Undergraduate Texts in Physics

Series Editors

Kurt H. Becker, NYU Polytechnic School of Engineering, Brooklyn, USA

Jean-Marc Di Meglio, Matière et Systèmes Complexes, Université Paris Diderot, Bâtiment Condorcet, Paris, France

Sadri D. Hassani, Department of Physics, Loomis Laboratory, University of Illinois at Urbana-Champaign, Urbana, USA

Morten Hjorth-Jensen, Department of Physics, Blindern, University of Oslo, Oslo, Norway

Michael Inglis, Patchogue, USA

Bill Munro, NTT Basic Research Laboratories, Optical Science Laboratories, Atsugi, Kanagawa, Japan

Will Raven, Department of Physics, Smith College, Northampton, MA, USA

Susan Scott, Department of Quantum Science, Australian National University, Acton, Australia

Martin Stutzmann, Walter Schottky Institute, Technical University of Munich, Garching, Germany

Undergraduate Texts in Physics (UTP) publishes authoritative texts covering topics encountered in a physics undergraduate syllabus. Each title in the series is suitable as an adopted text for undergraduate courses, typically containing practice problems, worked examples, chapter summaries, and suggestions for further reading. UTP titles should provide an exceptionally clear and concise treatment of a subject at undergraduate level, usually based on a successful lecture course. Core and elective subjects are considered for inclusion in UTP.

UTP books will be ideal candidates for course adoption, providing lecturers with a firm basis for development of lecture series, and students with an essential reference for their studies and beyond.

Anders Malthe-Sørenssen

Elementary Electromagnetism Using Python

A Modern Course Combining Analytical and Numerical Techniques

Anders Malthe-Sørenssen
Department of Physics
University of Oslo
Oslo, Norway

ISSN 2510-411X ISSN 2510-4128 (electronic)
Undergraduate Texts in Physics
ISBN 978-3-032-19875-4 ISBN 978-3-032-19876-1 (eBook)
https://doi.org/10.1007/978-3-032-19876-1

© The Editor(s) (if applicable) and The Author(s), under exclusive license to Springer Nature Switzerland AG 2026

This work is subject to copyright. All rights are solely and exclusively licensed by the Publisher, whether the whole or part of the material is concerned, specifically the rights of translation, reprinting, reuse of illustrations, recitation, broadcasting, reproduction on microfilms or in any other physical way, and transmission or information storage and retrieval, electronic adaptation, computer software, or by similar or dissimilar methodology now known or hereafter developed.
The use of general descriptive names, registered names, trademarks, service marks, etc. in this publication does not imply, even in the absence of a specific statement, that such names are exempt from the relevant protective laws and regulations and therefore free for general use.
The publisher, the authors and the editors are safe to assume that the advice and information in this book are believed to be true and accurate at the date of publication. Neither the publisher nor the authors or the editors give a warranty, expressed or implied, with respect to the material contained herein or for any errors or omissions that may have been made. The publisher remains neutral with regard to jurisdictional claims in published maps and institutional affiliations.

This Springer imprint is published by the registered company Springer Nature Switzerland AG
The registered company address is: Gewerbestrasse 11, 6330 Cham, Switzerland

If disposing of this product, please recycle the paper.

Preface

This book was developed as a textbook for the course Fys1120—Electromagnetism at the University of Oslo, Norway, from 2018 and onwards. The book was written to provide students with an integrated text that includes both analytical and computational methods in a seamless manner and that interfaces with how students learn to use computational methods in associated courses in mathematics.

I wrote the book as I was making a major transformation on both the content and learning methods used in the course, where I wanted to focus on student agency—teaching students to become active participants in their own learning process and trying to push students towards higher levels in Bloom's taxonomy, towards creativity, exploration and expression. This was done by introducing students to computational essays in collaboration with Tor Ole Odden. You can find examples of computational essays here.[1]

This book is written in the spirit of Hans Petter Langtangen, who was my great mentor in textbook writing and the integration of computing across the sciences. The first year I was teaching this course it was with the help of Henrik Andersen Sveinsson, and several of the exercises have been developed by him. The text is also inspired by the Norwegian text written by Johannes Skaar, "Elektromagnetisme" (Skaar 2019a), and the lecture series by Johannes Skaar, which both are inspiringly concise and have inspired both exposition and examples in this book. In addition, the second chapter on circuits is inspired by Johannes Skaar, "Kretsanalyse basert på elektromagnetisme" (Skaar 2019b). This book has also benefited from several students who have helped develop examples and exercises and from Learning Assistants who have helped improve text and examples, including Kine Ødegaard Hanssen and Even Nordhagen.

Oslo, Norway
August 2025

Anders Malthe-Sørenssen

[1] https://uio-ccse.github.io/computational-essay-showroom/intro.

References

Johannes Skaar. *Elektromagnetisme*. University of Oslo, 2019a.
Johannes Skaar. *Kretsanalyse basert pa elektromagnetisme*. University of Oslo, 2019b.

Competing Interests The author has no competing interests to declare that are relevant to the content of this manuscript.

Introduction

In this book, we will introduce the fundamental laws of electromagnetism and learn to use them to understand nature and to develop and understand technology. The field of electromagnetism is rich and beautiful, and the theory of electromagnetism is one of the most beautiful theories we know of—described concisely by Maxwell's four equations. However, the theory by itself is of little use if we do not know how to harness it to understand and master nature. This text therefore focuses on making you a proficient user of electromagnetism by providing you with a solid understanding of the fundamentals, a good understanding of how to model physical systems with simplified physics-based models, and the tools to solve the models analytically and computationally. The book has a unique approach where the computational methods are deeply integrated into exposition, examples and exercises. The book has companion material in the form of computational essay examples and projects that help students become active participants in their own learning and to start asking their own questions, posing their own problems, and discovering how to pursue these problems, analyse them and communicate the results effectively. The book also has extensive sets of analytical and computational exercises with answers and solutions in a dynamic web-page setting that facilitate good learning. My goal is for you as a student to succeed by making electromagnetism a part of you, so that physics in general and electromagnetism in particular shape how you think, and help you think, discover, innovate and communicate better—letting physics become an extended part of you.

Electromagnetism

Electromagnetic interactions are everywhere. Electromagnetism is a fundamental theory of interactions that are both hidden and evident in the world around you. In mechanics, we have addressed gravitational forces and various types of contacts forces such as normal forces, friction and air resistance. All the macroscopic forces beyond gravity are electromagnetic forces in various forms. All contact forces are due to electromagnetic interactions. When we treat atoms classically, by for example integrating their equations of motion to discover the behavior of many-particle systems, the forces between atoms are typically modelled as various forms of electrostatic forces. The attractive interaction between surfaces due to van der Waals interactions is due to an induced dipole-dipole interaction—which we will discuss in detail in this book. Many aspects of cell mechanics, in particular for nerve cells, are best understood in terms of electromagnetic concepts, and of course modern technology depends on our mastery of electromagnetism for circuits, light, sensors, and communications. Your cellphone is a marvel of engineering, based on our understanding of electromagnetism as well as other aspects of physics.

An elegant theory. Electromagnetism is also an elegant theory. Maybe this is your first encounter with a theory that is so beautiful that you are only left to marvel. All the electromagnetic phenomena you have learned about in various contexts so far in your education will be covered by Maxwell's four equations. However, this compactness and beauty relies on a well-developed mathematical machinery. This book assumes that you have had an introductory course in vector calculus. However, it is fully possible to follow the text without this background knowledge. To help you if you are in this situation, we have provided you with a brief introduction to vector calculus.

A lasting theory. When you have read this book, you fully understand the last chapter, Maxwell's equations, and you have built your skills so that you can use them effectively, you have become a master of one of the cornerstones of our civilization. And you will have acquired lasting knowledge and skills. News, celebrities, and human glory are transient, but the fundamental theory of electromagnetism remains unchanged.

A complete theory. However, you should not be fooled by the beauty of the theory. It also means that this part of physics is largely finished. We understand the fundamentals and have been able to formulate the laws concisely. My advise to you is to pursue areas where our understanding is yet lacking, where there is no beautiful theory—yet—and be the one to create this theory. But whatever it is you will pursue—if it is to understand our brain or develop new green technologies and energy solutions—I am sure you will need a solid understanding of electromagnetism to succeed!

Computational Methods

Power of physics. This book has been written to provide a textbook where computational methods are deeply integrated into the exposition, theory, examples and problems. I want you to be able to use computational methods as naturally as analytical methods when you learn and apply physics. And I want to help you build a set of tools that you can use to address any physics problem, not only the ones that are analytically tractable. As you read this book and work through the examples and problems, you should gradually get the feeling of the *power of physics*. That you learn methods that are powerful and that can be applied to any problem. That there are robust solution methods that can be applied and that allows you to study phenomena and processes you are intrigued by.

Worked examples. I provide you with complete worked examples, where all aspects are addressed: The problem is specified and a physics-based model is introduced, the model is solved computationally, and sometimes we make a simplified theoretical model to understand the behavior we observe computationally. The full computer code is provided where feasible.

Building and implementing models. You will also see that many of the same computational methods are applied over and over. This will help you see mathematical and methodological similarities between different aspects of electromagnetism. The book has a focus on modeling, which is the process of simplifying and abstracting a physical system into a model system that we can address using the theoretical and computational tools we have available. Learning to model and to simplify a problem down to its core components is a very important part of becoming a physicist—indeed it may be argued that this is the core of physics as a practice. However, it is often difficult to learn these skills and to learn to trust your own intuition. I am sure you will often ask yourself if you made the model too simple or why you can ignore a particular aspect of a problem. Often the proof is in the pudding. You test your model to see if it reproduces the aspects you are interested in modeling and thus learn to trust it. Learning modeling is a long process, which requires that you are challenged by your teachers, peers and yourself. But it should also be a fun process with a feeling of adventure as you learn to see the world in its simplest/simplified form.

Learning Physics

Learning effectively. We have significant amounts of research-based knowledge on how to teach and how to learn physics. Most of educational research boils down to the following: To learn effectively you have to work on problems that are just beyond your current level of mastery, formulate your understanding, and receive constructive and immediate feedback. We know that learning is a social process—you learn better if you learn with other. It is also a strenuous activity—you should be tired after trying to learn. We also know that it is good to alternate between being highly focused, which is the strenuous activity, and defocused, where you allow you mind to wander more freely.

Learning by retrieval practice. Learning is most effective through *retrieval practice*. This is when you try to remember what you have learned. When you read something new—a new theory, a new example or a new methods—you place everything into your working memory. Unfortunately, your working memory can typically only store 4–5 chunks. Your goal is to imprint what is in your working memory into your long-term memory. This frees up room in working memory for you to think. And the best way to do this is by retrieval. One way to apply this to your studies is to:

(i) first read a section; (ii) close the book and write down the main points; (iii) then work through an example; (iv) close the page and write down how you remember that the example was solved by going through the whole process as you remember it; (v) then solve a problem without looking at the example—strain yourself hard to use the approach from the example as you remember it. The day after you start with a blank sheet. Write down the main points of the theory from memory. Go through how you solved a problem from memory. Then repeat after about a week and after about a month. Does this sound tiresome? It is. But this is how you learn. If you try this, you will learn, and you will gradually put more and more into your long-term memory, allowing you to draw on them when you think, and eventually allowing you to think creatively and make connections between theory, examples, methods and applications. This is where creativity actually comes from!

Using worked examples to learn. Research indicates that you learn physics well by going through worked examples before you solve the first problem. Use examples actively and solutions carefully. Solutions to problems are like Kryptonite. If you come too close to the solution, your powers of learning will weaken. I provide solutions to all problems in this text at the accompanying website. Be careful when you unroll them. Try hard first. Then look. Use the examples actively, but wisely. Try to close all books and notes when you attempt to solve a problem, thus using the power of recall to imprint concepts, facts and methods into long-term memory.

Methods are clearly described. I provide "Method" boxes throughout the text that provides you with step-by-step instructions extracted from the examples on how to solve a particular type of problem. Use these and learn them so that you can apply them when you need them, but also deviate from the methods when the situations require you to.

Exercises, answers and solutions. You find a basic set of exercises and answers in this book. You can find a full set of exercises, answers and solutions on the accompanying website.[2]

Structure of the Book

This book follows a traditional structure of the exposition. We start from electrostatics and magnetostatics, where charges and currents are stationary, before moving to electrodynamics, where charges can move and currents may change with time. Each chapter contains a few key worked examples and associated "Methods" sections. Many sections have been written to provide you with a solid toolbox to address problems: We introduce computational methods to find the electric and magnetic fields from any charge or current distribution, and we provide you with three tools to solve Laplace equation. The goal is not for you to learn all the details of the methods,

[2] https://malthent.github.io/electromagnetism/book/bookexercises.html.

but instead to feel confident in their use, and to learn their limitations so that you can evaluate the results you get in real situations. I wish you good fortune in your efforts to learn to master electromagnetism!

Formulas from Vector Calculus

Cartesian Coordinates

$$\mathrm{d}\mathbf{l} = \mathrm{d}x\,\hat{\mathbf{x}} + \mathrm{d}y\,\hat{\mathbf{y}} + \mathrm{d}z\,\hat{\mathbf{z}}$$

$$\nabla V = \frac{\partial V}{\partial x}\hat{\mathbf{x}} + \frac{\partial V}{\partial y}\hat{\mathbf{y}} + \frac{\partial V}{\partial z}\hat{\mathbf{z}}$$

$$\nabla \cdot \mathbf{A} = \frac{\partial A_x}{\partial x} + \frac{\partial A_y}{\partial y} + \frac{\partial A_z}{\partial z}$$

$$\nabla \times \mathbf{A} = \left(\frac{\partial A_z}{\partial y} - \frac{\partial A_y}{\partial z}\right)\hat{\mathbf{x}} + \left(\frac{\partial A_x}{\partial z} - \frac{\partial A_z}{\partial x}\right)\hat{\mathbf{y}} + \left(\frac{\partial A_y}{\partial x} - \frac{\partial A_x}{\partial y}\right)\hat{\mathbf{z}}$$

$$\nabla^2 V = \frac{\partial^2 V}{\partial x^2} + \frac{\partial^2 V}{\partial y^2} + \frac{\partial^2 V}{\partial z^2}$$

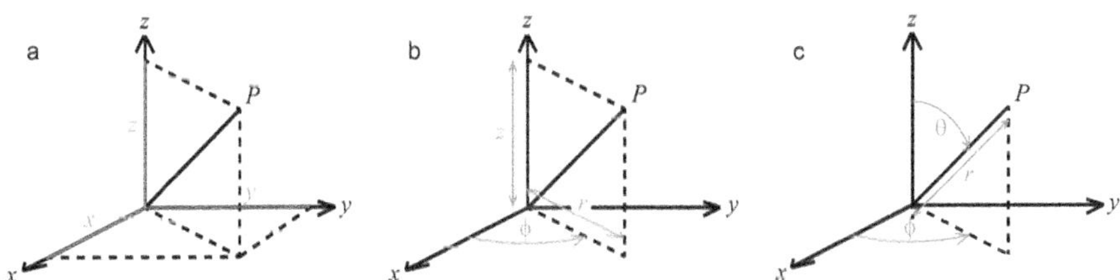

Axes for **a** Cartesian coordinates, **b** cylindrical coordinates, **c** spherical coordinates

Cylindrical Coordinates

$$\mathrm{d}\mathbf{l} = \mathrm{d}r\,\hat{\mathbf{r}} + r\,\mathrm{d}\phi\,\hat{\boldsymbol{\phi}} + \mathrm{d}z\,\hat{\mathbf{z}}$$

$$\hat{\mathbf{r}} = \cos\phi\,\hat{\mathbf{x}} + \sin\phi\,\hat{\mathbf{y}}$$

$$\hat{\boldsymbol{\phi}} = -\sin\phi\,\hat{\mathbf{x}} + \cos\phi\,\hat{\mathbf{y}}$$

$$\hat{\mathbf{z}} = \hat{\mathbf{z}}$$

$$\nabla V = \frac{\partial V}{\partial r}\hat{\mathbf{r}} + \frac{1}{r}\frac{\partial V}{\partial \phi}\hat{\boldsymbol{\phi}} + \frac{\partial V}{\partial z}\hat{\mathbf{z}}$$

$$\nabla \cdot \mathbf{A} = \frac{1}{r}\frac{\partial (rA_r)}{\partial r} + \frac{1}{r}\frac{\partial A_\phi}{\partial \phi} + \frac{\partial A_z}{\partial z}$$

$$\nabla \times \mathbf{A} = \left(\frac{1}{r}\frac{\partial A_z}{\partial \phi} - \frac{\partial A_\phi}{\partial z}\right)\hat{\mathbf{r}} + \left(\frac{\partial A_r}{\partial z} - \frac{\partial A_z}{\partial r}\right)\hat{\boldsymbol{\phi}} + \left(\frac{\partial (rA_\phi)}{\partial r} - \frac{\partial A_r}{\partial \phi}\right)\frac{\hat{\mathbf{z}}}{r}$$

$$\nabla^2 V = \frac{1}{r}\frac{\partial}{\partial r}\left(r\frac{\partial V}{\partial r}\right) + \frac{1}{r^2}\frac{\partial^2 V}{\partial \phi^2} + \frac{\partial^2 V}{\partial z^2}$$

Spherical Coordinates

$$\mathrm{d}\mathbf{l} = \mathrm{d}r\,\hat{\mathbf{r}} + r\sin\theta\,\mathrm{d}\phi\,\hat{\boldsymbol{\phi}} + r\,\mathrm{d}\theta\,\hat{\boldsymbol{\theta}}$$

$$\hat{\mathbf{r}} = \sin\theta\cos\phi\,\hat{\mathbf{x}} + \sin\theta\sin\phi\,\hat{\mathbf{y}} + \cos\theta\,\hat{\mathbf{z}}$$

$$\hat{\boldsymbol{\phi}} = -\sin\phi\,\hat{\mathbf{x}} + \cos\phi\,\hat{\mathbf{y}}$$

$$\hat{\boldsymbol{\theta}} = \cos\theta\cos\phi\,\hat{\mathbf{x}} + \cos\theta\sin\phi\,\hat{\mathbf{y}} - \sin\theta\,\hat{\mathbf{z}}$$

$$\nabla V = \frac{\partial V}{\partial r}\hat{\mathbf{r}} + \frac{1}{r}\frac{\partial V}{\partial \theta}\hat{\boldsymbol{\theta}} + \frac{1}{r\sin\theta}\frac{\partial V}{\partial \phi}\hat{\boldsymbol{\phi}}$$

$$\nabla \cdot \mathbf{A} = \frac{1}{r^2}\frac{\partial (r^2 A_r)}{\partial r} + \frac{1}{r\sin\theta}\frac{\partial (\sin\theta\,A_\theta)}{\partial \theta} + \frac{1}{r\sin\theta}\frac{\partial A_\phi}{\partial \phi}$$

$$\nabla \times \mathbf{A} = \left(\frac{\partial (\sin\theta\,A_\phi)}{\partial \theta} - \frac{\partial A_\theta}{\partial \phi}\right)\frac{\hat{\mathbf{r}}}{r\sin\theta} + \left(\frac{1}{\sin\theta}\frac{\partial A_r}{\partial \phi} - \frac{\partial (rA_\phi)}{\partial r}\right)\frac{\hat{\boldsymbol{\theta}}}{r} + \left(\frac{\partial (rA_\theta)}{\partial r} - \frac{\partial A_r}{\partial \theta}\right)\frac{\hat{\boldsymbol{\phi}}}{r}$$

$$\nabla^2 V = \frac{1}{r^2}\frac{\partial}{\partial r}\left(r^2\frac{\partial V}{\partial r}\right) + \frac{1}{r^2\sin\theta}\frac{\partial}{\partial \theta}\left(\sin\theta\frac{\partial V}{\partial \theta}\right) + \frac{1}{r^2\sin^2(\theta)}\frac{\partial^2 V}{\partial \phi^2}$$

Gradient Operators

$$\nabla(f + g) = \nabla f + \nabla g$$

$$\nabla(fg) = f\nabla g + g\nabla f$$

$$\nabla(\mathbf{A}\cdot\mathbf{B}) = \mathbf{A}\times(\nabla\times\mathbf{B}) + \mathbf{B}\times(\nabla\times\mathbf{A}) + (\mathbf{A}\cdot\nabla)\mathbf{B} + (\mathbf{B}\cdot\nabla)\mathbf{A}$$

$$\nabla\cdot(f\mathbf{A}) = f(\nabla\cdot\mathbf{A}) + \mathbf{A}\cdot(\nabla f)$$

$$\nabla\cdot(\mathbf{A}\times\mathbf{B}) = \mathbf{B}\cdot(\nabla\times\mathbf{A}) - \mathbf{A}\cdot(\nabla\times\mathbf{B})$$

$$\nabla\times(f\mathbf{A}) = f(\nabla\times\mathbf{A}) + (\nabla f)\times\mathbf{A}$$

$$\nabla\times(\mathbf{A}\times\mathbf{B}) = (\mathbf{B}\cdot\nabla)\mathbf{A} - (\mathbf{A}\cdot\nabla)\mathbf{B} + \mathbf{A}(\nabla\cdot\mathbf{B}) - \mathbf{B}(\nabla\cdot\mathbf{A})$$

$$\nabla\cdot(\nabla\times\mathbf{A}) = 0$$

$$\nabla\cdot(\nabla f) = \nabla^2 f$$

$$\nabla\times(\nabla f) = 0$$

Integral Equations

$$\int_v \nabla \cdot \mathbf{A}\, \mathrm{d}v = \oint_S \mathbf{A} \cdot \mathrm{d}\mathbf{S} \quad \text{(divergence theorem)}$$

$$\int_S \nabla \times \mathbf{A} \cdot \mathrm{d}\mathbf{S} = \oint_C \mathbf{A} \cdot \mathrm{d}\mathbf{l} \quad \text{(Stokes' theorem)}$$

Contents

Chapter 1
Vector Calculus

1.1 Vectors

Electromagnetism may be the first time in physics where you need to study systems in three dimensions. You may have studied three-dimensional systems in mechanics, but it is only for some rotational systems that the three-dimensional nature of the problem is important. In electromagnetism, many of the systems are inherently three dimensional as illustrated in Fig. 1.1. It is therefore necessary to introduce the mathematical tools we need to describe systems in three dimensions. We first repeat what you probably know about vectors as well as scalar and vector functions. Then we describe how our tools of calculus, our basic mathematical tools in physics, can be extended to three dimensions and to both scalar and vector functions. This will provide you with the tools you need to address electromagnetism in the form of vector calculus. If you are already familiar with these tools, you may proceed to the next chapter. However, if you do not have experience with visualization techniques for two- and three-dimensional systems, we advise you to look at the examples provided here as we will apply these techniques in later chapters.

Vector Algebra

You have an intuition about vectors from mechanics where we used vectors to describe displacement, velocity, acceleration and force. A **vector** has both a direction and a length, whereas a **scalar** is a single number. We use the notation $\mathbf{a}$ for a vector and a for a scalar.

We add two vectors, $\mathbf{a}$ and $\mathbf{b}$, as we would add two displacements by placing vector $\mathbf{b}$ at the end of vector $\mathbf{a}$ and finding the sum graphically as illustrated in Fig. 1.2a. We can scale a vector by multiplication with a scalar. For example, the vector $2\mathbf{b}$ is twice as long as the vector $\mathbf{b}$ as illustrated in Fig. 1.2b. Vectors obey the **distributive**

© The Author(s), under exclusive license to Springer Nature Switzerland AG 2026

A. Malthe-Sørenssen, *Elementary Electromagnetism Using Python*, Undergraduate Texts in Physics, https://doi.org/10.1007/978-3-032-19876-1_1

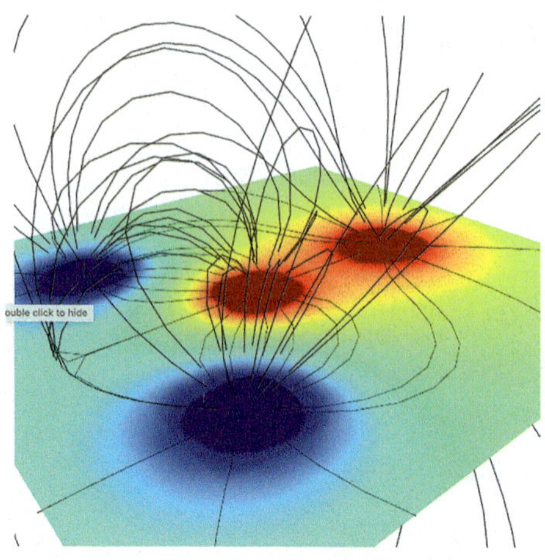

Fig. 1.1 In electromagnetic systems we cannot avoid the three dimensional nature of the systems. Here, illustrated with the field lines around a charge distribution

rule: $c(\mathbf{a}+\mathbf{b}) = c\mathbf{a} + c\mathbf{b}$. And the sequence in which we add vectors is arbitrary: $\mathbf{a}+\mathbf{b} = \mathbf{b}+\mathbf{a}$. We call the length of a vector the vector **norm** and denote it as $|\mathbf{a}|$. We call a vector that is of unit length, that is a length of 1, a **unit vector** and use the following notation for unit vectors $\hat{\mathbf{n}}$.

Vectors on component form. We describe a coordinate system by the position of the origin and by unit vectors along each axis. For a Cartesian coordinate system the unit vectors are $\hat{\mathbf{x}}$, $\hat{\mathbf{y}}$, $\hat{\mathbf{z}}$ and we write a vector $\mathbf{a}$ on **component form** as

$$\mathbf{a} = a_x\hat{\mathbf{x}} + a_y\hat{\mathbf{y}} + a_z\hat{\mathbf{z}}, \tag{1.1}$$

where a_x, a_y, a_z are the components along the x, y, and z-axes respectively as illustrated in Fig. 1.2c. In physics we often use the **coordinate form** $\mathbf{a} = (a_x, a_y, a_z)$.

Vector addition and scalar multiplication. Vector addition and scalar multiplication occurs by adding and multiplying the components:

$$(a_x, a_y, a_z) + (b_x, b_y, b_z) = (a_x + b_x, a_y + b_y, a_z + b_z), \tag{1.2}$$

$$c(a_x, a_y, a_z) = (ca_x, ca_y, ca_z). \tag{1.3}$$

Norm of a vector on coordinate form. The norm of a vector on coordinate form is

$$|\mathbf{a}| = |(a_x, a_y, a_z)| = \sqrt{a_x^2 + a_y^2 + a_z^2}. \tag{1.4}$$

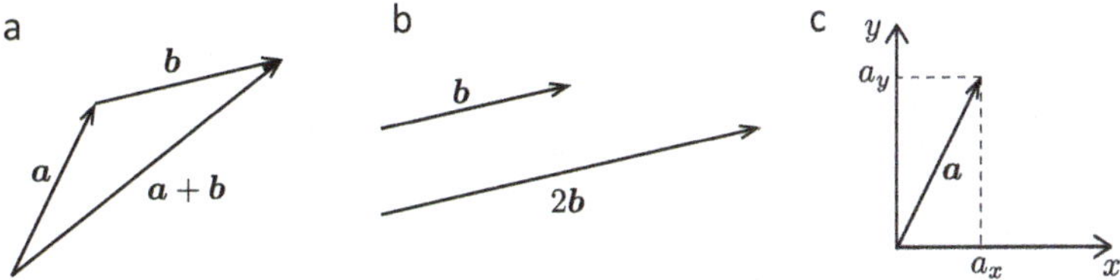

Fig. 1.2 **a** Graphical illustration of vector addition, **b** Graphical illustration of scalar multiplication, **c** The components of a vector

Numerical representation. In Python we represent vectors using an `array` in `numpy`. The addition of two vectors is done by:

```
import numpy as np
a = np.array([1,2,3])
b = np.array([-1,0,0])
print(a+b)
```

```
Out: [0 2 3]
```

and the norm is found by

```
a = np.array([3,4,0])
print(np.linalg.norm(a))
```

```
Out: 5.0
```

Dot Product

Definition of dot product. The **dot product** or inner product between two vectors **a** and **b** is defined as

$$\mathbf{a} \cdot \mathbf{b} = |\mathbf{a}|\,|\mathbf{b}|\cos\theta, \tag{1.5}$$

where θ is the angle between the vectors as illustrated in Fig. 1.3. Notice that the dot product is the product of two vectors and produces a *scalar*. The dot product satisfies the *commutative law*: $\mathbf{a} \cdot \mathbf{b} = \mathbf{b} \cdot \mathbf{a}$ and the *distributive law*: $\mathbf{a} \cdot (\mathbf{b} + \mathbf{c}) = \mathbf{a} \cdot \mathbf{b} + \mathbf{a} \cdot \mathbf{c}$. We usually define the norm of a vector through the dot product, $|\mathbf{a}| = \sqrt{\mathbf{a} \cdot \mathbf{a}}$.

Orthogonal vectors. The condition for two vectors to be *orthogonal* (normal) is that the dot product of the two vectors is zero. That is, the two vectors **a** and **b** are orthogonal if and only if $\mathbf{a} \cdot \mathbf{b} = 0$.

Geometric interpretation. The most common geometric interpretation of the dot product is that the dot product of a vector and a unit vector is the projection of the

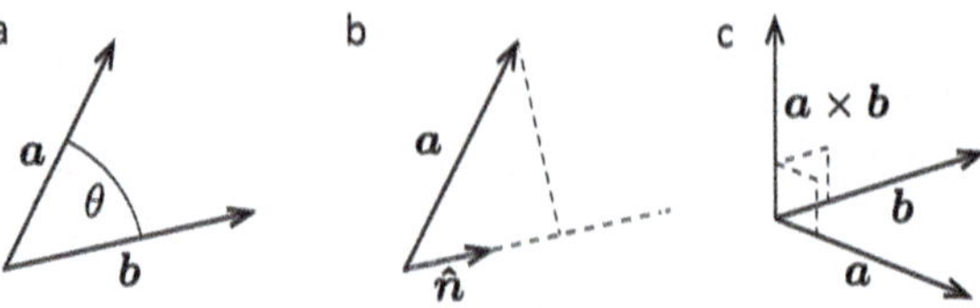

Fig. 1.3 **a** Illustration of the dot product. **b** The projection of vector **a** onto the direction given by the unit vector $\hat{\mathbf{n}}$. **c** Illustration of the cross-product

vector on an axis along the unit vector as illustrated in Fig. 1.3. That is, for a unit vector $\hat{\mathbf{n}}$, the length of the vector **a** along $\hat{\mathbf{n}}$ is $\mathbf{a} \cdot \hat{\mathbf{n}}$.

Coordinate form. The dot product on coordinate form is the sum of the products of the coordinates. That is, for vectors $\mathbf{a} = (a_x, a_y, a_z)$ and $\mathbf{b} = (b_x, b_y, b_z)$ the dot product is

$$(a_x, a_y, a_z) \cdot (b_x, b_y, b_z) = a_x b_x + a_y b_y + a_z b_z, \tag{1.6}$$

which can be generalized to any dimension. You can prove this from the definition by writing out the vectors on component form.

Unit vectors. The unit vectors $\hat{\mathbf{x}}$, $\hat{\mathbf{y}}$ and $\hat{\mathbf{z}}$ in a Cartesian coordinate system are unit vectors that are normal to each other so that $\hat{\mathbf{x}} \cdot \hat{\mathbf{x}} = \hat{\mathbf{y}} \cdot \hat{\mathbf{y}} = \hat{\mathbf{z}} \cdot \hat{\mathbf{z}} = 1$, but $\hat{\mathbf{x}} \cdot \hat{\mathbf{y}} = \hat{\mathbf{x}} \cdot \hat{\mathbf{z}} = \hat{\mathbf{y}} \cdot \hat{\mathbf{z}} = 0$.

Numerical implementation. The dot product is computed numerically using the `dot` function in `numpy`:

```python
import numpy as np
a = np.array([1,2,3])
b = np.array([-1,0,0])
print(np.dot(a,b))
```

```
Out: -1
```

Notice that using the multiplication operator `*` in Python does not return the dot product, but instead returns a vector consisting of the component-wise multiplications of the components, that is, `a*b` returns `[a_x*b_x,a_y*b_y,a_z*b_z]`, which should not be confused with the dot product.

Cross Product

Definition of cross product. The cross product is defined as $\mathbf{a} \times \mathbf{b} = |\mathbf{a}|\,|\mathbf{b}|\,\sin\theta\,\hat{\mathbf{n}}$ where θ is the angle between the two vectors and $\hat{\mathbf{n}}$ is a vector that is normal to both vectors and points in the positive direction according to the right-hand rule as illustrated in Fig. 1.3. The cross product *does not commute* because $\mathbf{a} \times \mathbf{b} = -\mathbf{b} \times \mathbf{a}$. The order of multiplication is important because the right hand rule gives the opposite direction for the cross products $\mathbf{a} \times \mathbf{b}$ and $\mathbf{b} \times \mathbf{a}$. Notice that the result of a cross product is a *vector* and the cross product is normal to each of the vectors in the product. The cross product obeys the *distributive law*: $\mathbf{a} \times (\mathbf{b} + \mathbf{c}) = \mathbf{a} \times \mathbf{b} + \mathbf{a} \times \mathbf{c}$. Notice also that $\mathbf{a} \times \mathbf{a} = 0$.

Geometric interpretation. A common geometrical interpretation of the cross product of two vectors **a** and **b** is that the magnitude of the product is the area of the parallelogram spanned by the two vectors. In addition, the cross product forms a vector normal to both vectors. We therefore often use the cross product to find the normal vector to a plane spanned by two vectors.

Unit vectors. The cross product of two orthonormal unit vectors, that is unit vectors that are of unit length and orthogonal to each other, results in the third unit vector (with the sign depending on the order of the product). For example, $\hat{\mathbf{x}} \times \hat{\mathbf{y}} = \hat{\mathbf{z}}$ because the coordinate system is positively oriented.

Coordinate form. The cross product in coordinate form is a bit more complicated than the dot product, but you can teach yourself various schemes to memorize the rules:

$$(a_x, a_y, a_z) \times (b_x, b_y, b_z) = (a_y b_z - a_z b_y, a_z b_x - a_x b_z, a_x b_y - a_y b_x). \tag{1.7}$$

These expressions can be derived by writing the vectors in component form and using the cross products of the unit vectors. A common trick is to remember the determinant form, where the cross product is given as the determinant of

$$\mathbf{a} \times \mathbf{b} = \begin{vmatrix} \hat{\mathbf{x}} & \hat{\mathbf{y}} & \hat{\mathbf{z}} \\ a_x & a_y & a_z \\ b_x & b_y & b_z \end{vmatrix} = \hat{\mathbf{x}} \begin{vmatrix} a_y & a_z \\ b_y & b_z \end{vmatrix} - \hat{\mathbf{y}} \begin{vmatrix} a_x & a_z \\ b_x & b_z \end{vmatrix} + \hat{\mathbf{z}} \begin{vmatrix} a_x & a_y \\ b_x & b_y \end{vmatrix} \tag{1.8}$$

Numerical implementation. The cross product is computed using the `cross` function in `numpy`:

```
import numpy as np
a = np.array([1,2,3])
b = np.array([-1,0,0])
print(np.cross(a,b))
```

```
Out: [ 0 -3  2]
```

1.2 Multivariate Functions and Vector Fields

Physical quantities may vary in space. For example, the temperature $T(x, y, z)$ or the pressure $p(x, y, z)$ may vary with the spatial position (x, y, z). We call such functions of more than one variable multivariate functions. You are already well versed in single-valued functions from mechanics and you may indeed have experience with multivariate functions such as the potential energy $U(x, y, z)$ of an object subject to the Sun's gravity. We often use the word *field* to describe functions that vary in space, that is, functions of the (three) spatial coordinates (x, y, z). If the function returns a single number, we call it a *scalar field*. If the function returns a vector, we call the function a *vector field*.

Scalar Fields

An example of a scalar field is the height of a terrain, $h(x, y)$, which we can illustrate as a three-dimensional surface, as a map with colors that indicate the height, or as a set of contours, which describe curves where the height is a specific value. A contour corresponds to the points (x, y) where $h(x, y) = h_0$. All three methods are illustrated for the surface $h(x, y) = -(3x^2 + 2y^2)$ in Fig. 1.4.

Visualizing a scalar field. How do we visualize a scalar field $h(x, y)$ in Python? In order to visualize it, we need to calculate the field at specific positions (x_i, y_i) in space, for example on a regular grid of points, and then draw a surface, a color, or a set of contours to illustrate these values. We will do this in three steps: (1) We will generate a grid of regularly spaced points (x_i, y_j) on which we will calculate the field; (2) We will calculate the field at these positions and store the values in an array; (3) We will visualize this discretized version of the field in various ways.

Grid geometry. We want to generate a set of points, (x_i, y_j), so that they form a square grid as illustrated in Fig. 1.5. We want the grid to consist of $N_x + 1$ values from x_0 to x_1 along the x-axis and $N_y + 1$ values from y_0 to y_1 along the y-axis. This means that we divide the line from x_0 to x_1 into N_x line elements that each have a length $\Delta x = (x_1 - x_0)/N_x$. Notice the small, but important, counting difference: There are $N_x + 1$ points, but there are N_x line elements connecting these points. Similarly, along the y-axis, the distance between the points are $\Delta y = (y_1 - y_0)/N_y$. We therefore want to generate a set of points $(x_{i,j}, y_{i,j})$ for $i = 0, \ldots, N_x$ and $j = 0, \ldots, N_y$ so

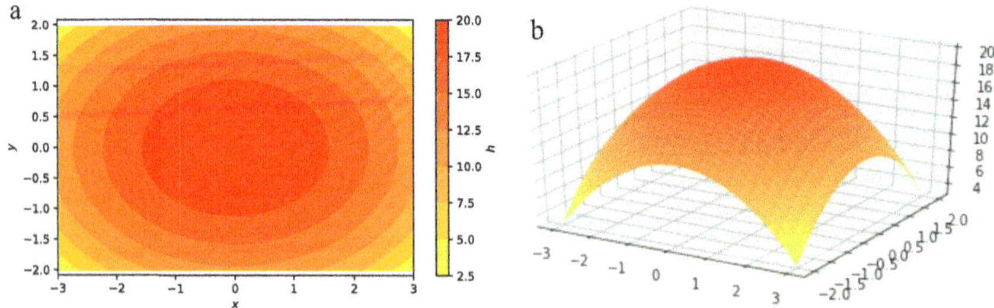

Fig. 1.4 **a** Plot of the function $h(x, y)$ as an image with contours. **b** Plot of $h(x, y)$ as a surface plot

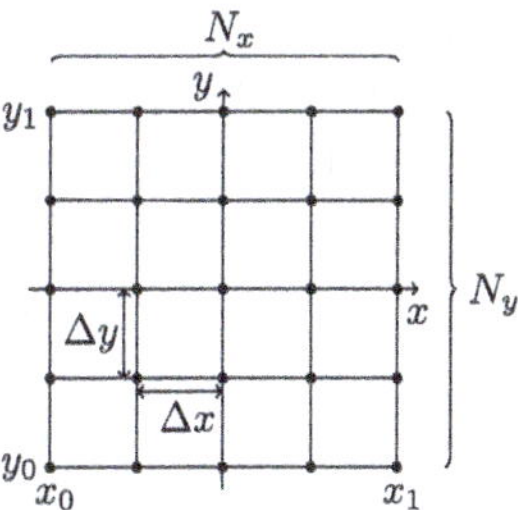

Fig. 1.5 Illustration of a grid used to calculate the values $h(x_{i,j}, y_{i,j})$

that $x_{i,j} = x_0 + i\Delta x$ and $y_{i,j} = y_0 + j\Delta y$. Notice again the small, but important, counting detail: There are $N_x + 1$ points, the first for $i = 0$ and the last for $i = N_x$, and similarly in the y-direction.

Generating a grid in Python. Fortunately, there is a function in Python that automatically generates such a grid. We first need to define the points we would like to have along the x- and y-directions using for example `linspace` and then we create the grid using the `meshgrid` function. We make a grid with $N_x = N_y = 5$ points and $x_0 = -L_x = -3$, $x_1 = +L_x = 3$, $y_0 = -L_y = -2$. $y_1 = L_y = 2$:

```
import numpy as np
Lx, Ly = 3,2
Nx, Ny = 5,4
x = np.linspace(-Lx,Lx,Nx+1)
y = np.linspace(-Ly,Ly,Ny+1)
rx,ry = np.meshgrid(x,y,indexing="ij")
```

Let us look at the `rx` and `ry` arrays generated:

```
print(rx)
```

```
Out = [[-3.   -3.   -3.   -3.   -3. ]
       [-1.8 -1.8 -1.8 -1.8 -1.8]
       [-0.6 -0.6 -0.6 -0.6 -0.6]
       [ 0.6  0.6  0.6  0.6  0.6]
       [ 1.8  1.8  1.8  1.8  1.8]
       [ 3.   3.   3.   3.   3. ]]
```

```
print(ry)
```

```
Out = [[-2. -1.  0.  1.  2.]
       [-2. -1.  0.  1.  2.]
       [-2. -1.  0.  1.  2.]
       [-2. -1.  0.  1.  2.]
       [-2. -1.  0.  1.  2.]
       [-2. -1.  0.  1.  2.]]
```

How can we understand these values? The function `meshgrid` returns two arrays, `rx` and `ry`, that correspond to $x_{i,j}$ and $y_{i,j}$ respectively. When you generate these arrays using `meshgrid` we were careful to provide the option `indexing="ij"`. This tells Python to use an indexing scheme which is similar to the i, j scheme we have introduced here. The element `rx[i,j]` therefore corresponds to the value $x_{i,j}$. We test this by checking that $x_{1,0}$, which should be $x_0 + 1\Delta x = -3.0 + 16/5 = -1.8$, indeed corresponds to the value of `rx[1,0]`.

```
print(rx[1,0])
```

```
Out: -1.8
```

Notice that if we did not specify `indexing="ij"`, Python would have used its default indexing scheme, which is called `indexing="xy"`, in which the element `rx[j,i]` corresponds to the value $x_{i,j}$. This creates a lot of confusion for us when we do physics, so we need to remember to use the `ij` scheme.

Defining the function $h(x, y)$. We define a function $h(x, y)$ that takes an array `r` corresponding to a vector $\mathbf{r} = (x, y)$ as an input and returns the height $h(\mathbf{r}) = h(x, y)$:

```
def h(r):
    return -(3*r[0]**2 + 2*r[1]**2)
```

Calculating values on the grid. We now calculate the values of the function h at the positions $(x_{i,j}, y_{i,j})$. First, we create an empty array `hij` of the same size as `rx` and `ry` to store the values. Then we make two nested loops over $i = 0, 1, \ldots, N_x$ and $j = 0, 1, \ldots, N_y$ to calculate the value of $h(x, y)$ at each point and store it in the `hij` array:

```
hij = np.zeros((Nx+1,Ny+1))
for i in range(Nx+1):
    for j in range(Ny+1):
        r = np.array([rx[i,j],ry[i,j]])
        hij[i,j] = h(r)
```

Notice how the first index `i` corresponds to x-values and the second index `j` corresponds to y-values.

Visualizing the scalar field as a flat image. Finally, we visualize the scalar field. We can plot it as a two-dimensional image with contours where colors indicate the value of h in a particular point. This is done using the `contourf` function:

```
import matplotlib.pyplot as plt
plt.contourf(rx,ry,hij)
plt.colorbar(label="$h$")
plt.xlabel("$x$")
plt.ylabel("$y$")
plt.axis("equal")
```

The resulting plot is shown in Fig. 1.4a.

Visualizing the scalar field as a surface. Alternatively, we can visualize the scalar field as a surface in a three-dimensional plot. This is done by the following program:

```
from mpl_toolkits import mplot3d
fig = plt.figure(figsize =(7, 4))
ax = plt.axes(projection ="3d")
ax.plot_surface(rx, ry, hij, cmap="autumn_r")
plt.show()
```

And the resulting plot is shown in Fig. 1.4b. Notice that we have increased the values of N_x and N_y to make the surface smoother in the plot.

Vector Fields

A vector field is a function that returns a vector instead of a scalar value. Typical examples are the gravitational force field or a velocity field in the Earth's atmosphere. You know that the gravitational force on a object from for example the Sun depends on the position of the object relative to the Sun and that the force is a vector. This means that the gravitational force $\mathbf{G}$ is a function of the position (x, y, z) in space. Similarly, the wind velocity in the atmosphere is a vector $\mathbf{v}(x, y, z)$ that varies with the position (x, y, z). Vector fields are essential in electromagnetism as we use them to describe electric and magnetic forces and how they vary in space.

Visualizing a vector field. To visualize a vector field we will follow the same approach as for scalar fields and find the field at discrete positions $(x_{i,j}, y_{i,j})$. Here, we demonstrate the method for three simple vector fields, $\mathbf{F}_1(x, y, z) = F_0\hat{\mathbf{x}}$ (a uniform field), $\mathbf{F}_2(x, y, z) = (x, 0, 0)$ and $\mathbf{F}_3(x, y, z) = (-y, 0, 0)$. We use the same program as above, but modify it slightly by flattening the two-dimensional arrays to one-dimensional arrays and then looping through all the elements in the flattened one-dimensional arrays to only use a single loop:

```
Lx, Ly = 3,2
Nx, Ny = 10,8
x = np.linspace(-Lx,Lx,Nx+1)
y = np.linspace(-Ly,Ly,Ny+1)
rx,ry = np.meshgrid(x,y,indexing="ij")

F2x = rx.copy()
F2y = rx.copy()
for i in range(len(F2x.flat)):
    r = np.array([rx.flat[i],ry.flat[i]])
    F2x.flat[i] = r[0]
    F2y.flat[i] = 0.0

F3x = rx.copy()
F3y = rx.copy()
for i in range(len(F3x.flat)):
    r = np.array([rx.flat[i],ry.flat[i]])
```

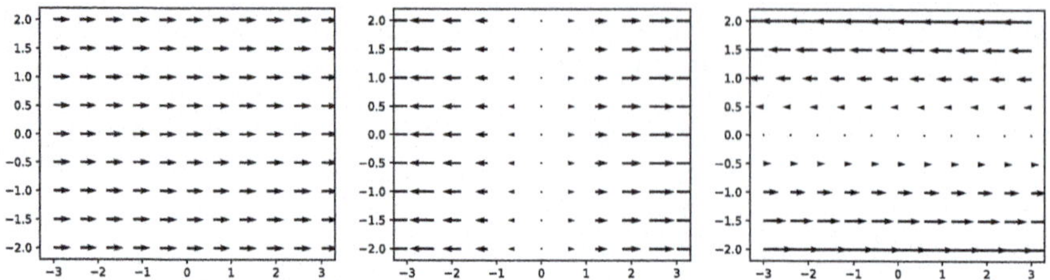

Fig. 1.6 Plots of three vector fields $\mathbf{F}_1(x, y)$ (left), $\mathbf{F}_2(x, y)$ (middle), $\mathbf{F}_3(x, y)$ (right)

```
    F3x.flat[i] = -r[1]
    F3y.flat[i] = 0
```

Plotting the field with arrows. We illustrate a vector field by plotting small arrows that represent the direction and magnitude of the vector $\mathbf{F}(x, y, z)$ at the grid positions $(x_{i,j}, y_{i,j}, 0)$ using the function `quiver`:

```
plt.quiver(rx,ry,F2x,F2y)
```

The resulting plots for the three vector fields are shown in Fig. 1.6. It may be worth experimenting with the options to the `quiver` function to for example scale or color the arrows. We provide examples on this in the following chapters.

1.3 Integration

In this section we will extend the concept of integration to scalar and vector fields and demonstrate how you can solve integrals analytically and numerically.

Volume Integrals

What is the mass of an object with a mass density, ρ, which varies in space? A simple solution would be to divide the object into many small pieces at positions $\mathbf{r}_i$ with small volumes $\mathrm{d}v_i$, for example, by dividing the object into small cubes of sides $\mathrm{d}x$, $\mathrm{d}y$ and $\mathrm{d}z$ so that the volume is $\mathrm{d}v = \mathrm{d}x\,\mathrm{d}y\,\mathrm{d}z$ as shown in Fig. 1.7. The mass of a small volume element at $\mathbf{r}_i$ is $\mathrm{d}m = \rho(\mathbf{r}_i)\,\mathrm{d}v$. The total mass is the sum of all the small elements, which approaches the integral over the volume v in the limit when the small volumes approach zero:

$$M = \sum_i \rho(\mathbf{r}_i)\,\mathrm{d}v \rightarrow \int_v \rho(\mathbf{r})\,\mathrm{d}v. \tag{1.9}$$

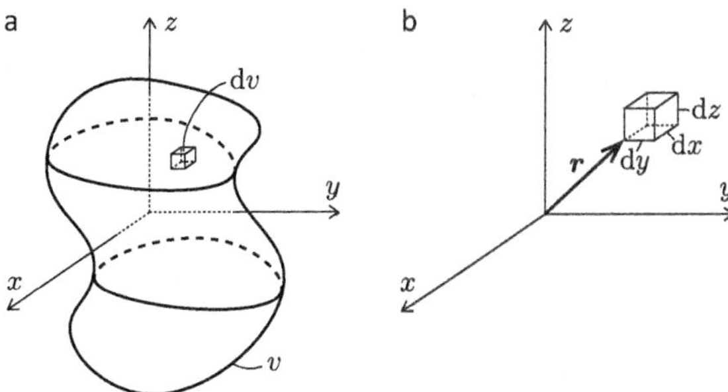

Fig. 1.7 **a** A small volume element dv inside a volume v. **b** A small integration volume dv = dx dy dz

In Cartesian coordinates we often write out the integral with respect to each of the coordinates:

$$M = \int_v \rho(\mathbf{r})\,\mathrm{d}v = \int\int\int \rho(x, y, z)\,\mathrm{d}x\,\mathrm{d}y\,\mathrm{d}z. \tag{1.10}$$

Notice that you need to ensure that you integrate over the specific volume by including the constraint of the volume in the integration limits.

Example: Mass of a Cube and a Triangular Prism

Mass of a cube. What is the mass of a cube from $x = 0$ to $x = a$, $y = 0$ to $y = a$ and $z = 0$ to $z = a$ with a mass density $\rho(x, y, z) = \rho_0(x/a)$ as illustrated in Fig. 1.8a.

We find the mass by integrating the density over the volume of the cube:

$$M = \int_v \rho\,\mathrm{d}v = \int_0^a \int_0^a \int_0^a \rho_0 \frac{x}{a}\,\mathrm{d}x\,\mathrm{d}y\,\mathrm{d}z. \tag{1.11}$$

The integrand does not depend on y or z. We can therefore first integrate with respect to y and z, where each integration gives a factor a, before integrating with respect to x:

$$\int_0^a \int_0^a \int_0^a \rho_0 \frac{x}{a}\,\mathrm{d}x\,\mathrm{d}y\,\mathrm{d}z = \int_0^a a^2 \rho_0 \frac{x}{a}\,\mathrm{d}x = \frac{1}{2}\rho_0 a^3. \tag{1.12}$$

Mass of triangular prism. What is the mass of the triangular prism with mass density $\rho(x, y, z) = \rho_0(x/a)$ as illustrated in Fig. 1.8b?

The integration is now from $z = 0$ to $z = a$, from $x = 0$ to $x = a$, but for each x-value, we need to integrate from $y = 0$ to $y = a - x$ to stay within the triangle. Notice that we unpack the integrals from the innermost integral and outwards, so that the innermost integral is over z, the next integral is over y and the outmost integral is over x:

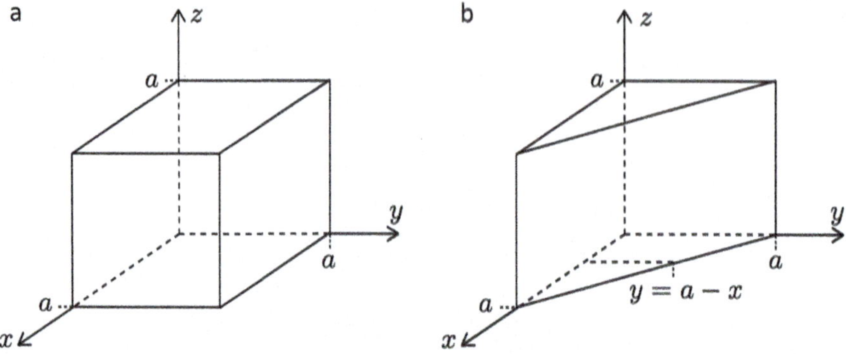

Fig. 1.8 **a** A cube with sides a. **b** A triangular prism with sides a

$$
\begin{aligned}
M &= \int_0^a \int_0^{a-x} \int_0^a \rho_0 \frac{x}{a}\, \mathrm{d}z\, \mathrm{d}y\, \mathrm{d}x = \int_0^a \int_0^{a-x} a\rho_0 \frac{x}{a}\, \mathrm{d}y\, \mathrm{d}x \\
&= \int_0^a a(a-x)\rho_0 \frac{x}{a}\, \mathrm{d}x = \rho_0 \left[\frac{1}{2} a x^2 - \frac{1}{3} x^3 \right]_0^a = \frac{1}{6} \rho_0 a^3 .
\end{aligned}
\tag{1.13}
$$

Example: Numerical Calculation of a Volume Integral

What is the mass of an object with mass density $\rho = \rho_0(z/a)$ underneath the surface $z = h(x, y) = |\sin(x/a)\cos(y/a)|$ from $x = 0$ to $x = a$ and $y = 0$ to $y = a$?

We integrate by summing up the values for points (x_i, y_i, z_i) with N points along each of the axis so that the grid distances are $\Delta x = \Delta y = \Delta z = a/N$. The integral is then approximated by

$$
I = \sum_i \rho(x_i, y_i, z_i)\, \Delta x\, \Delta y\, \Delta z . \tag{1.14}
$$

where the sum is over all points that are inside the object. We perform this sum by the following program:

```
import numpy as np
def h(x,y,a):
    return np.abs(np.sin(x/a)*np.cos(y/a))
N = 100
a, rho0 = 1.0, 1.0 dx = a/N mass = 0 for ix in range(N):
    x = ix*dx
    for iy in range(N):
        y = iy*dx
        for iz in range(N):
            z = iz*dx
```

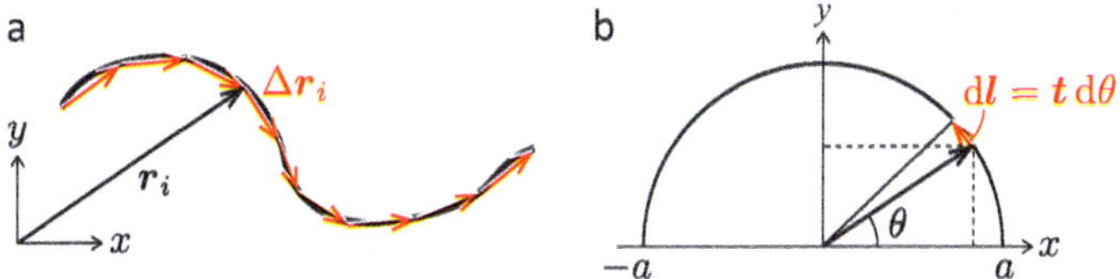

Fig. 1.9 **a** A small element along a curve. **b** A circular curve is parameterized with θ

```
            if (z<=h(x,y,a)): # Check if inside volume
                mass += rho0*(z/a)*dx*dx*dx
print("Integral = ",mass)
```

Notice that you need the loop in the z-direction to have an upper limit large enough for z to reach the maximum value of the $h(x, y)$ function.

Line Integrals

You may have encountered line integrals in mechanics to calculate the work of a force for motion along a curve. In this case, the curve is described by a sequence of displacements, $\Delta \mathbf{r}_i$, along a curve C as illustrated in Fig. 1.9a. The work performed by a force, $\mathbf{f}(x, y, z)$, during motion from $\mathbf{r}_i$ to $\mathbf{r}_{i+1} = \mathbf{r}_i + \Delta \mathbf{r}_i$ is $W_i = \mathbf{f}(\mathbf{r}_i) \cdot \Delta \mathbf{r}_i$. The total work performed along the curve C is

$$W = \sum_i \mathbf{f}(\mathbf{r}_i) \cdot \Delta \mathbf{r}_i \tag{1.15}$$

In the limit where the length of the steps goes to zero, this becomes the **curve integral** or the **line integral** along the curve C:

$$\int_C \mathbf{f} \cdot \mathrm{d}\mathbf{l}. \tag{1.16}$$

Here we use the notation $\mathrm{d}\mathbf{l}$ for the infinitesimal step along the curve, because the notation $\mathrm{d}\mathbf{r}$ may be confusing. If the curve is a closed curve, that is, if is starts and ends in the same point, we write the integral as

$$\oint_C \mathbf{f} \cdot \mathrm{d}\mathbf{l}. \tag{1.17}$$

This integral usually depends on the curve. However, for some vector fields that we call **conservative** vector fields, the integral does not depend on the path, but only on the position of the end-points.

How to calculate the curve integral. In order to calculate the curve integral, we need to describe the curve C as a parameterized curve. For example, we could describe a curve along a semi-circle from $(a, 0)$ to $(-a, 0)$ with a curve described as $\mathbf{r} = a(\cos\theta, \sin\theta)$ where θ goes from 0 to π as illustrated in Fig. 1.9b. In this case, the curve is parameterized by the parameter θ. The line element $\mathbf{dl}$ points along the curve. What does this mean? It means that it points in the direction of the *tangent vector* in each point along the curve. The length of $\mathbf{dl}$ when we increase the parameter by a value $d\theta$ is

$$\mathbf{dl} = \frac{d\mathbf{r}}{d\theta}\, d\theta, \tag{1.18}$$

where the vector $\mathbf{t} = \frac{d\mathbf{r}}{d\theta} = a(-\sin\theta, \cos\theta)$ is a tangent to the curve, that is, it points along the curve. Notice that you always need to create a parameterization of the curve in order to calculate the curve integral. With this parameterization, the line integral of a vector field $\mathbf{f}(\mathbf{r})$ is

$$\int_C \mathbf{f} \cdot \mathbf{dl} = \int_0^{\pi} \mathbf{f}(a(\cos\theta, \sin\theta)) \cdot a(-\sin\theta, \cos\theta)\, d\theta, \tag{1.19}$$

For a vector field $\mathbf{f} = (x/a)\hat{\mathbf{y}}$, this parameterization results in the integral

$$\int_C \mathbf{f} \cdot \mathbf{dl} = \int_0^{\pi} (0, \cos\theta) \cdot a(-\sin\theta, \cos\theta)\, d\theta = \int_0^{\pi} a\cos^2\theta\, d\theta = \pi a/2. \tag{1.20}$$

Notice that the result of the integral will always be independent of the choice of parameterization of the curve, so you are free to choose a parameterization that you find convenient. Sometimes we use the arc length, s, as the parameterization variable. In this case, $d\mathbf{r}/ds$ becomes the unit tangent vector along the curve.

Example: Line Integral Along a Straight Line

A vector field $\mathbf{f} = (f_0/a)\mathbf{r}$ points away from the origin. What is the curve integral of this vector field for a straight line from $(a, 0)$ to $(2a, a)$?

To find the integral, we need to make a parameterization of the curve. A straight line from $\mathbf{r}_0$ to $\mathbf{r}_1$ can be parameterized as $\mathbf{r}(t) = \mathbf{r}_0 + t\,(\mathbf{r}_1 - \mathbf{r}_0)$ for t from 0 to 1. The derivative of this parameterized curve with respect to t is

$$\frac{d\mathbf{r}}{dt} = (\mathbf{r}_1 - \mathbf{r}_0) = (2a, a) - (a, 0) = (a, a), \tag{1.21}$$

The line integral is therefore

$$
\begin{aligned}
\int_C \mathbf{f} \cdot \mathrm{d}\mathbf{l} &= \int_0^1 \mathbf{f}(\mathbf{r}(t)) \cdot \frac{\mathrm{d}\mathbf{r}}{\mathrm{d}t}\,\mathrm{d}t = \int_0^1 \frac{f_0}{a}\mathbf{r}(t) \cdot (a, a)\,\mathrm{d}t \\
&= \frac{f_0}{a}\int_0^1 ((a, 0) + t(a, a)) \cdot (a, a)\,\mathrm{d}t \\
&= \frac{f_0}{a}\int_0^1 ((a, 0) \cdot (a, a) + t(a, a) \cdot (a, a))\,\mathrm{d}t \\
&= \frac{f_0}{a}\int_0^1 \left(a^2 + ta^2 + ta^2\right)\,\mathrm{d}t \\
&= \frac{f_0}{a}\left(a^2 + a^2\right) = 2f_0 a
\end{aligned}
\tag{1.22}
$$

Notice that we can prove that this vector field is conservative. In general, the line integral along any curve C from $\mathbf{r}_0$ to $\mathbf{r}_1$ can be parameterized by some curve $\mathbf{r}(t)$ from $t = 0$ to $t = 1$. The integral is therefore

$$
\begin{aligned}
\int_C \mathbf{f} \cdot \mathrm{d}\mathbf{l} &= \int_C \frac{f_0}{a}\mathbf{r} \cdot \frac{\mathrm{d}\mathbf{r}}{\mathrm{d}t}\,\mathrm{d}t = \frac{f_0}{a}\int_0^1 \frac{\mathrm{d}}{\mathrm{d}t}\left(\frac{1}{2}\mathbf{r} \cdot \mathbf{r}\right)\,\mathrm{d}t \\
&= \frac{f_0}{2a}(\mathbf{r}_1 \cdot \mathbf{r}_1 - \mathbf{r}_0 \cdot \mathbf{r}_0)
\end{aligned}
\tag{1.23}
$$

This result is independent of the curve C. The line integral is therefore independent of the curve and the vector field is conservative.

Surface Integrals and the Flux of a Vector Field

For a vector field $\mathbf{f}$ we define $\mathbf{f} \cdot S\,\hat{\mathbf{n}}$ as the **flux** of the vector field through the surface with area S and surface normal $\hat{\mathbf{n}}$. Notice that we describe a surface by both an area, S, and a normal vector $\hat{\mathbf{n}}$. Often we write this as $\mathbf{S}$ for simplicity.

The flux of a fluid through a surface. That was a very abstract definition. We can make it easier to understand by looking at a fluid flowing with a velocity field $\mathbf{v}$. How much fluid flows through a small surface $\Delta\mathbf{S}$ per unit time, where we assume that the surface is so small that the velocity field $\mathbf{v}$ does not vary significantly across the surface? First, let us assume that $\mathbf{v}$ is in the direction of the surface normal $\hat{\mathbf{n}}$. In this case, the volume of fluid flowing through the surface in a short time interval Δt is the volume of the prism with base ΔS and length $v\Delta t$ as illustrated in Fig. 1.10a, that is, the volume is $\Delta S v \Delta t$ and the volume of fluid flowing through the surface per unit time, which is what we call the flux of the fluid, is $\Delta S v$. Since the velocity field and the surface normal points in the same direction this is also $\mathbf{v} \cdot \Delta\mathbf{S}$.

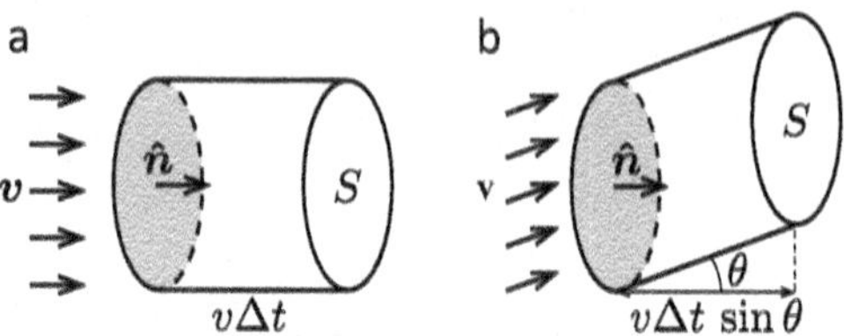

Fig. 1.10 **a** The flux through a surface S. **b** The flux through a surface S when $\mathbf{v}$ and $\hat{\mathbf{n}}$ are not parallell

The flux when the surface and the field are not parallel. This dot-product expression also makes sense even if the velocity field is not parallell to the surface normal. In this case, the volume of the fluid passing through the surface in a time interval Δt is illustrated in Fig. 1.10b. The volume of this prism with an inclined side is given by the base area of the prism, which is ΔS, multiplied with the height of the prism normal to the base area, which is $\Delta S v \Delta t \cos\theta$, which is $\Delta \mathbf{S} \cdot \mathbf{v}\Delta t$. We can also interpret this as the height of the prism being the projection of $\mathbf{v}\Delta t$ onto the surface normal $\hat{\mathbf{n}}$. The volume flowing through the surface per unit time is therefore $\mathbf{v} \cdot \Delta \mathbf{S}$, which is the flux.

A more complex surface. For a surface S where the field $\mathbf{f}$ varies across the surface or the surface has a more complex shape, we may approximate the flux of the field through the surface by dividing the surface into small elements $\Delta \mathbf{S}_i$ and summing up the contributions from all the elements. In the limit when the elements become small, this approaches what we call the **surface integral** of the field across the surface S, which we somewhat sloppily may define as:

$$\lim_{\Delta S \to 0} \sum_i \mathbf{f}(\mathbf{r}_i) \cdot \Delta \mathbf{S}_i = \int_S \mathbf{f} \cdot d\mathbf{S}. \tag{1.24}$$

In general, the normal vector may point in two possible directions for each surface element. However, we must choose one direction as the positive direction of the surface, which we call the orientation of the surface. A surface on which there is a well-defined orientation in each point is called an *orientable surface*. The surface integral is only defined for orientable surfaces. If the surface S is closed, the tradition is to define the outward direction as positive and to write a circle on the integral, $\oint_S \mathbf{f} \cdot d\mathbf{S}$. We typically only write one integral sign for the surface integral even though a surface is two-dimensional.

The **flux**, Φ, of a vector field $\mathbf{f}(\mathbf{r})$ through a surface S is defined as the surface integral

$$\Phi = \int_S \mathbf{f} \cdot d\mathbf{S}. \tag{1.25}$$

The flux is a *scalar*.

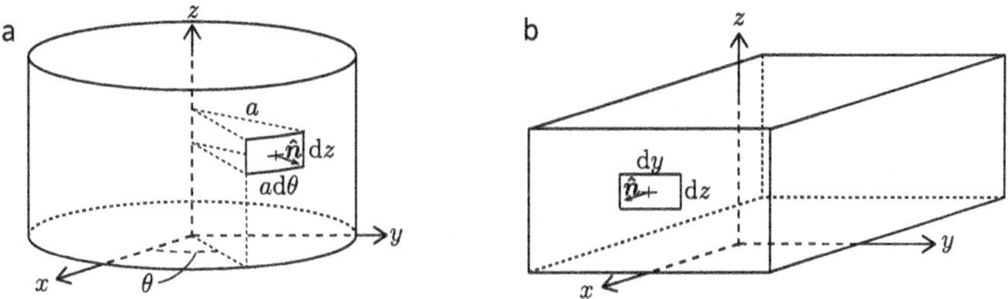

Fig. 1.11 **a** A cylindrical surface. **b** A surface with flat faces

Example: A Parameterized Surface

To calculate the surface integral, we need to describe the surface as a parameterized surface, that is, we need to describe the position of each point, $\mathbf{r}(t, s)$, on the surface S using two parameters, t, s, and with a surface normal $\hat{\mathbf{n}}$ and a surface element $\mathrm{d}S$ in each point. How this is done is best demonstrated by an example. Figure 1.11 illustrates two surfaces: a cylinder with radius a and height h in the z-direction and a surface consisting of 4 flat faces of length $2a$ and height L. How can we parameterize these surfaces?

A cylinder surface. A point on the cylinder surface is uniquely described by two coordinates, the angle θ with the x-axis and the position z along the z-axis. The point has coordinates $\mathbf{r}(\theta, z) = (a\cos\theta, a\sin\theta, z)$. The surface element has a normal in the same direction, $\hat{\mathbf{n}} = (\cos\theta, \sin\theta, 0)$ and the area of a surface element is $\mathrm{d}S = a\,\mathrm{d}\theta\,\mathrm{d}z$. The surface integral of a vector field $\mathbf{f} = (f_0/a)\mathbf{r}$ across this surface is then

$$\begin{aligned}\int_S \mathbf{f}\cdot\hat{\mathbf{n}}\,\mathrm{d}S &= \int_0^h \int_0^{2\pi} \frac{f_0}{a}(a\cos\theta, a\sin\theta, z)\cdot(\cos\theta, \sin\theta, 0)a\,\mathrm{d}\theta\,\mathrm{d}z \\ &= \frac{f_0}{a}h\int_0^{2\pi} a(\cos^2\theta + \sin^2\theta)a\,\mathrm{d}\theta = \frac{f_0}{a}ha^2 2\pi = f_0 2\pi ah\end{aligned} \tag{1.26}$$

A flat surface. A point on the flat surface at $x = a$ from $y = -a$ to $y = a$ is given by two coordinates: $\mathbf{r} = (a, y, z)$ where $-a < y < a$ and $0 < z < L$. The surface normal points in the positive x-direction, $\hat{\mathbf{n}} = \hat{\mathbf{x}}$. The area of a surface element is $\mathrm{d}S = \mathrm{d}y\,\mathrm{d}z$. The surface integral of a vector field $\mathbf{f} = (f_0/a)\mathbf{r}$ across this surface is then:

$$\begin{aligned}\int_S \mathbf{f}\cdot\hat{\mathbf{n}}\,\mathrm{d}S &= \int_0^L \int_{-a}^a \frac{f_0}{a}\mathbf{r}\cdot(1,0,0)\,\mathrm{d}y\,\mathrm{d}z = L\int_{-a}^a \frac{f_0}{a}(a, y, z)\cdot(1,0,0)\,\mathrm{d}y \\ &= \int_{-a}^a \frac{f_0}{a}La\,\mathrm{d}y = 2aLf_0.\end{aligned} \tag{1.27}$$

1.4 Differentiation

Partial Derivates and the Gradient

The derivative of a function, $f(x)$, of a single variable x, tells us how quickly the function is varying as we increase x: $f(x+\Delta x) \simeq f(x) + \frac{\mathrm{d}f}{\mathrm{d}x}\Delta x$. How can we generalize this to multivariable functions such as $f(x, y, z)$? In this case, we are interested in how the function changes when we make a change in both x, y and z, that is, a small change $\mathbf{dl} = \hat{\mathbf{x}}\,\mathrm{d}x + \hat{\mathbf{y}}\,\mathrm{d}y + \hat{\mathbf{z}}\,\mathrm{d}z$. The change $\mathrm{d}f$ is given by the theorem of partial derivatives:

$$\mathrm{d}f = \frac{\partial f}{\partial x}\mathrm{d}x + \frac{\partial f}{\partial y}\mathrm{d}y + \frac{\partial f}{\partial z}\mathrm{d}z \tag{1.28}$$

This can also be expressed as a dot product with the change $\mathbf{dl}$ in position:

$$\mathrm{d}f = \left(\frac{\partial f}{\partial x}\hat{\mathbf{x}} + \frac{\partial f}{\partial y}\hat{\mathbf{y}} + \frac{\partial f}{\partial z}\hat{\mathbf{z}}\right) \cdot \left(\hat{\mathbf{x}}\,\mathrm{d}x + \hat{\mathbf{y}}\,\mathrm{d}y + \hat{\mathbf{z}}\,\mathrm{d}z\right) \tag{1.29}$$

$$= (\nabla f) \cdot \mathbf{dl}, \tag{1.30}$$

where

$$\nabla f = \frac{\partial f}{\partial x}\hat{\mathbf{x}} + \frac{\partial f}{\partial y}\hat{\mathbf{y}} + \frac{\partial f}{\partial z}\hat{\mathbf{z}}. \tag{1.31}$$

We call ∇f the **gradient** of f. Notice that ∇f is a vector, even though the notation may not show it.

Interpretation of the gradient. What does the gradient tell us about the function $f(x, y, z)$? We know that if we move from $\mathbf{r} = (x, y, z)$ to $\mathbf{r} + \mathbf{dl}$, then the change in f is given as $\mathrm{d}f = \nabla f \cdot \mathbf{dl} = |\nabla f|\,|\mathbf{dl}|\cos\theta$, where θ is then angle between the gradient and $\mathbf{dl}$. If we keep the length of $\mathbf{dl}$ the same and vary the direction θ, then the maximum change in f occurs when ∇f and $\mathbf{dl}$ points in the same direction. This means that ∇f points in the direction in which the function f increases the most. Similarly, the function f decreases most rapidly in the direction of $-\nabla f$. Notice that ∇f is a function of the spatial position, depending on the function f. Also, notice that there is no change in the function when $\theta = \pi/2$ or $\theta = -\pi/2$, because then $\cos\theta = 0$. This means that the function does not change in the direction normal to ∇f. We call surfaces (in 3d, curves in 2d) along which the function f does not change, contours or equisurfaces. The gradient, ∇f, is therefore normal to the contour or equisurface.

The **gradient** ∇f of a function $f(x, y, z)$ is a vector that is defined as

$$\nabla f = \frac{\partial f}{\partial x}\hat{\mathbf{x}} + \frac{\partial f}{\partial y}\hat{\mathbf{y}} + \frac{\partial f}{\partial z}\hat{\mathbf{z}}. \tag{1.32}$$

It points in the direction of maximum increase of the function f. The magnitude of the gradient tells how rapidly the function is changing.

The gradient is a useful concept that you will frequently use, just as you are using the derivative. For example, a cell that always moves in the direction of increasing concentration of food, moves along the gradient of the food concentration. Alternatively, if you want to find out where a good smell is coming from, you would move along the gradient of the concentration of smell molecules. If you want to move uphill as quickly as possible, you would move along the gradient of the height. If you want to descend as quickly as possible, you would move opposite to the gradient.

Gradient descent. If we want to find the minimum of a multivariate function, we could start in a given position and then move to lower values of the function, that is, in the direction opposite to the gradient. We would call such an algorithm a gradient descent algorithm. It would not work if the gradient was zero. This is analogous to the one-dimensional situation, where a zero derivative indicates that the function is a maximum, a minimum or an inflection point. We may conclude the same for higher dimensional functions. Minima or maxima will occur where the gradient is zero.

Example: Finding the Gradient

What is the gradient of a height field, $f = z$, or the distance to the origin, $g = r = (x^2 + y^2 + z^2)^{1/2}$?

The gradient of the height field is

$$\nabla(z) = \frac{\partial z}{\partial x}\hat{\mathbf{x}} + \frac{\partial z}{\partial y}\hat{\mathbf{y}} + \frac{\partial z}{\partial z}\hat{\mathbf{z}} = 0\hat{\mathbf{x}} + 0\hat{\mathbf{y}} + 1\hat{\mathbf{z}} = \hat{\mathbf{z}}. \tag{1.33}$$

Notice that the result is a vector, as it should be, and it points upwards. This means that the function $f = z$ increases most rapidly in the positive z-direction. This is indeed reasonable.

The gradient of the distance to the origin, $g = r = (x^2 + y^2 + z^2)^{1/2}$, is

$$\nabla r = \frac{\partial r}{\partial x}\hat{\mathbf{x}} + \frac{\partial r}{\partial y}\hat{\mathbf{y}} + \frac{\partial r}{\partial z}\hat{\mathbf{z}}. \tag{1.34}$$

Here, we find that

$$\frac{\partial}{\partial x}\left(x^2+y^2+z^2\right)^{1/2}=\frac{\frac{1}{2}2x}{\left(x^2+y^2+z^2\right)^{1/2}}=\frac{x}{r}. \tag{1.35}$$

We will get similar results for the y and z components, so that:

$$\nabla r=\frac{x}{r}\hat{\mathbf{x}}+\frac{y}{r}\hat{\mathbf{y}}+\frac{z}{r}\hat{\mathbf{z}}=\frac{\mathbf{r}}{r}=\hat{\mathbf{r}}, \tag{1.36}$$

which is the unit vector in the radial direction. This means that the distance to the origin, g or r, increases most rapidly in the radial direction $\hat{\mathbf{r}}$, as we expected.

The Del Operator

It is common to introduce the simplified notation ∇f to describe the gradient using the *del operator*:

$$\nabla f=\frac{\partial f}{\partial x}\hat{\mathbf{x}}+\frac{\partial f}{\partial y}\hat{\mathbf{y}}+\frac{\partial f}{\partial z}\hat{\mathbf{z}}=\underbrace{\left(\frac{\partial}{\partial x}\hat{\mathbf{x}}+\frac{\partial}{\partial y}\hat{\mathbf{y}}+\frac{\partial}{\partial z}\hat{\mathbf{z}}\right)}_{=\nabla}f, \tag{1.37}$$

The **del operator** is defined as

$$\nabla=\left(\hat{\mathbf{x}}\frac{\partial}{\partial x}+\hat{\mathbf{y}}\frac{\partial}{\partial y}+\hat{\mathbf{z}}\frac{\partial}{\partial z}\right) \tag{1.38}$$

Notice that ∇ is an operator that acts on what is to the right of it. We must therefore be careful with the order in which we write the operator and the objects that it acts on. Notice also that ∇ is a vector and we will address three ways that it acts on objects: (i) It can act on a scalar function ∇f; (ii) It can act on a vector field through the dot product $\nabla\cdot\mathbf{f}$; (iii) It can act on a vector field through the cross product $\nabla\times\mathbf{f}$. The first case is the gradient, which we have discussed. In the following we will discuss the other two operations.

Divergence

The divergence of a vector field $\mathbf{f}(\mathbf{r})$ describes the flux out of each point in the vector field. The flux out of a surface S is given by the surface integral $\Phi = \oint_S \mathbf{f} \cdot \mathrm{d}\mathbf{S}$. To define the flux out from of point, we make a small surface ΔS around the point and then let the surface area go to zero. As the volume Δv that is confined by the surface becomes smaller and smaller, we define the divergence as the ratio of the flux out of the volume to the volume:

$$\mathrm{div}\, \mathbf{f} = \lim_{\Delta v \to 0} \frac{\oint_{\Delta S} \mathbf{f} \cdot \mathrm{d}\mathbf{S}}{\Delta v}. \tag{1.39}$$

This definition gives us an intuitive understanding of what the divergence is. If the divergence is positive in a point, it means that there is a flux out of this point. If the divergence is negative, it means that there is a flux into the point. However, the definition is not that practical to use to calculate the divergence for a given vector field.

The divergence in Cartesian coordinates. Let us see if we can find a more practical expression for the divergence by addressing a small, cubic volume at (x, y, z) with sides $\mathrm{d}x, \mathrm{d}y, \mathrm{d}z$ as illustrated in Fig. 1.12. First, let us look at the x-direction. For the surface at $x + \mathrm{d}x$ the normal vector is in the $\hat{\mathbf{x}}$ direction. The flux through this surface is therefore $\mathbf{f}(x + \mathrm{d}x, y, z) \cdot \hat{\mathbf{x}}\, \mathrm{d}S = f_x(x + \mathrm{d}x, y, z)\, \mathrm{d}S$. Similarly, for the surface at x, the normal vector is in the $-\hat{\mathbf{x}}$ direction and the flux through this surface is $\mathbf{f}(x, y, z) \cdot (-\hat{\mathbf{x}})\, \mathrm{d}S = -f_x(x, y, z)\, \mathrm{d}S$, where $\mathrm{d}S = \mathrm{d}y\, \mathrm{d}z$. The contribution to the flux along the x-axis is therefore $\mathrm{d}\Phi_x = (f_x(x + \mathrm{d}x, y, z) - f_x(x, y, z))\, \mathrm{d}y\, \mathrm{d}z$. We will get similar expressions in the y and the z directions. The surface integral out of the cubic surface S is therefore:

$$\begin{aligned} \oint_S \mathbf{f} \cdot \mathrm{d}\mathbf{S} = &\, (f_x(x + \mathrm{d}x, y, z) - f_x(x, y, z))\, \mathrm{d}y\, \mathrm{d}z \\ &+ \big(f_y(x, y + \mathrm{d}y, z) - f_y(x, y, z)\big)\, \mathrm{d}x\, \mathrm{d}z \\ &+ (f_z(x, y, z + \mathrm{d}z) - f_z(x, y, z))\, \mathrm{d}x\, \mathrm{d}y. \end{aligned} \tag{1.40}$$

The volume of the cube is $\mathrm{d}x\, \mathrm{d}y\, \mathrm{d}z$. In the limit when $\mathrm{d}x$, $\mathrm{d}y$ and $\mathrm{d}z$ becomes very small, we find the divergence to be:

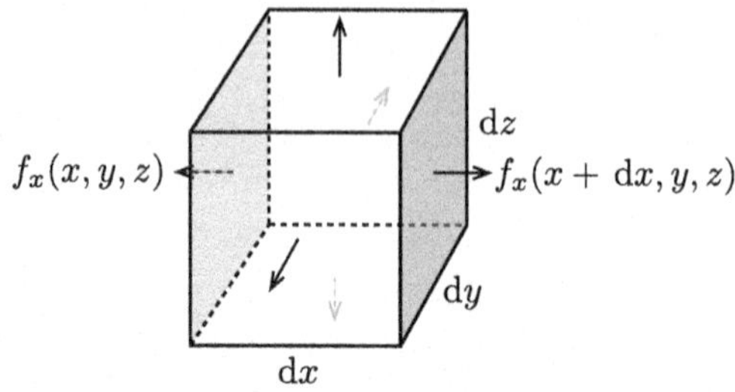

Fig. 1.12 The flux of a vector field **f** out through a small cubic volume

$$
\begin{aligned}
\operatorname{div} \mathbf{f} &= \frac{\oint_C \mathbf{f} \cdot \mathrm{d}\mathbf{s}}{\Delta v} \\
&= \frac{(f_x(x + \mathrm{d}x, y, z) - f_x(x, y, z))\, \mathrm{d}y\, \mathrm{d}z}{\mathrm{d}x\, \mathrm{d}y\, \mathrm{d}z} \\
&+ \frac{\big(f_y(x, y + \mathrm{d}y, z) - f_y(x, y, z)\big)\, \mathrm{d}x\, \mathrm{d}z}{\mathrm{d}x\, \mathrm{d}y\, \mathrm{d}z} \\
&+ \frac{(f_z(x, y, z + \mathrm{d}z) - f_z(x, y, z))\, \mathrm{d}x\, \mathrm{d}y}{\mathrm{d}x\, \mathrm{d}y\, \mathrm{d}z} \\
&= \frac{\partial f_x}{\partial x} + \frac{\partial f_y}{\partial y} + \frac{\partial f_z}{\partial z} = \left(\hat{\mathbf{x}}\frac{\partial}{\partial x} + \hat{\mathbf{y}}\frac{\partial}{\partial y} + \hat{\mathbf{z}}\frac{\partial}{\partial z}\right) \cdot \mathbf{f} = \nabla \cdot \mathbf{f}
\end{aligned}
\tag{1.41}
$$

The **divergence** of a vector field $\mathbf{f}(\mathbf{r})$ quantifies the flux out from each point in the vector field. In Cartesian coordinates the divergence is

$$
\operatorname{div} \mathbf{f} = \nabla \cdot \mathbf{f} = \frac{\partial f_x}{\partial x} + \frac{\partial f_y}{\partial y} + \frac{\partial f_z}{\partial z}. \tag{1.42}
$$

The divergence is a *scalar*.

Example: Divergence of Three Fields

Figure 1.13 illustrates three vector fields: $\mathbf{f} = f_0\, \hat{\mathbf{x}}$, $\mathbf{f} = (f_0/a)x\, \hat{\mathbf{x}}$, and $\mathbf{f} = -(f_0/a)\, y\, \hat{\mathbf{x}}$. Find the divergence of these fields.

We find the divergence by applying the formula $\nabla \cdot \mathbf{f}$ in each case. When $\mathbf{f} = f_0\hat{\mathbf{x}}$ the vector field is uniform, that is, it does not depend on the position. All the partial derivatives are therefore zero and the divergence is zero. For the case $\mathbf{f} = (f_0/a)x\, \hat{\mathbf{x}}$ we find that

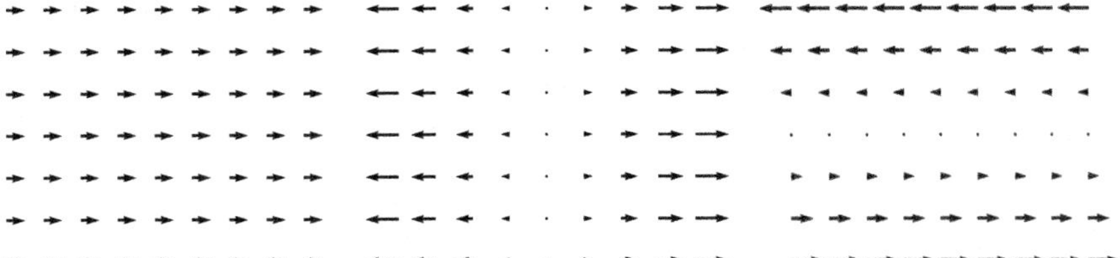

Fig. 1.13 Illustration of three vector fields: $\mathbf{f} = f_0\hat{\mathbf{x}}$, $\mathbf{f} = (f_0/a)x\hat{\mathbf{x}}$, and $\mathbf{f} = -(f_0/a)y\hat{\mathbf{x}}$

$$\nabla \cdot \mathbf{f} = \frac{\partial (f_0/a)x}{\partial x} + 0 + 0 = \frac{f_0}{a}. \tag{1.43}$$

And for the case $\mathbf{f} = -(f_0/a)y\,\hat{\mathbf{x}}$ we see that $f_y = f_z = 0$ and $f_x = -(f_0/a)y$ so that

$$\nabla \cdot \mathbf{f} = \frac{\partial - (f_0/a)y}{\partial x} = 0. \tag{1.44}$$

The Divergence Theorem

The divergence theorem relates the divergence of a vector field in a volume to the flux of the vector field out from this volume:

Divergence theorem. For any volume v which is enclosed by the closed surface S, we have that for any vector field $\mathbf{f}(\mathbf{r})$:

$$\int_v \nabla \cdot \mathbf{f}\, \mathrm{d}v = \oint_S \mathbf{f} \cdot \mathrm{d}\mathbf{S}. \tag{1.45}$$

This theorem is also known as Gauss' theorem.

Relation to fundamental theorem of calculus. This theorem is in form similar to the fundamental theorem of calculus, which states that the integral of the derivative of a function is given by the value of the function at the boundaries:

$$\int_a^b \frac{\mathrm{d}f}{\mathrm{d}x}\,\mathrm{d}x = f(b) - f(a). \tag{1.46}$$

This is similar to the divergence theorem, which states that the integral of the derivative of a vector field, in the form of the divergence, over a volume is equal to the integral of the flux of the vector field over the surface that forms the boundary of the volume.

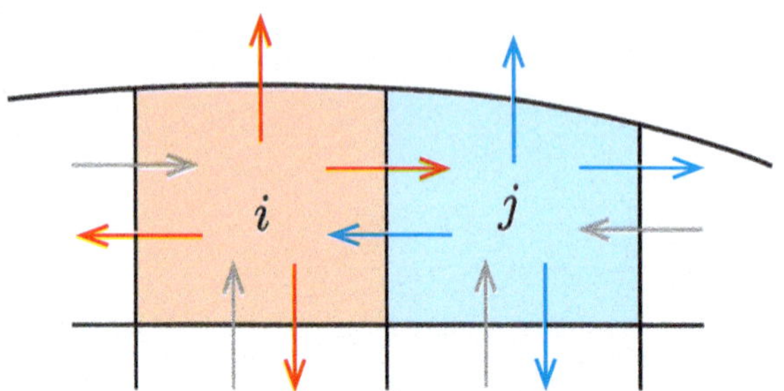

Fig. 1.14 A volume is divided into small volume elements. The arrows indicates the fluxes out of one element and into the neighboring elements. If we sum all the fluxes, all the internal fluxes come in pairs that cancel because they are equal in size but are in opposite directions

Motivation for the divergence theorem. The divergence theorem states that the flux out through the outer surface S is the sum of the divergences in all the volume elements enclosed by the surface. Formulated like this the theorem appears to be similar to a conservation law: the divergence from a small volume element is the flux out through surfaces of the volume element. The sum of all these fluxes is the flux out through the outer surface, because all the internal fluxes between the volume elements cancel each other out. Let us be more precise: We divide the volume v into small pieces Δv_i as illustrated in Fig. 1.14. For each such small volume element, the definition of divergence gives that

$$\frac{\oint_{S_i} \mathbf{f} \cdot d\mathbf{S}}{\Delta v_i} \simeq \nabla \cdot \mathbf{f} \Rightarrow \nabla \cdot \mathbf{f}\, \Delta v_i \simeq \oint_{S_i} \mathbf{f} \cdot d\mathbf{S}. \tag{1.47}$$

Now, if we sum all these contributions together, we will have a sum of surface integrals:

$$\sum_i \nabla \cdot \mathbf{f}\, \Delta v_i = \sum_i \oint_{S_i} \mathbf{f} \cdot d\mathbf{S}. \tag{1.48}$$

The flux through surface S_i is the sum of the fluxes through the surfaces that make up the surface S_i as shown in Fig. 1.14. For all the *inner* surfaces, there will be two such terms in the sum over all surfaces S_i: One contribution from the flux from i to j and one contribution from the flux from j to i. The surface from i to j and to j to i are the same, but they are oriented in the opposite directions. Thus the two contributions will sum to zero. This is the case for all inner surfaces. But for the outer surfaces, there is no element on the other side. Thus the sum of the all the surface integrals over all surfaces S_i is the same as the sum of the surface integrals over the outer surface. Thus the sum on the right-hand side in Eq. (1.48) is the integral over the outer surface S, whereas the left-hand side is the integral over all the volume elements in the limit when the elements become small. Thus we have demonstrated the theorem.

Curl

While the divergence characterizes the flux out of a point, we now introduce the **curl** which characterizes the rotation of a field around a point. A rotation must be defined relative to an axis. The curl is therefore a vector. How can we define how much a vector field rotates? We can see how much of the field is aligned with a path around the axis. To measure the rotation around the z-axis, we make a small path C in a plane normal to the z-axis, that is, in a plane parallel to the xy-plane. We also orient the curve so that the direction along the curve is in the positive direction around the z-axis according to the right-hand rule. We then define the curl of the vector field in a point as the limit of the line integral relative the area enclosed by the curve as the area of becomes small:

$$(\text{curl}\,\mathbf{f})_z = \lim_{\Delta S \to 0} \frac{\oint_C \mathbf{f} \cdot \mathrm{d}\mathbf{l}}{\Delta S}, \tag{1.49}$$

where the area ΔS is the area of a surface enclosed by the curve C and the curve C is in a plane normal to the z-axis. The definition of the curl in the two other directions, x and y, is similar.

Curl in Cartesian coordinates. Finding the curl of an actual vector field using this definition is not practical, so let us instead see if we can find a general formula by starting with a square curve with sides $\mathrm{d}x$ and $\mathrm{d}y$ in a plane parallel to the xy-plane as illustrated in Fig. 1.15. The line integral for this system is the sum of the contributions from each of the four lines of length $\mathrm{d}x$ and $\mathrm{d}y$. The contributions are

$$\begin{aligned}\oint_C \mathbf{f} \cdot \mathrm{d}\mathbf{l} &= f_x(x, y, z)\,\mathrm{d}x + f_y(x + \mathrm{d}x, y, z)\,\mathrm{d}y \\ &\quad + f_x(x, y + \mathrm{d}y, z)(-\mathrm{d}x) + f_y(x, y, z)(-\mathrm{d}y).\end{aligned} \tag{1.50}$$

We divide by the area $\mathrm{d}S = \mathrm{d}x\,\mathrm{d}y$ and find that

$$\begin{aligned}\frac{\oint_C \mathbf{f} \cdot \mathrm{d}\mathbf{l}}{\mathrm{d}S} &= -\frac{f_x(x, y + \mathrm{d}y, z) - f_x(x, y, z)}{\mathrm{d}y} \\ &\quad + \frac{f_y(x + \mathrm{d}x, y, z) - f_y(x, y, z)}{\mathrm{d}x}.\end{aligned} \tag{1.51}$$

We recognize that in the limit of small $\mathrm{d}x$ and $\mathrm{d}y$ the right-hand side is the partial derivatives of f_x with respect to y and f_y with respect to x, giving us

$$(\text{curl}\,\mathbf{f})_z = \frac{\partial f_y}{\partial x} - \frac{\partial f_x}{\partial y} \tag{1.52}$$

Now, by comparing this expression with $\nabla \times \mathbf{f}$, we see that we have found the z-component $(\nabla \times \mathbf{f})_z$. Similarly, you can find the two other components of the curl. Together we find that

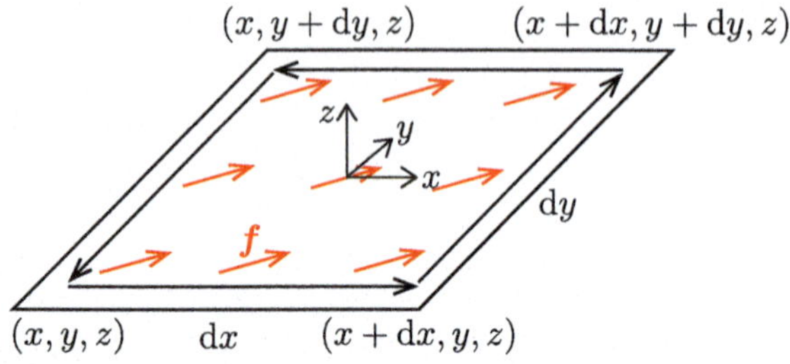

Fig. 1.15 An illustration of a small curve C in the xy-plane around the z-axis

The **curl** of a vector field $\mathbf{f}(x, y, z)$ is given as

$$\text{curl}\,\mathbf{f} = \nabla \times \mathbf{f}, \tag{1.53}$$

which in Cartesian coordinates is

$$\nabla \times \mathbf{f} = \left(\frac{\partial f_z}{\partial y} - \frac{\partial f_y}{\partial z}, \frac{\partial f_x}{\partial z} - \frac{\partial f_z}{\partial x}, \frac{\partial f_y}{\partial x} - \frac{\partial f_x}{\partial y} \right). \tag{1.54}$$

The curl is a *vector*.

Example: Curl of Three Fields

Figure 1.13 illustrates three vector fields: $\mathbf{f} = f_0\hat{\mathbf{x}}$, $\mathbf{f} = (f_0/a)x\hat{\mathbf{x}}$, and $\mathbf{f} = -(f_0/a)y\hat{\mathbf{x}}$. Find the curl of these fields.

We find the curl by applying the formula $\nabla \times \mathbf{f}$ in each case. When $\mathbf{f} = f_0\hat{\mathbf{x}}$ the vector field does not depend on x, y or z and all partial derivatives are zero. The curl is therefore zero. For the case $\mathbf{f} = (f_0/a)x\hat{\mathbf{x}}$ we find that the only partial derivative that is non-zero is of f_x with respect to x. Thus all the off-diagonal partial derivatives are zero and the curl is zero. For the case $\mathbf{f} = -(f_0/a)y\hat{\mathbf{x}}$ we see that the only non-zero partial derivative is of f_x with respect to y. This is an off-diagonal element, which enters in the z-component of the curl, which is

$$(\nabla \times \mathbf{f})_z = \frac{\partial f_y}{\partial x} - \frac{\partial f_x}{\partial y} = \frac{f_0}{a}. \tag{1.55}$$

We see that the curl is in the z-direction and positive, which corresponds to the observed orientation of the rotation of the field in Fig. 1.13.

Stokes' Theorem

We found that the divergence theorem was analogous to the fundamental theorem of calculus for the divergence. What is the corresponding theorem for the curl? It is called Stokes' theorem and it states that

Stokes' theorem. For any curve C that encloses a surface S we have that:

$$\oint_C \mathbf{f} \cdot d\mathbf{l} = \int_S (\nabla \times \mathbf{f}) \cdot d\mathbf{S}. \tag{1.56}$$

Motivation for Stokes' theorem. We can motivate the theorem using a similar approach to the divergence theorem. We demonstrate the theorem for the z-component of the curl. We choose a curve C that encloses a surface S in the xy-plane. We divide the surface into smaller pieces on a square grid as illustrated in Fig. 1.16. The idea is that we can replace the line integral around the outer boundary C with the sum of the line integrals along the enclosing curves C_i around each of the small surfaces ΔS_i:

$$\oint_C \mathbf{f} \cdot d\mathbf{l} = \sum_i \oint_{C_i} \mathbf{f} \cdot d\mathbf{l}. \tag{1.57}$$

In the sum on the right-hand side, there will be two contributions from each internal border between two surfaces. For example, for the two surfaces i and j, there will be one contribution to the line integral around surface S_i along the common border between the two, but an identical contribution in the opposite direction from the integral around the surface S_j. These two contributions will cancel out in the sum. However, all the contributions along the outer border do not have a similar contribution from the other side of the border. These contributions will therefore add up and in total they will be the line integral along the external border C of the surface S.

For each small surface ΔS_i, we use the definition of the curl to relate the integral to the curl:

$$(\nabla \times \mathbf{f})_z \, \Delta S_i = \oint_{C_i} \mathbf{f} \cdot d\mathbf{l}. \tag{1.58}$$

The line integral around the curve C that encloses S is therefore the sum of the line integrals around all the surfaces ΔS_i, which again is equal to the sum of the curls:

$$\oint_C \mathbf{f} \cdot d\mathbf{l} = \sum_i \oint_{C_i} \mathbf{f} \cdot d\mathbf{l} = \sum_i (\nabla \times \mathbf{f})_z \, \Delta S_i. \tag{1.59}$$

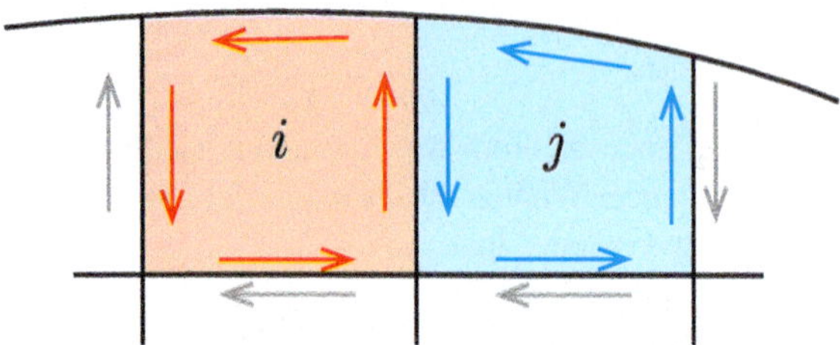

Fig. 1.16 A surface S is divided into small surface elements ΔS_i. The arrows indicates the curves C_i along the borders of each of the elements. Along each internal boundary between two surface elements i and j there are two path elements in opposite directions. The sum of the line integral along two such paths is zero

When the size of the surfaces ΔS_i goes to zero, the right-hand side is the definition of the surface integral of the curl of the field. We have therefore motivated that

$$\oint_C \mathbf{f} \cdot \mathrm{d}\mathbf{l} = \int_S (\nabla \times \mathbf{f}) \cdot \mathrm{d}\mathbf{S}, \tag{1.60}$$

which is Stokes' theorem.

1.5 Coordinate Systems

We are used to describe positions in space using the Cartesian coordinate system. In this case, we describe a position in terms of three orthogonal unit vectors $\hat{\mathbf{x}}$, $\hat{\mathbf{y}}$ and $\hat{\mathbf{z}}$ along the x, y and z axes. However, in many cases it is practical to use coordinates that correspond to particular symmetries in the system we study. The two most commonly used coordinate systems are *cylindrical coordinates* and *spherical coordinates*.

Cylindrical Coordinates

In a cylindrical coordinate system we describe a position in space relative to the z-axis by the position along the z-axis, the distance r from the axis and the angle ϕ around the z-axis measured in the positive direction from the x-axis as illustrated in Fig. 1.17. We can go between descriptions in Cartesian and cylindrical coordinates by realizing that a position (r, ϕ, z) in cylindrical coordinates corresponds to (x, y, z) in Cartesian coordinates where

$$x = r\cos\phi \quad y = r\sin\phi \quad z = z. \tag{1.61}$$

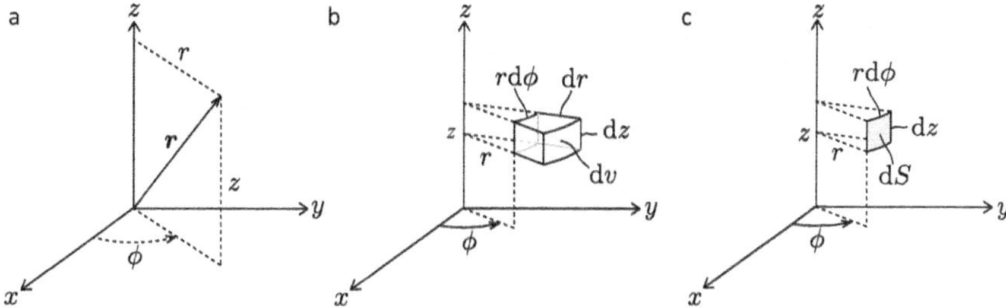

Fig. 1.17 Illustration of a position in a cylindrical coordinate system where a position is described by the (r, ϕ, z) coordinates

We call this a transformation from the cylindrical to the Cartesian coordinate system. We can also perform the transformation in the reverse. For a position (x, y, z) in the Cartesian coordinate system, we can find the corresponding coordinates in the cylindrical coordinate system by $r = (x^2 + y^2)^{1/2}$ and $z = z$. To find ϕ we note that $x = r \cos\phi$ and $y = r \sin\phi$. This can be used to find ϕ through $\phi = \arctan(y/x)$ and a consideration of the signs of x and y.

Vector fields in cylindrical coordinates. For Cartesian coordinates, we describe a vector field using the three unit vectors $\hat{\mathbf{x}}$, $\hat{\mathbf{y}}$, and $\hat{\mathbf{z}}$: $\mathbf{E}(x, y, z) = E_x(x, y, z)\hat{\mathbf{x}} + E_y(x, y, z)\hat{\mathbf{y}} + E_z(x, y, z)\hat{\mathbf{z}}$. Similarly, we would like to describe a vector field in cylindrical coordinates using unit vectors $\hat{\mathbf{r}}$, $\hat{\boldsymbol{\phi}}$ and $\hat{\mathbf{z}}$ so that a vector field can be written as

$$\mathbf{E} = E_r(r, \phi, z)\hat{\mathbf{r}} + E_\phi(r, \phi, z)\hat{\boldsymbol{\phi}} + E_z(r, \phi, z)\hat{\mathbf{z}}. \tag{1.62}$$

To transform from cylindrical coordinates to Cartesian coordinates, we rewrite the cylindrical unit vectors in Cartesian coordinates. They are

$$\hat{\mathbf{r}} = \hat{\mathbf{x}}\cos\phi + \hat{\mathbf{y}}\sin\phi \quad \hat{\boldsymbol{\phi}} = -\hat{\mathbf{x}}\sin\phi + \hat{\mathbf{y}}\cos\phi \quad \hat{\mathbf{z}} = \hat{\mathbf{z}}. \tag{1.63}$$

Similarly, to transform from Cartesian coordinates to cylindrical coordinates, we rewrite the Cartesian unit vectors in cylindrical coordinates as

$$\hat{\mathbf{x}} = \hat{\mathbf{r}}\cos\phi - \hat{\boldsymbol{\phi}}\sin\phi \quad \hat{\mathbf{y}} = \hat{\mathbf{r}}\sin\phi + \hat{\boldsymbol{\phi}}\cos\phi \quad \hat{\mathbf{z}} = \hat{\mathbf{z}}. \tag{1.64}$$

The field $\mathbf{E} = E_0\hat{\mathbf{r}} + E_1\hat{\boldsymbol{\phi}}$ in cylindrical coordinates is therefore $\mathbf{E} = E_0(\hat{\mathbf{x}}\cos\phi + \hat{\mathbf{y}}\sin\phi) + E_1(-\hat{\mathbf{x}}\sin\phi + \hat{\mathbf{y}}\cos\phi) = \hat{\mathbf{x}}(E_0\cos\phi - E_1\sin\phi) + \hat{\mathbf{y}}(E_0\sin\phi + E_1\cos\phi)$ in Cartesian coordinates.

Cylindrical symmetry. We will often use this type of decomposition of vector fields in cylindrical coordinates when we study systems that have cylindrical symmetry. What does it mean to have cylindrical symmetry? A physical system has cylindrical symmetry if the system does not change when the system is rotated around the z-axis or when it is translated along the z-axis. An infinitely long cylinder will have cylindrical symmetry. A finite cylinder has rotational symmetry around the z-axis, but not the translational symmetry along the z-axis. However, it is often still useful to describe a finite cylinder using cylindrical coordinates to include the rotational symmetry of the system.

Derivatives in cylindrical coordinates. The gradient in cylindrical coordinates is

$$\nabla f = \frac{\partial f}{\partial r}\hat{\mathbf{r}} + \frac{1}{r}\frac{\partial f}{\partial \phi}\hat{\boldsymbol{\phi}} + \frac{\partial f}{\partial z}\hat{\mathbf{z}}. \tag{1.65}$$

The divergence in cylindrical coordinates is

$$\nabla \cdot \mathbf{f} = \frac{1}{r}\frac{\partial (r f_r)}{\partial r} + \frac{1}{r}\frac{\partial f_\phi}{\partial \phi} + \frac{\partial f_z}{\partial z}. \tag{1.66}$$

The curl in cylindrical coordinates is

$$\nabla \times \mathbf{f} = \left(\frac{1}{r}\frac{\partial f_z}{\partial \phi} - \frac{\partial f_\phi}{\partial z}\right)\hat{\mathbf{r}} + \left(\frac{\partial f_r}{\partial z} - \frac{\partial f_z}{\partial r}\right)\hat{\boldsymbol{\phi}} + \left(\frac{\partial (r f_\phi)}{\partial r} - \frac{\partial f_r}{\partial \phi}\right)\frac{1}{r}\hat{\mathbf{z}}. \tag{1.67}$$

Volume integration in cylindrical coordinates. A small volume element $\mathrm{d}v$ in cylindrical coordinates is represented as a piece of a cylinder shell with thickness $\mathrm{d}z$ along the z-axis, extending from r to $r + \mathrm{d}r$ in the radial direction and from ϕ to $\phi + \mathrm{d}\phi$ in the azimuthal (angular) direction. The volume of such an element is $\mathrm{d}v = \mathrm{d}r\, r\, \mathrm{d}\phi\, \mathrm{d}z$.

Surface integration in cylindrical coordinates. A small surface element on the outside of the curved cylinder surface has a surface normal in the $\hat{\mathbf{r}}$ direction and an area $\mathrm{d}S = r\, \mathrm{d}\phi\, \mathrm{d}z$ so that $\mathrm{d}\mathbf{S} = r\, \mathrm{d}\phi\, \mathrm{d}z\, \hat{\mathbf{r}}$.

Spherical Coordinates

In a spherical coordinate system we describe a position in space as we would describe the position on a spherical surface. The radius of the surface, r, tells us how far out the spherical surface is. In addition, we need two angles to describe the position on the surface, just as we need to angles to describe positions on the surface of the Earth. We use the angle ϕ to describe rotation around the z-axis and the angle θ to describe the angular position relative to the z-axis as shown in Fig. 1.18. A position (r, ϕ, θ) in spherical coordinates can be represented in Cartesian coordinates as $(r \sin\theta \cos\phi, r \sin\theta \sin\phi, r \cos\theta)$.

Unit vectors for spherical coordinates. The unit vectors in spherical coordinates are $\hat{\mathbf{r}}$, $\hat{\boldsymbol{\phi}}$ and $\hat{\boldsymbol{\theta}}$, which depend on the position. In Cartesian coordinates the unit vectors are $\hat{\mathbf{r}} = \hat{\mathbf{x}} \sin\theta \cos\phi + \hat{\mathbf{y}} \sin\theta \sin\phi + \hat{\mathbf{z}} \cos\theta$, $\hat{\boldsymbol{\phi}} = -\hat{\mathbf{x}} \sin\phi + \hat{\mathbf{y}} \cos\phi$ and

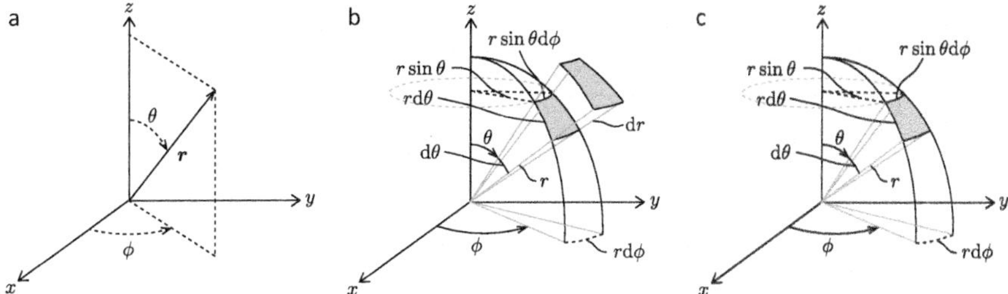

Fig. 1.18 Illustration of a position in a spherical coordinate system where a position is described by the (r, θ, ϕ) coordinates

$\hat{\boldsymbol{\theta}} = \hat{\mathbf{x}} \cos\theta \cos\phi + \hat{\mathbf{y}} \cos\theta \sin\phi - \hat{\mathbf{z}} \sin\theta$. A vector field in a position (r, ϕ, θ) is described in terms of its components along the unit vectors, $\mathbf{f} = f_r \hat{\mathbf{r}} + f_\phi \hat{\boldsymbol{\phi}} + f_\theta \hat{\boldsymbol{\theta}}$.

Spherical symmetry. We will often use this type of decomposition of vector fields in spherical coordinates when we study systems that have spherical symmetry. What does it mean to have spherical symmetry? A physical system has spherical symmetry if the system does not change when the system is rotated around any axis. The most typical example would be a sphere where the distribution of properties only depends on the distance to the center of the sphere and not on the direction.

Derivatives in spherical coordinates. The gradient in spherical coordinates is

$$\nabla f = \frac{\partial f}{\partial r}\hat{\mathbf{r}} + \frac{1}{r}\frac{\partial f}{\partial \theta}\hat{\boldsymbol{\theta}} + \frac{1}{r\sin\theta}\frac{\partial f}{\partial \phi}\hat{\boldsymbol{\phi}}. \tag{1.68}$$

The divergence in spherical coordinates is

$$\nabla \cdot \mathbf{f} = \frac{1}{r^2}\frac{\partial\left(r^2 f_r\right)}{\partial r} + \frac{1}{r\sin\theta}\frac{\partial\left(\sin\theta f_\theta\right)}{\partial\theta} + \frac{1}{r\sin\theta}\frac{\partial f_\phi}{\partial\phi}. \tag{1.69}$$

The curl in spherical coordinates is

$$\begin{aligned}\nabla \times \mathbf{f} = &\frac{1}{r\sin\theta}\left(\frac{\partial\left(\sin\theta f_\phi\right)}{\partial\theta} \quad \frac{\partial f_\theta}{\partial\phi}\right)\hat{\mathbf{r}} \\ &+ \frac{1}{r}\left(\frac{1}{\sin\theta}\frac{\partial f_r}{\partial\phi} - \frac{\partial\left(r f_\phi\right)}{\partial r}\right)\hat{\boldsymbol{\theta}} + \frac{1}{r}\left(\frac{\partial(r f_\theta)}{\partial r} - \frac{\partial f_r}{\partial\theta}\right)\hat{\boldsymbol{\phi}}.\end{aligned} \tag{1.70}$$

Surface integration in spherical coordinates. A small surface element $\mathrm{d}S$ in spherical coordinates is shown in Fig. 1.18. The surface consists of a small piece of the sphere spanned by the circle around the z-axis with radius $r\sin\theta$ and the circle with radius r and an angle ϕ with the x-axis as illustrated. The area of this surface element is approximately $\mathrm{d}S = r\,\mathrm{d}\theta\, r\sin\theta\,\mathrm{d}\phi = r^2\sin\theta\,\mathrm{d}\theta\,\mathrm{d}\phi$.

Volume integration in spherical coordinates. A small volume element $\mathrm{d}v$ in spherical coordinates is represented as a piece of a spherical shell with thickness $\mathrm{d}r$ along the r-axis, extending over angles $\mathrm{d}\theta$ and $\mathrm{d}\phi$ as illustrated in Fig. 1.18. The thickness of the shell is $\mathrm{d}r$ and the surface area of the shell is $\mathrm{d}S = (r \sin\theta\, \mathrm{d}\phi)\,(r\, \mathrm{d}\theta)$ so that the volume is $\mathrm{d}v = r^2 \sin\theta\, \mathrm{d}r\, \mathrm{d}\theta\, \mathrm{d}\phi$.

Summary

Dot product. The dot product between two vectors $\mathbf{a}$ and $\mathbf{b}$ is defined as $\mathbf{a}\cdot\mathbf{b} = |\mathbf{a}||\mathbf{b}|\cos\theta$, where θ is the angle between the two vectors. In Cartesian coordinates the dot product is $(a_x, a_y, a_z)\cdot(b_x, b_y, b_z) = a_x b_x + a_y b_y + a_z b_z$. The dot product between two vectors is zero if and only if the two vectors are orthogonal.

Cross product. The cross product between two vectors $\mathbf{a}$ and $\mathbf{b}$ is defined as $\mathbf{a}\times\mathbf{b} = \hat{\mathbf{n}}|\mathbf{a}||\mathbf{b}|\sin\theta$, where θ is the angle between the two vectors and $\hat{\mathbf{n}}$ is a unit vector in a direction normal to both vectors and the direction of $\hat{\mathbf{n}}$ is chosen according to the right-hand-rule.

Scalar and vector fields. We define a scalar field as a multivariate function $f(\mathbf{r})$, which depends on the position $\mathbf{r} = (x, y, z)$. A vector field is a vector function $\mathbf{f}(\mathbf{r})$ which depends on the position $\mathbf{r}$.

Integrals. We introduced three different integral types: The volume integral $\int_v f(\mathbf{r})\,\mathrm{d}v$, the line integral $\int_C \mathbf{f}\cdot\mathrm{d}\mathbf{l}$, and the surface integral $\int_S \mathbf{f}\cdot\mathrm{d}\mathbf{S}$. To calculate the line and surface integrals we must parameterize the curve or the surface along which we integrate.

The gradient. The gradient of a scalar field

$$\operatorname{grad} f = \nabla f = \hat{\mathbf{x}}\frac{\partial f}{\partial x} + \hat{\mathbf{y}}\frac{\partial f}{\partial y} + \hat{\mathbf{z}}\frac{\partial f}{\partial z},$$

points in the direction in which the scalar field increases the most. The gradient is normal to the equisurfaces, that is, surfaces with the same value of f.

The Del operator. We often use the Del operator to write differential operations in shorthand. The Del operator is

$$\nabla = \hat{\mathbf{x}}\frac{\partial}{\partial x} + \hat{\mathbf{y}}\frac{\partial}{\partial y} + \hat{\mathbf{z}}\frac{\partial}{\partial z}.$$

Divergence. The divergence of a vector field describes the flux of the vector field out from a point. The divergence is defined as

$$\operatorname{div}\mathbf{f} = \lim_{\Delta S\to 0}\frac{\int_{\Delta S}\mathbf{f}\cdot\mathrm{d}\mathbf{S}}{\Delta v},$$

Fig. 1.19 Two vectors

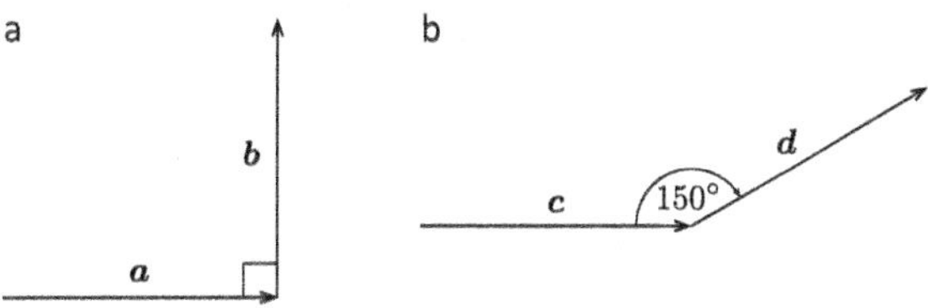

where the integral is over a small surface ΔS enclosing a volume Δv around a point $\mathbf{r}$. In Cartesian coordinates the divergence is:

$$\operatorname{div}\mathbf{f} = \nabla \cdot \mathbf{f} = \frac{\partial f_x}{\partial x} + \frac{\partial f_y}{\partial y} + \frac{\partial f_z}{\partial z}.$$

Curl. The curl of a vector field describes how much the vector field is curling around a point. The curl around an axis in the z-direction is defined as

$$(\operatorname{curl}\mathbf{f})_z = \lim_{C\to 0} \frac{\oint_C \mathbf{f} \cdot \mathbf{dl}}{\Delta S},$$

where the curve C encloses a surface ΔS around the point $\mathbf{r}$ in a plane normal to the z-axis. In Cartesian coordinates the curl is:

$$\nabla \times \mathbf{f} = \left(\frac{\partial f_z}{\partial y} - \frac{\partial f_y}{\partial z}, \frac{\partial f_x}{\partial z} - \frac{\partial f_z}{\partial x}, \frac{\partial f_y}{\partial x} - \frac{\partial f_x}{\partial y} \right).$$

Exercises

Discussion Exercises

1.1 Manhattan-norm. We have introduced the norm of a vector (a_x, a_y, a_z) as $\sqrt{a_x^2 + a_y^2 + a_z^2}$. We call this the Euclidean norm or the $L2$ norm. In general we can introduce the Lk norm for an integer k as

$$|\mathbf{a}|_k = \left(|a_x|^k + |a_y|^k + |a_z|^k\right)^{1/k} \tag{1.71}$$

Why do you think that the $L1$ norm is often called the Manhattan norm? How would you interpret the $L0$ norm? And what does the $L\infty$-norm correspond to?

Exercises

1.2 Vector sums. Figure 1.19a shows two vectors **a** and **b** each of length 1m.
(**a**) Draw the vector $\mathbf{A} = \mathbf{a} + \mathbf{b}$.

(b) Find $\mathbf{a} \cdot \mathbf{b}$.
(c) Find the length of $\mathbf{a} + \mathbf{b}$.

Figure 1.19b shows two vectors **c** and **d** each of length 1m.
(d) Draw the vector $\mathbf{c} + \mathbf{d}$.
(e) Find $\mathbf{c} \cdot \mathbf{d}$.
(f) Find the length of $\mathbf{c} + \mathbf{d}$

Assume that the unit vectors for a coordinate system are $\hat{\mathbf{x}} = \mathbf{a}/|\mathbf{a}|$ and $\hat{\mathbf{y}} = \mathbf{b}/|\mathbf{b}|$ and that $\mathbf{c} = \mathbf{a}$.
(g) What is $\mathbf{c} \cdot \hat{\mathbf{x}}$, $\mathbf{c} \cdot \hat{\mathbf{y}}$, $\mathbf{d} \cdot \hat{\mathbf{x}}$, and $\mathbf{d} \cdot \hat{\mathbf{y}}$?
(h) Write **c** and **d** in Cartesian coordinates, that is, on the form (x, y).

1.3 Decomposition. We have two vectors in a Cartesian coordinate system: $\mathbf{a} = (2, 1)$ and $\mathbf{b} = (2, 4)$.
(a) Find $\mathbf{a} \cdot \mathbf{b}$.
(b) Explain why we can interpret $\mathbf{b} \cdot \mathbf{a}/|\mathbf{a}|$ as the length of **b** along a.
(c) We call the vector $\mathbf{b_a} = \mathbf{b} \cdot (\mathbf{a}/|\mathbf{a}|)\,(\mathbf{a}/|\mathbf{a}|)$ the projection of **b** onto **a**. Find the projection of **b** onto **a**.
(d) Show that $\mathbf{b} - \mathbf{b_a}$ is orthogonal to **a**.

1.4 Cross products of a unit vector. The Cartesian unit vectors in a three-dimensional system are $\hat{\mathbf{x}} = (1, 0, 0)$, $\hat{\mathbf{y}} = (0, 1, 0)$, $\hat{\mathbf{z}} = (0, 0, 1)$.
(a) A set of vectors are called orthonormal if they all are of unit length and they are orthogonal to each other. Show that the set of unit vector $\{\hat{\mathbf{x}}, \hat{\mathbf{y}}, \hat{\mathbf{z}}\}$ are orthonormal.
(b) A sequence of three orthonormal unit vectors $(\hat{\mathbf{x}}, \hat{\mathbf{y}}, \hat{\mathbf{z}})$ form a positively orientated coordinate system if $\hat{\mathbf{x}} \times \hat{\mathbf{y}} = \hat{\mathbf{z}}$. Show that this is true for $(\hat{\mathbf{x}}, \hat{\mathbf{y}}, \hat{\mathbf{z}})$.
(c) What other sequences of the unit vectors form a positively oriented coordinate system?

1.5 Sketching a vector field. A vector field is given as $\mathbf{E}(\mathbf{r}) = \mathbf{r}/|\mathbf{r}|^3$.
(a) Make a sketch of the vector field in the xy-plane.
(b) Write a program to plot the vector field in the xy-plane.
(c) A vector field is given as $\mathbf{E} = \frac{3(\mathbf{r}\cdot\mathbf{p})\mathbf{r}}{r^5} - \frac{\mathbf{p}}{r^3}$ where $\mathbf{p} = (1, 0)$ and $r = |\mathbf{r}|$. Write a program to visualize the vector field in the xy-plane.

1.6 Fluxes. Figure 1.20 shows six different surfaces S with surface normals indicated by the red vector. There is a vector field $\mathbf{E} = E_0\hat{\mathbf{z}}$ everywhere in space. Find the flux Φ of the vector field through the different surfaces.
(a) Surface S_a has vertices at $(1, 1, 0)$, $(-1, 1, 0)$, $(1, -1, 0)$ and $(-1, -1, 0)$.
(b) Surface S_b has vertices at $(1, 1, 1)$, $(-1, 1, 1)$, $(1, -1, 1)$ and $(-1, -1, 1)$.
(c) Surface S_c has vertices at $(1, 1, 1)$, $(-1, 1, 1)$, $(1, -1, 0)$ and $(-1, -1, 0)$.
(d) Surface S_d has vertices at $(1, 1, 0)$, $(-1, 1, 0)$, $(1, -1, 1)$ and $(-1, -1, 1)$.
(e) Surface S_e has vertices at $(1, 0, 1)$, $(-1, 0, 1)$, $(1, 0, -1)$ and $(-1, 0, -1)$.
(f) Surface S_f has vertices at $(1, 1, 0)$, $(-1, 1, 0)$, $(1, -1, 1)$ and $(-1, -1, 1)$.

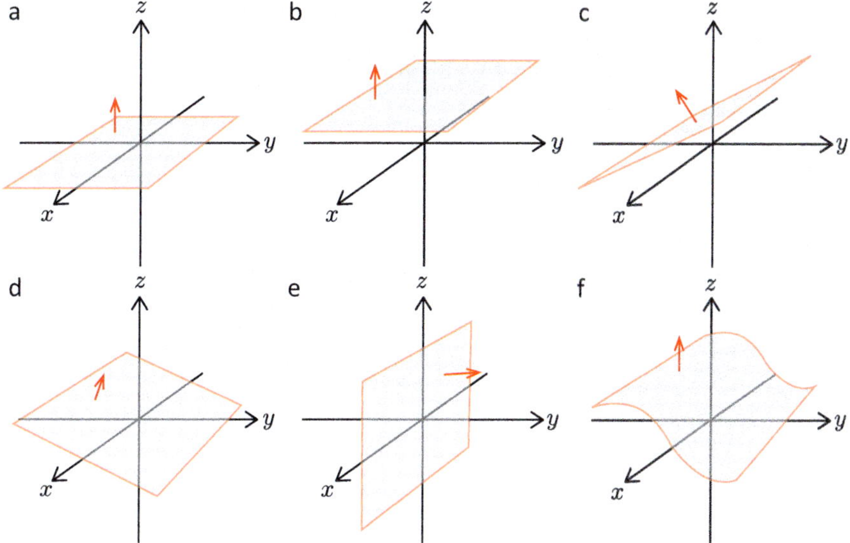

Fig. 1.20 Illustration of surfaces S_a to S_f.

1.7 Gradients. Find the gradients of the following scalar fields
(a) $V(x, y) = V_0 x$
(b) $V(x, y) = V_0 \sin x$
(c) $V(x, y, z) = V_0(x + y - z)$
(d) $V(r) = r^2 = (x^2 + y^2 + z^2)$
(e) $V(r) = r = (x^2 + y^2 + z^2)^{1/2}$
(f) $V(r) = V_0/r$

1.8 Divergence and curl. Find the divergence and curl of the following vector fields
(a) $\mathbf{E} = (1, 0, 0)$
(b) $\mathbf{E} = \mathbf{r} = (x, y, z)$
(c) $\mathbf{E} = (x, y, -2z)$
(d) $\mathbf{E} = (-y, x, 0)$
(e) $\mathbf{E} = \nabla V_0 e^{-y} \sin x$
(f) $\mathbf{E} = \mathbf{r}/r^3$

Tutorials

1.9 The field in a tornado. You have just seen the movie Twisters and start to discuss the wind velocities inside a tornado.
(a) Sketch the velocity field inside a tornado with its center in the origin.
(b) What is the behavior in the middle of the tornado? And far away?

(c) Could you describe this velocity field, $\mathbf{v}(x, y)$, as a vector field? Write it down as a function on a form that you find reasonable.
(d) How can you check what this field looks like in the xy-plane? Visualize the field in the xy-plane.

1.10 Eye of the storm. A vector field is given as

$$\mathbf{f}(x, y) = \frac{1}{\sqrt{x^2 + y^2)}}(-y, x)$$

(a) Sketch the field in the xy-plane.
(b) What happens to the field when $r = \sqrt{x^2 + y^2}$ approaches zero? Is the field defined here? Why/why not?

1.11 Along a line or along a curve?. A vector field has the shape $\mathbf{f} = (y + 2, 2)$ in the xy-plane.
(a) Define a parameterized curve along the x-axis and calculate the integral from $x = 0$ to $x = 5$ using this parameterization.
(b) Find the line integral along a line $y = 2x$ from $(0, 0)$ to $(1, 2)$.
(c) Are there different ways to parameterize this line? Does the choice of parameterization change the values of the line integral? Why/why not?

1.12 Against the current. You need to fill a bucket of water from a stream. Assume that the velocity field in the stream is uniform: $\mathbf{v} = v_0\hat{\mathbf{x}}$. Also assume that you will not change the current of water with the bucket.
(a) You want to fill the bucket as quickly as possible. How should you then orient it? Describe the orientation of the surface S, which is the opening of the bucket.

The surface integral $\Phi = \int_S \mathbf{v} \cdot d\mathbf{S}$, where S is the opening of the bucket, tells how much water that flows into the bucket per unit time. We call this the flux.
(b) Find Φ for the bucket with the orientation you found in (a).
(c) What happens if you instead orient the opening of the bucket along the current, that is along the x-axis. What is Φ in this case?
(d) Your friend Q watch you turn the bucket and says that it does not matter which way you orient the bucket. The flux only depends on the cross sectional area of the bucket. And the size of the opening of the bucket does not change even if you turn it. So the flux should become the same. Do you agree with Q? If not, what is wrong with Q's argument?
(e) What is the positive orientation of the surface S? What does this mean for Φ? It is meaningful to describe the surface with a vector, $\mathbf{S}$? What is the physical interpretation of changing the sign of the surface $\mathbf{S}$ you chose in part (a)?
(f) What if the surface S was not a plane, but was shaped like a half sphere. What is then the flux Φ through the surface with the orientation you chose in part (a)?

1.13 Down from the mountain. The scalar field $h(x, y)$ in Fig. 1.21 varies only along the x-axis.
(a) Sketch the gradient, ∇h, in the xy-plane.

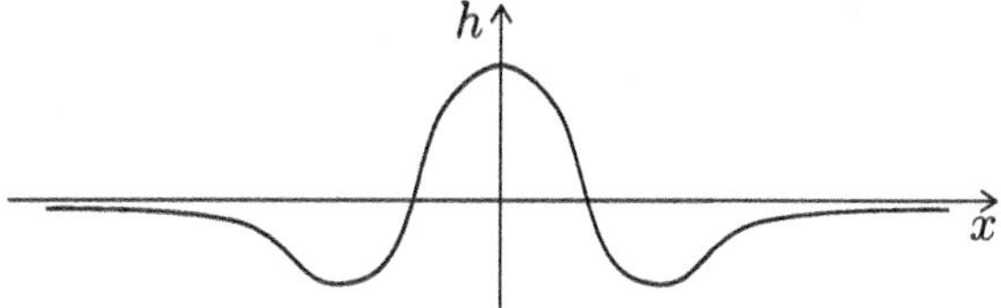

Fig. 1.21 Illustration of a height field $f(x, y)$ plotted along the x-axis

(b) Where is minimum and maximum of the magnitude of the gradient?
(c) How can you use the gradient to find the lowest point?
(d) Can the gradient be zero in a point, in which the surface if not locally a plane in this point?

1.14 Do not unleash the divergence. Your friend Q want to describe the flow of water on a table. The water flows down on the table from a water outlet above the origin and flows out over the table. You observe that the water film on the table has approximately the same thickness everywhere and that this thickness does not change over time. Q has made a model for the velocity field of the water: $\mathbf{v} = C(x, y)$ where C is a constant.
(a) Sketch the field.
(b) Find the divergence of the field.
(c) Can this field describe the physical system? (Try to understand what the result you got for the divergence tells you about the system).
(d) Can you propose a better model for the velocity field of the water?

Homework

1.15 Exploring a scalar field. The scalar field $h(x, y) = e^{-y} \sin x$ is defined in the region $0 < x < \pi$ and $0 < y < \pi$.
(a) Write a Python program to visualize the scalar field both as a heat map (an image) and as a surface.
(b) The gradient of the scalar field is a vector field, ∇h. Calculate ∇h analytically.
(c) Write a Python program to calculate and visualize ∇h by plotting your analytical result for ∇h.
(d) Write a Python program to calculate and visualize ∇h by taking the numerical derivative your calculation of $h(x, y)$ found in the first exercise.

1.16 Volume integrals.
(a) What is the integral of $\exp(-x)$ over a cube $0 < x < a, 0 < y < a, 0 < z < a$?
(b) What is the integral of $\rho(\mathbf{r}) = \rho_0$, where ρ_0 is a constant, over a spherical volume centered in the origin with radius a?

1.17 Line integrals. A vector field is defined as $\mathbf{E} = E_0\hat{\mathbf{x}}$ for all $\mathbf{r}$.
(a) Make a sketch of the vector field.

(b) Calculate the line integral along a line from $(-1, 0)$ to $(1, 0)$. Draw the curve into your sketch.
(c) Calculate the line integral along a line from $(0, -2)$ to $(0, 2)$. Draw the curve into your sketch.
(d) Calculate the line integral along a line from $(-1, -2)$ to $(1, 2)$
(e) Calculate the line integral along a circular path in the positive direction from $(1, 1)$ to $(-1, -1)$.

1.18 Fluxes of curved surfaces.
(a) A cylindrical surface along the z-axis has a radius a and a length L. The surface is oriented with its surface normal pointing outward. What is the flux of the field $\mathbf{E} = E_0\hat{\mathbf{r}}$, where $\hat{\mathbf{r}}$ is the radial unit vector in cylindrical coordinates.
(b) Assume instead that the field depends on the distance r to the z-axis, $\mathbf{E} = E(r)\hat{\mathbf{r}}$. What functional form should $E(r)$ have for the flux to be independent of the radius a of the surface?
(c) A spherical surface centered at the origin has a radius a and a surface normal pointing outward. What is the flux of the field $\mathbf{E} = E_0\hat{\mathbf{r}}$, where $\hat{\mathbf{r}}$ is a radial unit vector $\hat{\mathbf{r}} = \mathbf{r}/|\mathbf{r}|$.
(d) Assume instead that the field depends on the distance r to the origin, $\mathbf{E} = E(r)\hat{\mathbf{r}}$. What functional form should $E(r)$ have for the flux to be independent of the radius a of the surface?

1.19 Cylindrical coordinates.
(a) A scalar field $V(r, \phi, z) = V_0 r$ in cylindrical coordinates. Find the volume integral of V over a volume between radius $r = a$ and $r = b$ from $z = 0$ to $z = L$.
(b) A vector field is $\mathbf{E} = E_0\hat{\boldsymbol{\phi}}$. What is the flux of this field through a cylindrical surface with radius a from $z = 0$ to $z = L$.
(c) A vector field is $\mathbf{E} = E_0\hat{\boldsymbol{\phi}}$. What is the curve integral of this field along a circular path of radius a in the positive direction around the z-axis.
(d) A scalar field in cylindrical coordinates is $V(r, \phi, z) = V_0 \ln r$. What is the gradient in cylindrical coordinates?
(e) A vector field is $\mathbf{E} = (-y, x, 0)$ in Cartesian coordinates. What is it in cylindrical coordinates?

Chapter 2
Electric Field

The interactions between a hemoglobin molecule and a water molecule as shown in Fig. 2.1 depends on the electrostatic forces between the charges in the two molecules. Negative and positive charges are distributed in each of the molecules, and the sum of the interactions between all the charges determines the forces acting between them. While this net interaction is complex, because the distribution of charges is complex, the fundamental principles are simple: The interactions are due to Coulomb's law for the force between two charges. In this chapter we will introduce Coulomb's law and the corresponding field called the electric field. You will learn to find the electric field that arises from individual point charges and charge distributions. This will allow you to find forces from complex charge distributions in the hemoglobin molecule, in a thunder cloud, or in the solar wind. We will address methods to calculate and visualize the electric field analytically and numerically. Thus by the end of this chapter, you will be armed with the tools to address the interactions between microscopic charged objects such as atoms or molecules or macroscopic charged objects such as moon dust and astronauts.

2.1 Coulomb's Law

If I rub a balloon against my hair and push it towards a wall, it will stick to the wall as illustrated in Fig. 2.2. Why? We know from mechanics that gravity will pull the balloon down. Since it does not slide down, there must be another force acting in the opposite direction of gravity. What is this force?

Forces between charged objects. The force that acts from the wall on the balloon is an electrostatic force. It is a fundamental force, just like gravity, acting between all *charged* objects. The balloon becomes charged when we rub it against our hair. Electrostatic forces are along with gravity one of the four fundamental forces of

© The Author(s), under exclusive license to Springer Nature Switzerland AG 2026

A. Malthe-Sørenssen, *Elementary Electromagnetism Using Python*, Undergraduate Texts in Physics, https://doi.org/10.1007/978-3-032-19876-1_2

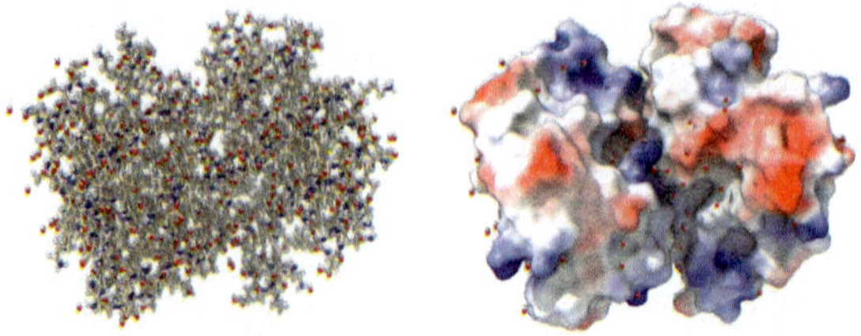

Fig. 2.1 Illustration of a hemoglobin molecule (left) and the charge distribution around it

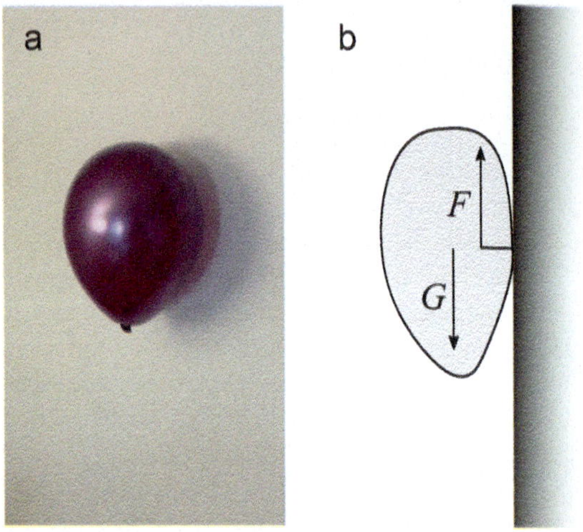

Fig. 2.2 **a** Illustration of a balloon hanging on a wall. **b** Force diagram for the balloon, indicating gravity, **G**, and the contact force, **F**, due to electrostatic interactions

nature. In our everyday lives we mostly have experience from gravity and electrostatic forces. Often, we do not realize that forces around us have an electrostatic origin. Instead, we often give these forces other names. For example, the normal force between you and the ground is due to electrostatic forces between the atoms on your skin and the surface of the ground. Indeed, all interatomic forces can be considered to be electrostatic forces that are due to the distribution of charges around atoms.

The electrostatic force law. The force that acts between all charged objects is formulated in *Coulomb's* law. It is a fundamental law like gravity, and the force law has a similar form to that of gravity. The force between two point objects with charges q and Q at positions $\mathbf{r}_q$ and $\mathbf{r}_Q$ respectively is given by Coulomb's law (see Fig. 2.3):

Coulomb's law

The force *on* a point-particle with charge q at a position $\mathbf{r}_q$ *from* a point-particle with charge Q at position $\mathbf{r}_Q$ is:

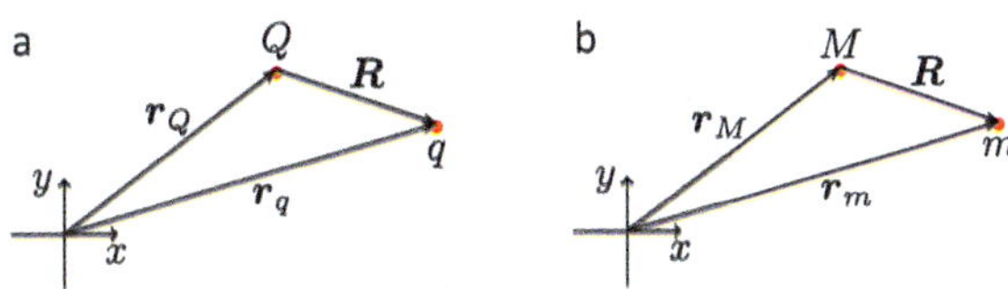

Fig. 2.3 The force from a charge Q at $\mathbf{r}_Q$ on a charge q at $\mathbf{r}$. The force from a mass M at $\mathbf{R}_M$ on a mass m at $\mathbf{r}_m$

$$\mathbf{F} = \frac{qQ}{4\pi\epsilon_0}\frac{\hat{\mathbf{R}}}{R^2} = \frac{qQ}{4\pi\epsilon_0}\frac{\mathbf{R}}{R^3}, \tag{2.1}$$

where we have introduced $\mathbf{R} = \mathbf{r}_q - \mathbf{r}_Q$, $\hat{\mathbf{R}} = \mathbf{R}/R$, and $R = |\mathbf{R}|$.

Notice the two different ways to write the law: Either with the unit vector $\hat{\mathbf{R}}$ divided by R^2 or by the vector $\mathbf{R}$ divided by R^3. The two ways are identical because $\hat{\mathbf{R}} = \mathbf{R}/R$.

Coulomb's law is an experimentally established law. It is correct as far as we can measure.

Similarity to gravity. Coulomb's law looks very much like Newton's law of gravity. We recall that the gravitational force on a mass m at position $\mathbf{r}_m$ from a mass M at a position $\mathbf{r}_M$ is

$$\mathbf{F}_G = -GmM\frac{\hat{\mathbf{R}}}{R^2}, \tag{2.2}$$

where $\mathbf{R} = \mathbf{r}_m - \mathbf{r}_M$, $\hat{\mathbf{R}} = \mathbf{R}/R$, and $R = |\mathbf{R}|$. You may already have experience with Newton's law of gravitation: You know how to find its potential energy, you know how to solve the motion of objects affected by this force, and you know how to solve such problems analytically and numerically. We will build on this intuition to develop our understanding of Coulomb's law.

Charges versus masses. We see that Newton's law of gravity has the opposite sign of Coulomb's law: For gravity, two objects always attract each other because the masses m and M are always positive. Whereas for Coulomb's law, two objects will repel each other if qQ is positive and attract each other if qQ is negative. You already have experience and intuition about masses. You know that mass is a fundamental property of matter and that masses only can be positive. What about charge q? We need to build a similar understanding and intuition for charge.

Electrons, protons and the conservation of charge. Charge is also a fundamental property of matter. The fundamental building blocks of charges are the charges of electrons and protons: Electrons have a charge of $-e$ and protons have a charge $+e$. We will therefore often refer to the atomic model of matter when we build our intuition about charged objects. Charges are measured in units of coulomb, where $e = 1.602 \times 10^{-19}$C is the charge of the proton. Charge is a *conserved quantity*. The net charge of a system does not change unless we add or remove charges from it. For example, a hydrogen atom has zero net charge. If it is split into an electron and a proton, the net charge is $-e + e = 0$. The net charge is still zero. In the example of the balloon above, we changed the net charge of the balloon when we rubbed it against our hair. Electrons were transferred from the hair to the balloon, making the balloon negatively charged. As a result the balloon had a net charge different from zero and a net force was acting between the wall and the balloon.[1]

Adding masses and charges. Because charges can be positive and negative, the *net charge* of an object can be zero even if there is internal structure in the charges. For example, the net charge of a hydrogen atom is zero: It has one proton with charge e and one electron with charge $-e$. This is a significant difference from gravity, because masses only add up. We will gradually learn the consequences of this difference, but an important effect is that since most objects have approximately zero net charge, the effective electrostatic force falls off more rapidly with distance for electromagnetic forces than for gravity. Gravitational forces add up. The gravitational force from the Sun is the sum of all the masses in the Sun, which is non-negligible. Because electrostatic forces both add and subtract, the electrostatic force from the Sun is the net sum of all the attractive and repulsive forces from the individual charges in the Sun, which is typically small since the net charge of the Sun is close to zero.

Comparing the magnitude of gravity and electrostatic forces. How large are the electrostatic forces compared to gravity? We can compare the electrostatic and gravitational forces between an electron and a proton. The mass of the electron is $m_e = 9.10938356 \times 10^{-31}$ kg, the mass of the proton is $m_p = 1.6726219 \times 10^{-27}$ kg, the gravitational constant is $G = 6.67430 \times 10^{-11}$ m^{3}kg$^{-1}s^{-2}$. The gravitational force is $F_G = G m_e m_p / R^2$ and the electrostatic force is $F_C = e^2/(4\pi\epsilon_0 R^2)$. Here ϵ_0 is called the permittivity of vacuum and its value is $\epsilon_0 = 8.85 \times 10^{-12}$C^{2}N$^-$1m^{-2}. Both forces depend on R^2. If we look at the ratio, the Rs will cancel out:

$$\frac{F_G}{F_C} = \frac{G m_e m_p}{e^2/(4\pi\epsilon_0)} = \frac{G m_e m_p 4\pi\epsilon_0}{e^2} \tag{2.3}$$

```
import numpy as np
e = 1.602e-19
me = 9.10938356e-31
mp = 1.6726219e-27
G = 6.67430e-11
epsilon0 = 8.85e-12
ratio = G*me*mp*4*np.pi*epsilon0/(e*e)
print("ratio = ",ratio)
```

```
Out: ratio =  4.406772215446942e-40
```

[1] However, the wall would typically have zero net charge before it comes in contact with the balloon. To understand what happenes in this case, we need to develop our understanding of electrostatic systems further. We will return to this case when we have developed the necessary concepts.

The gravitational force between an electron and a proton is therefore very much smaller than the electrostatic force—irrespective of distance. If we for simplicity assume that all matter consists of electrons and protons (forgetting about neutrons for the argument), this should also hold for the Earth and the Sun. But we must then remember that the *net* charge often is approximately zero, as discussed above. Masses add up, but charges tend to (almost) cancel out.

The permittivity in vacuum. The quantity ϵ_0 is called the permittivity of vacuum. We have written the constant in front of Coulomb's law as $1/(4\pi\epsilon_0)$. This seems to be a strange choice. Why not simply call it $K = 1/(4\pi\epsilon_0)$ in analogy to the G in the gravitational law? It turns out that this way of writing it is practical, but the reason for this will first become apparent in Chap. 4.

Point charges and extended objects. We recall from mechanics that we started by describing objects as point objects before we moved on to address extended objects as consisting of many of point objects. We will pursue a similar approach in electromagnetism. We will start by addressing *point charges*, which are charged bodies where the dimensions of the body is much smaller than typical distances between bodies of interest. Electrons and protons are considered point charges, but we will use this concept also as simplified models of macroscopic bodies. To find the net effect of many point charges, we need to apply the superposition principle of forces.

Superposition principle. The superposition principle for forces also holds for Coulomb's force law: The force *on* a point charge q at $\mathbf{r}$ *from* point charges Q_1 and Q_2 at $\mathbf{r}_1$ and $\mathbf{r}_2$, respectively, is:

$$\mathbf{F} = \mathbf{F}_1 + \mathbf{F}_2 = \frac{qQ_1}{4\pi\epsilon_0}\frac{\hat{\mathbf{R}}_1}{R_1^2} + \frac{qQ_2}{4\pi\epsilon_0}\frac{\hat{\mathbf{R}}_2}{R_2^2}, \tag{2.4}$$

where $\mathbf{R}_1 = \mathbf{r} - \mathbf{r}_1$ and $\mathbf{R}_2 = \mathbf{r} - \mathbf{r}_2$. The superposition principle can be extended to any number of point charges. The principle will be used as both a theoretical and as a practical tool in this text—so spend some time to ensure you master the principle now.

No self interaction. Just as for the gravitational force from a single mass particle, a point charge does not interact with itself. However, just like for gravitational systems, the point charges making up an extended body may interact with other point charges in the same body, but not with themselves.

Test your understanding

The net force on a system due to all the electrostatic forces between charges in the system is zero. Is this statement true or false? Explain your answer.

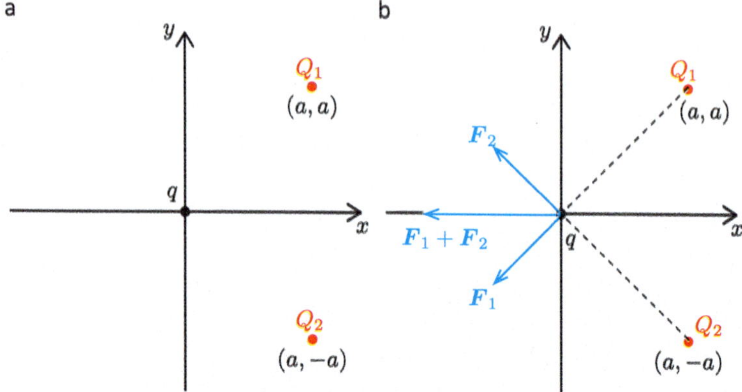

Fig. 2.4 A system with two charges, Q_1, Q_2, near a charge q in the origin. Forces acting on q are illustrated when $qQ_1 > 0$ and $qQ_2 > 0$

Test your understanding

What is the ratio between the force F_A on charge A and F_B on charge B?[2]

$\boldsymbol{F}_A$ A B $\boldsymbol{F}_B$

$3q$ q

The Force from Two Charges

Let us look at three ways to use the superposition principle to find the net force on a charge q in the origin. Figure 2.4 shows a system of two charges Q_1 and Q_2 at positions (a, a) and $(a, -a)$ respectively.

Graphical vector addition. If all the charges are the same ($Q_i = Q$), we can use graphical vector addition to find the solution. As illustrated in Fig. 2.4b the sum of the forces only has a component along the x-axis.

Analytical vector addition. The force *from* charge Q_1 *on* charge q depends on $\mathbf{R}_1 = \mathbf{r} - \mathbf{r}_1 = (0, 0) - (a, a) = (-a, -a)$, where we have used that $\mathbf{r} = (0, 0)$ and $\mathbf{r}_1 = (a, a)$:

$$\mathbf{F}_1 = \frac{qQ}{4\pi\epsilon_0}\frac{\mathbf{R}_1}{R_1^3} = \frac{qQ_1}{4\pi\epsilon_0}\frac{(-a, -a)}{\left(\left(a^2 + a^2\right)^{1/2}\right)^3} = \frac{qQ}{4\pi\epsilon_0 2a^2}\frac{(-1, -1)}{\sqrt{2}}. \tag{2.5}$$

[2] 1.

For charge Q_2 we have $\mathbf{R}_2 = \mathbf{0} - \mathbf{r}_2 = (-a, a)$ and the force is:

$$\mathbf{F}_2 = \frac{qQ_2}{4\pi\epsilon_0}\frac{\mathbf{R}_2}{R_2^3} = \frac{qQ}{4\pi\epsilon_0}\frac{(-a, a)}{\left(a^2 + a^2\right)^{3/2}} = \frac{qQ}{4\pi\epsilon_0 2a^2}\frac{(-1, 1)}{\sqrt{2}}. \tag{2.6}$$

If we assume $Q_1 = Q_2 = Q$, we find the net force as the sum by component-wise addition:

$$\begin{aligned}\mathbf{F} = \mathbf{F}_1 + \mathbf{F}_2 &= \frac{qQ}{4\pi\epsilon_0 2a^2}\frac{(-1, -1)}{\sqrt{2}} + \frac{qQ}{4\pi\epsilon_0 2a^2}\frac{(-1, 1)}{\sqrt{2}} \\ &= \frac{qQ}{4\pi\epsilon_0 2a^2}\frac{(-2, 0)}{\sqrt{2}} = -\frac{qQ}{4\pi\epsilon_0\sqrt{2}a^2}\hat{\mathbf{x}}.\end{aligned} \tag{2.7}$$

Numerical vector addition. We can also calculate the sum numerically in Python, by first calculating the forces and then adding them using vector addition. When we calculate Coulomb's law numerically, it is common to use the following form:

$$\mathbf{F} = \frac{qQ}{4\pi\epsilon_0}\frac{\mathbf{R}}{R^3}. \tag{2.8}$$

Let us assume that the charges $Q_1 = 1\mu$C and $Q_2 = 2\mu$C and that $a = 1$cm.

```
import numpy as np
import scipy.constants as sc
epsilon0 = sc.epsilon_0
q = 1e-6 # in units of C
a = 1e-2 # in units of meters
r = np.array([0,0])
Q1 = 1e-6 # in units of C
R1 = r - np.array([a,a])
F1 = (q*Q1)/(4*np.pi*epsilon0)*R1/np.linalg.norm(R1)**3
Q2 = 2e-6 # in units of C
R2 = r - np.array([a,-a])
F2 = (q*Q2)/(4*np.pi*epsilon0)*R2/np.linalg.norm(R2)**3
F = F1+F2
print("F = ",F," N")
```

```
Out: F =  [-95.32738228  31.77579409] N
```

Notice that we use `np.linalg.norm` to find the length of the $\mathbf{R}_i$ vectors. Also notice the use of `scipy.constants` for physical constants. The result is a force, which is measured in units of Newtons (N = kgms^{-2}). Notice that Python returns many digits—always remember to reduce the number of significant digits to correspond to values provided.

Test your understanding

A charge Q is placed in $(a, 0)$. You want to find the force from Q on a charge q in $(0, a)$. (a) What is the **R**-vector? (b) What is the force **F** from Q on q?[3]

2.2 Electric Field

For a given set of charges, such as the charges in a molecule, a thunder cloud or a solar wind, we can find the force acting on a charge q anywhere in space. If all the other charges are kept in place, the force would depend on the position of the charge as well as the magnitude and sign of the charge. The force is a *vector field*. It is a vector function that depends on the position in space—just like the forces on a spacecraft from the Sun and all the planets in the Solar system depends on the position of the spacecraft. We can calculate the force using Coulomb's law. It is the sum of the forces from each of the individual charges. Figure 2.5 schematically illustrates such a situation where a set of charges Q_i are in positions $\mathbf{r}_i$. What is the net force from this set of charges on a charge q at the position $\mathbf{r}$? It is

$$\mathbf{F} = \sum_i \frac{q Q_i}{4\pi\epsilon_0} \frac{\mathbf{R}_i}{R_i^3} . \tag{2.9}$$

We notice that the charge q is a factor in all the forces, and can be placed outside the sum:

$$\sum_i \frac{q Q_i}{4\pi\epsilon_0} \frac{\mathbf{R}_i}{R_i^3} = q \underbrace{\sum_i \frac{Q_i}{4\pi\epsilon_0} \frac{\mathbf{R}_i}{R_i^3}}_{\mathbf{E}} = q\mathbf{E}. \tag{2.10}$$

The sum now only depends on the charges Q_i, their positions $\mathbf{r}_i$, and the position $\mathbf{r}$, where the force acts on the charge q. We call this sum the *electric field*, $\mathbf{E}(\mathbf{r};\, Q_i, \mathbf{r}_i)$. If the charge distribution is given and does not change, the sum $\mathbf{E}$ is a function of only the position $\mathbf{r}$.

We can calculate and study this vector field independently of the charge q. Electrostatics, which is the subject we are now starting to study, is largely about the properties of electric fields from charge distributions. Our task is to calculate and understand the electric field for a given charge distribution.

[3] (a) $\mathbf{R} = (0, a) - (a, 0) = (-a, a)$; (b) $\mathbf{F} = qQ/(4\pi\epsilon_0 a^2)\,(-1, 1)/2^{3/2}$.

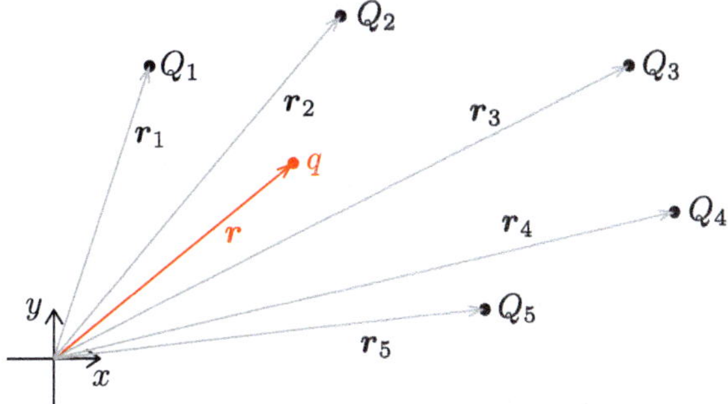

Fig. 2.5 Illustration of a charge distribution Q_i. The force on a charge q depends on the position $\mathbf{r}$ of the charge

Electric field

The electric field $\mathbf{E}(\mathbf{r})$ in a point $\mathbf{r}$ is defined as:

$$\mathbf{E} = \frac{\mathbf{F}_q}{q} \quad (q \to 0)\,, \tag{2.11}$$

where $\mathbf{F}_q$ is the force on a test charge q from other charges in space. We assume q is a small[a] test charge. The force from the electric field $\mathbf{E}$ on a charge q at $\mathbf{r}$ is

$$\mathbf{F} = q\mathbf{E}. \tag{2.12}$$

We call the point $\mathbf{r}$ the **observation point**. This is where we *observe* (or measure) the electric field.

[a] Why do we say that the test charge is small? This is to make clear that the test charge q does not affect the distribution of charges, Q_i, that set up the electric field. The test charge is so small, that we do not have to think about how q would affect the charge distribution. This becomes more important when we address conductors and ideal conductors. Conventionally, the test charge is assumed to be positive.

The electric field is a vector field. The electric field varies with the position $\mathbf{r}$ in space. It is a vector field or a vector function: For each position $\mathbf{r}$, the function returns a vector value, $\mathbf{E}(\mathbf{r})$. We can calculate the vector function for a given set of charges Q_i and position $\mathbf{r}_i$. The function may be complicated, but it is in principle straight forward to calculate it and to visualize it in space.

The electric field is analogous to the gravitational field. You already have intuition about a similar concept: the gravitational field. You may recall that the gravitational force on an object of mass m is $m\mathbf{g}$. When the object is close to the Earth's surface, we know that $\mathbf{g}$ is approximately constant. When the object is further away, the gravitational field $\mathbf{g}(\mathbf{r})$ depends on the position $\mathbf{r}$ of the object relative to the Earth. And in any point in the Solar system, the gravitational field $\mathbf{g}(\mathbf{r})$ depends on the positions and masses of all the other object in the Solar system. Just like the gravitational field is set up by the distribution of masses in space, the electric field is set up the distribution of charges in space.

The Electric Field from a Point Charge

What does the electric field from a single point charge Q_1 look like? The force from Q_1 on q can be found from Coulomb's law. We assume the charge Q_1 is in $\mathbf{r}_1$. According to Coulomb's law, the force on a charge q in $\mathbf{r}$ is:

$$\mathbf{F} = \frac{qQ}{4\pi\epsilon_0}\frac{\mathbf{R}}{R^3}, \tag{2.13}$$

where $\mathbf{R}$ is the vector *from* the charge Q_1 at $\mathbf{r}_1$ *to* the charge q at $\mathbf{r}$: $\mathbf{R} = \mathbf{r} - \mathbf{r}_1$:

$$\mathbf{F} = \frac{qQ}{4\pi\epsilon_0}\frac{\mathbf{r} - \mathbf{r}_1}{|\mathbf{r} - \mathbf{r}_1|^3}, \tag{2.14}$$

and the electric field is:

$$\mathbf{E} = \frac{\mathbf{F}}{q} = \frac{Q}{4\pi\epsilon_0}\frac{\mathbf{r} - \mathbf{r}_1}{|\mathbf{r} - \mathbf{r}_1|^3}. \tag{2.15}$$

Electric field in Cartesian coordinates. Let us make this more specific. Assume that the charge Q is a $(0, 0, a)$. What is the electric field in the position $\mathbf{r} = (x, y, z)$? We insert these values in $\mathbf{R} = \mathbf{r} - \mathbf{r}_1$, getting $\mathbf{R} = (x, y, z) - (0, 0, a) = (x, y, z - a)$ and $R = (x^2 + y^2 + (z - a)^2)^{1/2}$. We again insert this in (2.15), and find

$$\mathbf{E} = \frac{Q}{4\pi\epsilon_0}\frac{(x, y, z - a)}{\left(x^2 + y^2 + (z - a)^2\right)^{3/2}}. \tag{2.16}$$

Notice that the electric field indeed depends on the position (x, y, z) in space. The formula is rather complex, even though this problem is about the simplest problem we could think of. We therefore need to develop effective methods to calculate electric fields—both analytically and numerically.

Visualization of the electric field. It is customary to visualize the electric field by drawing small arrows to illustrate the vector field. The arrows indicate the magnitude and direction of the field in a given point in space, which usually corresponds to where the arrow is starting. We draw a sufficient number of arrows to provide a good illustration of the field. (What you believe to be a sufficient number of arrows is up to your judgment). We have illustrated the field from a single charge in $(0, 0, a)$ in Fig. 2.6.

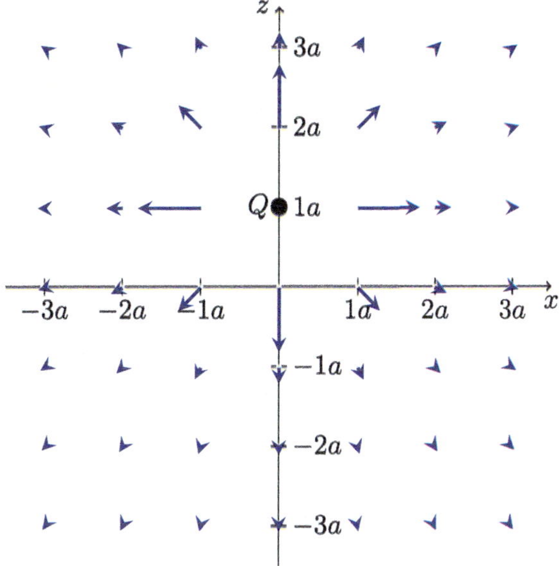

Fig. 2.6 Drawing of the electric field in the xz-plane from a point charge at $(0, 0, a)$

The R-vector

The vector $\mathbf{R}$ or $\mathbf{R}_i$ plays a special role in the expression for the electric field. We often write only $\mathbf{R}_i$, but we should remember that this is really a vector pointing from a charge Q_i at a position $\mathbf{r}_i$ to the observation point $\mathbf{r}$. That is, we must always remember that $\mathbf{R}_i$ really is given as $\mathbf{R}_i = \mathbf{r} - \mathbf{r}_i$. For example, in the expression for the field from a set of charges Q_i at positions $\mathbf{r}_i$

$$\mathbf{E} = \sum_i \frac{Q_i}{4\pi\epsilon_0} \frac{\hat{\mathbf{R}}_i}{R_i^2}, \tag{2.17}$$

each of the vectors $\mathbf{R}_i$ are $\mathbf{R}_i = \mathbf{r} - \mathbf{r}_i$. A common mistake is to insert $\mathbf{r}_i$ or $\mathbf{r}$ instead of $\mathbf{R}_i$. (This becomes even more important when we address continuous charge distributions later on).

Notice that the $\mathbf{R}_i$-vector points *from* the charge Q_i *to* the position $\mathbf{r}$. It is common to call the position $\mathbf{r}$ the *observation point*. The vector $\mathbf{R}_i$ therefore points from the charge (Q_i) to the observation point.

Test your understanding

We place a charge Q in the origin. We want to find the electric field in the point $\mathbf{r}$. (a) What is the $\mathbf{R}$-vector? (b) What is the force $\mathbf{F}$ on a charge q in $\mathbf{r}$? (c) What is the electric field $\mathbf{E}$ in $\mathbf{r}$?[4]

Superposition Principle for the Electric Field

The superposition principle can be used to find the net force on a charge q at $\mathbf{r}$ from several charges Q_i at $\mathbf{r}_i$:

$$\mathbf{F} = \sum_i \mathbf{F}_i = \sum_i \frac{qQ_i}{4\pi\epsilon_0}\frac{\mathbf{R_i}}{R_i^3} = q\sum_i \underbrace{\frac{Q_i}{4\pi\epsilon_0}\frac{\mathbf{R_i}}{R_i^3}}_{\mathbf{E}_i} = q\sum_i \mathbf{E}_i = q\mathbf{E}. \tag{2.18}$$

The superposition is therefore also valid for electric fields:

$$\mathbf{E} = \sum_i \mathbf{E}_i = \sum_i \frac{Q_i}{4\pi\epsilon_0}\frac{\mathbf{R_i}}{R_i^3}. \tag{2.19}$$

Electric field from a set of point charges

For a set of charges Q_i in positions $\mathbf{r}_i$, the net electric field in a position $\mathbf{r}$ is the sum of the electric fields from each of the charges:

$$\mathbf{E}(\mathbf{r}) = \sum_i \mathbf{E}_i(\mathbf{r}) = \sum_i \frac{Q_i}{4\pi\epsilon_0}\frac{\hat{\mathbf{R}}_i}{R_i^2} = \sum_i \frac{Q_i}{4\pi\epsilon_0}\frac{\mathbf{R}_i}{R_i^3}. \tag{2.20}$$

where $\mathbf{E}_i(\mathbf{r})$ is the field from charge Q_i, $\mathbf{R}_i = \mathbf{r} - \mathbf{r}_i$, and $R_i = |\mathbf{R}_i|$. Inserting $\mathbf{R}_i = \mathbf{r} - \mathbf{r}_i$ gives the explicit expression:

$$\mathbf{E}(\mathbf{r}) = \sum_i \frac{Q_i}{4\pi\epsilon_0}\frac{(\mathbf{r} - \mathbf{r}_i)}{|\mathbf{r} - \mathbf{r}_i|^3}. \tag{2.21}$$

[4] (a) $\mathbf{R} = \mathbf{r}$; (b) $\mathbf{F} = qQ/(4\pi\epsilon_0)\,\mathbf{r}/r^3$; (c) $\mathbf{E} = Q/(4\pi\epsilon_0)\,\mathbf{r}/r^3$.

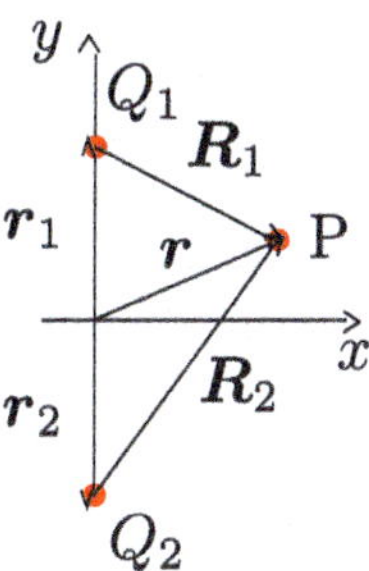

Fig. 2.7 Illustration of two charges Q_1 and Q_2. The electric field at a position $\mathbf{r}$ (the observation point P) is the sum of the contributions from each of the two charges

Example: Electric Field from Two Charges

What is the electric field from two charges Q_1 and Q_2 at $\mathbf{r}_1 = (0, a, 0)$ and $\mathbf{r}_2 = (0, -a, 0)$ respectively?
Approach: We plan to use the superposition principles to add the fields from the charges at the point $\mathbf{r} = (x, y, z)$, the observation point, P, as illustrated in Fig. 2.7.
Solve: First, we find an expression for the electric field $\mathbf{E}_1(\mathbf{r})$ from the charge Q_1 using Coulomb's law:

$$\mathbf{E}_1(\mathbf{r}) = \frac{1}{4\pi\epsilon_0}\frac{Q_1}{R_1^2}\hat{\mathbf{R}}_1 = \frac{1}{4\pi\epsilon_0}\frac{Q_1}{R_1^2}\frac{\mathbf{R}_1}{R_1} = \frac{1}{4\pi\epsilon_0}\frac{Q_1}{R_1^3}\mathbf{R}_1 . \tag{2.22}$$

This is still a general result. We now insert the specific position $\mathbf{r}_1 = (0, a, 0)$ and $\mathbf{r} = (x, y, z)$, finding $\mathbf{R}_1 = \mathbf{r} - \mathbf{r}_1 = (x - 0, y - a, z - 0) = (x, y - a, z)$, and insert this in (2.22), getting:

$$\mathbf{E}_1(x, y, z) = \frac{1}{4\pi\epsilon_0}\frac{Q_1\,(x, y - a, z)}{\left(x^2 + (y - a)^2 + z^2\right)^{3/2}} . \tag{2.23}$$

We find a similar result for Q_2, but now $\mathbf{R}_2 = (x, y, z) - (0, -a, 0) = (x, y + a, z)$:

$$\mathbf{E}_2(x, y, z) = \frac{1}{4\pi\epsilon_0}\frac{Q_2\,(x, y + a, z)}{\left(x^2 + (y + a)^2 + z^2\right)^{3/2}} . \tag{2.24}$$

From the superposition principles, the electric field is the sum of the two fields:

$$\begin{aligned}\mathbf{E} &= \mathbf{E}_1 + \mathbf{E}_2 \\ &= \frac{1}{4\pi\epsilon_0}\frac{Q_1\,(x, y - a, z)}{\left(x^2 + (y - a)^2 + z^2\right)^{3/2}} + \frac{1}{4\pi\epsilon_0}\frac{Q_2\,(x, y + a, z)}{\left(x^2 + (y + a)^2 + z^2\right)^{3/2}} .\end{aligned} \tag{2.25}$$

Discuss: This result cannot be further simplified. We see that it is straight-forward to find this expression, but that the formula is a bit complex. This is typical for problems in electromagnetism. You will have to be careful in your calculations, but we can often follow a systematic procedure.

Test your understanding

Two charges $3q$ and q are placed on the x-axis at $(-a, 0)$ and $(a, 0)$ as shown in the figure. (a) Draw a vector diagram of the electric field along the x-axis. (b) Sketch the electric field $E_x(x)$ as a function of x along the x-axis.

2.3 Visualizing Electric Fields

To help you build intuition for electric fields, we use various tools to visualize the fields. Here, we will introduce you to two main tools: visualization through arrows to represent the field and visualization through field lines.

Direct Visualization Using Arrows

An electric field is typically visualized by drawing arrows to illustrate the field at various points in space. This is cumbersome to do by hand. Instead, we would like to visualize the field in Python. We will use the approach introduced in Sect. 1.2 to find the electric field on a grid of regularly spaced (x, y)-points by: (1) writing a function to calculate the contribution to the electric field from a single point charge; (2) generating a grid of regularly spaced (x, y)-points on which to calculate the electric field; (3) calculating the field at these positions using superposition; and (4) visualizing the vectors values as arrows or stream lines. We developing the methods using an example of two charges $Q_1 = -1\mu\text{C}$ at $\mathbf{r}_1 = (-a, 0)$ and $Q_2 = +1\mu\text{C}$ at $\mathbf{r}_2 = (+a, 0)$, where $a = 1$ mm.

Function for the field from a single point charge. We want to write a function `E(r,q0,r0)` that calculates the electric field at a position $\mathbf{r}$ from a charge Q_0 at position $\mathbf{r}_0$ using Coulomb's law:

$$\mathbf{E} = \frac{Q_0}{4\pi\epsilon_0}\frac{\mathbf{R}}{R^3}, \tag{2.26}$$

where $\mathbf{R} = \mathbf{r} - \mathbf{r}_0$. The code to do this directly follows the mathematical expression for the field: We first find $\mathbf{R}$, which we call `R`, then the length $R = |\mathbf{R}|$, which we call `Rnorm`, and then we calculate the electric field $\mathbf{E}$. Notice that we import the value of ϵ_0 from the `scipy` module:

```
import numpy as np
import matplotlib.pyplot as plt
import scipy.constants as sc
def E(r,q0,r0):
    # Find E at r from a charge q at position r0
    R = r-r0
    Rnorm = np.linalg.norm(R)
    return q0*R/(4*np.pi*sc.epsilon_0*Rnorm**3)
```

We test the code by checking the value for $Q_0 = 1$ C, $r = (2, 0)$ m and $r_0 = (0, 0)$ m, which should give $\mathbf{E}\, 4\pi\epsilon_0 = 1\,\text{C}\,(2, 0)/2^3\,\text{m}^{-2} = (0.25, 0)\,\text{C/m}^2$:

```
Q0 = 1.0
r0 = np.array([0,0])
r = np.array([2,0])
print(E(r,Q0,r0)*4*np.pi*sc.epsilon_0)
```

```
Out: [0.25 0.  ]
```

This is indeed correct. It is always smart to check the validity of a function with a few simple tests.

Generating a regular grid of points. We recall from Sect. 1.2 that we generate a regular grid of points using the `meshgrid` function. Here, we generate a grid on the scale given by a, from $-3a$ to $+3a$ in the x and y directions:

```
Lx , Ly, Nx, Ny = 3*a, 3*a, 21, 21
x = np.linspace(-Lx,Lx,Nx)
y = np.linspace(-Ly,Ly,Ny)
rx,ry = np.meshgrid(x,y,indexing="ij")
```

We recall the use of the `indexing` option to get the correct ordering of the elements so that $x_{i,j}$ corresponds to `rx[i,j]`.

Creating the field with a loop. Now, we calculate the electric field at each point $\mathbf{r}_{i,j}$ on the grid by looping through all the positions in the grid. First, we generate arrays to store the x and y components of the electric field, `Ex` and `Ey` respectively. These arrays are of the same dimensions as `rx` and `ry` since they store one component of the field at each position in the `rx` and `ry` arrays.

```
a = 1e-3 # in meters
Ex = rx.copy(); Ey = ry.copy()
r1 = np.array([-a,0])
Q1 = -1.0e-6 # -1 micro coulomb
r2 = np.array([+a,0])
Q2 = +1.0e-6 # -1 micro coulomb
for i in range(Nx):
    for j in range(Ny):
        r = np.array([rx[i,j],ry[i,j]])
        Ex[i,j],Ey[i,j] = E(r,Q1,r1) + E(r,Q2,r2)
```

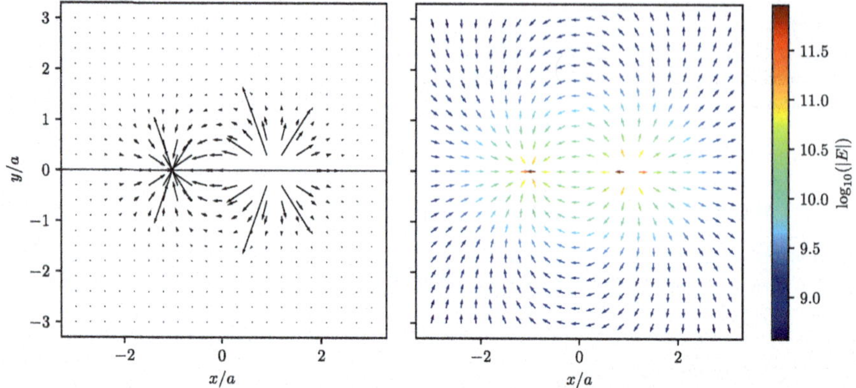

Fig. 2.8 Visualization of the field from two charges using arrows, where the length of the arrow indicates the magnitude of the field (left), and where the color of the arrow indicates the magnitude of the field

It is possible to simplify this by flattening the arrays to one-dimensional arrays and then using a single counter to loop through all the elements using the built-in method `flat` as in the following loop:

```
for i in range(len(rx.flat)):
    r = np.array([rx.flat[i],ry.flat[i]])
    Ex.flat[i],Ey.flat[i] = E(r,Q1,r1) + E(r,Q2,r2)
```

We visualize the field using the `quiver` plot, which plots arrows in each point on the grid. To make the plots clearer, we measure lengths in units of $a = 1$mm, by dividing `rx` and `ry` by `a`. The resulting plot is shown in Fig. 2.8.

```
plt.quiver(rx/a,ry/a,Ex,Ey)
plt.xlabel("$x/a$"); plt.ylabel("$y/a$")
plt.axis("equal")
```

Complete program. The complete program so far is now

```
import numpy as np
import matplotlib.pyplot as plt
import scipy.constants as sc
def E(r,q0,r0):
    # Find E at r from a charge q at position r0
    R = r-r0
    Rnorm = np.linalg.norm(R)
    return q0*R/(4*np.pi*sc.epsilon_0*Rnorm**3)
# Generate grid
a = 1e-3 # in meters
Lx , Ly, Nx, Ny = 3*a, 3*a, 21, 21
x = np.linspace(-Lx,Lx,Nx)
y = np.linspace(-Ly,Ly,Ny)
rx,ry = np.meshgrid(x,y,indexing="ij")
```

```
# Find electric field
Ex = rx.copy(); Ey = ry.copy()
r1 = np.array([-a,0])
Q1 = -1.0e-6 # -1 micro coulomb
r2 = np.array([+a,0])
Q2 = +1.0e-6 # +1 micro coulomb
for i in range(Nx):
    for j in range(Ny):
        r = np.array([rx[i,j],ry[i,j]])
        Ex[i,j],Ey[i,j] = E(r,Q1,r1) + E(r,Q2,r2)
# Visualization
plt.quiver(rx/a,ry/a,Ex,Ey)
plt.xlabel("$x/a$"); plt.ylabel("$y/a$")
plt.axis("equal")
```

Vector visualization using colors. For electric fields from point charges, the magnitude of the field diverges close to the charges. This means that the arrows used to visualize the field become very large close to the charges and small far away from the charges. It is therefore sometimes useful to separate the visualization of the direction and the magnitude of the field. This can be done by drawing unit vectors to indicate the direction of the field and then illustrate the magnitude with a color-scheme. This is done by the following short program. The resulting visualization is shown in Fig. 2.8.

```
# Calculate the magnitude of the field
Emag = np.sqrt(Ex**2 + Ey**2)
# Calculate unit vectors in the fields direction
nEx = Ex / Emag
nEy = Ey / Emag
# Visualize
plt.quiver(rx/a,ry/a,nEx,nEy,np.log10(Emag),cmap="jet")
plt.xlabel("$x/a$"); plt.ylabel("$y/a$")
plt.colorbar(label="$log_{10}(|E|)$")
```

Here, we have also used the logarithm of the magnitude in order to better visualize the range of values because the range of magnitudes for the field varies by orders of magnitude from far away to close to the charges. Experiment with these methods and various colormaps to find a visualization scheme you like. Example values for the colormap may be `jet`, `copper`, `plasma`, `viridis`, or `inferno`.

Visualization Using Field Lines

Another practical and insightful way to visualize the electric field is through *stream lines* and *field lines*. A stream line is a continuous curve constructed in such a way that the tangent in any point along the line points in the direction of the electric field in that point. For field lines, the local density of stream lines is proportional to the local magnitude of the electric field: When the lines are closely spaced, the field is strong, and when they are sparsely spaced, the field is weaker. Stream lines and field lines radiate from positive charges and terminate on negative charges. However, we may have to consider charges at infinity to make a system neutral. Stream lines cannot intersect other stream lines and the same is the case for field lines.

You will often find visualizations of field lines in textbooks or illustrations of electromagnetic phenomena. But how can we construct and visualize stream lines and field lines?

Vector visualization using streamlines. Python comes with a built-in function to visualize stream lines in the form of the `streamplot` function. The function visualizes the local electric field by drawing curves in such a way that the electric field is tangent to any point along the line and the curve is directed in the positive direction of the electric field. The `streamplot` function strives to make a constant density of stream lines. We visualize the electric field by:

```
plt.streamplot(rx.T/a,ry.T/a,Ex.T,Ey.T)
plt.xlabel("$x/a$"); plt.ylabel("$y/a$")
```

Notice that we need to transpose both the coordinates and the field by adding the `.T` at the end of `rx`, `ry`, `Ex` and `Ey`. This is because we have used the `indexing="ij"` to create and order the arrays with the `meshgrid` function. If you decide to use `indexing="xy"`, which is the default in Python, you should not use the transpose here. We consider this a small inconvenience for the advantage of having what we feel is a natural ordering of the indicies of arrays.

The resulting plot for the electric field from Fig. 2.8 is shown in Fig. 2.9a. Notice that the density of stream lines is approximately constant and *does not* represent the density of the field. Also notice that the stream lines do not all start at positive charges and terminate at negative charges. The plot does therefore not show the intensity of field lines, which gives information about the strength of the field, but still provides a useful visualization of the local direction of the field. Thus stream lines do not correspond to what we call field lines in electromagnetism, which is one of the most common ways to visualize electric fields.

Understanding field lines. What are field lines? We can think of them as the trajectories of positive charges with no inertia moving in the field. (This is like the motion of a particle in a highly viscous fluid or the motion of a submicron particle in water). Newton's second law for this particle would be $q\mathbf{E} - D\mathbf{v} = m\mathbf{a} = 0$, where the D-term is due to viscosity and we assume that the mass is very small (no inertia). Therefore, $\mathbf{v} = (q/D)\mathbf{E}$, that is, the velocity points in the direction of the electric field!

This provides us with a nice intuitive notion about field lines. We also realize that field lines cannot cross each other. Why? Because the direction of a field line out of a point depends on the direction of $\mathbf{E}$. If two field lines were to cross, there would have to be two field lines out of a point, which would mean that the field would have to point in two different directions in the same point.

To visualize the field lines, we start with a given density of field lines, that is a given number of field lines starting from each charge q. For a small sphere with

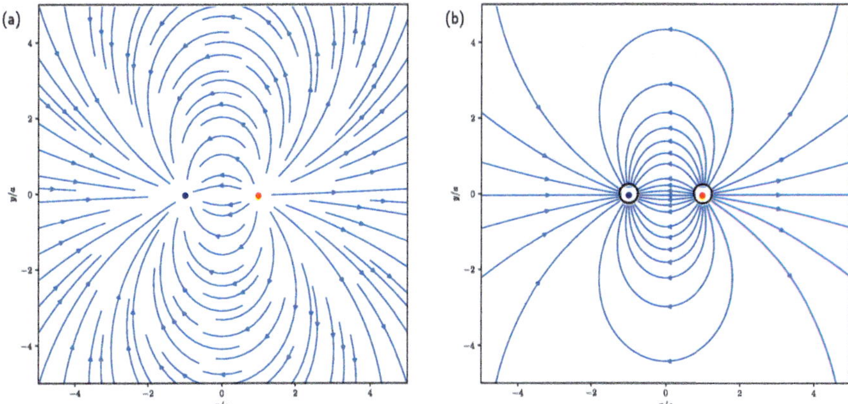

Fig. 2.9 Visualization of the field from two charges. **a** Using stream lines. **b** Using field lines. The black dots show where the field lines were started

radius a around this charge, the density of field lines would be $N/(4\pi a^2)$, where N is the number of field lines. If we are close to the charge, we assume that the electric field is dominated by the field from the charge, which is $q/(4\pi\epsilon_0 a^2)$. We therefore see that if we choose N so that it is proportional to the charge q, then the density of field lines is proportional to the magnitude of the electric field.

We visualize the field by initiating a number of field lines proportional to the charge at each positive charge. Similarly, we will terminate field lines at negative charges. With this construction, we ensure that the density of field lines will correspond to the magnitude of the electric field close to the charges. However, this is true everywhere in space—the density of field lines is proportional to the magnitude of the field.

Notice that this argument is only correct in three dimensions. If we make this construction in two dimensions instead, the density of lines passing through a circle of radius r would be $N/(2\pi r)$, which does not go like $1/r^2$.

Finding field lines using `streamplot`. The function `streamplot` *can* be used to find field lines, but then we need to specify the starting points of the field lines. We can start field lines from a small circle of radius r around the charges, where the number of points on the circle is proportional to the charge. (Note the comment on two and three dimensional systems above). We need to create a list of these starting points and submit them to `streamplot`. First, we generate lists of the points in the array `start_xy` for the electric field visualized in Fig. 2.8. Notice the use of the length a since the charges are at positions $\mathbf{r}_i = (\pm a, 0, 0)$.

```
numlines = 32
start_x = []
start_y = []
Ri = np.array([-a,0])
rad = 0.25*a
for ni in range(numlines):
    ang = 2*np.pi*ni/numlines
    start_x.append(Ri[0]+rad*np.cos(ang))
    start_y.append(Ri[1]+rad*np.sin(ang))
Ri = np.array([a,0])
rad = 0.25*a
for ni in range(numlines):
    ang = 2*np.pi*ni/numlines
    start_x.append(Ri[0]+rad*np.cos(ang))
    start_y.append(Ri[1]+rad*np.sin(ang))
start_xy = np.column_stack([start_x/a, start_y/a])
```

We then call `streamplot` with this list as input:

```
plt.streamplot(rx.T/a,ry.T/a,Ex.T,Ey.T,start_points=start_xy,density=100)
plt.plot(start_x, start_y, ".k")
```

The resulting plot is shown in Fig. 2.9b. Notice again the use of the transpose whenever we use `streamplot` as explained above. Notice also that we need to specify a high number for `density` when calling `streamplot`. You should explore this parameter, but if you choose too small a number, you will not get a correct visualization. You may also explore what happens if you only include points around the positive charges—can you explain what happens?

2.4 Continuous Distributions of Charge

We have learned that we can use the superposition principle to find the net electric field from many charges. But what if the charges are distributed continuously in space? Fig. 2.10 illustrates a set of charges Q_i. The net electric field from these charges is

$$\mathbf{E} = \sum_i \frac{Q_i}{4\pi\epsilon_0} \frac{\mathbf{R}_i}{R_i^3} . \tag{2.27}$$

When there are many, many small charges it becomes simpler not to think of individual point charges, but instead introduce a volume density of charges, a charge density. If we divide the system in Fig. 2.10 into small boxes of volume Δv_j containing a net charge ΔQ_j, then the total electric field is approximately:

$$\mathbf{E} = \sum_j \frac{\Delta Q_j}{4\pi\epsilon_0} \frac{\mathbf{R}_j}{R_j^3} , \tag{2.28}$$

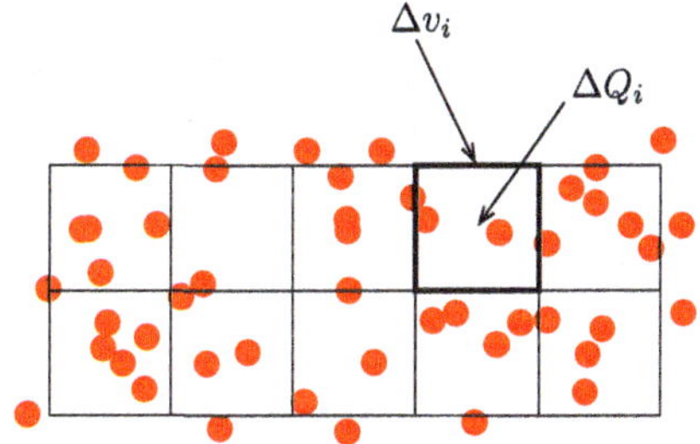

Fig. 2.10 Illustration of a set of charges divided into small volumes of size ΔV_i. Volume i has a net charge Q_i, which is the sum of the charges inside volume ΔV_i

where the sum is over all the boxes j and $\mathbf{R}_j = \mathbf{r} - \mathbf{r}_j$ refers to the center $\mathbf{r}_j$ of box j. As the boxes become smaller and smaller, this approaches an integral over the volume elements Δv_i:

$$\mathbf{E} = \sum_j \frac{\Delta Q_j}{\Delta v_j} \frac{\Delta v_j}{4\pi\epsilon_0} \frac{\mathbf{R}_j}{R_j^3} = \sum_j \rho_j \frac{\Delta v_j}{4\pi\epsilon_0} \frac{\mathbf{R}_j}{R_j^3} \rightarrow \int_v \frac{\rho_v(\mathbf{r}')\,\mathrm{d}v}{4\pi\epsilon_0} \frac{\mathbf{R}}{R^3}, \tag{2.29}$$

where $\rho_j = \Delta Q_j/\Delta v_j$. In the limit when Δv_j becomes small, ρ_j approaches the charge density, $\rho(\mathbf{r}_j)$. The sum is over all $\mathbf{r}_j$ in some volume region. When we rewrite this as an integral, the integral is over the positions $\mathbf{r}'$, which correspond to the center of a box of volume $\mathrm{d}v$, thus $\mathbf{r}'$ corresponds to $\mathbf{r}_j$. In the integral, we therefore write the charge density as $\rho(\mathbf{r}')$ and we interpret $\mathbf{R}$ as $\mathbf{R} = \mathbf{r} - \mathbf{r}'$. We should therefore write the integral as

$$\mathbf{E}(\mathbf{r}) = \int_v \frac{\rho(\mathbf{r}')}{4\pi\epsilon_0} \frac{\mathbf{r} - \mathbf{r}'}{|\mathbf{r} - \mathbf{r}'|^3}\,\mathrm{d}v', \tag{2.30}$$

Note that the integral is *not* over $\mathbf{r}$, but over $\mathbf{r}'$! We will therefore sometimes use the notation $\mathrm{d}v'$ to make this distinction clear. We discuss this distinction in more detail below.

Charge density

The volume charge density is:

$$\rho = \rho(\mathbf{r}) = \rho_v(\mathbf{r}) = \frac{\mathrm{d}Q}{\mathrm{d}v}, \tag{2.31}$$

with unit $\mathrm{C/m^3}$.

The net charge in a volume. The charge density is generally a function of position, $\rho(\mathbf{r})$. The net charge in a volume v is the sum of all the charges inside the volume, which is given by the integral over the volume v. We write a v below the integral

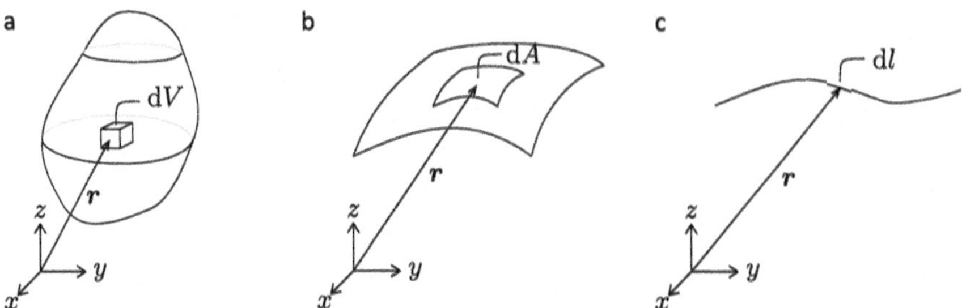

Fig. 2.11 Illustration of **a** volume charge distribution, **b** surface charge distribution, and **c** line charge distribution

sign to indicate that the integral is over a specific volume v in space:

$$Q_v = \int_v \rho(\mathbf{r})\,\mathrm{d}v. \tag{2.32}$$

Tricks of notation. Sometimes we use a compressed notation when we describe distributions of charge. Instead of specifying the volume of a small element as a differential, $\mathrm{d}v$, we instead specify the (net) charge of the small volume by writing $\mathrm{d}q$. We still think of this as the charge of a small element in a particular point $\mathbf{r}$ in space. Sometimes we will write

$$Q = \int_Q \mathrm{d}q. \tag{2.33}$$

What we really mean by this is that the integral is over a volume element $\mathrm{d}v$ and that the charge in this volume element is $\mathrm{d}q = \rho(\mathbf{r})\,\mathrm{d}v$ as illustrated in Fig. 2.11a. However, this notation also makes it easier to think of other charge elements, such as charges distributed over a surface (a surface charge) or charged distributed along a curve (a line charge).

Surface charge density. Following the same reasoning as for a volume charge density, we can describe charges that are distributed on a surface with a *surface charge density*, $\rho_s(\mathbf{r})$ as illustrated in Fig. 2.11b. In the literature, you may sometimes find the notation σ used for a surface charge density instead of ρ_s, but we will use ρ_s here. This is the charge per unit surface area

$$\rho_s = \frac{\mathrm{d}Q}{\mathrm{d}A} \tag{2.34}$$

with unit $\mathrm{C/m^2}$. The charge in a surface with area S is then given by an integral over the area S:

$$Q_S = \int_S \rho_s(\mathbf{r})\,\mathrm{d}S. \tag{2.35}$$

Line charge density. Finally, we also describe charges that are distributed on a line with a line charge density ρ_l (or often λ) as illustrated in Fig. 2.11c. This is the charge per unit length:

$$\rho_l = \lambda = \frac{\mathrm{d}Q}{\mathrm{d}l}, \tag{2.36}$$

with unit C/m. The charge in a length L is then given by an integral over the curve:

$$Q_L = \int_C \rho_l(\mathbf{r})\,\mathrm{d}l. \tag{2.37}$$

Test your understanding

(a) A sphere with radius R is filled with a material with a uniform volume charge density ρ. What is the total charge Q of the sphere? (b) There is a uniform volume charge density ρ in the region $0 < x < a$, $0 < y < b$, $0 < z < c$. Write down the integral for the total charge. (c) What is the value of this integral?[5]

Finding the Electric Field from the Charge Density

How can we use the volume charge density to find the electric field? We will do this by summing up the contributions from all the small charge elements $\mathrm{d}q$ making up the system. We need to be very precise with our notation. We want to find the electric field $\mathbf{E}(\mathbf{r})$ in a point $\mathbf{r}$. We call this point the *observation point*. We want to include the contribution to this field from all the charges. For discrete charges, the charges were placed in positions $\mathbf{r}_i$ and we summed over all i. For a continuous distribution of charges, we sum the contributions of all charge elements $\mathrm{d}q = \rho(\mathbf{r}')\,\mathrm{d}v'$ at $\mathbf{r}'$ to the electric field $\mathbf{E}(\mathbf{r})$ at $\mathbf{r}$. The contribution from one such element is:

$$\mathrm{d}\mathbf{E} = \frac{\rho(\mathbf{r}')\,\mathrm{d}v'}{4\pi\epsilon_0}\frac{\mathbf{r}-\mathbf{r}'}{|\mathbf{r}-\mathbf{r}'|^3}. \tag{2.38}$$

The contributions from all charges in a volume v is then found by the integral:

$$\mathbf{E} = \int_v \frac{\rho(\mathbf{r}')\,\mathrm{d}v'}{4\pi\epsilon_0}\frac{\mathbf{r}-\mathbf{r}'}{|\mathbf{r}-\mathbf{r}'|^3}. \tag{2.39}$$

Notice that the integral is over $\mathrm{d}v'$. What does this mean? It may become more obvious in Cartesian coordinates: In this case $\mathbf{r}' = (x', y', z')$ and the integral is over $\mathrm{d}v' = \mathrm{d}x'\,\mathrm{d}y'\,\mathrm{d}z'$. Notice that the *integration variable* is $\mathbf{r}'$, and that $\mathbf{r}$ is a *constant*

[5] (a) $Q = 4\pi\rho R^3/3$, (b,c) $Q = \int_0^a \int_0^b \int_0^c \rho\,\mathrm{d}z\,\mathrm{d}y\,\mathrm{d}x = \rho abc$.

in this integration! We use the triple integral notation if we want to write the integral explicitly in Cartesian coordinates:

$$\mathbf{E} = \iiint_v \frac{\rho(\mathbf{r}')}{4\pi\epsilon_0} \frac{\mathbf{r} - \mathbf{r}'}{|\mathbf{r} - \mathbf{r}'|^3} \, \mathrm{d}x' \, \mathrm{d}y' \, \mathrm{d}z'. \tag{2.40}$$

Using the R notation. It is common and often practical to introduce the vector $\mathbf{R} = \mathbf{r} - \mathbf{r}'$. Notice that the $\mathbf{R}$ goes *from* the element $\mathrm{d}v'$ at $\mathbf{r}'$ *to* the observation point at $\mathbf{r}$. We can then rewrite the integral for the electric field as:

$$\mathbf{E} = \int_v \frac{\rho(\mathbf{r}')}{4\pi\epsilon_0} \frac{\hat{\mathbf{R}}}{R^2} \, \mathrm{d}v'. \tag{2.41}$$

Here, it is important to remember that $\mathbf{R}$ depends on the integration variable $\mathbf{r}'$. You can therefore generally not put $\mathbf{R}$, $\hat{\mathbf{R}}$ or R outside the integral! They are generally not constants. However, there may be special cases where for example the magnitude R is a constant in the integral, while the direction $\hat{\mathbf{R}}$ is not. We will use the $\mathbf{R}$ notation because it is compact, but we will be careful to unpack its meaning in each case. A common mistake is to assume that $\mathbf{R}$ is a constant in the integral or to assume that $\mathbf{R}$ (or $\mathbf{r}$) is the integration variable instead of $\mathbf{r}'$.

Electric fields from surface and line charge densities. We can make exactly the same arguments for surface and line charge densities.

For a *surface charge density* ρ_s, the electric field from a surface S is

$$\mathbf{E} = \int_S \frac{\rho_s(\mathbf{r}') \, \mathrm{d}S'}{4\pi\epsilon_0} \frac{\hat{\mathbf{R}}}{R^2}. \tag{2.42}$$

For a *line charge density* ρ_l, the electric field from a line L is

$$\mathbf{E} = \int_L \frac{\rho_l(\mathbf{r}') \, \mathrm{d}l'}{4\pi\epsilon_0} \frac{\hat{\mathbf{R}}}{R^2}. \tag{2.43}$$

Test your understanding

The figure shows a line element of length $\mathrm{d}l$ of a material with a uniform line charge density ρ_l. (a) What is the contribution $\mathbf{dE}$ from the line element $\mathrm{d}l$ at x' to the electric field at the origin? (b) What is the contribution $\mathbf{dE}$ from the line element $\mathrm{d}l$ at the origin to the electric field at x?[6]

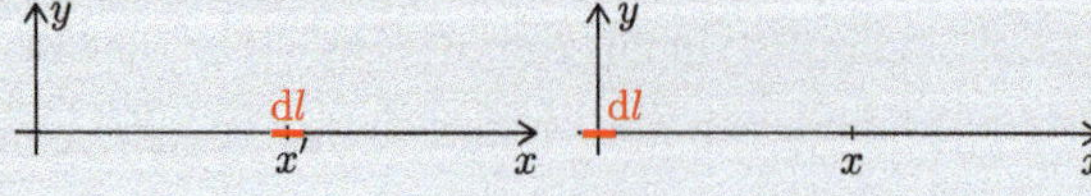

[6] (a) $\mathbf{dE} = \rho_l \, \mathrm{d}l/(4\pi\epsilon_0)\,(-x')/|x'|^3 \, \hat{\mathbf{x}}$; (b) $\mathbf{dE} = \rho_l \, \mathrm{d}l/(4\pi\epsilon_0)\, x/|x|^3 \, \hat{\mathbf{x}}$.

Method for finding the electric field

The electric field $\mathbf{E}(\mathbf{r})$ from a set of discrete charges or a continuous charge distribution can be found using the following procedure:

- Model the problem: Choose a coordinate system to describe the system that reflects the symmetries of the system.
- Make a drawing: Draw a sketch of the system where you indicate the charges and the observation point $\mathbf{r}$.

Discrete charges:

- Find the vector $\mathbf{R}_i$ from the charge Q_i at $\mathbf{r}_i$ to the observation point $\mathbf{r}$: $\mathbf{R}_i = \mathbf{r} - \mathbf{r}_i$ and $R_i = |\mathbf{R}_i|$ for each charge Q_i.
- Sum the contributions from all charges Q_i:

$$\mathbf{E}(\mathbf{r}) = \sum_i \frac{Q_i}{4\pi\epsilon_0} \frac{\mathbf{R}_i}{R_i^3} = \sum_i \frac{Q_i}{4\pi\epsilon_0} \frac{\mathbf{r} - \mathbf{r}_i}{|\mathbf{r} - \mathbf{r}_i|^3}. \tag{2.44}$$

Continuous charges:

- Describe a small element $\mathrm{d}q$ at a position $\mathbf{r}'$. Relate $\mathrm{d}q$ to a volume; surface; line element $\mathrm{d}q = \rho\, \mathrm{d}v'$; $\rho_s\, \mathrm{d}S'$; $\rho_l\, \mathrm{d}l'$.
- Find the vector $\mathbf{R}$ from the small element $\mathrm{d}q$ at $\mathbf{r}'$ to the observation point $\mathbf{r}$: $\mathbf{R} = \mathbf{r} - \mathbf{r}'$ and $R = |\mathbf{R}|$.
- Integrate over all $\mathbf{r}'$ that make up the body of interest analytically or numerically. Remember that the integration variable is $\mathbf{r}'$ and not $\mathbf{r}$! For a volume charge density the integral is:

$$\mathbf{E}(\mathbf{r}) = \int_v \frac{\rho\, \mathrm{d}v'}{4\pi\epsilon_0} \frac{\mathbf{R}}{R^3} = \int_v \frac{\rho\, \mathrm{d}v'}{4\pi\epsilon_0} \frac{\mathbf{r} - \mathbf{r}_i}{|\mathbf{r} - \mathbf{r}_i|^3}. \tag{2.45}$$

Example: Electric Field from a Ring Charge

A ring of radius a in the xy-plane with center in the origin has a total charge Q which is uniformly distributed along the ring. Find the electric field from the ring.

Model. We develop a more precise model of the system. We assume that the length of the ring is $L = 2\pi a$. Because the charge is uniformly distributed along the ring, the line charge density is $\rho_l = Q/L = Q/(2\pi a)$. We plan to find the electric field in the point $\mathbf{r}$ by summing the contributions from all parts of the ring. A small part of length

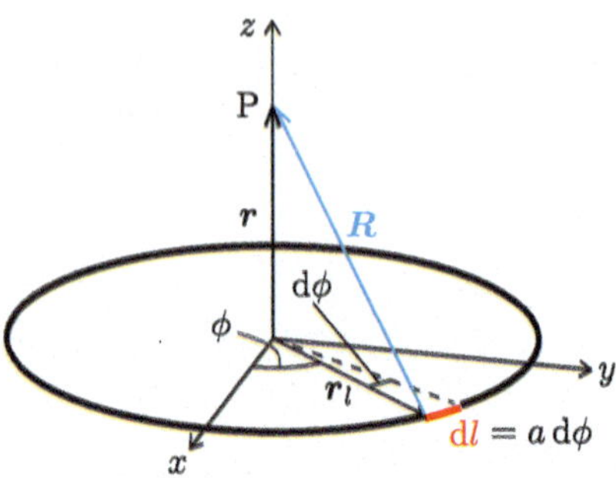

Fig. 2.12 An illustration of a ring with a charge Q

dl at a position ϕ has a charge $dq = \rho_l \, dl$. We will sum up all such contributions. The system is illustrated in Fig. 2.12 where **r** is chosen along the z-axis.

Analytical solution along the z -axis. We start by finding the electric field at a point $\mathbf{r} = z\hat{\mathbf{z}}$ along the z-axis. We choose this observation point because the system has rotational symmetry around the z-axis, and we therefore expect the solution to be simpler along this axis. We describe the position of an element along the ring using the angle ϕ. A small length dl along the ring is then equal to $dl = a \, d\phi$, and the charge of this element is $dq = \rho_l a \, d\phi$.

The position of the element dl is $\mathbf{r}' = (a \cos\phi, a \sin\phi, 0)$. We find the field by summing up all contributions through the integral

$$\mathbf{E} = \int_Q \frac{dq}{4\pi\epsilon_0} \frac{\mathbf{R}}{R^3} = \int_l \frac{\rho_l(\mathbf{r}')\, dl'}{4\pi\epsilon_0} \frac{\mathbf{R}}{R^3}, \tag{2.46}$$

where the vector $\mathbf{R} = \mathbf{r} - \mathbf{r}' = (0, 0, z) - (a\cos\phi, a\sin\phi, 0) = (-a\cos\phi, -a\sin\phi, z)$. We see that

$$R^2 = a^2\left(\cos^2\phi + \sin^2\phi\right) + z^2 = a^2 + z^2, \tag{2.47}$$

where we have used the identity $\cos^2\phi + \sin^2\phi = 1$. This means that R^2 does not depend on ϕ. We rewrite the integral to be an integral over ϕ:

$$\mathbf{E} = \int_l \frac{\rho_l \, dl'}{4\pi\epsilon_0} \frac{\mathbf{R}}{R^3} = \int_0^{2\pi} \frac{\rho_l a \, d\phi}{4\pi\epsilon_0} \frac{\mathbf{R}}{R^3} = \frac{\rho_l a}{4\pi\epsilon_0} \int_0^{2\pi} \frac{\mathbf{R}}{R^3} \, d\phi. \tag{2.48}$$

We insert the expression for **R** and R, getting:

$$\mathbf{E} = \frac{\rho_l a}{4\pi\epsilon_0} \int_0^{2\pi} \frac{(-a\cos\phi, -a\sin\phi, z)}{\left(a^2 + z^2\right)^{3/2}} \, d\phi. \tag{2.49}$$

The integrals over $\cos\phi$ and $\sin\phi$ are zero. The resulting field **E** along the z-axis therefore only has a component in the z-direction:

$$\mathbf{E}(0,0,z) = \frac{\rho_l a}{4\pi\epsilon_0} 2\pi \frac{z}{\left(a^2+z^2\right)^{3/2}} \hat{\mathbf{z}} = \frac{Q}{4\pi\epsilon_0} \frac{z}{\left(a^2+z^2\right)^{3/2}} \hat{\mathbf{z}}. \tag{2.50}$$

It is good practice always to check this result in the limits. Limits typically means for very large or very small values of z. Here, the limit of large z corresponds to the case when $z \gg a$ meaning that we are far away from the ring compared to the size of the ring. In this case, $a^2 + z^2 \simeq z^2$ and the electric field approaches:

$$\mathbf{E} \to \frac{Q}{4\pi\epsilon_0} \frac{z}{z^3}, \quad z \gg a. \tag{2.51}$$

Is this result reasonable? Yes, we recognize this as the electric field from a point charge with charge Q placed at the origin. When we are far away from the ring, we expect that we can approximate it as a point charge, and this is indeed what we have found.

Symmetry argument. Let us see how we can solve this problem using symmetry instead of brute force. The symmetry argument may be more elegant, but it does require that you are good at finding and using symmetries to solve the problem. This takes some practice, but is a useful skill in physics.

In this case, we notice that for each element $\mathrm{d}\phi$ at a position ϕ, there is also an element on the opposite side of the ring (at $\phi + \pi$), and that the xy contributions from these two components cancel, thus leaving only the z-component. Thus, we can limit ourselves to finding the z-component. This simplifies the integral. We can instead formulate an integral in terms of $\mathrm{d}l$ and not use the explicit parameterization of the curve through the angle ϕ. In this case, the integral for the z-component of $\mathbf{E}$, E_z, becomes:

$$E_z = \int_l \frac{\rho_l \,\mathrm{d}l}{4\pi\epsilon_0} \frac{R_z}{R^3}, \tag{2.52}$$

where R_z is the z-component of $\mathbf{R}$. We see from the geometry that R_z is simply z and that $R^2 = a^2 + z^2$ from Pytagoras. Hence the integral becomes

$$E_z = \int_l \frac{\rho_l \,\mathrm{d}l}{4\pi\epsilon_0} \frac{z}{R^3} = \int_l \frac{\rho_l \,\mathrm{d}l}{4\pi\epsilon_0} \frac{z}{\left(a^2+z^2\right)^{3/2}}, \tag{2.53}$$

where there are no parts inside the integral that depend on the position along the ring. This gives

$$\begin{aligned} E_z &= \int_l \frac{\rho_l \,\mathrm{d}l}{4\pi\epsilon_0} \frac{z}{\left(a^2+z^2\right)^{3/2}} = \frac{\rho_l}{4\pi\epsilon_0} \frac{z}{\left(a^2+z^2\right)^{3/2}} \int_l \mathrm{d}l \\ &= \frac{\rho_l 2\pi a}{4\pi\epsilon_0} \frac{z}{\left(a^2+z^2\right)^{3/2}} = \frac{Q}{4\pi\epsilon_0} \frac{z}{\left(a^2+z^2\right)^{3/2}}, \end{aligned} \tag{2.54}$$

where we have used that $\int_l \, dl = 2\pi a$ and $\rho_l = Q/(2\pi a)$. This is indeed the same result as we found above.

Solution outside the z-axis. If we want to find the solution outside the z-axis, we need to evaluate the integral numerically or to discretize the ring into finite linear segments and sum of the contributions from each of these segments.

Numerical integration through piecewise linear discretization. We can take a physically motivated approach to the discretization of the ring. We divide the ring into small pieces of length Δl placed along the circle, and then sum the contributions from each of these elements to the electric field at a position $\mathbf{r}$.

If we divide the ring into N elements, then the charge of an element is Q/N. The (line) element starts at $\mathbf{r}_i = a(\cos\phi_i, \sin\phi_i, 0)$ and ends at $\mathbf{r}_{i+1} = a(\cos\phi_{i+1}, \sin\phi_{i+1}, 0)$ where $\phi_i = i2\pi/N$ for $i = 0, 1, \ldots, N-1$. We may consider the position of the element to be the midpoint of this line, which would be at $(\mathbf{r}_i + \mathbf{r}_{i+1})/2$. However, we do not make a large mistake if we instead use $\mathbf{r}_i$ as the position of element i. (Indeed, the error becomes smaller as the number of elements N increases.)

We find the total electric field $\mathbf{E}$ as a sum of the contributions from each of the i elements:

$$\mathbf{E}(\mathbf{r}) = \sum_i \mathbf{E}_i = \sum_i \frac{Q_i}{4\pi\epsilon_0} \frac{\mathbf{R}_i}{R_i^3}, \tag{2.55}$$

where $\mathbf{R}_i = \mathbf{r} - \mathbf{r}_i$.

This method can be directly implemented in Python. To store all the charges and their positions we construct two lists, a list of charges, `Qlist`, with elements $Q_i = Q/N$, and a list of positions, `rlist`, with elements $\mathbf{r}_i$ and then we sum the contributions from all these charges to find the total field. First, we create the lists

```
import numpy as np
import matplotlib.pyplot as plt
import scipy.constants as sc
# Function to find E at r from a charge q at position r0
def E(r,q0,r0):
    R = r-r0
    Rnorm = np.linalg.norm(R)
    return q0*R/(4*np.pi*sc.epsilon_0*Rnorm**3)
# Set up lists of charges for ring charge
a = 1.0 # radius of ring
N = 20 # number of discrete elements in ring
q = 1.0 # charge of ring
rcenter = np.array([0,0,0]) # center of circle
Qlist = [] # Empty list for charges
rlist = [] # Empty list for positions
deltaphi = 2*np.pi/N
Qi = q/N
for i in range(N):
    phi_i = i*deltaphi
    r_i = np.array([a*np.cos(phi_i),a*np.sin(phi_i),0])+rcenter
    Qlist.append(Qi)
    rlist.append(r_i)
```

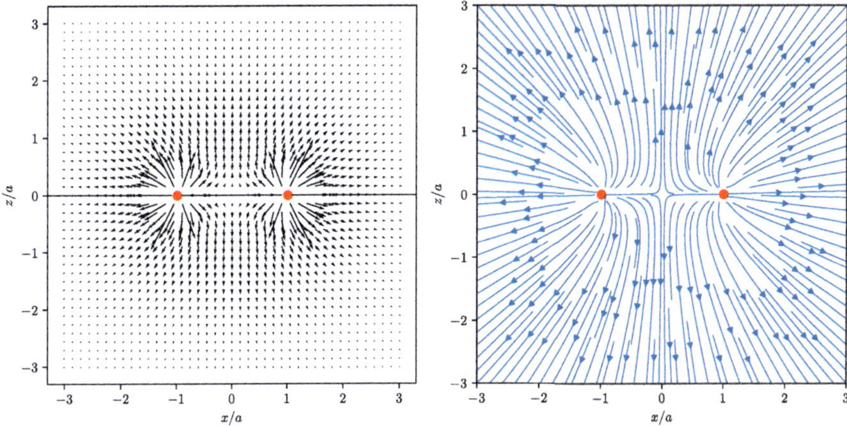

Fig. 2.13 Plot of the calculated field in the xz-plane

Then we calculate the field on a grid and visualize the results. Because the system is rotationally symmetric around the z-axis—the ring is identical if we rotate it around the z-axis—we expect the field to have the same symmetry. It is therefore sufficient to calculate and visualize the field in the xz-plane. Any other plane, such as the yz-plane, will be identical.

```
# Set up grid for field calculation
Lx , Lz, Nx, Nz = 3*a, 3*a, 41, 41
x = np.linspace(-Lx,Lx,Nx)
z = np.linspace(-Lz,Lz,Nz)
rx,rz = np.meshgrid(x,z,indexing="ij")
Ex = np.zeros(rx.shape); Ey = 0; Ez = np.zeros(rz.shape)
# Calculate electric field
for i in range(Nx):
    for j in range(Nz):
        r = np.array([rx[i,j],0,rz[i,j]])
        for ic in range(len(Qlist)):
            Ex[i,j],Ey,Ez[i,j] = np.array([Ex[i,j],Ey,Ez[i,j]]) \
               + E(r,Qlist[ic],rlist[ic])
# Plot results
plt.quiver(rx,rz,Ex,Ez); plt.axis("equal")
plt.xlabel("$x/a$"); plt.ylabel("$z/a$")
```

Notice the use of the dummy variable `Ey` when we call the `E`-function. When a three-dimensional vector is sent as input to this function, it will also return a three-dimensional vector, but we are only interested in the x- and z-coordinates. The resulting plot can be seen in Fig. 2.13.

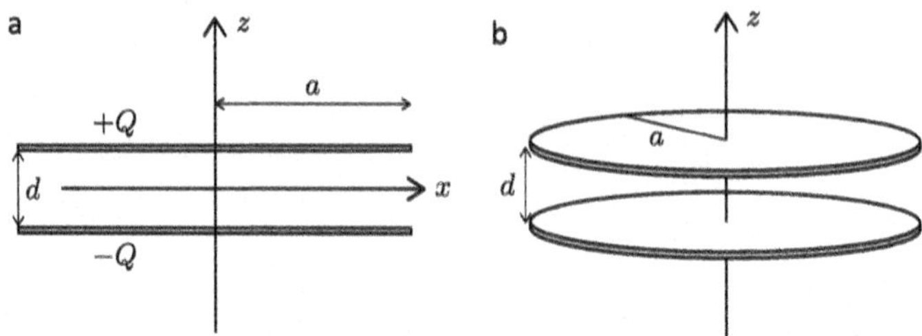

Fig. 2.14 Illustration of a system of two circular plates

Example: Two Charged Plates

Two charged thunder clouds or the two sides of a cell membrane can be simplified as two charged, planar surfaces with opposite charges. Address the electric field from such a simplified model in the form of two circular, thin plates of radius a that a placed a distance d apart. One plate has a charge Q and one plate has a charge −Q. Find the electric field close to the plates and far away from the plates.

Model. A model is suggested, but we need to make it more precise. We place the two plates in a coordinate system parallel to the xy plane. We assume that the plates are infinitely thin and that the charges are uniformly distributed on the plates. We place the positive plate at $z = d/2$ and the negative plate at $z = -d/2$, so that they are nicely symmetric around the xy-plane. (We like symmetries in physics). We make a drawing of the system as shown in Fig. 2.14. We should also be more precise about what to calculate: We will calculate the electric field numerically close to the plates. To validate our numerical solution we will find the field analytically along the z-axis. Finally, we will make a model for the field far away, that is, at distances much larger than d and a.

Electric field along z-axis. Due to the rotational symmetry around the z-axis, it is possible to find the electric field along the z-axis using methods similar to what we did for the ring charge. We leave this as an exercise. Here, we simply use the result, which is that the electric field along the z-axis from a circular plate with radius a and charge Q in the xy-plane is given as:

$$E_z = \frac{z}{|z|}\frac{Q}{2\pi\epsilon_0 a^2}\left(1 - \frac{|z|}{\sqrt{a^2+z^2}}\right), \tag{2.56}$$

where $z/|z|$ gives the sign of z, which we often write as sign(z). We use this result and the superposition principle to find the net field along the z-axis from two plates, one with charge Q at $z = d/2$ and one with charge $-Q$ at $z = -d/2$. For the field from the charge at $z = d/2$, we replace z with $z - d/2$ in the expression for the field and for the charge at $z = -d/2$, we replace z with $z + d/2$. The expression becomes a bit complex, but should be doable:

$$E_z = \text{sign}(z - \frac{d}{2})\frac{Q}{2\pi\epsilon_0 a^2}\left(1 - \frac{|z - d/2|}{\sqrt{a^2 + (z - d/2)^2}}\right)$$
$$- \text{sign}(z + \frac{d}{2})\frac{Q}{2\pi\epsilon_0 a^2}\left(1 - \frac{|z + d/2|}{\sqrt{a^2 + (z + d/2)^2}}\right), \tag{2.57}$$

Electric field in the xz-plane. Now, let us solve the full problem by finding the electric field in the xz-plane by summing the contributions from many small charges, dq. This can be done in many ways, for example, we can (i) construct an integral over elements $r\, d\phi\, dr$; (ii) construct a sum over charges placed on a regular grid in the planes $z = d/2$ and $z = -d/2$, or (iii) on a regular, circularly shaped grid. We will pursue option (i) and (iii) as exercises. Here, we will pursue approach (ii). We will place charges in positions $(x_{i,j}, y_{i,j}, \pm d/2)$ on a grid with N elements from $-a$ to a in both the x and y directions. However, only positions inside a distance a from the z-axis will be used. We loop through points $x_{i,j} = -a + i\Delta x$, where $\Delta x = 2a/(N-1)$, and similarly $y_{i,j} = -a + j\Delta x$. Notice that we divide by $(N-1)$ because there are N points and $N-1$ line elements and that the length from $-a$ to $+a$ is $2a$. It is useful to check the end points: for $i = 0$ we get $x_{0,j} = -a$ and for $i = N-1$ we get $x_{N-1,j} = a$, as we should.

First, we write a function to create a list of charges for a disk of radius a:

```
# Function to create discretized charges for disk
def makedisk(r0,a,q,N):
    # r0, a, q, N: center, radius, charge and number of elements
    r_list = []
    q_list = []
    ncharge = 0 # number of charges in each disk
    for i in range(N):
        for j in range(N):
            x_ij = -a + i/(N-1)*(2*a)
            y_ij = -a + j/(N-1)*(2*a)
            rad = np.sqrt(x_ij**2 + y_ij**2)
            if (rad<=a):
                r1 = r0 + np.array([x_ij,y_ij,0]) # +Q
                r_list.append(r1)
                ncharge += 1
    dq = q/ncharge
    for i in range(ncharge):
        q_list.append(dq) # q/n
    return r_list,q_list
```

Notice how we found the charge dq for each element: While adding charges to the list `r_list`, we count the total number of charges in the variable `ncharge`. Each charge `dq` should therefore be `q/ncharge`. We place these values in the corresponding list `q_list`.

We are then ready to set up the system:

```
import numpy as np
import matplotlib.pyplot as plt
```

```
import scipy.constants as sc
# Function to find E at r from a charge q at position r0
def E(r,q0,r0):
    R = r-r0
    Rnorm = np.linalg.norm(R)
    return q0*R/(4*np.pi*sc.epsilon_0*Rnorm**3)
# Create disks
a = 1.0 # radius of disk
d = 1.0 # distance between disks
N = 20 # number of elements in x and y directions
Q = 1.0 # charge of disk
# Positive disk
r0 = np.array([0,0,+d/2])
rlist1,Qlist1 = makedisk(r0,a,+Q,N)
# Negative disk
r0 = np.array([0,0,-d/2])
rlist2,Qlist2 = makedisk(r0,a,-Q,N)
rlist = rlist1 + rlist2
Qlist = Qlist1 + Qlist2
```

Finally, we calculate the electric field on a grid of points in the xz-plane and visualize the results:

```
# Set up grid for field calculation
Lx , Lz, Nx, Nz = 2*a, 2*a, 40, 40
x = np.linspace(-Lx,Lx,Nx)
z = np.linspace(-Lz,Lz,Nz)
rx,rz = np.meshgrid(x,z,indexing="ij")
Ex = rx.copy(); Ey = 0; Ez = rz.copy()
# Calculate electric field
for i in range(Nx):
    for j in range(Nz):
        r = np.array([rx[i,j],0,rz[i,j]])
        for ic in range(len(Qlist)):
            Ex[i,j],Ey,Ez[i,j] = np.array([Ex[i,j],Ey,Ez[i,j]]) \
               + E(r,Qlist[ic],rlist[ic])
# Plot results
plt.subplot(1,2,1)
plt.quiver(rx,rz,Ex,Ez); plt.axis("equal")
plt.xlabel("$x/a$"); plt.ylabel("$z/a$")
plt.subplot(1,2,2)
plt.streamplot(rx.T,rz.T,Ex.T,Ez.T,density=2)
plt.xlabel("$x/a$"); plt.ylabel("$z/a$")
```

The resulting plots are shown in Fig. 2.15.

Comparing with the exact solution along the z-axis. We know the exact solution along the z-axis. We compare the results from the numerical model and the exact solution by plotting the results in the same plot, that is, by plotting E_z as a function of z. First, we calculate the values from $z = -3a$ to $z = 3a$ using the exact solution

```
# Field from a single disk
z = np.linspace(-3*a,3*a,1000)
Ez_exact1 =  np.sign(z-d/2)*Q/(2*np.pi*sc.epsilon_0*a**2)* \
```

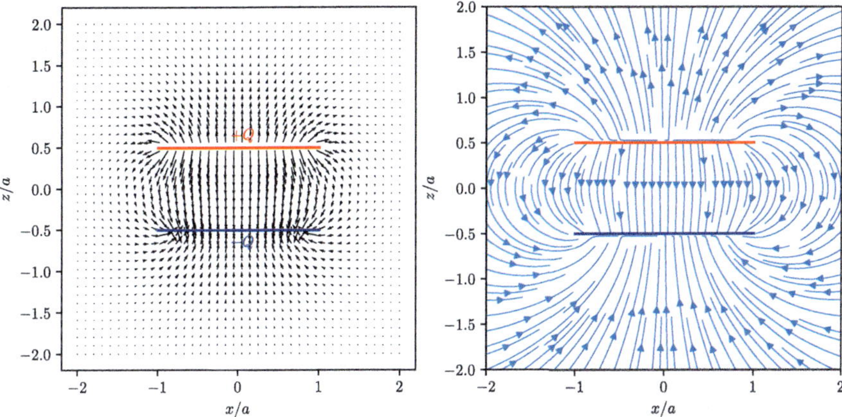

Fig. 2.15 Plot of the electric field in the xz-plane for a system of two charges, circular plates

```
         (1-np.abs(z-d/2)/(np.sqrt(a**2 + (z-d/2)**2)))
Ez_exact2 = -np.sign(z+d/2)*Q/(2*np.pi*sc.epsilon_0*a**2)* \
         (1-np.abs(z+d/2)/(np.sqrt(a**2 + (z+d/2)**2)))
Ez_exact = Ez_exact1 + Ez_exact2
```

Second, we find the corresponding values for E_z using the numerical method we have developed:

```
def findEz(z):
    Ez_num = np.zeros(z.shape)
    for i in range(len(z)):
        r = np.array([0,0,z[i]])
        for ic in range(len(Qlist)):
            dEx,dEy,dEz = E(r,Qlist[ic],rlist[ic])
            Ez_num[i] = Ez_num[i] + dEz
    return Ez_num
Ez_num = findEz(z)
```

For comparison, we plot the two results in the same plot, as shown in Fig. 2.16a. We notice that the two results correspond reasonably well, indicating that the numerical method provides a good approximation to the exact result.

```
plt.plot(z,Ez_exact,label="exact")
plt.plot(z,Ez_num,label="num")
plt.xlabel("$z/a$"); plt.ylabel("$E_z (V/m)$")
plt.legend()
```

The electric field far away from the plates. What happens far away from the two plates? It may be tempting to assume that since the net charge inside a sphere around the origin (with a large enough radius) is zero, then the effective field will be negligible far away. We see from Fig. 2.16a that the electric field decays quickly towards zero. But how quickly does it fall to zero? We know from Coulomb's law that the electric

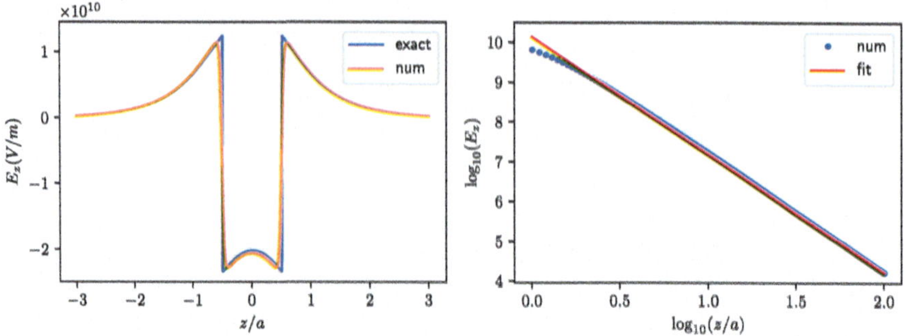

Fig. 2.16 **a** Plot of the electric field E_z along the z axis for the exact and the numerical solution. **b** Double-logarithmic plot of the electric field E_z along the z-axis

field from a single charge decays as $1/r^2$ with distance. In Fig. 2.16b we have plotted E_z as a function of z on logarithmic scales to see how quickly the field decays. If the field has the functional form $E_z = C/z^k$, then $\log_{10}(E_z) = \log_{10} C - k \log_{10} z$, which means that $\log_{10} E_z$ would be a linear function of $\log_{10} z$ and that the slope gives the power-law exponent k. In Fig. 2.16b we have fitted a linear function to the logarithms using

```
zlarge = np.linspace(a,100*a,1000)
Ez_large = findEz(zlarge)
lz = np.log10(zlarge)
lEz = np.log10(Ez_large)
p = np.polyfit(lz,lEz,1)
print("p = ",p)
```

```
Out: p =  [-2.9642887 10.1935955]
```

This means that the powerlaw exponent k is approximately -3, that is, that $E_z \propto z^{-3}$. This means that the electric field decays faster than for a single charge.

Modeling the electric field far away from the plates. What would be a reasonable simplified model of the system that would give such a behavior far away? A possible simple model would be to represents each plate as a single charge: a charge $+Q$ at $z = d/2$ and a charge $-Q$ at $z = -d/2$. The electric field along the z-axis is then:

$$
\begin{aligned}
E_z &= \frac{Q}{4\pi\epsilon_0(z-d/2)^2} + \frac{-Q}{4\pi\epsilon_0(z+d/2)^2} \\
&= \frac{Q}{4\pi\epsilon_0}\frac{(z+d/2)^2-(z-d/2)^2}{(z-d/2)^2(z+d/2)^2} \\
&= \frac{Q}{4\pi\epsilon_0}\frac{2zd}{(z-d/2)^2(z+d/2)^2} .
\end{aligned}
\tag{2.58}
$$

When $z \gg d$, we can approximate $z - d/2$ and $z + d/2$ both as z, getting:

$$E_z = \frac{Q}{4\pi\epsilon_0}\left(\frac{2zd}{(z-d/2)^2(z+d/2)^2}\right) \simeq \frac{1}{4\pi\epsilon_0}\frac{2dQ}{z^3} . \quad (2.59)$$

This is indeed the observed behavior! Such a system consisting of two opposite charges is called a *dipole*. The dipole represents the simplest form of internal charge distribution in a system that has net zero charge, and we will therefore return to the dipole many times in this book.

Comments. We have seen that for two charged plates, the electric field between the plates is approximately constant, the field around the plates is more complex, but far away from the plates it resembles the field from a dipole. We have also demonstrated how we can compare exact and numerical solutions and how to analyze the behavior far away from the systems. These are fundamental aspects of *modeling* of physical system.

Method: Modeling a system

An important aspect of our analysis of electromagnetic systems is what we call *modeling*. Modeling is to take a physical situation and convert it into a simplified, cartoon system that represents the essential physical properties we need to address the system.

The difficult part is to figure out what physical aspects need to be included in our model, and how to model these aspects. This takes experience and a bit of courage. In physics, we want to create a model that is as simple as possible, but not simpler. And we want to model the system using components that we can understand and use in computations.

For example, let us analyze a balloon close to a wall as illustrated in the figure part **a**. **b** The balloon has a charge q. **c** We simplify it as a point charge q. **d** We model the wall as displaced charges. **e** The distance from the balloon to the wall is R_0. **f** We model the displacement of the charges in the wall as a charge Δq displaced by a distance d, where $d\Delta q$ is proportional to the electric field from q at the surface of the wall: $d\Delta q \simeq Cq/R_0^2$. This is similar to a charge Δq attached to a small spring with a deflection proportional to the force. We could continue to refine the model, but stop here. This allows us to explain the phenomenon, and maybe even measure C and R_0?

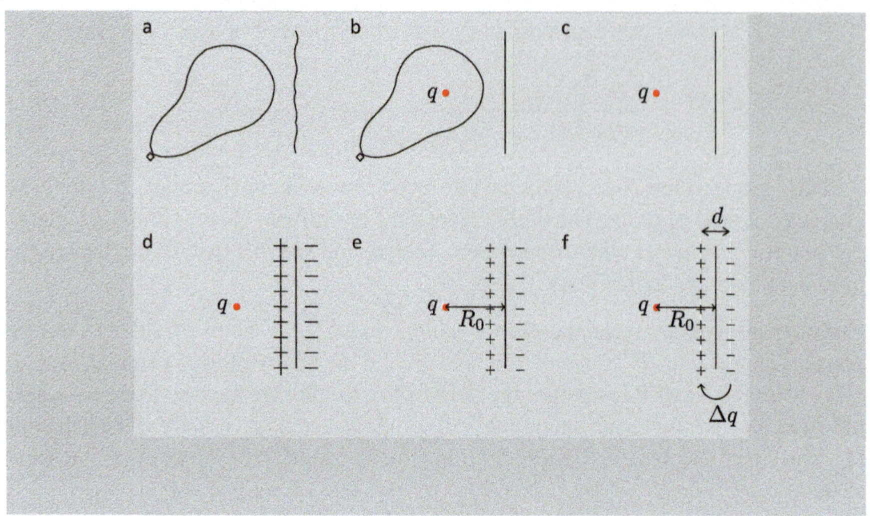

Summary

Coulomb's law. The force on a point charge q at $\mathbf{r}$ from a point charge Q at $\mathbf{r}'$ is

$$\mathbf{F} = \frac{qQ}{4\pi\epsilon_0}\frac{\mathbf{R}}{R^3} = \frac{qQ}{4\pi\epsilon_0}\frac{\mathbf{r}-\mathbf{r}'}{|\mathbf{r}-\mathbf{r}'|^3}. \tag{2.60}$$

Electric field. The electric field $\mathbf{E}$ at a point $\mathbf{r}$ due to a set of charges is defined as

$$\mathbf{E} = \frac{\mathbf{F}}{q}, \tag{2.61}$$

where $\mathbf{F}$ is the net force on a test charge q due to the set of charges.

Properties of the electric field.

- The electric field from a point charge Q at $\mathbf{r}'$ is

$$\mathbf{E}(\mathbf{r}) = \frac{Q}{4\pi\epsilon_0}\frac{\mathbf{R}}{R^3} = \frac{Q}{4\pi\epsilon_0}\frac{\mathbf{r}-\mathbf{r}'}{|\mathbf{r}-\mathbf{r}'|^3}. \tag{2.62}$$

- The vector $\mathbf{R}$ is from the charge Q at $\mathbf{r}'$ to the observation point at $\mathbf{r}$: $\mathbf{R} = \mathbf{r} - \mathbf{r}'$.
- The electric field obeys the **superposition principle** : $\mathbf{E} = \sum_i \mathbf{E}_i$.

Charge distributions. We describe a continuous distribution of charges using a volume; surface; or line charge density with charge elements $dq = \rho_v\, dv$; $\rho_s\, dS$; $\rho_l\, dl$. The electric field from a:

- volume charge density is $\mathbf{E} = \int_v \frac{\rho(\mathbf{r}')\, dv'}{4\pi\epsilon_0} \frac{\mathbf{r}-\mathbf{r}'}{|\mathbf{r}-\mathbf{r}'|^3}$
- surface charge density is $\mathbf{E} = \int_S \frac{\rho_s(\mathbf{r}')\, dS'}{4\pi\epsilon_0} \frac{\mathbf{r}-\mathbf{r}'}{|\mathbf{r}-\mathbf{r}'|^3}$
- line charge density is $\mathbf{E} = \int_C \frac{\rho_l(\mathbf{r}')\, dl'}{4\pi\epsilon_0} \frac{\mathbf{r}-\mathbf{r}'}{|\mathbf{r}-\mathbf{r}'|^3}$

Exercises

Discussion Exercises

2.1 Zero field. Draw a diagram with two point charges so that the electric field is zero somewhere and show where that position is. For a system of two charges, is there always a point where the electric field field is zero?

2.2 Field and force. What is the relationship between the terms "field" and "force"? What are their units?

2.3 Constant field. How would you illustrate an electric field that is a constant (in space and time)? How does this field affect a positive and a negative charge at $(0, 0)$ and $(1, 1)$? Make drawings to illustrate your answers.

2.4 Peeling tape. If you peel two strips of transparent tape off the same roll and immediately let them hang near each other, they will repel each other. If you then stick the sticky side of one to the shiny side of the other and rip them apart, they will attract each other. Give a plausible explanation.

2.5 Mechanisms for charge transfer. If a glass rod is rubbed on a silk cloth, the rod becomes positively charged, but if you rub the rod on fur, it becomes negatively charged. How can you explain the mechanisms of charge transfer?

Tutorials

2.6 Field from a single charge. We place a charge Q in the point $(a, 0)$ as shown in Fig. 2.17.
(a) Sketch the direction and magnitude of the electric field in the points marked with an x in the figure.
(b) Draw the **R**-vector your would use to find the electric field from the charge Q in the points marked in the figure.

2.7 Field and R-vector for a single charge. A charge $-Q$ is placed in the origin.

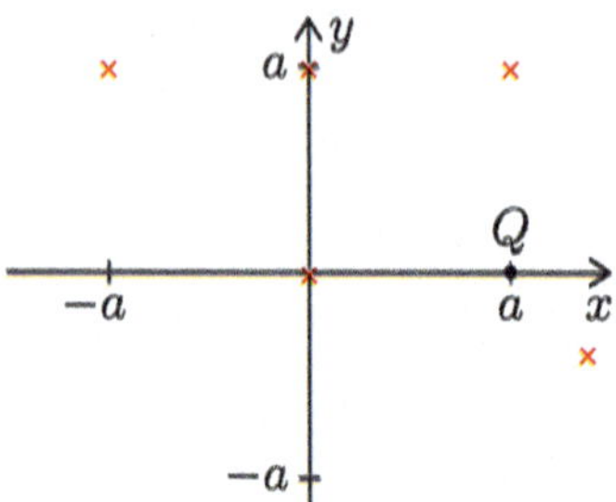

Fig. 2.17 A single charge in $(a, 0)$

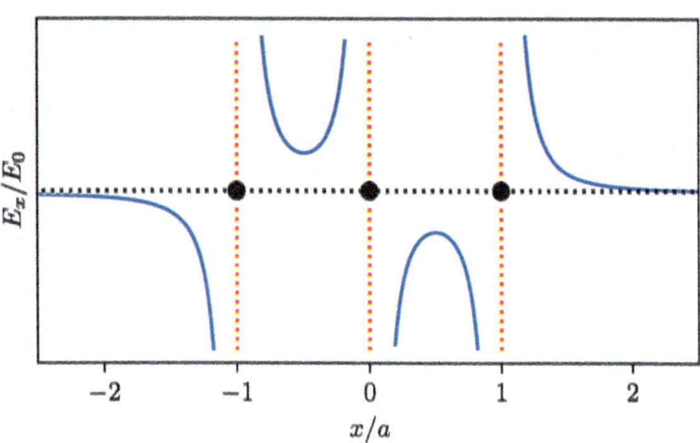

Fig. 2.18 Plot of $E_x(x)$ for three charges

(a) Draw the **R**-vector used to find the electric field in the point $(a, 0)$.
(b) Draw the **R**-vector used to find the electric field in the point $(0, a)$.
(c) Find the electric field in the point (a, a).

2.8 Field from two charges. A charge $+Q$ is placed in the point $(0, -a)$ and a charge $-Q$ is placed in the point $(0, a)$.
(a) Make a drawing of the system.
(b) Draw the **R**-vectors used to find the electric field in the point $(a, 0)$.
(c) Draw the magnitude and direction of the electric field in the 9 points (x_i, y_j) where $x_i = -a, 0, a$ and $y_j = -a, 0, a$.

2.9 Three unknown charges. Three charges q_1, q_2 and q_3 lie along the x-axis in the points $(-a, 0)$, $(0, 0)$ and $(a, 0)$ respectively. All the charges have the same magnitude ($|q_1| = |q_2| = |q_3|$). Figure 2.18 shows the electric field $E_x(x)$ along the x-axis. What are the signs of the three charges?

2.10 Net charge of a charge distribution. A volume charge density ρ is given as

$$\rho(\mathbf{r}) = \begin{cases} C & \text{for } r < a \\ 0 & \text{for } r \geq a \end{cases}$$

(a) Make a sketch of the charge density.
(b) How much charge, $Q(r)$, is inside a radius r from the origin?

2.11 Balloon. A spherical balloon with radius a has a charge $+Q$.
(a) How would you model the charged balloon? Can you describe the charge with a charge distribution?
(b) What do you think the electric field is at the center of the balloon?

2.12 A long line of charge. A thin, long rod is oriented parallel to the x-axis and intersects the y-axis at a distance a from the x-axis. A length L of the rod has a charge Q. You can assume that the rod is infinitely long and that it has a uniformly distributed charge.
(a) Draw the system.
We want to find the electric field in a point $\mathbf{r} = (x, 0, 0)$ on the x-axis.
(b) Without making a calculation, which way do you think that the electric field will be directed in the point $(x, 0, 0)$?
(c) Your friend Q says that if you want to find the electric field in the position $(x, 0, 0)$, you could instead find it it in $(0, 0, 0)$, because the result is the same. Is Q right? Provide an argument for your answer.
(d) Draw a small line element dl of the rod in the position $(x', a, 0)$. What is the charge of this line element?
(e) What is the **R**-vector your must use to find the contribution from the line element dl to the electric field in the point $(0, 0, 0)$?
(f) What is the contribution $d\mathbf{E}$ to the **E**-field at $(0, 0, 0)$ from this line element?
(g) Write down the integral to find the **E**-field. (It is sufficient to write down the integral for now).

Exercises

2.13 Electric dipole. An electric dipole consists of two particles: particle 1 with a charge Q at $\mathbf{r}_1 = (a, 0, 0)$ and particle 2 with a charge $-Q$ at $\mathbf{r}_2 = (-a, 0, 0)$.
(a) Make a sketch of the system. What is the direction of the electric field **E** for a point on the y-axis?
(b) What are the two **R**-vectors you used in this argument? (The **R**-vector appears in the expression for the electric field from a single charge: $\mathbf{E} = Q/(4\pi\epsilon_0)\mathbf{R}/R^3$.)
(c) For a point in a plane through $x = a$ parallel to the yz-plane, that is for points $\mathbf{r} = (a, y, z)$, what are the two **R**-vectors needed to find the electric field?
(d) Find an expression for the electric field $\mathbf{E}(a, y, z)$.
(e) What type of coordinate system would you use to take advantage of the symmetry of the problem? Provide an argument for your choice.
(f) What is the electric field **E** expressed in this coordinate system in the point $\mathbf{r} = (x, y, z)$?

2.14 Line charge. A rod of length L has a charge Q. We place the rod along the x-axis with its center at the origin.
(a) Make a drawing of the system. What assumptions would you need to make to approximate this as a line charge with a line charge density ρ_l? Find ρ_l.

(b) A small piece of the rod from x to $x + \mathrm{d}x$ has a length $\mathrm{d}x$. What is the charge of this piece?
(c) We want to calculate the electric field $\mathbf{E}(0, y, 0)$ along the y-axis. What is the contribution to the field from the piece of length $\mathrm{d}x'$ at x' expressed using the **R**-vector? Write down an explicit expression for the **R**-vector.
(d) Write down an expression for the electric field $\mathbf{E}(0, y, 0)$ in terms of an integral. Explain what variable you integrate over. (Ensure that your result is a vector!)
(e) Explain how you can use the symmetry of the problem to simplify the calculation. What type of symmetry do we have in this problem. Use this symmetry to explain where you know the field if you calculate it in the point $(0, y, 0)$.

2.15 Plane charge. An infinite plane with a homogeneous charge density ρ_s is placed at $z = 0$.
(a) What does the word homogeneous mean in this context?
(b) What is the total charge of the plane?
(c) We want to find the field in a point $\mathbf{r} = (x, y, z)$. Use a symmetry argument to find the direction of the electric field in this point.
(d) What is the contribution $\mathrm{d}\mathbf{E}$ to the electric field at $\mathbf{r} = (x, y, z)$ from a small piece $(x', x' + \mathrm{d}x'), (y', y' + \mathrm{d}y')$ at $(x', y', 0)$? First, make a drawing of the system. Explain why we use x' and not x. What is the **R**-vector? Find an expression for $\mathrm{d}\mathbf{E}$.
(e) Find an integral-expression for the electric field $\mathbf{E}$ in a point $\mathbf{r} = (x, y, z)$ from charges in the region $0 < x' < a, 0 < y' < b, z' = 0$. (You do not need to solve the integral).

2.16 Angle between suspended charges. Two charges of identical mass m with charges $q_1 = q$ and $q_2 = 2q$ as suspended from the same point in strings of length L. You can assume that the gravitational force is much larger than the electrostatic force on each charge.
(a) Draw a force diagram for each charge.
(b) Find an approximate expression for the angles θ_1 and θ_2 that each charge makes with respect to the vertical.
(c) Why did you have to assume that the electrostatic force was small compared with gravity to solve this problem. How can you solve this problem if you do not make this assumption?

2.17 Adding and subtracting using superposition. We use superposition to find the net electric field on a charge from a distribution of individual charges or from a continuous distribution of charges. However, superposition is often used as a "trick" when solving problems.
(a) Six identical charges Q are placed at the vertices of a hexagon with edge length L. What is the net force on a test charge q placed at the center of the hexagon?
(b) We now remove one of the charges. There are now 5 equal charges present at five of the vertices of the hexagon. What is the net force on the test charge now?
(c) How can you generalize your result to the case where you place n equidistant charges Q on a circle of radius r? What is the electric field in the center of the circle? What is the electric field in the center of the circle if you remove one charge at $(r, 0, 0)$?

(d) The electric field at a distance r from the center of a uniformly charged sphere of radius a and volume charge density ρ is

$$\mathbf{E} = \begin{cases} \frac{\rho}{3\epsilon_0} r \hat{\mathbf{r}} & r < a \\ \frac{\rho}{3\epsilon_0} \frac{a^3}{r^2} \hat{\mathbf{r}} & r \geq a \end{cases} .$$

Explain how you can use superposition to find the electric field inside a spherical shell of uniform charge density ρ with inner radius b and outer radius a. (Hint: A spherical shell with outer radius a and inner radius b plus a sphere of radius b is a sphere of radius a.)(Hint: $\mathbf{E}_a = \mathbf{E}_{\text{shell}} + \mathbf{E}_b$.)

Homework

2.18 Point charges. A point charge $q_1 = 4.0$ nC is on located the x-axis at $x = 2.0$ m. Another point charge, $q_2 = -6.0$ nC, is located on the y-axis at $y = 1.0$ m.
(a) Draw a figure of the charges in a coordinate system and sketch the forces acting on the two charges.
(b) Calculate the force $\mathbf{F}_2$ on the charge q_2 and the magnitude $|\mathbf{F}_2|$ of the force on the charge q_2.
(c) What is the force $\mathbf{F}_1$ on the charge q_1?

2.19 The electric field from groups of point charges. We address a system with a single charge q in $\mathbf{r} = 0$.
(a) Use Python to make an arrow plot that shows the electric field $\mathbf{E}(x, y)$ in the xy-plane and a plot that show the stream lines in the xy-plane.
(b) Plot $E(r) = |\mathbf{E}(r)|$, where $r = |\mathbf{r}|$. Explain how to use a plot of $E(r)$ to obtain the functional form of $E(r)$ and show that it is $E(r) \propto r^{-2}$.

We will then address a system of two charges: a charge q in $\mathbf{r} = (a, 0)$ and a charge $-q$ in $\mathbf{r} = (-a, 0)$, where a is a characteristic length.
(c) Make an arrow plot of the electric field in the xy-plane and a plot that shows the stream lines in the xy-plane.
(d) Plot $E(r)$. From the plot, find the functional form, $E(r)$, in the limit of large r ($r \gg a$).
(e) Will these results change significantly if both charges have the same sign? How?

We will then address a system of four charges—a quadrapole—with q in $\mathbf{r} = (\pm a, 0)$ and $-q$ in $\mathbf{r} = (0, \pm a)$.
(f) Make an arrow plot of the electric field in the xy-plane and a plot of the stream lines in the xy-plane.
g) Using the method you have developed to find the functional form of $E(r)$ in the limit of large r ($r \gg a$).

2.20 The electric field above a finite disk. In this exercise we study the electric field along the z-axis (the symmetry axis) from a disk of radius a and constant surface charge density ρ_A.

Fig. 2.19 Half-circle shaped line charge

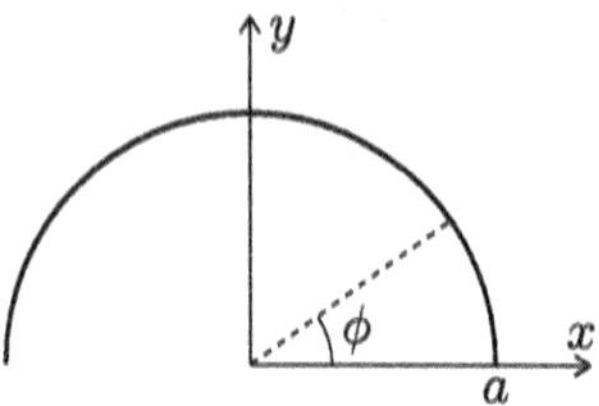

(a) What is the contribution to the electric field in a position z from an element from r to $r + dr$ and from θ to $\theta + d\theta$. (You can assume $z > 0$).
(b) What is the electric field in a position z (for $z > 0$)?
(c) Find the electric field **E** in the limit where z is small ($z \to 0$ from above). Interpret the result.
(d) Sketch $|\mathbf{E}|$ as a function of z for $z > 0$.

2.21 Work on point charges. Three point charges with the same charge Q are placed at each corner A, B, C in an equilateral triangle with side lengths a. One charge is displaced along the dashed line from A to the midpoint of A' on the side BC. The charges in B and C are kept in place throughout the displacement. Calculate the work done to carry out this displacement.

2.22 Plate with a hole. Find $|\mathbf{E}|$ at a height z above an infinitely large, plane surface with a hole with radius a. The surface has a constant surface charge density ρ_s. (Hint: Superposition.)

2.23 Half circle. We will here study a half-circle shaped line charge with radius a as illustrated in Fig. 2.19.
(a) Assume that the line has a uniform charge density with a total charge Q. Find the electric field **E** in the center of the half circle, that is, in the origin of the figure.
(b) Let as assume that the half circle is *not* uniformly charged, but that the line charge density is given as

$$\rho_l(\phi) = \frac{Q}{2a} \sin\phi .$$

Sketch the line charge density, and show that the total charge for the half circle is still Q.
(c) Find the magnitude and direction of the electric field in the origin for the charge distribution in **b**. Which of the charge distributions (from **a** or **b**) gives the largest magnitude for the electric field?

2.24 The electric field above an infinite plane. What is the electric field above an infinite charged plane? This is a result we will return to in many situations, because we often can approximate a part of the system we are studying as an infinite plane. For example, many electric components called capacitors often consists of two parallel planes. The goal of this exercise is to build intuition for the simple form of the electric field above (or below) a charged plane.

We place an infinite plane with a uniform surface charge density ρ_S in the xy-plane.

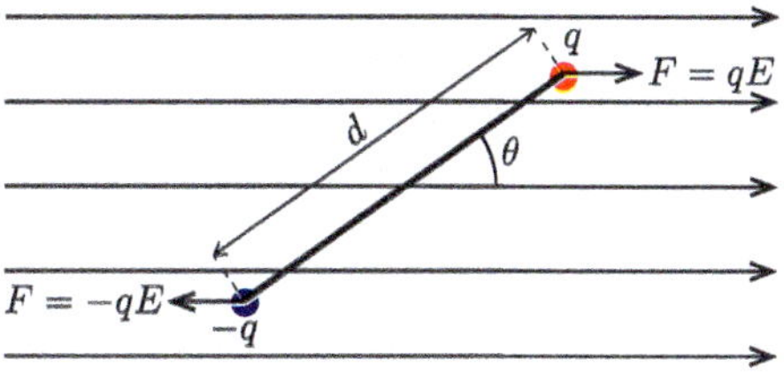

Fig. 2.20 Dipole in a uniform electric field

(a) Argue for why the electric field only can depend on the z-coordinate.
(b) We use cylindrical coordinates. What is the charge in a region from r to $r + \mathrm{d}r$ and from θ to $\theta + \mathrm{d}\theta$?
(c) What is the contribution $\mathrm{d}\mathbf{E}$ to the electric field at $(0, 0, z)$ from the charge $\mathrm{d}q$ in a position $(r\cos\theta, r\sin\theta, 0)$? (You can assume that $z > 0$).
(d) Find the contribution to the electric field at $(0, 0, z)$ from a circular region with radius r by integrating over θ.
(e) Find the electric field E_z at a position z by integrating over all r.

2.25 Dipole in a uniform electric field. Figure 2.20 shows an electric dipole in a uniform electric field. The distance between the two charges in the dipole is d as shown in the figure.
(a) What is the net force on the dipole?
(b) Find the torque around the center of mass of the dipole in terms of the dipole moment $\mathbf{p} = Q\mathbf{d}$ and the electric field $\mathbf{E}$. (Hint: The torque is $\boldsymbol{\tau} = \mathbf{r} \times \mathbf{F}$, where $\mathbf{r}$ is the vector going from the center of mass to where the force $\mathbf{F}$ is applied.)
(c) For what angles θ do we have stable and unstable equilibrium? And when do we have maximum torque?
(d) Show the dipole will act as a simple harmonic oscillator for small by using the small angle approximation ($sin(\theta) \approx \theta$) and $\tau = I\alpha$, where I is the moment of inertia and α is the angular acceleration. (Hint: A simple harmonic oscillator has the differential equation $m(\mathrm{d}^2x/\mathrm{d}t^2) = -kx$ where m and k are constants and x describes the position. The solution is $x(t) = Acos(\omega t + \phi)$ where $\omega = \sqrt{k/m}$ and A and ϕ are constants.)
(e) Find the angular frequency ω of the oscillator.
(f) Assume that the dipole starts at rest at $t = 0$ with an angle θ_0 with the electric field. Solve the differential equation of motion to find $\theta(t)$.
(g) Now, let us compare the small angle solution with the full numerical solution. Write a program to find the numerical solution without using the small angle approximation. Plot the results over four periods and plot the approximate solution in the same plot. For these calculations you can use $q = e, d = 3 \cdot 10^{-9}$m, $m = 3 \cdot 10^{-5}$kg and $E = 1000$N/C.

2.26 Field from a hemisphere shell. In this exercise we are going to calculate the field from a hemisphere shell as shown in Fig. 2.21. The shell has constant surface charge density ρ and radius R. Calculate the electric field, $\mathbf{E}$, in the origin.

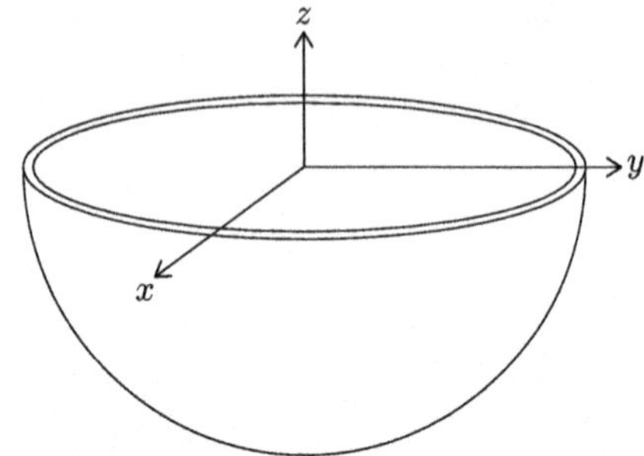

Fig. 2.21 Illustration of a hemisphere shell. The origin is equidistant to the hemisphere and the hemisphere is oriented such that it has rotational symmetry along the z-axis

2.27 Visualize electric field. Here we are going to visualize the electric field from several particles in two dimensions. We are going to do this in parts.
(a) Consider a particle in (x_1, y_1) with charge Q, write down the general electric field from this in position (x, y).
(b) Write a Python program that takes in position and charge from a particle, a point in space, and returns the electric field vector in that point. Make a vector arrow plot of the electric field.
(c) Expand the function you made to take in an arbitrary number of particles and then returns the resulting electric field. Make a vector arrow plot.

2.28 Parallel lines.
(a) An infinitely long line charge with uniform line charge density ρ_l is placed along the x-axis. There are no other charges present and the system is in vacuum. Find the electric field $\mathbf{E}$ at $\mathbf{r} = (x, y, z)$.
(b) An infinitely long line charge with uniform line charge density ρ_l is instead placed on a line parallel to the x-axis through $y = d$. There are no other charges present and the system is in vacuum. Find the electric field $\mathbf{E}$ in the point $\mathbf{r} = (x, y, z)$.
(c) Let us assume that the system consists of two infinitely long line charges that are parallel with the x-axis. One line passes through $(0, 0, 0)$ and the other passes through $(0, d, 0)$. Both lines have uniform line charge densities ρ_l. Show that the electric field $\mathbf{E}(\mathbf{r})$ in the xy-plane is:

$$\mathbf{E} = \frac{\rho_l}{2\pi\epsilon_0}\left(\frac{1}{y} + \frac{1}{y-d}\right)\hat{y}.$$

We will now study a system that consists of a line along the x-axis and through the origin with a line charge density ρ_l, which is not uniform, but varies with x:

$$\rho_l(x) = \begin{cases} 0 & x < 0 \\ \rho_l(x) & 0 \le x \le L \\ 0 & x > L \end{cases}.$$

(d) Find an expression on integral form for the electric field $\mathbf{E}(\mathbf{r})$ in $\mathbf{r} = (x, y, z)$ from the line charge density $\rho_l(x)$ on the x-axis. (You do not need to solve the integral.)
(e) Assume that you may approximate the integral with a sum over N elements of width $dx_i = L/N$ at $x_i = (i + 1/2)dx_i$ for $i = 0, 1, \ldots, N - 1$. Every element contributes with a point charge $dq_i = \rho_l(x_i)dx_i$. Write a short program to find an approximate value for the electric field $\mathbf{E}(\mathbf{r})$ at $\mathbf{r}$ by summing the fields from the point charges dq_i.

2.29 Crossing lines. In this exercise we assume that you know that the electric field from an infinitely long line along the z-axis with line charge density Q/L is $\mathbf{E} = (Q/L)/(2\pi\epsilon_0 r)\,\hat{\mathbf{r}}$, where $\hat{\mathbf{r}}$ is a unit vector that points radialy outward from the z-axis and r is the distance to the z-axis.
(a) We place an inifinitely long line along the x-axis with line charge density $-Q/L$. What is the electric field in the xy-plane, $\mathbf{E}(x, y)$, from this line?
(b) We place a new inifinitely long line with line charge density Q/L parallel to the x-axis and through the point $(0, a)$. What is the electric field in the xy-plane for the system consisting of these two lines? Make a plot to illustrate the field.
(c) Instead, we place two infinite lines: One line with line charge density Q/L along the x-axis and one line with line charge density $-Q/L$ along the y-axis. What is the electric field in the xy-plane. Write a program to visualize the field.

2.30 Induced dipole interactions in one dimension. We started this chapter with a demonstration of a balloon being attracted to (and sticking to) a wall, but we did not describe the details of the physics involved. Another popular demonstration is to show that if we rub a balloon against our hair, it will attract an empty aluminum can. How can we understand this process? The aluminum can does not have a net charge, so why is it attracted to the charge of balloon? We call this effect an *induced dipole interaction*. This effect is very important and common on the atomic scale.

Figure 2.22 illustrates a situation where a charge q is placed near a neutral atom such as Argon. We assume that the core of the atom does not move, but that the distribution of charge (gray) around the atom may deform.
(a) Explain why, if q is positive, the electron cloud is slightly shifted toward q.

We assume that q is placed in the origin and that the positive core of the atom is at x. The effect of q is to move a charge $-\Delta q$ a small distance d. Since the atom is neutral, this means that the atom will be negative on the left side and positive on the right side, which we model as a charge $-\Delta q$ at the position $x - d$ and a charge Δq at x. We assume that the displacement d is much smaller than x. We say that the charge q *induces* a dipole of strength $2d\Delta q$.
(b) How large is $2d\Delta q$? We will make a simplified model for this: We will assume that the response is linear, that is, that $d\Delta q$ is proportional to the electric field from the charge q at the position x. Show that the dipole moment is

$$2d\Delta q = C'\frac{1}{4\pi\epsilon_0}\frac{1}{x^2} = C/x^2,$$

where $C = C'/(4\pi\epsilon_0)$ is a constant.

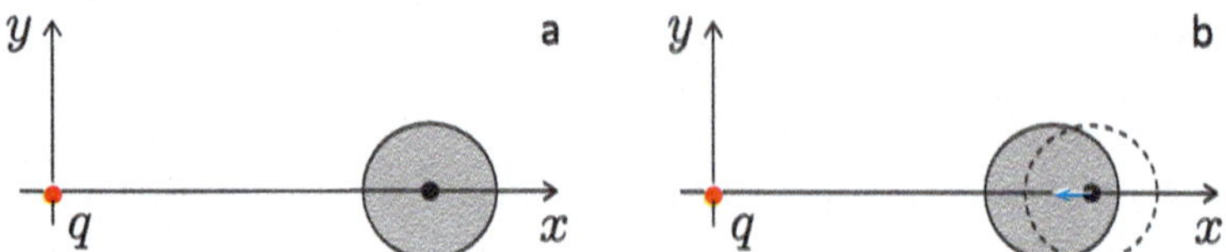

Fig. 2.22 Illustration of an induced dipole. We place a charge q near an atom with a center and a charge cloud. **a** Initial situation. **b** Situation after the charge distribution around the atom has equilibrated to a new configuration. (This usually happens very fast)

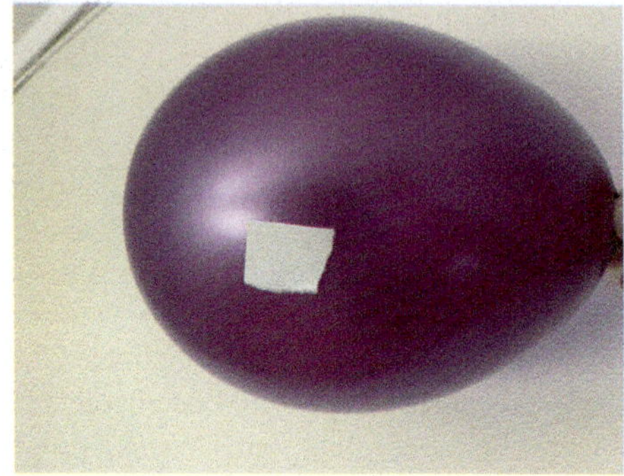

Fig. 2.23 A piece of paper attached to a balloon

(c) Show that the force from the induced dipole on the charge q is

$$F_x = \frac{qC}{4\pi\epsilon_0 x^5} .$$

(Hint: It may be simpler to find the force from the dipole on the charge.)
(d) Explain how you can use a similar argument for the soda can. Is the force between a charge and a soda can attractive or repulsive?

Modeling Projects

2.31 Paper on balloon. A small piece of paper attaches to a balloon so strongly that it does not fall down as shown in Fig. 2.23. Explain how you would model this system.

2.32 Wavy surface. If you cut into a material such as a mineral, the fresh surfaced may have effective surface charges.
(a) How will you model the electric field from a freshly cut mineral surface?
(b) If the surface is flat, what is the electric field outside the surface?

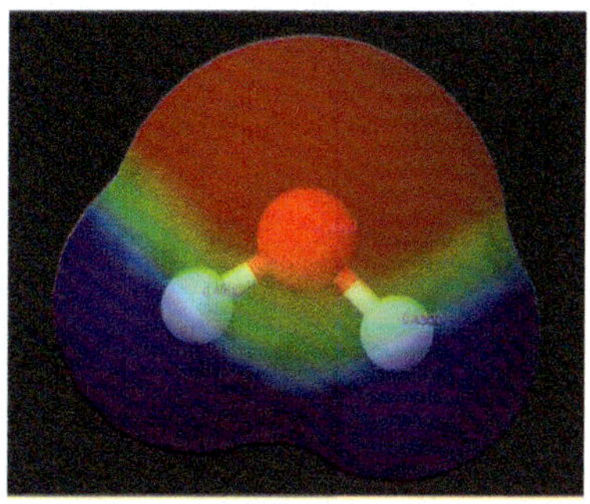

Fig. 2.24 Charge distribution around a water molecule calculated by `jmol` from `molview`. Here red are negative charges and blue are positive charges

(c) Assume that the surface is a fracture. We model the fracture as a sine-wave with amplitude A and wavelength λ. Write a Python program to find the electric field outside the surface. How can you use the result from the previous exercise to test your results?
(d) A fracture creates two sine-wave surfaces that are displaced a small distance d from each other. One surface has positive charge and the other negative charge. What is the electric field between the surfaces?
(e) Assume that the surface is carefully constructed to be have a rectangular wave shape with an amplitude A and a wavelength λ. Modify your Python program to find the electric field outside the surface. What are the main differences between the two systems?
(f) How would the charged surfaces affect a water molecule? You can assume that you can model the water molecule as a dipole.
(g) A real fracture surface has a more complicated shape called a self-affine fractal. An example of a self-affine fractal is a random walk. (Although the scaling behavior of the random walk is different from most fracture surfaces, they both are examples of self-affine fractals). You can generate a random walk of length L with the following script

```
import numpy as np
import matplotlib.pyplot as plt
L = 100
y = np.cumsum(np.random.randn(L))
x = np.arange(L)
y = y - (y[-1]-y[0])/L*x
plt.plot(x,y)
plt.axis("equal")
```

Use this model of a fracture to study the electric field inside a fracture.

2.33 Water models. In this project we will develop and study an increasingly complex model for a water molecule. Figure 2.24 illustrates the charge distribution around a water molecule. The dipole moment of water is 1.85D, where D is a unit called Debye: $1\text{D} \simeq 3.33564 \cdot 10^{-30}$Cm. We will now build a model for a water molecule.

(a) How would you model a water molecule as a set of point charges based on the illustration in the figure. (You need to look up realistic numbers and make your own assumptions here).
(b) Write a Python script based on your model to find the electric field around the water molecule.
(c) For a dipole, the field far ($r \gg d$) from the dipole along the dipole axis is $2p/(4\pi\epsilon_0 r^3)$, where $p = dq$ is the dipole moment. How can you use this to estimate the dipole moment from the electric field?
(d) Estimate the dipole moment of your water model and compare with the $p = 1.85$D. Comment on this result.
(e) There are many different types of water models used for molecular-scale modeling where the water molecule is modelled as various set of point charges. Look up and describe the models called SPC and TIP4P. Calculate the electric field for these two models and compare them. Why do you think the TIP4P model was introduced? Do you think the SPC model an exact description of the system?

Chapter 3
Electric Potential

3.1 Electric Potential

You most probably have heard the term potential directly or indirectly. You may know that a battery has a voltage of 1.5 V. The term voltage refers to the electric potential between the two poles of the battery. Similarly, you may have heard about high voltage as found in power lines or in lightning strikes (Fig. 3.1). How does this concept of voltage correspond to the concept of the electric field? We will try to build this intuition here.

You probably have developed an intuition for the relation between (conservative) forces and potential energy in mechanics. The relation between the electric field and the electric potential (the voltage) is similar. Gravity is a conservative force field. It is a force field in the sense that it is a force acting everywhere in space on masses due to particles with masses. It is conservative in the sense that the work done by the force is independent on the path, and it is therefore possible to introduce a potential energy. You may recall that for a conservative force, **F**, the potential energy in a point A was defined as the line integral along an arbitrary path from the point A to an (arbitrary) reference point 0:

$$U(A) - U(0) = \int_A^0 \mathbf{F} \cdot d\mathbf{l} \tag{3.1}$$

We can therefore determine the potential energy from the force field. Similarly, we recall that $\mathbf{F} = -\nabla U$, so we can find the force field from the potential energy. The potential energy provided us with a different way to reason about the motion of gravitational forces—a supplement to reasoning with forces which proved useful in many situations.

We will build on this intuition when we introduce the electrical potential. We will first demonstrate that electrostatic forces are conservative and use this to introduce a potential for the electric field.

© The Author(s), under exclusive license to Springer Nature Switzerland AG 2026

A. Malthe-Sørenssen, *Elementary Electromagnetism Using Python*, Undergraduate Texts in Physics, https://doi.org/10.1007/978-3-032-19876-1_3

Fig. 3.1 Electric potentials in **a** A 1.5 V battery, **b** a 10 kV power line, **c** a 10 GV lightning strike

3.1.1 Electric Forces are Conservative

First, let us demonstrate that the force from an electric field is conservative, that is, that the work done by the force does not depend on the path taken. Coulomb's law describes the interactions between two charges: a charge Q_1 setting up the field and a charge q moving in the field. Let us place the origin of the coordinate system at Q_1. The force on q from the field from Q_1 is then:

$$\mathbf{F} = q\mathbf{E} = q\frac{Q_1}{4\pi\epsilon_0}\frac{\hat{\mathbf{R}}}{R^2}. \tag{3.2}$$

where $\mathbf{R} = \mathbf{r} - \mathbf{r}_1 = \mathbf{r}$, since $\mathbf{r}_1 = \mathbf{0}$. The work on a charge q as it moves along a path C from A to B is illustrated in Fig. 3.2 and is defined as:

$$W_{AB} = \int_A^B \mathbf{F}\cdot \mathrm{d}\mathbf{l} = q\frac{Q_1}{4\pi\epsilon_0}\int_A^B \frac{\hat{\mathbf{R}}\cdot \mathrm{d}\mathbf{l}}{R^2} \tag{3.3}$$

Let us show that the work does not depend on the path C: We approximate the path by dividing it into small pieces that follow a circular arc with Q_1 as its center, or a radial path directly towards Q_1. We have illustrated a few such segments in Fig. 3.2. The total integral is (approximately) the sum of the integrals along these smaller paths.

Along a **circular segment**, such as the segment from b to c, $\mathbf{R}$ points radially from Q_1, whereas the direction of the path is along the circle segment of radius r, $\mathrm{d}\mathbf{l} = r\,\mathrm{d}\phi\hat{\boldsymbol{\phi}}$, where $\hat{\boldsymbol{\phi}}$ is a unit vector along the circular arc. But the circular arc is normal to $\mathbf{R}$. Therefore, $\hat{\mathbf{R}}\cdot \mathrm{d}\mathbf{l}$ is zero for this segment. The same is true for all the segments that are along circular paths centered on Q_1. These segments will not contribute to the integral.

Along a **radial segment**, such as the segment from a to b, the displacement $\mathrm{d}\mathbf{l}$ is along the vector $\mathbf{R}$: $\mathrm{d}\mathbf{l} = \mathrm{d}l\hat{\mathbf{R}} = \mathrm{d}r\hat{\mathbf{R}}$, where $\mathrm{d}r$ is the change in the distance r from the origin where Q_1 is placed. This integral is therefore simply in the scalar distance r to Q_1. The same is true for all the segments that are along radial paths centered on Q_1. The total integral is therefore

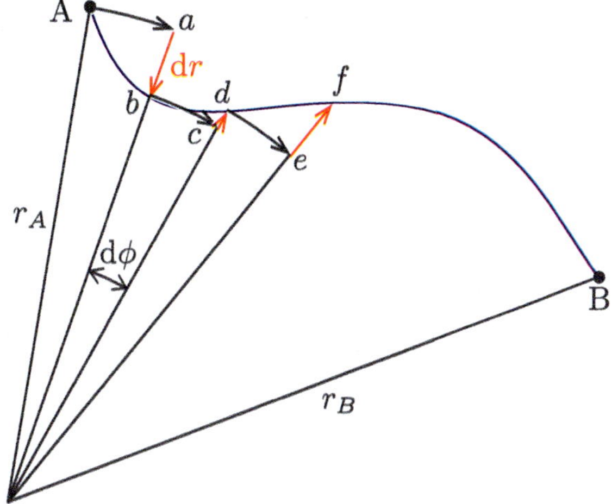

Fig. 3.2 Illustration of the work done on a charge q as it moves along the path C from A to B

$$W_{AB} = q\frac{Q_1}{4\pi\epsilon_0}\int_{r_A}^{r_B}\frac{\mathrm{d}r}{r^2} = -q\frac{Q_1}{4\pi\epsilon_0}\left(\frac{1}{r_B}-\frac{1}{r_A}\right). \tag{3.4}$$

This means that work does not depend on the path C, but only on the end points. We recall from mechanics that we call such a force field a *conservative* force field, and that we can introduce a potential energy for the force field from a single charge as the work needed to move the charge from a point A to a reference point:

$$U_A = W_{A,\mathrm{ref}} = U(r). \tag{3.5}$$

We can use the *superposition principle* to argue that if this is true for a single charge, it must also be true for any charge distribution: The total work is the sum of the work done by the individual forces $\mathbf{F}_i$ from each charge, Q_i:

$$U_A = W_{A,\mathrm{ref}} = \int_A^{\mathrm{ref}}(\mathbf{F}_1+\mathbf{F}_2+\cdots)\cdot\mathrm{d}\mathbf{l} \tag{3.6}$$

$$= \int_A^{\mathrm{ref}}\mathbf{F}_1\cdot\mathrm{d}\mathbf{l}+\int_A^{\mathrm{ref}}\mathbf{F}_2\cdot\mathrm{d}\mathbf{l}+\cdots \tag{3.7}$$

$$= W_1+W_2+\cdots = U_{A,1}+U_{A,2}+\cdots \tag{3.8}$$

This is the potential energy for a test particle with charge q. We can rewrite this in terms of the electric field instead, since $\mathbf{F} = q\mathbf{E}$:

$$U_A = \int_A^{\mathrm{ref}}\mathbf{F}\cdot\mathrm{d}\mathbf{l} = q\int_A^{\mathrm{ref}}\mathbf{E}\cdot\mathrm{d}\mathbf{l} = qV_A, \tag{3.9}$$

where we have introduced V_A, which we call the electric potential. The electric potential is therefore related to the potential energy, just as the electric field is related to the force: We find the potential energy of a charge by multiplying the potential with the charge.

Test your understanding

A positive charge moves in different trajectories around a positively charged rod. Is the work done by the force from the rod on the charge positive, negative or zero for the three cases? (a) A circular path with constant radius; (b) A helix-shaped path with constant radius; (c) A spiral-shaped path with diminishing radius.[1]

3.2 Properties of the Electric Potential

Electric potential

We define the *electric potential* $V(\mathbf{r})$ for an electric field $\mathbf{E}(\mathbf{r})$ as

$$V(\mathbf{r}) = \int_{\mathbf{r}}^{\text{ref}} \mathbf{E} \cdot \mathbf{dl}, \tag{3.10}$$

where the integral is a line integral along any curve from $\mathbf{r}$ to a reference point for which $V = 0$. It is common to choose the reference point infinitely far away.

The electric potential has the following properties:

- *The electric potential integral does not depend on the path, only on the end points.*
- *The electric potential obeys the superposition principle.*

The electric potential in a point A from a set of charges Q_i is the sum of the electric potential for each of the charges:

[1] (a) 0; (b) 0; (c) negative.

$$V_A = V_{A,1} + V_{A,2} \cdots = \sum_i V_{A,i} \tag{3.11}$$

- *The electric potential is measured in joule/coulomb. Which is called volt, V.*

The unit of the electric field is then V/m, volt per meter.

- *The integral of the electric field around a closed path is zero.*

Because the integral $W_{AB}/q = \int_C \mathbf{E} \cdot \mathbf{dl}$ is the same for all paths C from A to B, the integral for any closed loop is zero. Why? As illustrated in Fig. 3.3a we can always divide a closed path into two paths, one from A to B and one from B to A. The total integral along the path is the sum of these two integrals:

$$\oint_C \mathbf{E} \cdot \mathbf{dl} = \int_{C_1} \mathbf{E} \cdot \mathbf{dl} + \int_{C_2} \mathbf{E} \cdot \mathbf{dl} = 0, \tag{3.12}$$

where the second integral is the negative of the first because it goes in the opposite direction. Thus the total integral around a closed loop is zero.

- *The curl of* **E** *is zero.*

Because the integral around any closed path is zero, we can apply Stokes' theorem:

$$\oint_C \mathbf{E} \cdot d\mathbf{l} = \iint_S \nabla \times \mathbf{E} \cdot \mathbf{dS} = 0, \tag{3.13}$$

where S is a surface enclosed by the closed curve C as illustrated in Fig. 3.3b. Since this is true for any surface (and any corresponding closed curve), we find that

$$\nabla \times \mathbf{E} = 0. \tag{3.14}$$

This is true for **static** fields. We will later find modifications to this for time-varying fields.

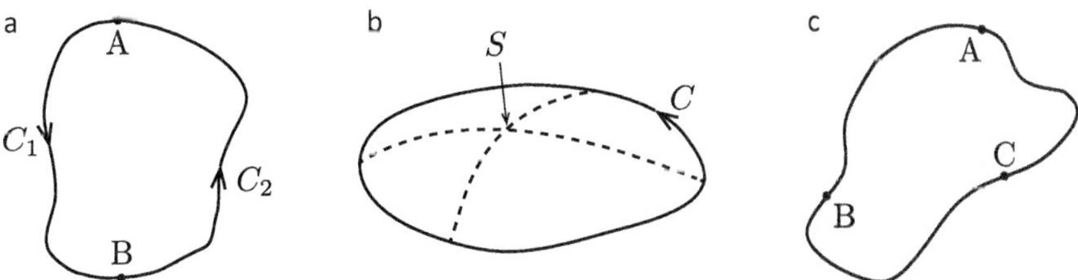

Fig. 3.3 **a** Illustration of how a closed loop can be divided into two curves C_1 from A to B and C_2 from B to A. **b** Illustration of a closed curve C and a corresponding surface S enclosed by C. **c** A circuit with three points A, B and C

- *The electric potential is only defined relative to a reference point.*

Notice that for finite charge distributions it is common to choose the reference point to be at infinity. A consequence is that the difference in potentials, V_{AB}, between two points A and B does not depend on the reference point:

$$V_{AB} = V_A - V_B = \int_A^{\text{ref}} \mathbf{E} \cdot d\mathbf{l} - \int_B^{\text{ref}} \mathbf{E} \cdot d\mathbf{l} \tag{3.15}$$

$$= \int_A^{\text{ref}} \mathbf{E} \cdot d\mathbf{l} + \int_{\text{ref}}^B \mathbf{E} \cdot d\mathbf{l} \tag{3.16}$$

$$= \int_A^B \mathbf{E} \cdot d\mathbf{l}. \tag{3.17}$$

- *The electric potential obeys Kirchhoff's voltage law: The sum of potential differences along a closed loop (such as a circuit) is zero.*

This is a consequence of the path integral of the electric field around a closed path being zero. For the case in Fig. 3.3c we find:

$$V_{AB} + V_{BC} + V_{CA} = \oint \mathbf{E} \cdot d\mathbf{l} = \int_A^B \mathbf{E} \cdot d\mathbf{l} + \int_B^C \mathbf{E} \cdot d\mathbf{l} + \int_C^A \mathbf{E} \cdot d\mathbf{l} = 0 \tag{3.18}$$

Test your understanding

The figure shows the electric potential at different points along a closed path. Is this situation possible?[2]

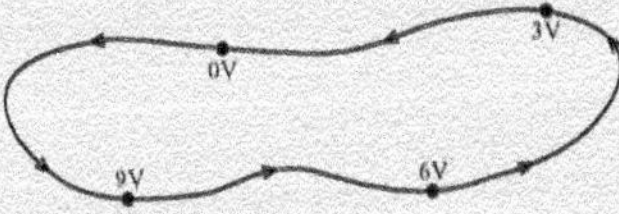

3.3 Electric Potentials from Charge Distributions

Let us now see how to find the electric potential $V(\mathbf{r})$ at a position $\mathbf{r}$ due to various charge distributions: a single charge; a set of charges; and charge densities for lines, surfaces and volumes.

[2] Yes. If we go a complete round, we are back where we started. It may be confusing that the potential increases from 0V to 9V. This may correspond to a battery in a circuit.

Single charge Q at the origin

We find the electric potential by applying the definition. We choose the reference point to be at infinity, and integrate from the point A at $\mathbf{r}$ to infinity:

$$V_A = \int_A^{\infty} \mathbf{E} \cdot d\mathbf{r}. \tag{3.19}$$

For a single charge Q at the origin, the electric field is:

$$\mathbf{E} = \frac{Q}{4\pi\epsilon_0} \frac{\mathbf{r}}{r^3}, \tag{3.20}$$

which gives

$$V(\mathbf{r}) = \int_{\mathbf{r}}^{\infty} \frac{Q}{4\pi\epsilon_0} \frac{\mathbf{r} \cdot d\mathbf{r}}{r^3} = \int_r^{\infty} \frac{Q}{4\pi\epsilon_0} \frac{dr}{r^2} = \frac{Q}{4\pi\epsilon_0 r}. \tag{3.21}$$

Here, we have used that $\mathbf{r} \cdot d\mathbf{r} = r\,dr$. This result demonstrates that the electric potential decays as $1/r$, while the electric field decays as $1/r^2$—as we would guess because the potential is the integral of the field. This is a useful rule of thumb that can be extended also to e.g. dipoles where the field decays at $1/r^3$, while the potential decays as $1/r^2$.

Single charge Q at $\mathbf{r}_1$

How does this result change if the charge Q_1 is in the position $\mathbf{r}_1$ instead? We introduce a new coordinate system $\mathbf{r}' = \mathbf{r} - \mathbf{r}_1$. In this coordinate system, the charge is in the origin, hence the potential is

$$V(r') = \frac{Q_1}{4\pi\epsilon_0 r'}, \tag{3.22}$$

and we insert $\mathbf{r}' = \mathbf{r} - \mathbf{r}_1$, getting

$$V(r) = \frac{Q_1}{4\pi\epsilon_0 |\mathbf{r} - \mathbf{r}_1|}. \tag{3.23}$$

A set of charges Q_i at $\mathbf{r}_i$

For a set of charges we can use the superposition principle for the electric potential. The total potential is therefore

$$V = \sum_i V_i = \sum_i \frac{Q_i}{4\pi\epsilon_0 R_i}, \tag{3.24}$$

where $\mathbf{R}_i = \mathbf{r} - \mathbf{r}_i$. As a function of $\mathbf{r}$ we get:

$$V(\mathbf{r}) = \sum_i \frac{Q_i}{4\pi\epsilon_0 |\mathbf{r} - \mathbf{r}_i|}. \tag{3.25}$$

Continuous charge distributions

We can extend this also to a continuous distribution of charges:

For a *volume charge density* ρ_v, we get:

$$V(\mathbf{r}) = \int_v \frac{\rho_v \, \mathrm{d}v}{4\pi\epsilon_0 R}, \tag{3.26}$$

where R is the distance from the point $\mathbf{r}$ to the volume element $\mathrm{d}v$.

For a *surface charge density* ρ_s, we get:

$$V(\mathbf{r}) = \int_S \frac{\rho_s \, \mathrm{d}S}{4\pi\epsilon_0 R}, \tag{3.27}$$

where R is the distance from the point $\mathbf{r}$ to the surface element $\mathrm{d}S$.

For a *line charge density* ρ_l, we get:

$$V(\mathbf{r}) = \int_C \frac{\rho_l \, \mathrm{d}l}{4\pi\epsilon_0 R}, \tag{3.28}$$

where R is the distance from the point $\mathbf{r}$ to the line element $\mathrm{d}l$.

Test your understanding

A charge Q is in the point $(a, 0, 0)$. (a) What is the electric potential in the origin? (b) What is the electric potential infinitely far away?[3]

[3] (a) $Q/(4\pi\epsilon_0 a)$, (b) 0.

Test your understanding

The figure shows a uniform electric field **E** between two infinitely large charged plates. Rank the electric potentials V_A, V_B, V_C, V_D and V_E from the largest to the smallest.[4]

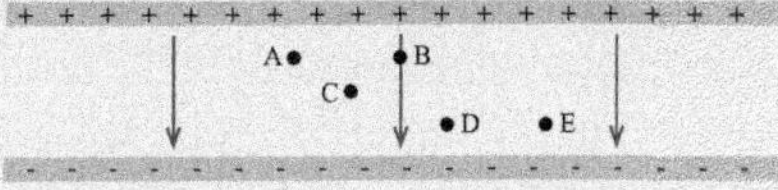

3.4 Relation Between Electric Potential and Electric Field

You may recall from mechanics that there is a relationship between the force **F** and the potential energy: $\mathbf{F} = -\nabla U$. This relation transfers directly to the electric potential:

$$\mathbf{E} = \frac{\mathbf{F}}{q} = -\frac{1}{q}\nabla U = -\nabla\left(\frac{U}{q}\right) \tag{3.29}$$

$$= -\nabla V = \left(-\frac{\partial V}{\partial x}, -\frac{\partial V}{\partial y}, -\frac{\partial V}{\partial z}\right). \tag{3.30}$$

Relation between the electric field and the electric potential

The electric potential $V(\mathbf{r})$ can be found from the electric field **E** by:

$$V(\mathbf{r}) = \int_r^{\text{ref}} \mathbf{E} \cdot d\mathbf{l}, \tag{3.31}$$

and the electric field $\mathbf{E}(\mathbf{r})$ can be found from the electric potential $V(\mathbf{r})$ by:

$$\mathbf{E} = -\nabla V. \tag{3.32}$$

Notice that in order to find the electric field from the potential, we need to know not only the electric potential in a single point. We must know how the potential varies around that point because the field depends on the (spatial) derivatives of the electric potential. Similarly, in order to find the electric potential from the field, we need to know not only the electric field in a single point or in a few points. We must

[4] $V_A = V_B > V_C > V_D = V_E$.

know how the electric field varies in space in order to calculate the path integral to determine the potential.

The compactness of the electric potential. It may be surprising that the electric field, which is a vector field with three components E_x, E_y, E_z, can be represented by the scalar electric potential, $V(x, y, z)$. This is because there are additional constraints for the electric field: It must, as we saw in (3.14), also satisfy $\nabla \times \mathbf{E} = 0$. This introduces three additional differential equations for the electric field. We could also have introduced this argument in the opposite order: Because the electric field can be written as the gradient of the potential, the curl of the electric field is zero, since the curl of a gradient always is zero: $\nabla \times (\nabla V) = 0$.

Proof of the relation between field and potential. If we look at the potential difference between A and a point B which is an infinitesimal displacement (dx, dy, dz) away, we get:

$$\mathrm{d}V = \frac{\partial V}{\partial x}dx + \frac{\partial V}{\partial y}dy + \frac{\partial V}{\partial z}dz = V_B - V_A \tag{3.33}$$

$$= -\int_A^B \mathbf{E} \cdot \mathrm{d}\mathbf{l} = -\left(E_x\,\mathrm{d}x + E_y\,\mathrm{d}y + E_z\,\mathrm{d}z\right), \tag{3.34}$$

Since this is valid for any choice of $\mathrm{d}x$, $\mathrm{d}y$, and $\mathrm{d}z$, we get that

$$E_x = -\frac{\partial V}{\partial x},\ E_y = -\frac{\partial V}{\partial y},\ E_z = -\frac{\partial V}{\partial z}. \tag{3.35}$$

Method for finding the electric potential

There is a robust method to find the electric potential $V(\mathbf{r})$ from a discrete or continuous charge distribution. Given the charges and their positions, we find the potential using the following procedure:

- Model the problem: Choose a coordinate system to describe the system that reflects the symmetries of the system.

Discrete charges:

- Find the vector $\mathbf{R}_i$ from a charge Q_i at $\mathbf{r}_i$ to the observation point $\mathbf{r}$, $\mathbf{R}_i = \mathbf{r} - \mathbf{r}_i$, for each charge.
- Sum the contributions from all charges Q_i:

$$V(\mathbf{r}) = \sum_i \frac{Q_i}{4\pi\epsilon_0|\mathbf{r} - \mathbf{r}_i|}. \tag{3.36}$$

Continuous charges:

- Describe a small element dq at a position $\mathbf{r}'$. Relate dq to a volume; surface; line element $dq = \rho\, dV'$; $\rho_s\, dS'$; $\rho_l\, dl'$.
- Find the vector $\mathbf{R}$ from the small element at $\mathbf{r}'$ to the observation point at $\mathbf{r}$: $\mathbf{R} = \mathbf{r} - \mathbf{r}'$.
- Integrate over all $\mathbf{r}'$ that make up the body of interest analytically or numerically. Remember that the integration variable is $\mathbf{r}'$ and not $\mathbf{r}$! For a volume charge density ρ_v this is:

$$V(\mathbf{r}) = \int_v \frac{\rho_v(\mathbf{r}')\, dv'}{4\pi\epsilon_0|\mathbf{r} - \mathbf{r}'|}. \tag{3.37}$$

Test your understanding

Which set of equipotential lines corresponds to the electric field in A?[5]

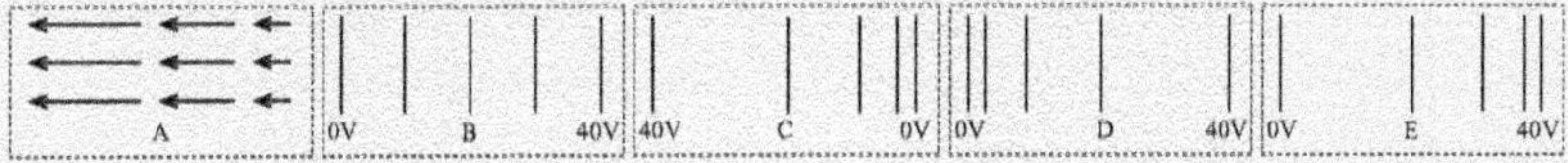

3.5 Applications of the Electric Potential

The relationship between the electric potential and the electric field opens for a new way to calculate the electric field: First we find the electric potential, and then we find the field from the potential. In many cases, it is mathematically simpler to find the potential than to find the field. This approach may also be used numerically.

Interpreting the electric potential

A linear potential. Figure 3.4 illustrates the electric potential V for three different cases. In Fig. 3.4a the electric potential decreases linearly from a high potential at $x = 0$ to a lower potential at $x = L$. How can we use the potential to gain insight into the electric field? We know that the electric field is $E_x = -\partial V/\partial x$ in the one-dimensional case. The electric field therefore points from a high potential toward lower potentials. In Fig. 3.4a the electric field points to the right as illustrated by the arrow. The field is proportional to the slope of the potential. Here, the slope is constant,

[5] D.

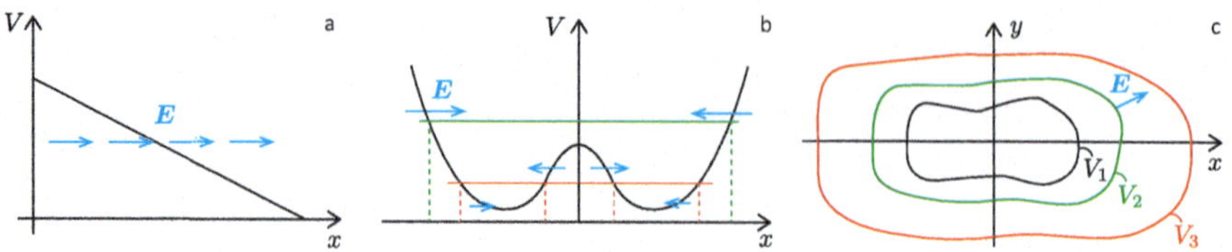

Fig. 3.4 Illustration of equipotential surfaces and the electric field for a linear potential **a**, a one-dimensional potential **b**, and a two-dimensional potential **c**

which means that the electric field is uniform. The electric field tells us in what direction a force on a positive charge acts. Positive charges will therefore move from high to low potential, whereas negative charges will move from low to high potential.

One-dimensional potential. Fig. 3.4b illustrates a more complicated one-dimensional electric potential. The electric field points in the direction the field is decreasing and its magnitude is illustrated by the length of the arrows.

Two-dimensional potential. Figure 3.4c illustrates a two-dimensional electric potential illustrated by its *equipotential surfaces*. An *equipotential surface* is a surface where the potential is a constant: $V = \text{const.}$. The gradient is normal to the equipotential surface and points in the direction that V increases the fastest. Since $\mathbf{E} = -\nabla V$, we see that the electric field is normal to the equipotential surface and points in the direction that V decreases the fastest. The electric field points from high potential values towards lower potential values. Figure 3.4c illustrates the equipotential curves for a two-dimensional electric field. How can you see from the equipotential curves where the gradient is steep? This is seen by regions where the equipotential surfaces are close: The gradient and hence the electric field has a larger magnitude along the y-axis than along the x-axis because the distance between the equipotential lines (surfaces) is longer along the x-axis than along the y-axis. In the following examples, we will calculate the electric potential and use equipotential curves and surfaces to visualize the potential.

Example: Dipole

The electric potential from a dipole. First, we find the electric potential from a dipole consisting of a charge Q at $y = a$ and a charge $-Q$ at $y = -a$. The system is illustrated in Fig. 3.5. We use the superposition principle, adding the contributions from the two charges, to find the potential in the point $\mathbf{r}$. The two charges are at $\mathbf{r}_1 = a\hat{\mathbf{y}}$ and $\mathbf{r}_2 = -a\hat{\mathbf{y}}$. The total potential is

$$V = V_1 + V_2 = \frac{Q}{4\pi\epsilon_0 R_1} + \frac{-Q}{4\pi\epsilon_0 R_2}, \tag{3.38}$$

where $\mathbf{R}_1 = \mathbf{r} - \mathbf{r}_1$ and $\mathbf{R}_2 = \mathbf{r} - \mathbf{r}_2$.

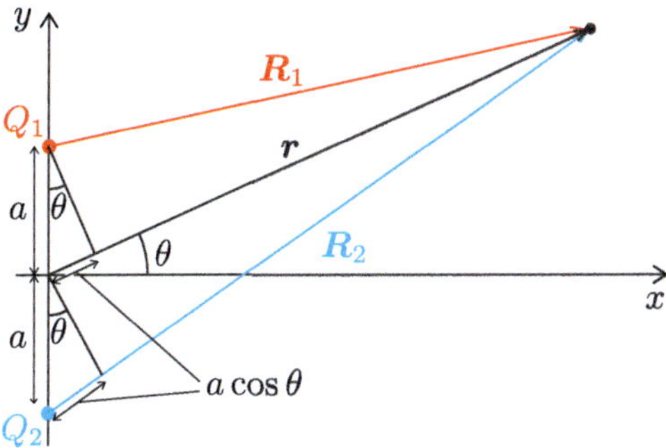

Fig. 3.5 Illustration of dipole

Electric potential for $r \gg a$. Then, we find an approximation for the electric field at $\mathbf{r}$ when $r \gg a$. However, for this we need to know a *trick*. I do not expect you to discover such a trick by yourself, but when you have seen it here, you may be able to use such a trick in another situation.

The trick is to rewrite the two lengths R_1 and R_2 using the angle θ from the axis of the dipole to the point $\mathbf{r}$. When $r \gg a$ we can assume that the triangles in the figure are approximately right triangles. We see from Fig. 3.5 that $R_1 = r - a\cos\theta$ and $R_2 = r + a\cos\theta$.

We can rewrite the potential as

$$V = \frac{Q}{4\pi\epsilon_0 R_1} + \frac{-Q}{4\pi\epsilon_0 R_2} = \frac{Q}{4\pi\epsilon_0}\left(\frac{R_2 - R_1}{R_1 R_2}\right). \tag{3.39}$$

We insert the expressions for R_1 and R_2 and also approximate $R_1 R_2 = r^2 - a^2\cos^2\theta \simeq r^2$ (when $r \gg a$), getting

$$V = \frac{Q}{4\pi\epsilon_0}\left(\frac{R_2 - R_1}{R_1 R_2}\right) \simeq \frac{Q}{4\pi\epsilon_0}\left(\frac{2a\cos\theta}{r^2}\right). \tag{3.40}$$

(Notice that the dipole decays as r^{-2}, whereas a single charge decays as r^{-1}.)

Again, we introduce a *trick* to rewrite the $\cos\theta$ expression as $2Qa\cos\theta = 2Qa\hat{\mathbf{y}} \cdot \hat{\mathbf{r}}$. Therefore, we can introduce the quantity $\mathbf{p} = Q2a\hat{\mathbf{y}}$, which we call the *dipole moment*, so that the electric potential becomes:

$$V(\mathbf{r}) = \frac{\mathbf{p} \cdot \mathbf{r}}{4\pi\epsilon_0 r^3} \tag{3.41}$$

Visualization of the dipole potential. We visualize the potential using the same approach as for the electric field. First, we introduce a function to find the potential in a point $\mathbf{r}$ from a charge Q_i at a position $\mathbf{r}_i$. We know that the potential from this charge is

$$V(\mathbf{r}) = \frac{Q_i}{4\pi\epsilon_0 R_i}, \tag{3.42}$$

where $\mathbf{R}_i = \mathbf{r} - \mathbf{r}_i$. We calculate this directly in Python:

```
import numpy as np
import matplotlib.pyplot as plt
import scipy.constants as sc

def Vfield(r,Qi,ri):
    # Find electric potential at r from charge Q at ri
    Ri = r - ri
    Rinorm = np.linalg.norm(Ri)
    V = Qi/(4.0*np.pi*sc.epsilon_0)*(1/Rinorm)
    return V
```

We follow the same procedure as for the electric field, and introduce a list of charges that contains the two charges in the dipole:

```
# Setup list of charges
qlist = []
rlist = []
a = 1.0 # Size of system in meters
r0 = np.array([a,0])
q0 = 1
rlist.append(r0)
qlist.append(q0)
r0 = np.array([-a,0])
q0 = -1
rlist.append(r0)
qlist.append(q0)
```

Then we construct a lattice of $\mathbf{r}_{ij}$-values and calculate the value $V_{ij} = V(\mathbf{r}ij)$ for each position on the lattice:

```
Lx, Ly, Nx, Ny = 3*a,3*a,21,21
x = np.linspace(-Lx,Lx,Nx)
y = np.linspace(-Ly,Ly,Ny)
rx,ry = np.meshgrid(x,y,indexing="ij")
# Set up electric potential
V = np.zeros((Nx,Ny),float)
# Calculate the potential
for i in range(Nx):
    for j in range(Ny):
        r = np.array([rx[i,j],ry[i,j]])
        for ic in range(len(qlist)):
            V[i,j] = V[i,j] + Vfield(r,qlist[ic],rlist[ic])
```

Finally, we visualize the resulting potential as a two-dimensional image:

```
plt.figure(figsize=(8,8))
plt.contourf(rx,ry,V,100,cmap="seismic")
plt.colorbar()
plt.axis("equal"), plt.axis("tight")
plt.xlabel("$x/a$"), plt.ylabel("$y/a$")
```

The resulting plot is shown in Fig. 3.6.

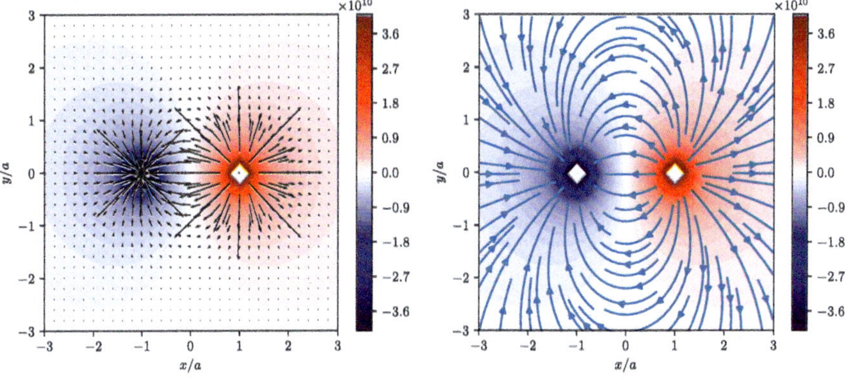

Fig. 3.6 Plot of the equipotential surfaces for a dipole

Numerical calculation of the field from the potential for a dipole. Based on these results, we want to estimate the electric field using $\mathbf{E} = -\nabla V$. The gradient operator, ∇, is provided by `gradient` in Python:

```
Ex,Ey = np.gradient(-V)
plt.figure(figsize=(16,6))
plt.subplot(1,2,1)
plt.contourf(rx,ry,V,100,cmap="seismic")
plt.colorbar()
plt.quiver(rx,ry,Ex,Ey)
plt.axis("equal"), plt.axis("tight")
plt.xlabel("$x/a$"), plt.ylabel("$y/a$")
plt.subplot(1,2,2)
plt.contourf(rx,ry,V,100,cmap="seismic")
plt.colorbar()
plt.streamplot(rx.T,ry.T,Ex.T,Ey.T)
plt.axis("equal"), plt.axis("tight")
plt.xlabel("$x/a$"), plt.ylabel("$y/a$")
```

The resulting plots are shown in Fig. 3.6.

Sometimes you would like to calculate the field with a high resolution, but then only draw arrows in some of the points you have calculated. This is done using the `slice` function in Python. The following only plots every 2nd arrow:

```
skip = (slice(None, None, 2), slice(None, None, 2))
plt.quiver(rx[skip],ry[skip],Ex[skip],Ey[skip])
```

Symbolic calculation of the field from a dipole. We can also find the electric field directly from the potential as $\mathbf{E} = -\nabla V$. We demonstrate how we can do this using Sympy. We use the package `CoordSys3D` to define a three-dimensional Cartesian coordinate system. The various unit vectors are then denoted `R.i`, `R.j` and `R.k` and the coordinates x, y, z are `R.x`, `R.y`, `R.z` respectively. The $\mathbf{r}$-vector is then `R.x*R.i + R.y*R.j + R.z*R.k`. We introduce the dipole moment $\mathbf{p} = (a, b, c)$ which is `a*R.i + b*R.j + c*R.k` and define the potential as $\mathbf{r} \cdot \mathbf{p}/r^3$, which in Sympy notation becomes `r.dot(p)/(r.dot(r))**(3/2)` in the following program (where we have removed the $1/(4\pi\epsilon_0)$ prefactor for simplicity):

```
import sympy as sy
from sympy.vector import CoordSys3D,gradient
R = CoordSys3D("R")
r = R.x*R.i + R.y*R.j + R.z*R.k
a = sy.symbols("a")
b = sy.symbols("b")
c = sy.symbols("c")
p = a*R.i + b*R.j + c*R.k
V = r.dot(p)/r.dot(r)**1.5
E = -gradient(V)
```

The result is:

$$\mathbf{E}' = \left(\frac{3.0 x_R (x_R a + y_R b + z_R c)}{(x_R^2 + y_R^2 + z_R^2)^{2.5}} - \frac{a}{(x_R^2 + y_R^2 + z_R^2)^{1.5}} \right) \hat{\mathbf{x}} \tag{3.43}$$

$$+ \left(\frac{3.0 y_R (x_R a + y_R b + z_R c)}{(x_R^2 + y_R^2 + z_R^2)^{2.5}} - \frac{b}{(x_R^2 + y_R^2 + z_R^2)^{1.5}} \right) \hat{\mathbf{y}} \tag{3.44}$$

$$+ \left(\frac{3.0 z_R (x_R a + y_R b + z_R c)}{(x_R^2 + y_R^2 + z_R^2)^{2.5}} - \frac{c}{(x_R^2 + y_R^2 + z_R^2)^{1.5}} \right) \hat{\mathbf{z}} \tag{3.45}$$

which we recognize can be simplified to:

$$\mathbf{E}' = \frac{3\mathbf{r}\,(\mathbf{r} \cdot \mathbf{p})}{r^5} - \frac{\mathbf{p}}{r^3}. \tag{3.46}$$

We reintroduce the prefactor $1/(4\pi\epsilon_0)$, getting,

$$\mathbf{E} = \frac{1}{4\pi\epsilon_0} \left(\frac{3\mathbf{r}\,(\mathbf{r} \cdot \mathbf{p})}{r^5} - \frac{\mathbf{p}}{r^3} \right). \tag{3.47}$$

Example: Charged Ring

Find the electric potential from a uniformly charged ring with charge Q and radius a.

Analytical solution along the axis of the ring. The system is illustrated in Fig. 3.7. The ring can be described as a line charge with a line charge density $\rho_l = Q/L = Q/(2\pi a)$. The contribution to the potential from an infinitesimal element $d\phi$ of charge $dq = \rho_l (a\, d\phi)$ is

$$dV = \frac{dq}{4\pi\epsilon_0 R}, \tag{3.48}$$

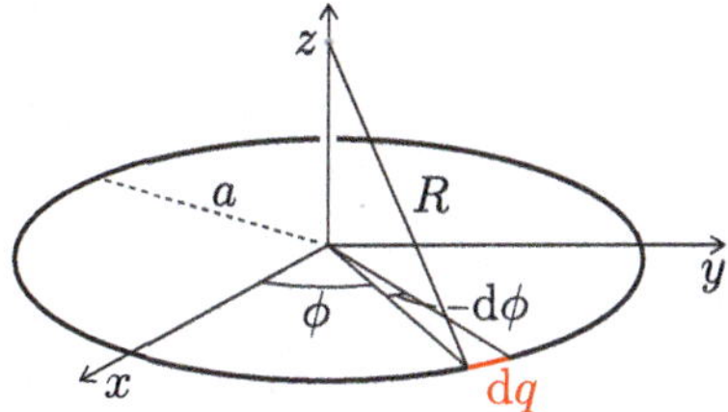

Fig. 3.7 Illustration of a charged ring

where the length R is the distance from the point $\mathbf{r} = (0, 0, z)$ on the axis to a point at the ring. We see from Fig. 3.7 that $R^2 = a^2 + z^2$ for all angles ϕ. The potential is therefore

$$\begin{aligned} V &= \int_0^{2\pi} \frac{\rho_l a \, \mathrm{d}\phi}{4\pi\epsilon_0 \left(a^2 + z^2\right)^{1/2}} = \frac{\rho_l a}{4\pi\epsilon_0 \left(a^2 + z^2\right)^{1/2}} \int_0^{2\pi} \mathrm{d}\phi \\ &= \frac{Q}{4\pi\epsilon_0} \frac{1}{\left(a^2 + z^2\right)^{1/2}}. \end{aligned} \tag{3.49}$$

We see that when $z \gg a$, the potential is the same as for a point charge Q located at the origin.

Numerical calculation of the potential at any point in space. In order to find the electric potential $V(\mathbf{r})$ in an arbitrary point $\mathbf{r}$, we need to sum up the contributions from all elements $\mathrm{d}\phi$ of the ring. We will first find the contribution from one element, and then integrate numerically to find the contributions from all elements.

A small element $\mathrm{d}\phi$ of the ring has a length $\mathrm{d}l = a\,\mathrm{d}\phi$, a charge $\mathrm{d}q = \rho_l\,\mathrm{d}l = \rho_l a\,\mathrm{d}\phi$, and is at a position $\mathbf{r}' = (a\cos\phi, a\sin\phi, 0)$, where ϕ is the angle with the x-axis. The contribution from this element to the electric potential is:

$$\mathrm{d}V = \frac{\mathrm{d}q}{4\pi\epsilon_0 R} = \frac{\rho_l a \; \mathrm{d}\phi}{4\pi\epsilon_0 R}, \tag{3.50}$$

where $\mathbf{R} = \mathbf{r} - \mathbf{r}'$.

To find the total electric field, we sum the contributions from $\phi = 0$ to $\phi = 2\pi$ by dividing the interval into N pieces of size $\mathrm{d}\phi = 2\pi/N$:

```
import numpy as np
import matplotlib.pyplot as plt
import scipy.constants as sc

def potentialfromring(r,a,rhol,N):
    # dV = \frac{\rho a d\phi}{4 \pi \epsilon_0 R}
    V = 0
    dphi = 2*np.pi/N
    for i in range(N):
        phi = i*dphi
        dl = a*dphi
```

```
        r0 = np.array([a*np.cos(phi),a*np.sin(phi),0])
        R = np.linalg.norm(r-r0)
        dV = rhol/(4.0*np.pi*sc.epsilon_0)*dl/R
        V = V+dV
    return V
```

We use this function to calculate the potential in the xz-plane with 30 points from $-3a$ to $3a$ for both x and z:

```
# Calculate potential in the xz plane
Nring = 100
a = 1.0
Q = 1.0
rhol = Q/(2*np.pi*a)
Lx, Lz, Nx, Nz = 3*a,3*a,30,30
x = np.linspace(-Lx,Lx,Nx)
z = np.linspace(-Lz,Lz,Nz)
rx,rz = np.meshgrid(x,z,indexing="ij")
# Set up electric potential
V = np.zeros((Nx,Nz),float)
# Calculate the potential
for i in range(Nx):
    for j in range(Nz):
        r = np.array([rx[i,j],0.0,rz[i,j]])
        V[i,j] = potentialfromring(r,a,rhol,Nring)
```

We can then visualize the result using the same methods as demonstrated above. The resulting visualization of the vector field and the stream lines are shown in Fig. 3.8.

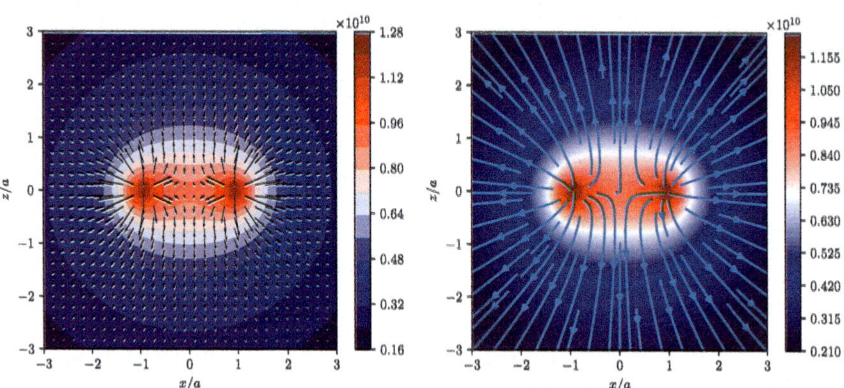

Fig. 3.8 Illustration of the electric potential and the electric field from a charged ring

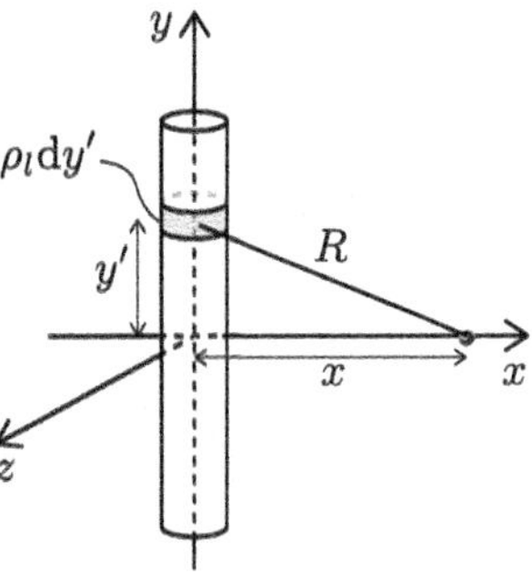

Fig. 3.9 Illustration of a charged line along the y-axis

Example: Charged Line Segment

Find the electric potential from a finite line charge, both analytically and numerically, compare and visualize the two.

First, let us specify how the system is modelled. We assume the line has a length L and a line charge density ρ_l. We place the line along the y-axis from $y = -L/2$ to $y = L/2$ as illustrated in Fig. 3.9. Then, let us find the contribution $\mathrm{d}V$ to the potential in a point $\mathbf{r}$ from a small element of length $\mathrm{d}y'$ at the position $\mathbf{r}' = (0, y', 0)$:

$$\mathrm{d}V = \frac{\mathrm{d}q}{4\pi \epsilon_0 R}, \tag{3.51}$$

where $\mathbf{R} = \mathbf{r} - \mathbf{r}'$. We will now use this to find the electric potential at a point on the x-axis analytically, and then find the electric potential for any point numerically.

Analytical solution on an axis normal to the line. For a point $\mathbf{r} = (x, 0, 0)$, the contribution to the electric potential from an element at y' is:

$$\mathrm{d}V = \frac{\mathrm{d}q}{4\pi \epsilon_0 R}, \tag{3.52}$$

where $\mathbf{R} = (x, 0, 0) - (0, y', 0) = (x, -y', 0)$ and $R = (x^2 + (y')^2)^{1/2}$. This gives

$$\mathrm{d}V = \frac{\rho_l(y')\,\mathrm{d}y'}{4\pi \epsilon_0 \left(x^2 + (y')^2\right)^{1/2}}. \tag{3.53}$$

We find the potential by summing up all these contributions from $y' = -L/2$ to $y' = L/2$. Notice that the integration variable is y' and not x!

$$V = \int_{-L/2}^{L/2} \frac{\rho_l(y')\,\mathrm{d}y'}{4\pi \epsilon_0 \left(x^2 + (y')^2\right)^{1/2}}. \tag{3.54}$$

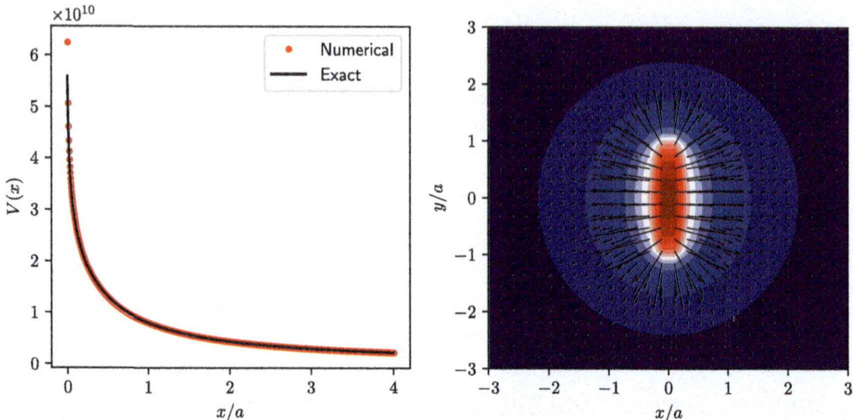

Fig. 3.10 Electric potential from a line charge from $y = -L/2$ to $y = L/2$. **a** Plot of $V(x, 0, 0)$. **b** Plot of the electric potential and the electric field in the xy-plane

How can we solve this integral? You may look it up in a table of integrals, use Sympy to solve it, or look it up with your favorite chatbot.

```
import sympy as sy
x = sy.Symbol("x")
y = sy.Symbol("y")
f = 1/sy.sqrt(x**2 + y**2)
V = sy.integrate(f,y)
print(V)
```

```
asinh(y/x)
```

We insert the end points $y' = -L/2$ and $y' = L/2$, getting

$$V = \frac{\rho_l}{4\pi\epsilon_0}\left(\operatorname{asinh}\left(\frac{L}{2x}\right) - \operatorname{asinh}\left(\frac{-L}{2x}\right)\right). \tag{3.55}$$

This function is plotted in Fig. 3.10.

Numerical solution for an arbitrary point. To find the solution at any point $\mathbf{r} = (x, y, z)$, we find the potential numerically. The contribution from a small element $\mathrm{d}y'$ at y' is:

$$\mathrm{d}V = \frac{\mathrm{d}q}{4\pi\epsilon_0 R}, \tag{3.56}$$

where $\mathrm{d}q = \rho_l \mathrm{d}y'$, $\mathbf{R} = (x, y, z) - (0, y', 0) = (x, y - y', z)$ and $R = (x^2 + (y - y')^2 + z^2)^{1/2}$. This gives

$$\mathrm{d}V = \frac{\rho_l(y')\,\mathrm{d}y'}{4\pi\epsilon_0\left(x^2 + (y - y')^2 + z^2\right)^{1/2}}. \tag{3.57}$$

We find the potential by summing up all these contributions from $y' = -L/2$ to $y' = L/2$. We divide the interval into N pieces of length $\mathrm{d}y' = L/N$ each and sum the contributions numerically:

```
import numpy as np
import matplotlib.pyplot as plt
import scipy.constants as sc
def potentialfromline(r,L,rhol,N):
    # dV = \frac{\rho dy}{4 \pi \epsilon_0 R}
    V = 0
    dl = L/N
    for i in range(N):
        y0 = -L/2+i*dl
        r0 = np.array([0,y0,0])
        R = np.linalg.norm(r-r0)
        dV = rhol/(4*np.pi*sc.epsilon_0)*dl/R
        V = V+dV
    return V
```

We visualize the result by calculating the field in the xy-plane and by comparing the numerical results with the exact results for $y = z = 0$. The resulting plot of $V(x, 0, 0)$ and the potential $V(x, y, 0)$ are shown in Fig. 3.10. We calculate the potential for a line of length `Lline` and charge `Q` using `Nline` elements to find the potential and visualize the potential on a grid of 30×30 points from `-Lx` to `Lx` and `-Ly` to `Ly`:

```
Lline, Q, Nline = 2.0, 1.0, 100
rhol = Q/Lline
Lx, Ly, Nx, Ny = 1.5*Lline,1.5*Lline,30,30
x = np.linspace(-Lx,Lx,Nx)
y = np.linspace(-Ly,Ly,Ny)
rx,ry = np.meshgrid(x,y,indexing="ij")
# Set up electric potential
V = np.zeros((Nx,Ny),float)
# Calculate the potential
for i in range(Nx):
    for j in range(Ny):
        r = np.array([rx[i,j],ry[i,j],0])
        V[i,j] = potentialfromline(r,Lline,rhol,Nline)
Ex,Ey = np.gradient(-V)
plt.contourf(rx,ry,V,15,cmap="seismic")
plt.quiver(rx,ry,Ex,Ey)
plt.axis("equal")
plt.xlabel("$x/L$"), plt.ylabel("$y/L$")
```

Summary

Electrical potential.

- The electrical potential $V(\mathbf{r})$ for an electrical field $\mathbf{E}(\mathbf{r})$ is defined as

$$V(\mathbf{r}) = \int_{\mathbf{r}}^{\mathrm{ref}} \mathbf{E} \cdot \mathrm{d}\mathbf{r},$$

where the integral is a path integral along any line from $\mathbf{r}$ to the reference point where $V = 0$.

- The electrical potential is defined relative to a reference point. It is only differences in electrical potential that have physical significance.

Electric field.

- The electric field is related to the electric potential through $\mathbf{E} = -\nabla V(\mathbf{r})$.
- The curl of a (static) electric field is 0: $\nabla \times \mathbf{E} = \mathbf{0}$.

Electric potential from a charge distribution.

- The electric potential in $\mathbf{r}$ from a single point charge Q in $\mathbf{r}'$ is

$$V(\mathbf{r}) = \frac{q}{4\pi\epsilon_0 R} = \frac{q}{4\pi\epsilon_0 |\mathbf{r} - \mathbf{r}'|}, \quad \mathbf{R} = \mathbf{r} - \mathbf{r}', \ R = |\mathbf{R}|.$$

- The electrical potential obeys the *superposition principle*:

$$V(\mathbf{r}) = \sum_i V_i(\mathbf{r}) = \sum_i \frac{Q_i}{4\pi\epsilon_0 R_i} = \sum_i \frac{Q_i}{4\pi\epsilon_0 |\mathbf{r} - \mathbf{r}_i|}$$

- The electrical potential from a volume charge density ρ_v is:

$$V(\mathbf{r}) = \int_v \frac{\rho_v \,\mathrm{d}v'}{4\pi\epsilon_0 R} = \int_v \frac{\rho_v(\mathbf{r}') \,\mathrm{d}v'}{4\pi\epsilon_0 |\mathbf{r} - \mathbf{r}'|}$$

- The electrical potential from a surface charge density ρ_s is:

$$V(\mathbf{r}) = \int_S \frac{\rho_s \,\mathrm{d}S'}{4\pi\epsilon_0 R} = \int_S \frac{\rho_s(\mathbf{r}') \,\mathrm{d}S'}{4\pi\epsilon_0 |\mathbf{r} - \mathbf{r}'|}$$

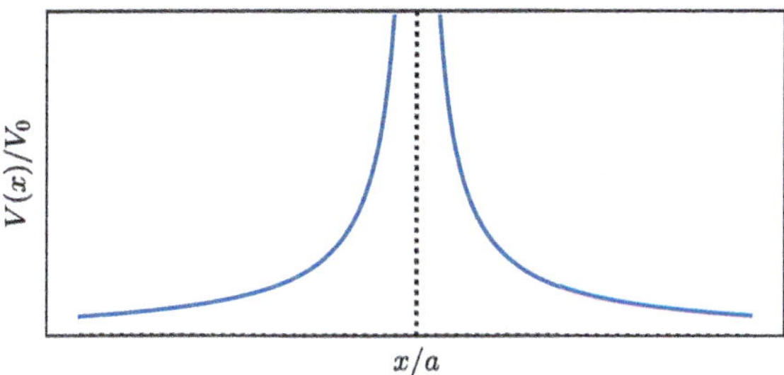

Fig. 3.11 Illustration of a simple potential

- The electrical potential from a line charge density ρ_l is:

$$V(\mathbf{r}) = \int_C \frac{\rho_l \, \mathrm{d}l'}{4\pi\epsilon_0 R} = \int_C \frac{\rho_l(\mathbf{r}') \, \mathrm{d}l'}{4\pi\epsilon_0 |\mathbf{r} - \mathbf{r}'|}$$

Exercises

Discussion exercises

3.1 Choosing your path. The potential (relative to a point at infinity) midway between two charges of equal magnitude and opposite sign is zero. Is it possible to bring a test charge from infinity to this midpoint in such a way that no work is done in any part of the displacement? If so, describe how it can be done. If it is not possible, explain why.

3.2 Finding the field from the potential. If the electric potential at a single point is known, can **E** at that point be determined? If so, how? If not, why not?

3.3 Zero field. If **E** is zero throughout a certain region of space, is the potential necessary also zero in this region? Why or why not? What *can* be said about the potential?

3.4 The zero-field path. If **E** is zero everywhere along a certain path the leads from point A to point B, what is the potential difference between those two points? Does this mean that **E** is zero everywhere along *any* path from A to B? Explain.

3.5 Motion in a simple field. Fig. 3.11 illustrates the scalar potential, $V(x)$, resulting from some charge configuration. Discuss what possible types of motion a positively and a negatively charged particle can have by comparing with the potential energy of a cart rolling down a hill. Illustrate the motions in an energy diagram ($U(x)$ for the particle and for the cart).

3.6 Potentials and sparks. A spark in air occurs where the electric field exceeds a maximum value beyond which the air becomes a conducting plasma. You may have experienced sparks between your finger and another object if you are charged with

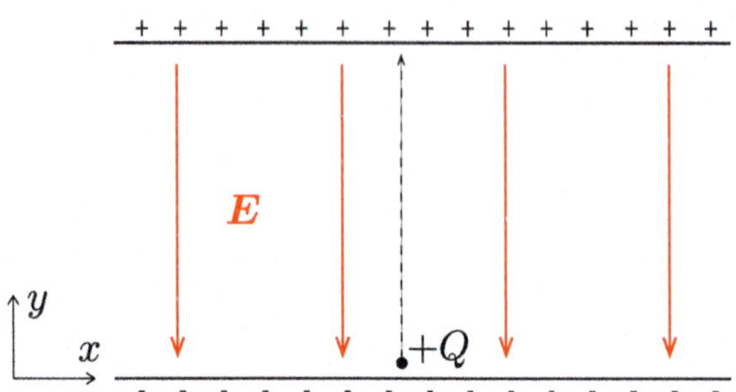

Fig. 3.12 Electric field between two charged plates

static electricity. How can you model this process by assuming that your finger has an electrical potential V and that for example your table lamp has potential zero. Explain what happens to the electric field as your finger approaches the lamp. (You can assume that the potential of your finger and the lamp does not change before a spark occurs).

Tutorials

3.7 Electric potential and potential energy. Two charged electric plates set up an electric field $\mathbf{E} = -E_0\hat{\mathbf{y}}$ between the plates as shown in Fig. 3.12. A proton with charge $Q = +e$ is lifted a distance d from the lower to the upper plate by an external force from an optical tweezer. (You may ignore the gravitational force on the proton).
(a) What is the direction of the force on the proton from the electric field?
(b) The work done by the electric field on the proton is: (A) Positive, (B) Negative, (C) Zero.
(c) The work done by the external force (the tweezer) is: (A) Positive, (B) Negative, (C) Zero.
(d) The change in potential energy is defined as $\Delta U = +W_{\text{ext}} = -W_{\text{field}}$. We define the potential energy of the proton to be $U_p(y = 0) = 0$ at the lower plate. What is then $U_p(y = d)$?
(e) Sketch the electric potential along the y-axis.
(f) We release the proton from rest at $y = d/2$. What is the electric potential in this point? Which way will the proton move—toward higher or lower electric potential?
(g) We remove the proton and instead place an electric at $y = d/2$. What is now the electric potential in this point? Which way will the electron move—toward higher or lower electric potential?

3.8 Optimal path. We move a charge $+Q$ along the paths (A, B, C) using an optical tweezer as shown in Fig. 3.13. Along which path does the optical tweezer perform the most work on the charge?

3.9 From potential to field. Fig. 3.14 illustrates electric potentials, $V(x)$, for different physical situations. First, sketch the corresponding electrical fields with arrows, then sketch the electric field $E_x(x)$.

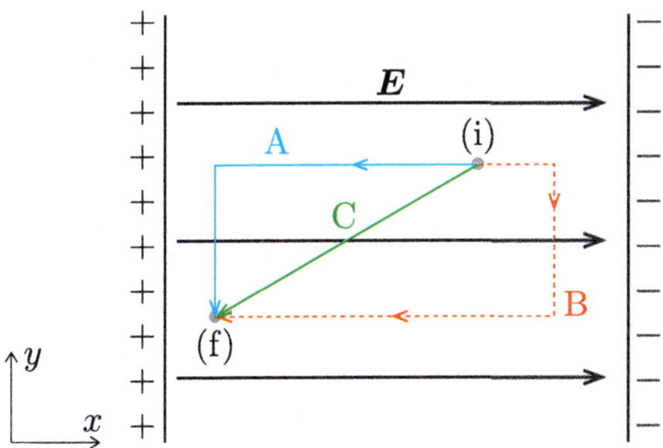

Fig. 3.13 Paths of motion through an electric field

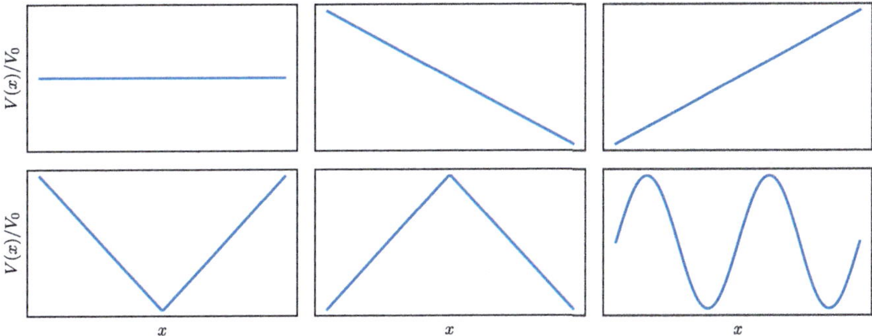

Fig. 3.14 Six electric potentials $V(x)$

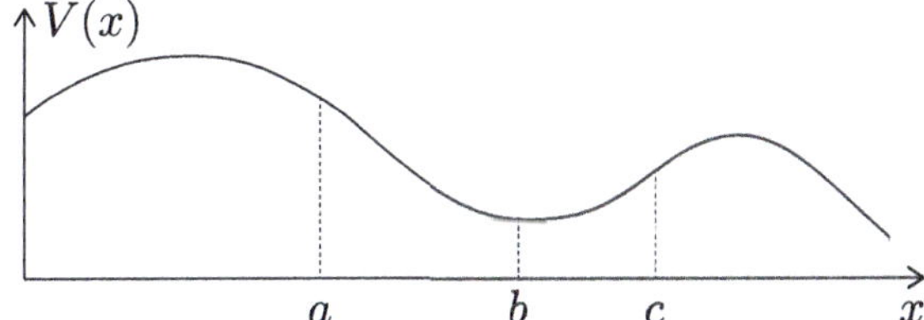

Fig. 3.15 Illustration of an electric potential $V(x)$ for a charge distribution

3.10 Potential, field and forces. Fig. 3.15 shows the electric potential $V(x)$ along the x-axis for a charge distribution. (We do not know the details of the charge distribution).

(a) Which way does the electric field point in the points a, b, and c ?

(b) Sketch the electric field $E_x(x)$.

(c) If we release a negative charge $-q$ in the points a, b and c, in which direction will the charge start moving?

3.11 Units for electric potential and energy. A shark has an advanced electrosensor called Lorenzini's ampullas. You may model this as a gel-filled cylinder. If the voltage difference between the gel in the cylinder and the interior of the cells around it exceeds a given threshold, ions will flow through the cell membrane. In this way, a shark may sense 10^{-6} volt over a centimeter of sea water.
(a) How long distance would we need to have between the two poles of a 1.5 V AA battery for a shark not to be able to sense the field from the battery?
(b) Chemists and material physicists often use the unit electronvolt for energy. It is the energy an electron receives when accelerated across a voltage difference of 1 V. How large is an electron volt (eV) in Joules?

3.12 Quadrupole. In electrostatics, the dipole plays a similar role to the harmonic oscillator in mechanics and quantum mechanics. Often, objects do not have a net charge, but the charges inside the object can be distributed in different ways. This is true from atoms or molecules to macroscopic objects such as a soda can, a balloon, a cloud or a star. The coarsest way to describe such a charge distribution is as a dipole. However, what happens if we add two dipoles together? We get a quadrupole. Here we will address the electric potential and the electric field around a quadrupole.

We will address a quadrupole consisting of four charges: $+Q$ in the points (a, a) and $(-a, -a)$ and $-Q$ in the points $(-a, a)$ and $(a, -a)$.
(a) Make a sketch of the system.
(b) Where do you think the electric potential is zero in this system? And where is the electric field zero?
(c) Sketch the electric potential along the x-axis.
(d) Write a python program to visualize the electric potential from this quadrupole system. You can use the program `epotlist`.
(e) Find and visualize the electric field (using the electric potential you already have calculated).
(f) Plot the electric potential as a function of the distance to the origin along a line ℓ which forms an angle $\pi/4$ with the x-axis.
(g) Plot the electric potential along the line ℓ for $r \gg a$. Use a double-logaritmic plot. What does this plot tell us about the behavior of the potential from a quadrupole far away from the center of the quadrupole? (Hint: A single charge behaves as $V \propto 1/r$ when $r \gg a$ while a dipole behaves as $V \propto 1/r^2$ when $r \gg a$.)
(h) We can characterize a charge distribution using the dipole moment of the distribution, which is defined as $\mathbf{p} = \sum_i Q_i \mathbf{r}_i$. If the charge distribution has zero net charge, i.e. it is neutral, then the dipole moment does not depend on the choice of origin. Find the dipole moment of the quadrupole.
(i) How should we instead characterize the quadrupole. Look up *multipole expansion* and discuss how you would characterize the quadrupole and what use such a multipole expansion may have.

Exercises

3.13 Two point charges. Two point charges $Q_1 = +3\text{nC}$ and $Q_2 = -5\text{nC}$ are separated by $d = 10\text{cm}$.

(a) What is the potential energy of the pair?
(b) What does the sign of your answer tell you about the system?
(c) What is the electric potential midway between the two charges?

3.14 Four charges. Four point charges, $Q = 1\text{nC}$ are placed on the corners of a square with side $d = 1\text{mm}$.
(a) What is the potential energy of the system? (Hint: Remember to include all the interactions.)
(b) What is the electric potential at the center of the square?
(c) What is the electric field and the center of the square?

3.15 Two charges again. Two point charges, $Q_1 = 1\text{nC}$ and $Q_2 = -4\text{nC}$ are separated by $d = 10\text{cm}$.
(a) Find where along a line passing through both charges that the electric potential is zero. (Hint: It may not only be in one point.)
(b) Find the electric field in this point.
(c) If the potential is zero, why is the electric field not zero?
(d) Write a program to visualize the electric potential and the electric field from this system in the xy-plane.

3.16 A distance away. At a given distance d from a charge Q the electric field is $E = |\mathbf{E}| = 100\text{V/m}$ and the electric potential is -1000V.
(a) What is the distance d from the charge?
(b) What is the sign and magnitude of the charge Q?

3.17 Potential in the plane. An electric potential has the functional form $V = Cxy$ in the xy-plane, where C is a constant.
(a) Find the electric field in the xy-plane.
(b) Show that for a small displacement $\mathrm{d}\mathbf{r} = (\mathrm{d}x, \mathrm{d}y)$ along an equipotential curve, the relation $\nabla V \cdot \mathrm{d}\mathbf{r} = 0$.
(c) Use this relation to find an equation for a tangent to an equipotential curve in the point (x_0, y_0).
(d) Write a program to visualize the electric potential and the electric field in the xy-plane. Check visually that the electric field is normal to the equipotential curves.

3.18 Accelerate an electron. An electron starts from rest and is accelerated through a potential difference of $\Delta V = 2\text{MV}$.
(a) What is the final kinetic energy of the electron?
(b) What is the rest-mass energy of the electron in eV (electron-volt)?
(c) Calculate v/c from the relativistic expression $K = m_0c^2\left[\left(1 - v^2/c^2\right)^{1/2} - 1\right]$.
(d) What would the value of v/c be if you instead used a classical calculation without including relativistic effects?

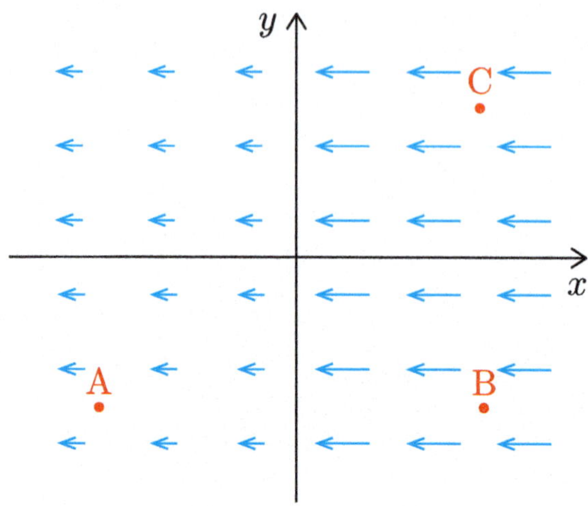

Fig. 3.16 Illustration of an electric field

3.19 Electric fields and electric potential. In Fig. 3.16 an electric field is illustrated by a vector plot, where the *length* of the vector is proportional to the magnitude of the electric field. The *direction* of the vector is in the direction of the electric field. The electric field is defined in all points in space, we just can't draw all the vectors. The three points A, B, C are at (−3mm, −2mm), (3mm, −2mm) and (3mm, 2mm) respectively. The magnitude of the electric field vector is 10^4 N/C for $x < 0$ and 2×10^4 N/C for $x > 0$.
(a) Suppose the reference point (zero) for the electric potential is at the origin. What are the electric potentials at points A, B, and C?
(b) With the same choice of reference point, plot the electric potential as a function of position along the line from A to B.
(c) You put a $+10\mu$C charge at point A and move it to B at a constant, slow velocity. What force vector $\mathbf{F}(x, y)$ do you need to apply to keep the velocity constant?
(d) Where is the potential energy zero?

3.20 Gradients. In this exercise we will practice basic skills to find the scalar potential, interpret gradients in the potential, and find the field from the potential. Two identical charges, q, are placed at $(a, 0, 0)$ and $(-a, 0, 0)$, where a is a given length.
(a) Find the electrical potential $V(x)$ everywhere along the x-axis. Is the potential negative when $x < a$? Is the potential zero at $x = 0$? Explain.
(b) Plot or sketch the potential along the x-axis and discuss the direction of the electric field.
(c) Use the sketch of the potential to draw equipotential points (at a constant spacing in potential).
(d) Find the potential $V(y)$ everywhere along the y-axis.
(e) Sketch equipotential curves in the xy-plane for this system.

3.21 Line charge. A rod of length L has a uniformly distributed charge Q. We place the rod along the x-axis with its center at the origin.
(a) We want to calculate the electric potential $V(0, y, 0)$ along the y-axis. What is the contribution to the potential from the piece of length dx at x?
(b) Write down an expression for the electric potential $V(0, y, 0)$ in terms of an integral. Explain what variable you integrate over.
(c) Anne argues that with this integral it is very simple to find the electric field $E_y = -\partial V/\partial y$: This is simply what is inside the integral, since derivation is the opposite of integration. Is Anne right? Explain.

3.22 Three-dimensional visualization. So far, we have addressed how to visualize the potential and the electric field in two dimensions. In this exercise we will extend the tools to visualize fields in three dimension. (For this you may need to install the `plotly` package in your python environment).
Generating a 3d potential field.
(a) First, we start by finding the potential from a single charge in three dimensions. Write a function to return the potential in a point `r` from a charge `Q` at a position `ri`, where both `r` and `ri` are vectors. Ensure that your function works for both two-dimensional and three-dimensional input vectors `r` and `ri`.
Dipole and ring charge: Second, we expand the methods we developed to compute the electric field from a set of charges to calculate the electric potential. We generate two lists, one list `rilist` which contains all the `ri` vectors and one list `qilist` which contains the corresponding charges. For a dipole this can be done by

```
qilist = []
rilist = []
ri = np.array([+1.0,0.0])
qi = +1
qilist.append(qi)
rilist.append(ri)
ri = np.array([-1.0,0.0])
qi = -1
qilist.append(qi)
rilist.append(ri)
```

(b) Generate a list of charges for a ring charge in the xy-plane with radius a and charge Q.
Calculating the potential in a cube. We use the following code to calculate the potential on a three-dimensional set of lattice points:

```
Lx, Ly, Lz = 3*a, 3*a, 3*a
Nx, Ny, Nz = 30, 30, 30
x = np.linspace(-Lx,Lx,Nx)
y = np.linspace(-Ly,Ly,Ny)
z = np.linspace(-Lz,Lz,Nz)
rx,ry,rz = np.meshgrid(x,y,z,indexing="ij")
# Set up electric potential
V = np.zeros((Nx,Ny,Nz),float)
# Calculate the potential
```

```
for ix in range(Nx):
    for iy in range(Ny):
        for iz in range(Nz):
            r = np.array([rx[ix,iy,iz],ry[ix,iy,iz],rz[ix,iy,iz]])
            for ic in range(len(qilist)):
                ri = rilist[ic]
                qi = qilist[ic]
                V[ix,iy,iz] = V[ix,iy,iz] + Vfield(r,qi,ri)
```

Visualization of cross-sections by slices. Now, we want to visualize a cross-section of the three-dimensional field. For example, we would like to visualize the field in the xz-plane in the middle of the cube for which we have calculated the field. That is, we want to pick out all the coordinates $\mathbf{r}_{ijk}$ and all the potential values $V(\mathbf{r}_{ijk})$ for which $j = N_y/2$. This is done by a *slice* through the three dimensional numpy-array. We slice the V array by writing V[:,int(Ny/2),:]. This generates a two-dimensional array. The : means that we include all values for the first and third coordinates, but only a specific value, int(Ny/2), for the second coordinate. We use int to ensure that the result is an integer value. We can use this to visualize the field in two dimensions:

```
# Visualize field in xz-plane
Ex,Ey,Ez = np.gradient(-V)
plt.figure(figsize=(6,6))
plt.contourf(rx[:,int(Ny/2),:],rz[:,int(Ny/2),:],V[:,int(Ny/2),:])
plt.quiver(rx[:,int(Ny/2),:],rz[:,int(Ny/2),:],\
    Ex[:,int(Ny/2),:],Ez[:,int(Ny/2),:])
plt.xlabel("$x/a$"), plt.ylabel("$z/a$"), plt.axis("equal")
```

(c) Use these methods to visualize the dipole field and the field from the ring charge in the xz-plane.

3d visualization of the potential. We can visualize the three dimensional field using plotly. Ensure that you have plotly installed. The following script provides a three-dimensional visualization of the potential:

```
import plotly.graph_objects as go
fig = go.Figure(data=go.Isosurface(
    x=rx.flatten(), y=ry.flatten(), z=rz.flatten(),
    value=V.flatten(), opacity=0.1, surface_count=20))
fig.show()
```

(d) Use this program to visualize the potential field from the dipole and from the ring charge.

3d visualization of vector field. We can also use plotly to visualize the three-dimensional electric field using the following script:

```
Ex,Ey,Ez = np.gradient(-V)
fig = go.Figure(data=go.Cone(
    x=rx.flatten(), y=ry.flatten(), z=rz.flatten(),
    u=Ex.flatten(), v=Ey.flatten(), w=Ez.flatten(),
    sizeref=2 ))
fig.show()
```

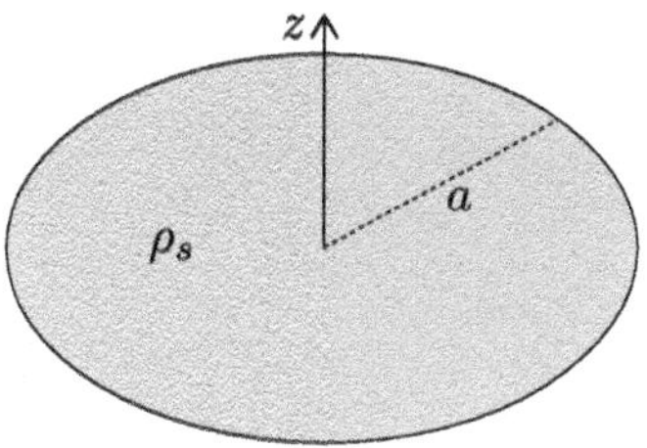

Fig. 3.17 Illustration of a disk with a radius a

(e) Visualize the electric fields from the dipole and the ring charge.

Homework

3.23 Dipole and gradient. Two identical charges, q, are placed in $(a, 0, 0)$ and $(-a, 0, 0)$, $a > 0$.
(a) Implement the Python programs from the text to find the electric potential in the xy-plane from the two charges. Illustrate the potential with contour lines.
(b) Find the electric field in the xy-plane numerically by taking the gradient of the electric potential and illustrate the field with a vector plot. Is the electric field normal to the contour lines?
(c) Let us make one of the charges negative, making the system a dipole. Repeat parts **a** and **b** for this new system.
(d) The vector plots are maybe not so pretty. Plot the same data using streamlines instead. What are the advantages and disadvantages of a streamline plot?

3.24 Half a line. In this exercise we will find the electric potential from half a line. We place a line charge with line charge density ρ_l along the x-axis from $x = 0$ to $x = -L$.
(a) What is the contribution $\mathrm{d}V$ to the electric potential in the point $\mathbf{r} = (x, 0)$ from a small line element $\mathrm{d}x'$ in the point x'? Make a drawing of the system and draw in the length R in the figure.
(b) Find the electric potential in the point $\mathbf{r} = (x, 0)$ for $x > 0$. Where is the potential zero?
(c) Find the electric field $E_x(x, 0)$ along the x-axis for $x > 0$.
(d) Explain why you cannot use the expression you have found for V to find the electric field $E_y(x, 0)$.

3.25 Potential and field above a disk. In this exercise we will address the electric potential and field along the z-axis above the center of a disk in the xy-plane. The disk has radius a and a constant surface charge density ρ_s as illustrated in Fig. 3.17.
(a) Find the contribution $\mathrm{d}V$ to the electric potential at a height z above the disk from an area element $\mathrm{d}\mathbf{S}$. You may assume $z > 0$. (Hint: Use cylinder coordinates, where $\mathrm{d}S = r\,\mathrm{d}r\,\mathrm{d}\phi$.)
(b) Find the electric potential V from the disk at the height z.
(c) Use the result from the previous exercise to find the E_z along the z-axis.

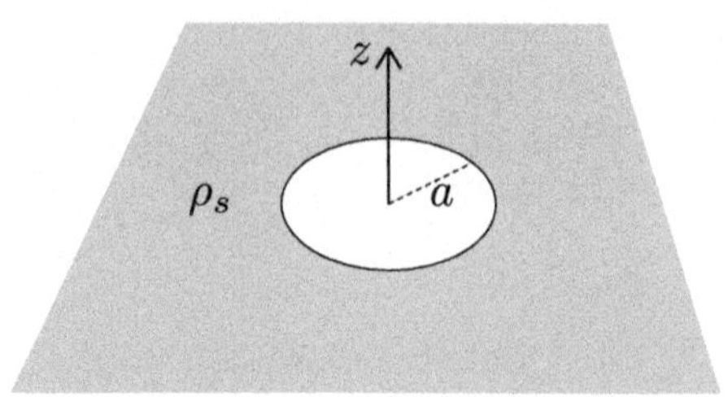

Fig. 3.18 Illustration of a plate with a circular hole

(d) Find the electric field $\mathbf{E}$ in the limit where z is small (that is $z \to 0$ from above). Interpret the result.
(e) Sketch/plot $|\mathbf{E}|$ as a function of z for $z > 0$.
(f) Show that $\mathbf{E}$ approaches the field from a point charge for large z. (Hint: Use the series expansion $(1 + x)^a \approx 1 + ax$ when $x \ll 1$.)

3.26 Far away from charges. Three point charges are placed along the x-axis: a charge $-Q$ at $x = -a$, a charge $+2Q$ at $x = 0$ and a charge $-Q$ at $x = a$. Find the potential along the x-axis in the limit when $|x| \gg a$.

3.27 Plate with hole. Fig. 3.18 illustrates an infinitely large, plane surface with a circular hole of radius a. The surface has a uniform charge density ρ_s and the center of the hole is in the origin. We want to find the electric field at a height z along the z-axis.

3.28 Dipole field. The electric potential at large distances from a dipole $\mathbf{p}$ at the origin is

$$V(\mathbf{r}) = \frac{\mathbf{p} \cdot \mathbf{r}}{4\pi \epsilon_0 r^3}.$$

Show that electric field at large distances is:

$$\mathbf{E}(\mathbf{r}) = \frac{1}{4\pi \epsilon_0} \left(-\frac{\mathbf{p}}{r^3} + \frac{3(\mathbf{p} \cdot \mathbf{r})\mathbf{r}}{r^5} \right).$$

3.29 Half spherical shell. Figure 3.19 shows half of a spherical shell of radius a and uniform surface charge density ρ_s.
(a) Find the electrical potential along the x-axis. (Hint: Use that $dq = \rho_s 2\pi a\, a \sin\theta \, d\theta$.)
(b) Show that the electrical potential is approximately $V(x) \simeq Q/(4\pi \epsilon_0 x)$ when $x \gg a$, where $Q = 2\pi a^2 \rho_s$.
(c) Find the electric field E_x along the x-axis for $x > 0$.

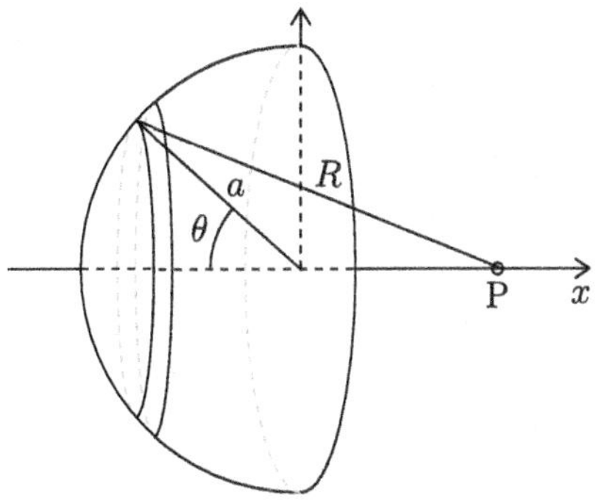

Fig. 3.19 Illustration of a half of a spherical shell

Modeling projects

3.30 Electric potential from an water molecule. In physics we *model*, that is, we take a physical system and develop a simplified model that we can address theoretically. When we model, we try to make reasonable assumptions. This requires courage and curiosity. You need to dare to make a simple model and explore the consequences.

In this exercise, we will model a water molecule as a set of point charges: a charge $Q_{\mathrm{O}} = -0.7e$ at position of the oxygen atom and a charge $Q_{\mathrm{H}} = 0.35e$ at the positions of the hydrogen atoms (1) (See e.g. Fyta[6] for data). We assume that the angle between the hydrogen atoms is $\theta = 104.5°$ and that the distance between the oxygen and the hydrogen atom is $a = 0.95$Å (2).

(a) We place the oxygen atom at the origin and then place the hydrogen atoms symmetrically around the y-axis. Show that the position of the two hydrogen atoms are then $\mathbf{r}_i = a(\pm \sin\theta/2, \cos\theta/2)$.

(b) Use the list method to set up lists `qilist` and `rilist` of the charges and positions of the charges in the water molecule. Write a Python script that generates the list for the water molecule. Measure all lengths in units of Angstrom and all charges in units of the electric charge.

(c) Write a Python program to find the electric potential of the water molecule with a reasonable resolution and visualize the results.

3.31 Dipole of two rings. In this exercise we will study the electric potential and field around a dipole consisting of two ringshaped charges. First, we will study a single ring charge, and then we will address two ring charges with opposite charge. We start by addressing a single ring charge with charge Q in the xy-plane. The ring charge consists of a thin ring with uniform charge density. It is centered in the origin with a radius a. You can assume the charge is in vacuum.

A single ring charge.

(a) Show that the electric potential along the z-axis is given as

---

[6] https://www2.icp.uni-stuttgart.de/~icp/mediawiki/images/3/35/SimmethodsII_ss13_lecture6_watermodels.pdf.

$$V(z) = \frac{Q}{4\pi\epsilon_0}\frac{1}{\sqrt{z^2 + a^2}}$$

(b) Find the z-component of the electric field along the z-axis, $E_z(z)$.
(c) Write a program to find the electric potential from the ring charge in the yz-plane and visualize this in a suitable region so that the shape of the potential is clearly shown.
(d) Check the results from your program by comparing with the exact solution you found in part **a**. Show using a plot that your computational results coincide with the theoretical results along the z-axis.

A dipole of two ring charges. We will now study a dipole that consists of two ring charges: A ring charge with charge $Q = 1\text{mC}$, radius a, and center in $(0, 0, a)$ and a ring charge with charge $Q = -1\text{mC}$, radius a, and center in $(0, 0, -a)$. Both rings are in planes parallel with the xy-plane.
(e) Write a program to find the electric potential in the yz-plane for this system and visualize the potential in a suitable region.
(f) The electric potential $V(\mathbf{r})$ in a point $\mathbf{r}$ from a dipole with a dipole moment $\mathbf{p} = q\mathbf{d}$ centered in the origin is approximately equal to $V(\mathbf{r}) = \mathbf{p} \cdot \hat{\mathbf{r}}/(4\pi\epsilon_0 r^2)$ where $\hat{\mathbf{r}} = \mathbf{r}/r, r = |\mathbf{r}|$ and $\mathbf{d}$ is a vector from the negative charge $-q$ to the positive charge q. Compare the electrical potential from the program with the approximate expression along a line that is parallel with the z-axis through $(2a, 0, 0)$ by plotting the two values for the potential in the same plot.
(g) Visualize the electric field from the two ring charges in the yz-plane.
(h) Use the program to find $E_z(x, 0, 0)$ for the system of two ring charges. (Hint: Find the electric potential immediately above and below the xy-plane along a line along the x-axis and use this to estimate E_z.)

References

C. Martin and H. Zipse. Charge distribution in the water molecule - a comparison of methods. *Journal of Computational Chemistry*, 26:97–105, 2005.
A. G. Csaszar, G. Czako, T. Furtenbacher, J. Tennyson, V. Szalay, S. V. Shirin, N. F. Zobov, and O. L. Polyansky. On equilibrium structures of the water molecule. *Journal of Chemical Physics*, 122:214305, 2005.

Chapter 4
Gauss' Law

4.1 Motivation for Gauss' Law

You are now familiar with Coulomb's law for the electric field from a single point charge in the origin

$$\mathbf{E} = \frac{Q}{4\pi\epsilon_0}\frac{\hat{\mathbf{r}}}{r^2}. \tag{4.1}$$

This law has a striking form, decaying like $1/r^2$ just like Newton's law of gravity. What does this mean in practice? As we move away from the charge, the field decreases while the area over which it acts increases. You can imagine a set of field lines that originate at the charge. The field lines point radially outward. As you move further away from the charge, the number of field lines at a given distance r remains the same. No new field lines are generated, since there are no additional charges. But the number of field lines per unit area decreases because the area increases. How does the area increase? The area of a sphere with a radius r is $S = 4\pi r^2$. The area therefore increases as r^2. This implies that if we multiply the magnitude of the electric field with the surface area, the result is a constant:

$$|\mathbf{E}|\, S = \frac{Q}{4\pi\epsilon_0 r^2} 4\pi r^2 = \frac{Q}{\epsilon_0}. \tag{4.2}$$

How can we interpret the product of the magnitude of the field and the surface area? If we address a small part of the surface $\mathrm{d}\mathbf{S}$, it has a magnitude $\mathrm{d}S$ and points outward, in the direction of the surface normal, which is in the $\hat{\mathbf{r}}$ direction. The field also points radially outward in the $\hat{\mathbf{r}}$ direction. The product $|\mathbf{E}|\,\mathrm{d}S$ is therefore the same as $\mathbf{E}\cdot\mathrm{d}\mathbf{S}$. This term is called the *flux* of the electric field. We can think of this as how much of the field is flowing through the surface. (We will go in depth on this concept in the next section.) Now, the product $|\mathbf{E}|\, S$ was for the whole surface S. We therefore need to add together the contributions from all surface elements $\mathrm{d}\mathbf{S}$ through an integral over the surface. This gives us

© The Author(s), under exclusive license to Springer Nature Switzerland AG 2026

A. Malthe-Sørenssen, *Elementary Electromagnetism Using Python*, Undergraduate Texts in Physics, https://doi.org/10.1007/978-3-032-19876-1_4

$$|\mathbf{E}|\, S = \oint_S \mathbf{E} \cdot \mathrm{d}\mathbf{S} = \frac{Q}{\epsilon_0}. \tag{4.3}$$

The flux out of a spherical surface S around a charge Q corresponds to the charge Q divided by ϵ_0. This beautiful relationship is a special case of Gauss' law. Further on, we prove that Gauss' law is true for any surface S as long as Q is the net charge within that surface. Now, let us formulate Gauss' law precisely, develop an intuition for the flux of the electric field, and learn how we can use Gauss' law to determine the electric field.

4.2 Gauss' Law in Integral Form

Gauss' law

Gauss' law states that the flux of the electric field through a *closed* surface S is proportional to the net total charge in the volume enclosed by the surface:

$$\oint_S \mathbf{E} \cdot \mathrm{d}\mathbf{S} = \frac{Q_S}{\epsilon_0} \tag{4.4}$$

Here, S is any closed surface and Q_S is the net charge in the volume inside the surface S. (We provide a proof of Gauss' law further down.)

- The surface S must be a *closed* surface. The charge Q_S is the net total charge inside the surface S. The net charge inside the surface S is either the sum of all the charges in the volume v enclosed by the surface S: $Q_S = \sum_i Q_{i,\mathrm{in}}$ or the integral of a volume charge density ρ_v over the volume v enclosed by the surface S: $Q_S = \int_v \rho_v \,\mathrm{d}v$.
- The integral $\oint_S \mathbf{E} \cdot \mathrm{d}\mathbf{S}$ is called the flux of the electric field through the surface S.

Electric Flux

The electric flux, $\mathrm{d}\Phi$, through a small surface $\mathrm{d}\mathbf{S}$ is defined as

$$\mathrm{d}\Phi = \mathbf{E} \cdot \mathrm{d}\mathbf{S}, \tag{4.5}$$

where $\mathrm{d}\mathbf{S}$ is an *oriented* surface element that has both an area and a direction. The direction is given by the unit normal vector for the surface, $\hat{\mathbf{n}}$, so that $\mathrm{d}\mathbf{S} = \hat{\mathbf{n}}\,\mathrm{d}S$. The directed surface element $\mathrm{d}\mathbf{S}$ points in the direction of the *positive* surface normal.

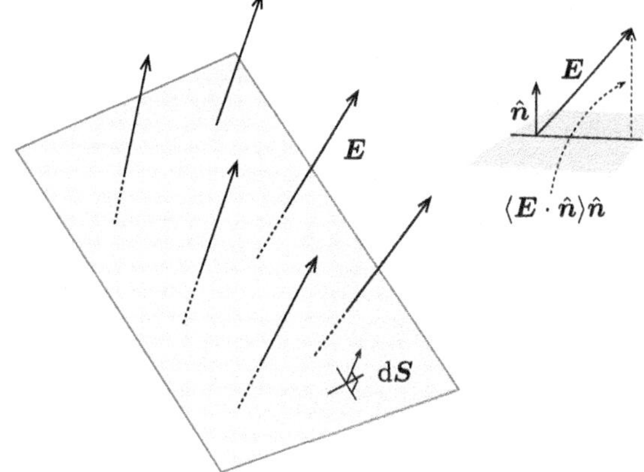

Fig. 4.1 Illustration of the electric field **E** and a small surface element with normal vector **n̂**

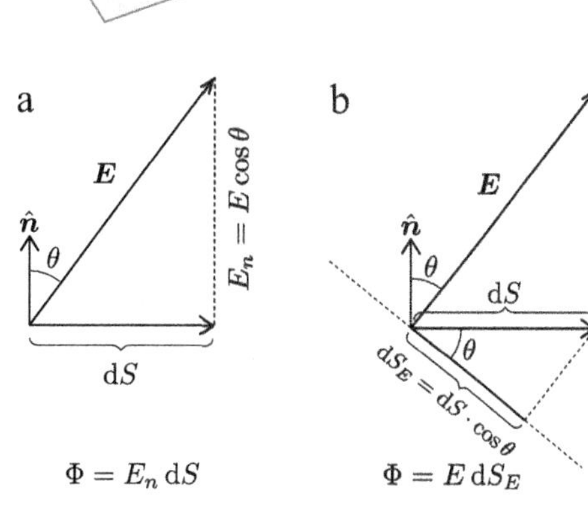

Fig. 4.2 Illustration of the electric field **E** and a small surface element with normal vector **n̂** showing two different ways of interpreting the flux

Flux and surface normal. The flux is given by the dot product of the field **E** and the surface element $\mathbf{dS} = \hat{\mathbf{n}}\, dS$. This is illustrated in Fig. 4.1. It is only the component of the field that is normal to the surface that contributes. Any component normal to the normal vector **n̂**, that is parallel to the surface, will not contribute to the flux.

There are two ways to think about this as illustrated in Fig. 4.2: (1) You can think of the flux as the projection of the electric field **E** onto the surface **dS** with surface normal **n̂**. (2) Or you can think of the flux as the projection of the surface area **dS** onto the electric field. In both cases $d\Phi = E\, dS \cos\theta$. In the first case, we interpret $E \cos\theta$ as the part of the electric field that is normal to the surface. It is only this part of the field that contributes to the flux. In the second case, we interpret $dS \cos\theta$ as the cross-sectional surface area, the projection of the area onto the direction of the electric field.

Only the normal component contributes to the flux. Figure 4.3 illustrates an electric field and two possible surfaces. For the case when the normal vector **n̂** for the surface is in the same direction as the electric field, **E**, the flux is maximum, Φ_{max}. For the case when the normal vector **n̂** for the surface is in a direction normal to the electric field, **E**, the flux is zero.

Fig. 4.3 Illustration of the electric field $\mathbf{E}$ and a small surface element with normal vector $\hat{\mathbf{n}}$ for two possible surfaces

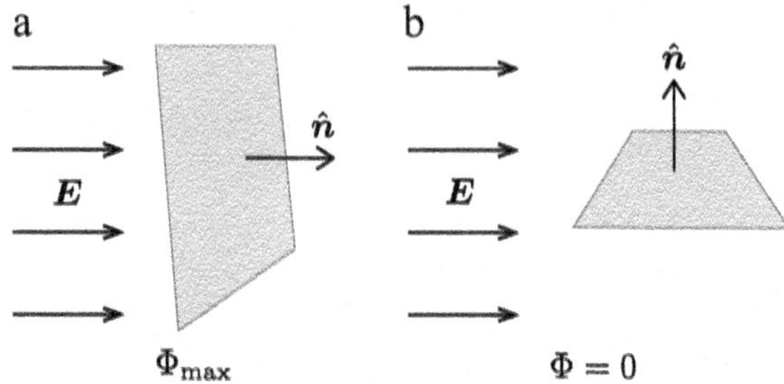

Direction of the surface normal. Notice that in the surface integral over the closed surface S:

$$\oint_S \mathbf{E} \cdot \mathrm{d}\mathbf{S} = \oint_S \mathbf{E} \cdot \hat{\mathbf{n}}\,\mathrm{d}S = \frac{Q_{\text{in}}}{\epsilon_0}, \tag{4.6}$$

the surface normal $\hat{\mathbf{n}}$, and the directed surface element $\mathrm{d}\mathbf{S}$, points *outward*. This is the flux *out of the volume* enclosed by S. When we calculate this surface integral we need to sum up the contributions from the part of the surface that make up the complete, closed surface.

Test your understanding
The figure shows a uniform field along the y-axis. Through which of the sides A, B or C is the flux, Φ, the largest?[1]

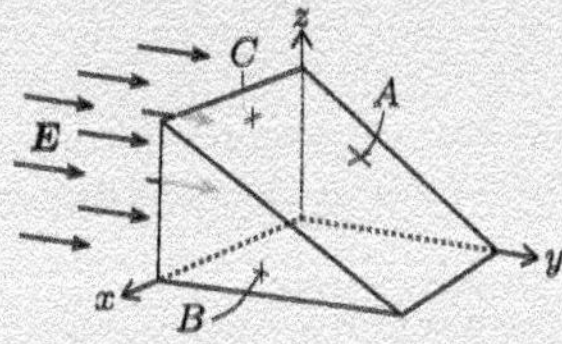

Example: Gauss' Law for a Single Charge

How we can use Gauss' law to find the electric field? We demonstrate an approach through an example, using Gauss' law to find the field from a single point charge in the origin.

For this system, Gauss' law states that for any closed surface that encloses the origin, the integral of the field over the surface is

---

[1] $\Phi_A = \Phi_C$ because the cross-sectional areas are the same, $\Phi_B = 0$.

$$\oint_S \mathbf{E} \cdot \mathrm{d}\mathbf{S} = \frac{Q}{\epsilon_0}. \tag{4.7}$$

If we know the electric field, we can show that this equation is correct. But how can we use this equation to *find* the electric field? This is not obvious, because the electric field is inside the integral. We need to find a functional form for the field so that it satisfies this integral equation. However, there is a *trick* that we will use. We select the surface carefully so that the inside of the integral, $\mathbf{E} \cdot \hat{\mathbf{n}}\, \mathrm{d}S$, is a constant during the integration. If we succeed with this, we can place $\mathbf{E} \cdot \hat{\mathbf{n}}$ outside the integral, and we can use the equation to find at least the component of the field that is normal to the surface.

To achieve this, we need to select the surface carefully taking into consideration the *symmetry* of the system. For a single charge in the origin, the system must have spherical symmetry: the system does not change if we rotate it. Consequently, the electric field must be the same for all possible rotations. We see the consequences of this symmetry if we describe the field in spherical coordinates:

$$\mathbf{E} = E_r\, \hat{\mathbf{r}} + E_\phi\, \hat{\boldsymbol{\phi}} + E_\theta\, \hat{\boldsymbol{\theta}}. \tag{4.8}$$

We realize that the field only can have a radial component. Otherwise the field would not be the same if we rotated it—it would violate the symmetry condition. Thus we expect the field to have the form

$$\mathbf{E} = E_r(r, \phi, \theta)\, \hat{\mathbf{r}}. \tag{4.9}$$

In addition, the field cannot depend on ϕ or θ, because this would also violate the symmetry. The radial component of the field must be the same in all directions, because the orientation of the coordinate system is arbitrary. Thus, we know that the field must have the form:

$$\mathbf{E} = E_r(r)\, \hat{\mathbf{r}}. \tag{4.10}$$

This describes the full symmetry of the field. The field points in the radial direction and only depends on the distance to the origin. The field is therefore constant on a surface that has a constant distance to the origin—on a sphere. And in this case the normal vector of the surface is $\hat{\mathbf{r}}$. We therefore choose this as our integration surface. We call an integration surface used in Gauss' law a *Gauss surface*.

We are now ready to apply Gauss' law on the Gauss surface, which is a spherical surface of radius r.

$$\oint_S \mathbf{E} \cdot \mathrm{d}\mathbf{S} = E_r(r) \oint_S \mathrm{d}S = E_r(r) 4\pi r^2 = \frac{Q}{\epsilon_0}. \tag{4.11}$$

Here the integral equals the area of the surface, $4\pi r^2$. Because the $E_r(r)$ is outside the integral, we can now solve for the field $E_r(r)$:

$$E_r(r) = \frac{Q}{4\pi\epsilon_0 r^2}, \tag{4.12}$$

and

$$\mathbf{E} = E_r(r)\,\hat{\mathbf{r}} = \frac{Q}{4\pi\epsilon_0 r^2}\,\hat{\mathbf{r}}. \tag{4.13}$$

Notice two things in this argument.

- First we see that we used *symmetry arguments* to simplify the description of the electric field.
- Second, this method only worked because *we chose a surface S where the electric field was constant*. It would not have worked if we chose a cubic surface enclosing the charge.

It is therefore essential to be able to recognize and use *symmetries* to use Gauss' law.

Method: Using Gauss' law to find the electric field

Based on this example, we can propose a general strategy for how to find the electric field for a charge distribution using Gauss' law.

- Find a set of surfaces that enclose a volume such that $\mathbf{E} \cdot \hat{\mathbf{n}}$ is constant on each such surface element. (It may be zero on some of the surfaces—zero is a constant!) We call these surfaces *Gauss surfaces* for the problem.
- This often requires that you find a simplified description of the field in a chosen coordinate system, such as $\mathbf{E} = E_r(r)\,\hat{\mathbf{r}}$.
- Find the flux integral.
- Use Gauss' law to find the electric field as a function of charge and position.
- Notice that the surface does not have to enclose all the charges—it is allowed and indeed often necessary to chose a surface that contains only some of the charges. However, the electric field must be a constant on the surfaces you have chosen.

Test your understanding

A spherical shell has a positive charge uniformly distributed on its surface. (a) There are no other charges present. What is the electric field inside the shell? (b) If we place a charge q outside the shell and keep the charges on the shell as they were, what is now the field inside the shell? Zero, non-zero, or is it not possible to tell if it is zero or not?[2]

[2] (a) Zero, (b) Non-zero.

Example: Electric Field from an Infinite Line Charge

Find the electric field from an infinite line charge.

Specifying the problem. The line is along the z-axis and we assume that it has a uniform line charge density, ρ_l.

Drawing the system. We illustrate this system in the drawing in Fig. 4.4.

Symmetry. We look for symmetries in this system. First, we notice that the line charge is along the z-axis. We therefore expect the field to be symmetric around the z-axis. First, let us use a cylindrical coordinate system. We can write $\mathbf{E}$ in cylindrical coordinates (r, ϕ, z), where r is the distance to the z-axis, ϕ is the angle around the z-axis with $\phi = 0$ along the x-axis, and z is the position along the z-axis. Using this coordinate system we have that $\mathbf{E} = E_r\,\hat{\mathbf{r}} + E_\phi\,\hat{\boldsymbol{\phi}} + E_z\,\hat{\mathbf{z}}$, where $\hat{\mathbf{r}}$, $\hat{\boldsymbol{\phi}}$ and $\hat{\mathbf{z}}$ are unit vectors.

We expect the field to depend on the distance from the z-axis, but not on the angle around the axis. The symmetry therefore implies that $E_r(r, \phi, z)$ only depends on r and z, $E_r = E_r(r, z)$.

Since the line is infinite along the z-axis, we expect the system to look the same no matter where we are along the z-axis. Thus we do not expect any z-dependence in the field, $E_r = E_r(r)$.

This means that the field will look like $\mathbf{E} = E_r(r)\hat{\mathbf{r}} + E_\phi(r)\hat{\boldsymbol{\phi}}$. Finally, what about the field in the ϕ-direction? Here, there are two arguments: Either we could argue that since the charge distribution is the same in both the positive and negative z-direction, we do not expect the ϕ-direction to break that symmetry. Alternatively, we can argue that since the electric field around any closed loop, $\oint \mathbf{E} \cdot \mathrm{d}\mathbf{l} = 0$, then we see that $\oint E_\phi\,\mathrm{d}l = 2\pi r E_\phi = 0 \Rightarrow E_\phi = 0$.

Finally, we have the symmetry of the problem written down in terms of how the electric field depends on the various variables. Usually, we do not need to be *this* systematic. I expect that you will be able to see such symmetries immediately and write down the corresponding form of the electric field.

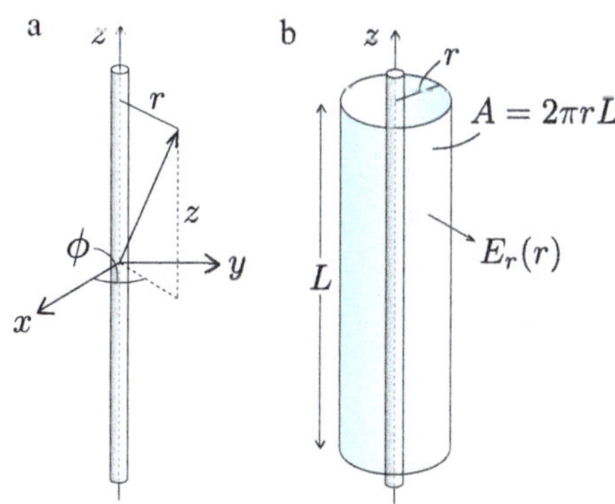

Fig. 4.4 Illustration of the electric field $\mathbf{E}$ from an infinite line charge along the z-axis

Applying Gauss' law using the found symmetry. With a cylindrical symmetry, we use a cylinder around the z-axis as the surface for Gauss' law. On the surface of this cylinder, the electric field will be constant, because $E_r(r)$ only depends on r. We then apply Gauss' law on the cylinder surface in Fig. 4.4:

$$\oint_S \mathbf{E} \cdot \mathrm{d}\mathbf{S} = \oint_S \mathbf{E} \cdot \hat{\mathbf{n}} \, \mathrm{d}S. \tag{4.14}$$

We divide the surface into three surfaces: The top and bottom edges of the cylinder and the curved cylinder surface with a radius r. On the *bottom* and *top* surfaces the electric field is radial and the surface normal is along the z-axis. Therefore $\mathbf{E} \cdot \hat{\mathbf{n}}$ is zero here. On the curved cylinder surface the electric field is constant $\mathbf{E} = E_r(r)\hat{\mathbf{r}}$ and directed along the surface normal $\hat{\mathbf{n}} = \hat{\mathbf{r}}$, which points outwards from the center of the cylinder. Therefore, $\mathbf{E} \cdot \hat{\mathbf{n}} = E_r(r)$, which is a constant on the whole surface. The integral is therefore simply $E_r(r)$ times the surface integral of the curved cylinder surface, $A = 2\pi r L$, where L is the length of the cylinder. We therefore find that the flux is

$$\oint_S \mathbf{E} \cdot \mathrm{d}\mathbf{S} = 2\pi r L E_r. \tag{4.15}$$

Gauss's law states that this is equal to the net charge inside the volume:

$$\oint_S \mathbf{E} \cdot \mathrm{d}\mathbf{S} = \frac{Q}{\epsilon_0}. \tag{4.16}$$

Here, the amount of charge inside is the charge line density times the length of the line: $Q = \rho_l L$. Gauss' law therefore gives:

$$\oint_S \mathbf{E} \cdot \mathrm{d}\mathbf{S} = 2\pi r L E_r = \frac{\rho_l L}{\epsilon_0} \Rightarrow E_r = \frac{\rho_l}{2\pi \epsilon_0 r}. \tag{4.17}$$

Example: Electric Field from a Spherical Charge

A sphere with radius a has a uniformly distributed charge Q. Find the electric field everywhere in space.

Specifying the problem. The problems sketches a sphere of radius a. Inside the sphere there is a constant volume charge density ρ_v. We find the charge density from:

$$Q = \int_v \rho_v \, dv = \rho_v \int_v \mathrm{d}v = \rho_v \frac{4}{3}\pi a^3 \Rightarrow \rho_v = \frac{3Q}{4\pi a^3}. \tag{4.18}$$

We can specify this further. The charge density is constant for $r < a$ and zero outside. This means that

$$\rho_v = \begin{cases} \rho_v & \text{when } r \leq a \\ 0 & \text{when } r > a \end{cases} \tag{4.19}$$

We want to find the electric field everywhere. This means that we want to find the electric field both for $r < a$ and for $r > a$!

Drawing the system. We illustrate this system in the drawing in Fig. 4.5.

Symmetry. What symmetries does this system have? We realize the system has spherical symmetry. We realize that the symmetry indicates the electric field is radial $\mathbf{E} = E\hat{\mathbf{r}}$. If this was not the case, we could have rotated the sphere half a turn around an axis through the center of the sphere, and the electric field would have changed. But this does not change the distribution of charges, therefore the field cannot change. This contradiction means that the field cannot have a component that is not radial. Similarly, we argue that E only depends on r and not on θ or ϕ. The field is therefore $\mathbf{E} = E(r)\hat{\mathbf{r}}$.

Applying Gauss' law. We can then apply Gauss' law using this symmetry. We choose to apply Gauss' law on a spherical surface centered on the center of the spherical charge distribution. In this case, the electric field $\mathbf{E} = E(r)\hat{\mathbf{r}}$ is always pointing in the direction of the outward-pointing surface normal $\hat{\mathbf{n}} = \hat{\mathbf{r}}$. The flux through a spherical surface is then

$$\Phi = \oint_S \mathbf{E} \cdot d\mathbf{S} = E(r) \oint_S dS = E(r)\, 4\pi r^2 = \frac{Q_{\text{in}}}{\epsilon_0}. \tag{4.20}$$

Now, what is the charge Q_{in}? It is the charge *inside* the surface. If $r > a$ the surface encloses the whole charge Q and $Q_{\text{in}} = Q$. In this case the field is

$$\mathbf{E} = \frac{Q}{4\pi\epsilon_0 r^2}\hat{\mathbf{r}} \quad r > a. \tag{4.21}$$

However, when $r < a$, the charge Q_{in} is only the charge that is inside the sphere of radius r. Inside the sphere, the volume charge density is constant. We can therefore

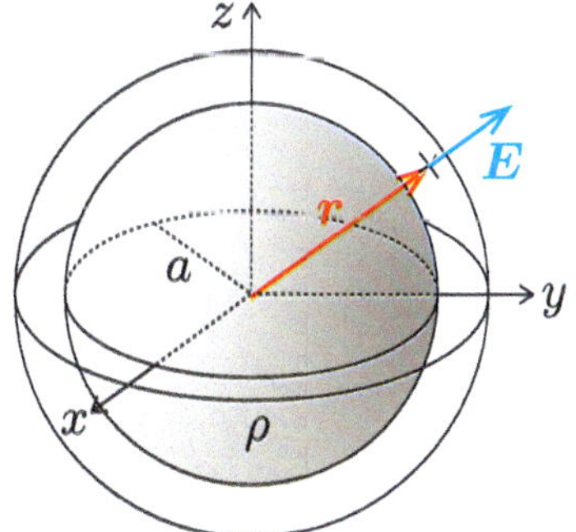

Fig. 4.5 Illustration of the electric field **E** from a spherical charge

find the charge Q_{in} as the integral over the volume enclosed by a sphere of radius r, v_r:

$$Q_{\text{in}} = \int_{v_r} \rho \, dv = \rho_v \int_{v_r} dv \tag{4.22}$$

$$= \rho_v \frac{4}{3}\pi r^3 = \frac{Q}{\frac{4}{3}\pi a^3}\frac{4}{3}\pi r^3 = Q\left(\frac{r}{a}\right)^3 . \tag{4.23}$$

We can then find the electric field from

$$E(r)\, 4\pi r^2 = \frac{Q_{\text{in}}}{\epsilon_0} = \frac{Q(r/a)^3}{\epsilon_0} \Rightarrow E(r) = \frac{Qr}{4\pi\epsilon_0 a^3}. \tag{4.24}$$

The electric field is therefore:

$$\mathbf{E} = \begin{cases} \frac{Q}{4\pi\epsilon_0 r^2}\hat{\mathbf{r}} & \text{when } r > a \\ \frac{Qr}{4\pi\epsilon_0 a^3}\hat{\mathbf{r}} & \text{when } r \le a \end{cases} . \tag{4.25}$$

4.3 Gauss' Law on Differential Form

Gauss' law as formulated so far is on *integral form*:

$$\oint_S \mathbf{E} \cdot d\mathbf{S} = \frac{Q_{\text{in}}}{\epsilon_0}. \tag{4.26}$$

This law is true for any surface S. In general, the net charge inside the surface is given as an integral over a volume charge density ρ:

$$Q_{\text{in}} = \int_v \rho \, dv, \tag{4.27}$$

where the volume v is the volume enclosed by the surface S.

We can rewrite the surface integral using the divergence theorem from vector calculus. This states that for a (continuous) vector field $\mathbf{f}$ we have that

$$\oint_S \mathbf{f} \cdot d\mathbf{S} = \int_v \nabla \cdot \mathbf{f} \, dv, \tag{4.28}$$

where the volume v is the volume enclosed by the surface S. We apply this to Gauss' law:

$$\oint_S \mathbf{E} \cdot d\mathbf{S} = \int_v \nabla \cdot \mathbf{E} \, dv = \frac{Q_{\text{in}}}{\epsilon_0} = \frac{1}{\epsilon_0}\int_v \rho \, dv. \tag{4.29}$$

Because this is true for any volume v, the two expressions inside the volume integrals are also identical:

$$\nabla \cdot \mathbf{E} = \frac{\rho}{\epsilon_0}. \tag{4.30}$$

This is called *Gauss' law on differential form*. This version of Gauss' law will be used frequently when we address electromagnetics waves and it is often the starting point for numerical computations of the electric field.

Gauss' law on differential form

Gauss' law on differential form states that the electric field $\mathbf{E}(\mathbf{r})$ is related to the local divergence of the volume charge density ρ:

$$\nabla \cdot \mathbf{E} = \frac{\partial E_x}{\partial x} + \frac{\partial E_y}{\partial y} + \frac{\partial E_z}{\partial z} = \frac{\rho}{\epsilon_0}. \tag{4.31}$$

4.4 Proof of Gauss' Law

Gauss law can be proved by proving it for a single point charge and then use the superposition principle to prove it for any charge distribution. We will follow the elegant approach from Johannes Skaar (1) where we use the divergence theorem on integral form and that the divergence of the electric field for a single point charge is zero everywhere except in the center of the charge. For a charge q inside a closed surface S, we know from the divergence theorem that the net flux of $\mathbf{E}$ through S is equal to the volume integral of the divergence of $\mathbf{E}$ over the volume v enclosed by S:

$$\oint_S \mathbf{E} \cdot \mathrm{d}\mathbf{S} = \int_v \nabla \cdot \mathbf{E}\, \mathrm{d}v. \tag{4.32}$$

As long as the charge q is inside the closed surface S and therefore also the volume v, we can divide the volume into two separate parts: a spherical volume v_s of a sphere centered on the charge q and the remaining volume, $v_r = v - v_s$. The volume integral is therefore:

$$\int_v \nabla \cdot \mathbf{E}\, \mathrm{d}v = \int_{v_s} \nabla \cdot \mathbf{E}\, \mathrm{d}v + \int_{v_r} \nabla \cdot \mathbf{E}\, \mathrm{d}v. \tag{4.33}$$

We will now show that the second integral is zero, because the divergence of **E** is zero everywhere away from the point charge, and that the first integral can be solved explicitly.

What is the divergence of the field from a single point charge? We place the origin in the point charge, so that the field from the point charge is

$$\mathbf{E} = \frac{q}{4\pi\epsilon_0 r^2}\hat{\mathbf{r}} = E_r\hat{\mathbf{r}}. \tag{4.34}$$

The divergence of a field in spherical coordinates, $\mathbf{E} = E_r\hat{\mathbf{r}} + E_\theta\hat{\boldsymbol{\theta}} + E_\phi\hat{\boldsymbol{\phi}}$, is

$$\nabla\cdot\mathbf{E} = \frac{1}{r^2}\frac{\partial\left(r^2E_r\right)}{\partial r} + \frac{1}{r\sin\theta}\frac{\partial\left(\sin\theta\, E_\theta\right)}{\partial\theta} + \frac{1}{r\sin\theta}\frac{\partial E_\phi}{\partial\phi} \tag{4.35}$$

For a single point charge $E_\theta = E_\phi = 0$. We see that since E_r is proportional to $1/r^2$, the term r^2E_r is a constant, and its derivative is zero. All the terms are therefore zero and the divergence is zero. Notice that this is only true when $r > 0$. But we have carefully selected the volume v_r so that $r > 0$ everywhere in this volume. Therefore, the volume integral of the divergence of **E** over v_r is zero:

$$\int_{v_r}\nabla\cdot\mathbf{E}\,\mathrm{d}v = 0. \tag{4.36}$$

Now, let us find the integral over the spherical volume v_s centered on the charge q. This is a sphere with radius a and center in the origin, where the charge q is. We apply the divergence theorem again and relate the volume integral to the surface integral:

$$\int_{v_s}\nabla\cdot\mathbf{E}\,\mathrm{d}v = \oint_{S_s}\mathbf{E}\cdot\mathrm{d}\mathbf{S}. \tag{4.37}$$

A surface element on this spherical surface of radius a is $\mathrm{d}\mathbf{S} = \mathrm{d}S\,\hat{\mathbf{r}}$ and at a distance $r = a$ we have that $\mathbf{E} = q/(4\pi\epsilon_0 a^2)\hat{\mathbf{r}}$ so that:

$$\oint_{S_s}\mathbf{E}\cdot\mathrm{d}\mathbf{S} = \oint_{S_s}\frac{q}{4\pi\epsilon_0 a^2}\hat{\mathbf{r}}\cdot\hat{\mathbf{r}}\,\mathrm{d}S = \frac{q}{4\pi\epsilon_0 a^2}\oint_{S_s}\mathrm{d}S = \frac{q}{4\pi\epsilon_0 a^2}4\pi a^2 = \frac{q}{\epsilon_0}. \tag{4.38}$$

We have therefore shown that when a charge q is inside a closed surface S:

$$\oint_S\mathbf{E}\cdot\mathrm{d}\mathbf{S} = \int_{v_s}\nabla\cdot\mathbf{E}\,\mathrm{d}v + \int_{v_r}\nabla\cdot\mathbf{E}\,\mathrm{d}v = 0 + \frac{q}{\epsilon_0} = \frac{q}{\epsilon_0}. \tag{4.39}$$

What if the charge q is not inside the surface, but outside the surface? In this case we do not need to subdivide the volume inside the surface into two parts, because the divergence will be zero everywhere in the volume inside the surface and the integral will be zero.

We have therefore demonstrated that Gauss' law in integral form is true for a single point charge q: the integral is q/ϵ_0 if the charge is inside the surface and zero if it is outside the surface. For a charge distribution with charges Q_i in positions $\mathbf{r}_i$ we therefore find that the flux through a closed surface S is

$$\oint_S \sum_i \mathbf{E}_i \cdot d\mathbf{S} = \sum_{i \text{ inside } S} \frac{Q_i}{\epsilon_0} = \frac{Q_{\text{inside}}}{\epsilon_0}. \tag{4.40}$$

Summary

Gauss' law in integral form. For a closed surface S:

$$\oint_S \mathbf{E} \cdot d\mathbf{S} = \frac{Q_{\text{in}}}{\epsilon_0},$$

where Q_{in} is the net charge inside the surface S.

Application of Gauss' law. We use Gauss' law to determine the electric field for systems with a high degree of symmetry, so that $\mathbf{E} \cdot d\mathbf{S}$ is constant E_0 on the surface. We can then find E_0 by applying Gauss' law and calculate the net charge inside the surface: $SE_0 = Q_{\text{in}}/\epsilon_0$, and therefore $E_0 = Q_{\text{in}}/(S\epsilon_0)$, where S is the area of the corresponding surface.

Gauss' law on differential form. Gauss' law on differential form is:

$$\nabla \cdot \mathbf{E} = \rho/\epsilon_0,$$

where ρ is the volume charge density.

Exercises

Discussion Exercises

4.1 Rubber balloon. We put a small charge inside a rubber balloon and inflate it. Does the flux of the electric field through the balloon depend on how inflated it is? Explain your reasoning.

4.2 Two charges. Figure 4.6 shows two charges and several surface A to D. Can you find the flux through each of the surfaces?

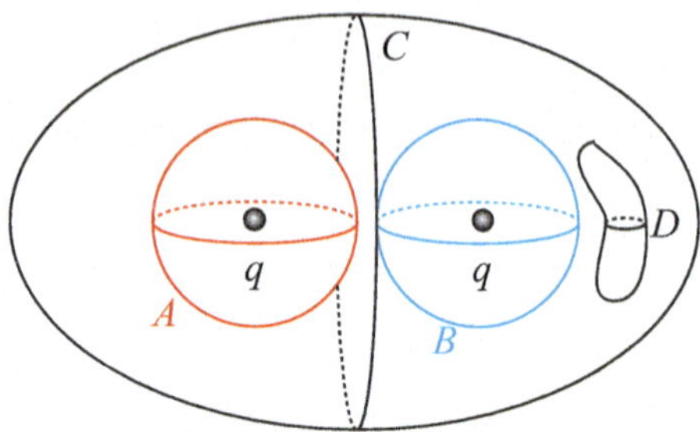

Fig. 4.6 Illustration of fluxes through surface A, B, C, and D

4.3 No charges, no field?. A closed surface S does not contain any charges. Does this imply that the electric field must be zero everywhere on the surface? Is the flux zero through all parts of the surface? What is the net flux through the surface?

4.4 Uniform charge density.
(a) The volume charge density ρ is uniform and positive in a region in space. Can the electric field $\mathbf{E}$ also be uniform in this region?
(b) Assume that this region, where ρ is uniform and positive, encloses a smaller closed region where $\rho = 0$. What can you say about the electric field in this region?

4.5 A different Coulomb's law. If the electric field from a point charge in the origin were proportional to $1/r^3$ instead of $1/r^2$, would Gauss' law still be valid? (Hint: Consider a spherical surface centered around a single point charge.)

Tutorials

4.6 Maximum flux. A uniform electric field is directed along the x-axis: $\mathbf{E} = E_0\hat{\mathbf{x}}$ as shown in Fig. 4.7. The left part show an oriented surface $\mathbf{S}$.
(a) What is the flux through the surface in **a**?
(b) Part **b** show an oriented surface $\mathbf{S}$, which is at an angle θ with the y-axis. For what angle θ is the flux through this surface maximal?
(c) For what angle is the flux minimal?
(d) Find an expression of the flux as a function of θ.

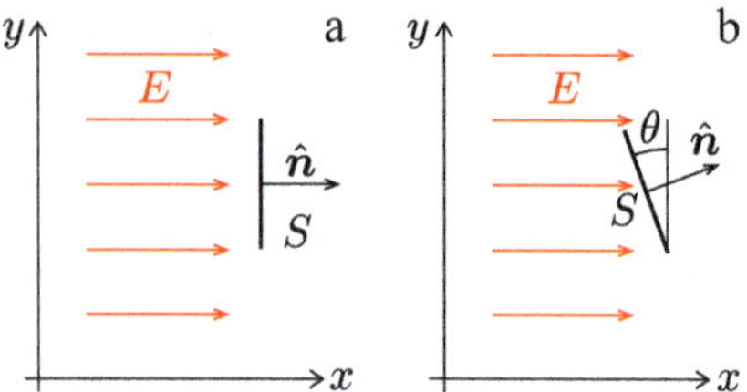

Fig. 4.7 The flux of a uniform field with two surfaces in **a** and **b**

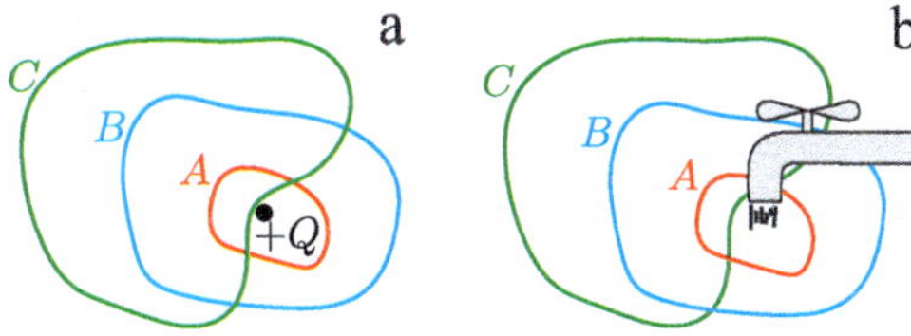

Fig. 4.8 **a** A charge and three closed surfaces. **b** A water faucet and three closed surfaces

4.7 Smartsurface. We address a uniform electric field $\mathbf{E} = E_0\hat{\mathbf{x}}$.
(a) Explain that the flux through a cube with side a always will be zero in this field (without using Gauss' law). Can you explain it with word and also show it mathematically?
(b) Your friend Q has designed a smart surface which is not closed, but still always has zero flux in a homogeneous field. The surface is a cube, where two opposite sides have been removed. Does Q's proposition make sense? Can you define a flux through a surface which is not closed?
(c) Is Q correct? Is the flux always zero through this surface in a uniform field?
(d) Q states that this is not only true for uniform fields, but this surface has zero flux for any field. Is Q right?

4.8 Waterflux. Figure 4.8a shows a charge $+Q$ and three closed surfaces A, B, and C. There are no other charges present.
(a) Out of which surface, A, B, or C, is the flux the largest?
(b) Figure 4.8b shows water flowing from a faucet onto a plane surface. You may assume that the water flows at a constant rate, the water has formed a water film with constant thickness, and that the faucet has been running for a long time so that the water is flowing at a constant velocity everywhere. Out of which of the surfaces, A, B or C, does the most water flow per unit time.
(c) What similarities do you see between the two systems? What would a negative charge correspond to in such an analogy?
(d) If we place a negative charge $-Q$ just inside a beside the positive charge. What is now the flux out through the three surfaces A, B, and C?

4.9 Unknown charge. An electric field $\mathbf{E} = E_0\hat{\mathbf{r}}$ is radiating from the origin.
(a) What is the flux of this field through a spherical surface with radius a?
(b) What is the net charge inside the spherical surface with radius a?

4.10 A plane charge. We will address an infinitely large, uniformly charged plane with surface charge density ρ_S in the xy plane.
(a) What does it mean that the plane is *uniformly* charged?
(b) What symmetries does this system have? List all you can think of—they may be useful when we describe the electric field.
(c) We will the describe the electric field using cylinder coordinates: $\mathbf{E} = E_r(r, \phi, z)\hat{\mathbf{r}} + E_\phi(r, \phi, z)\hat{\boldsymbol{\phi}} + E_z(r, \phi, z)\hat{\mathbf{z}}$. How can you use symmetry to argue that none of the components may depend on ϕ?

(d) How can you use symmetry to argue that none of the components can depend on r?
(e) This means that the electric field has the form: $\mathbf{E} = E_r(z)\hat{\mathbf{r}} + E_\phi(z)\hat{\boldsymbol{\phi}} + E_z(z)\hat{\mathbf{z}}$. How can you use symmetry to argue that E_r must be zero?
(f) How can you use symmetry to argue that E_ϕ must be zero? Can you also make an argument that does not depend on symmetry, but instead on physical laws?
(g) We are now left with the very simplified expression $\mathbf{E} = E_z(z)\hat{\mathbf{z}}$. We can find yet another symmetry. How is $E_z(z)$ related to $E_z(-z)$ and how can you argue for your answer?
(h) Now, we want to use Gauss' law to determine the electric field from this charged plane. What type of Gauss surface should we choose? (Hint: Remember that Gauss surface must be closed.)
(i) Find the charge inside the Gauss surface your have chosen, Q_{in}.
(j) Calculate the flux of the electric field through all the surfaces that together make up the Gauss surface.
(k) Now, use Gauss law to determine the electric field.
(l) Your friend Q wonder why you did not choose a spherical surface as the Gauss surface. A spherical surface also has rotational symmetry in the same way as a cylinder. Is it possible to use a spherical Gauss surface to solve this problem?
(m) Your friend Q thinks that hexagons are beautiful. Could you have used a hexagonal prism (a prism with a hexagon as the base) as a Gauss surface?
(n) If the surface charge density is not uniform, but depends on the distance r to the z-axis, what parts of the symmetry arguments do you need to change? Could we still use Gauss' law to find the electric field?

4.11 Multiple planes. We place three plane charges, all parallel to the xy-plane, above each other: A plane through $z = 0$ with surface charge density 2ρ, a plane through $z = h$ is surface charge density $-\rho$, and a plane through $z = -h$ with surface charge density $-\rho$.
(a) Sketch the system.
(b) Sketch the electric field as a function of z.
(c) Calculate the electric field as a function of z.

Exercises

4.12 Flux-trix. In this exercise you will learn to calculate fluxes for specific surfaces and to use Gauss' law to infer fluxes that may be difficult to calculate. We will study an electric field $\mathbf{E} = E_0\hat{\mathbf{y}}$ generated by charges that are far away. We will now study the flux through several surfaces illustrated in Fig. 4.9 (assume all lengths are in units of a). Surfaces A and B are planar, surface E is a cylindrical shell, and surfaces C and D are the top and bottom caps that ensure that the whole volume is closed.
(a) What are the surface normals for surfaces A, B, C, and D? How would you describe the surface normal for surface E?
(b) Find the flux through surfaces A, B, C, and D.

Fig. 4.9 Illustration of a surface

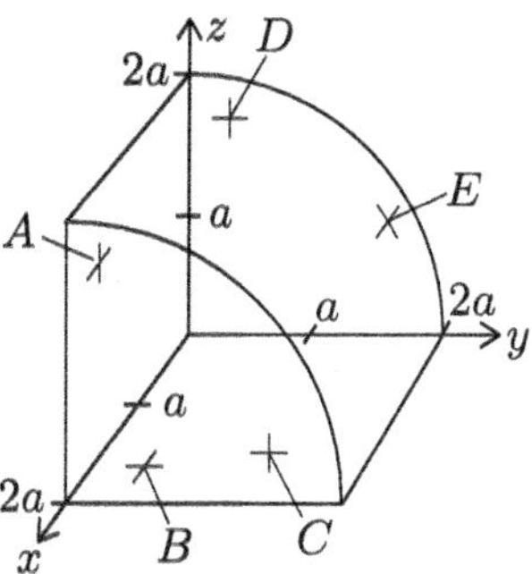

(c) Use Gauss' law to find the flux through surface E. How would you find this without using Gauss' law? Discuss how your methods can be generalized to other electric fields.
(d) Let us now use a similar principle to find the flux in another situation. A charge q is placed at the origin. This is the only charge in the system. The charge sets up an electric field $\mathbf{E}(\mathbf{r})$. What is the flux of $\mathbf{E}(\mathbf{r})$ through a triangular surface with corners at $(a, 0, 0)$, $(0, a, 0)$ and $(0, 0, a)$? Does the result depend on the value of a? (Hint: Make a sketch. Is there another surface you can easily find the flux through? How can you use Gauss' law?)

4.13 Overlapping clouds of charge.
(a) A charge is distributed uniformly throughout a sphere of radius a giving a uniform volume charge density ρ_0. Find the electric field inside and outside of the sphere.
(b) Using the result from (a), find the electric field from two oppositely charged spherical charge distributions, each with radius a, one with volume charge density ρ_0 and the other with uniform charge density $-\rho_0$. The two spheres overlap somewhat and their centers are separated by a distance d. Find the electric field in the overlapping region.
(c) Sketch the electric field in the overlapping region.
(d) Write a Python program to visualize the electric field in space. Place the center of the line connecting the two charges in the origin.

Homework

4.14 Fluxes from point charges. A point charge $q_1 = 4.0$ nC lies on the x-axis in $x = 2.0$ m. Another point charge $q_2 = -6.0$ nC lies on the y-axis in $y = 1.0$ m.
(a) A spherical surface S is centered in the origin with radius 0.5 m. What is the electric flux out of this surface, that is, $\Phi_E = \oint_S \mathbf{E} \cdot d\mathbf{S}$
(b) What is the flux out of the spherical surface if the radius is 1.5 m or 2.5 m?
(c) Would it be useful to use Gauss' law to find the electric field $\mathbf{E}(x, y, z)$ in this system?

4.15 Field from charge distributions. We have a total charge Q. Find the electric field $\mathbf{E}$ everywhere in space when:

(a) Q is a point charge.
(b) Q is uniformly distributed over the volume of a sphere with radius a so that the charge density ρ_v is:

$$\rho_v = \frac{Q}{4\pi a^3/3}. \tag{4.41}$$

(c) Q is uniformly distributed on a spherical shell with radius a so that the surface charge density ρ_s is

$$\rho_s = \frac{Q}{4\pi a^2}. \tag{4.42}$$

(d) Q is uniformly distributed over the volume of a sphere with radius a so that the charge density is proportional to the distance r from the center of the sphere, that is, $\rho_v = kr$ where k is a constant. (Hint: Determine k by calculating $Q = \int_v \rho \, dv$. The geometry in this subexercise indicates that spherical coordinates is a good choice. Remember to use the correct volume element in the integral.)

4.16 Two wires (A). Two very long wires lie parallel to each other in vacuum. The magnitude of their linear charge density ρ, is equal, although one is positive and the other is negative. The distance between the wires is x.
(a) Find the direction of the **E**-field half-way between the two wires without calculating it. (Hint: Compare the situation to that of finding the electric field strength in the half-way point between two equal but opposite charges. (Draw field lines))
(b) Calculate the field strength at this point. (Hint: Use cylindrical symmetry and Gauss' law)
(c) Without calculating, compare the **E**-field at a distance $x/3$ from the positive wire (and $2/3x$ from the negative) to the **E**-field at a distance $x/3$ from the negative wire (and $2/3x$ from the positive).
(d) Calculate the field strength at the locations in (c).

4.17 Two wires (B). Repeat the previous exercise, but with two positively charged wires.
(a) Find the direction of the **E**-field half-way between the two wires without calculating it.
(b) Calculate the field strength at this point.
(c) Without calculating, compare the **E**-field at a distance $x/3$ from one wire (and $2/3x$ from the other) to the **E**-field at a distance $x/3$ from the first wire (and $2/3x$ from the second).
(d) Calculate the field strength at the locations in (c).

4.18 Electrons in a box. A cubic box of sidelength $a = 1\text{m}$ contains N electrons in a vacuum. You may assume that the **E**-field is normal to the faces of the cube, and that the electric field strength is constant on all faces with a magnitude of $E = 6.03\mu\text{N/C}$. How many electrons are in the box? (Hint: Use Gauss' law on a Gaussian surface equal to the surface of the box.)

4.19 Two line charges.
(a) Find the electrical field from an infinitely long line charge directed along the z-axis through a point $(a, 0, 0)$ with uniform line charge density ρ_l.
(b) Find the electric field from two infinitely long lines charges directed along the z-axis: A line charge with line charge density ρ_l through $(a, 0, 0)$ and a line charge with line charge density $-\rho_l$ through $(-a, 0, 0)$.
(c) Find an approximate value for the electric field when $x^2 + y^2 \gg a^2$ and $y = 0$. Express the result in terms of the dipole moment per unit length, $\mathbf{p} = 2a\rho_l\hat{\mathbf{x}}$.

4.20 Cylinder. Find the electric field from an infinitely long cylinder with radius a and a uniform volume charge density ρ.

Reference

Johannes Skaar. *Elektromagnetisme*. University of Oslo, 2019.

Chapter 5
Polarization and Dielectrics

5.1 Dielectrics

A dielectric or an insulator is a material with very little free charges such as (pure) water, a ceramic, plastics, or glass. We discern between two types of dielectric: *polar* and *non-polar* as illustrated in Fig. 5.1. Polar systems consist of many small, permanent dipoles such as water molecules, that align with an external electric field, whereas a non-polar system consists of many small atoms with charge distributions that are displaced by an imposed electric field. What are the effects of these alignments or displacements?

Non-polar Dielectrics

In a non-polar dielectric without any electric field, the electrons are distributed symmetrically around the positive charges, so that there are no net dipoles. However, if we apply an external electric field, the electron clouds around atoms or molecules tend to be displaced: The negative electron cloud will be displaced in a direction opposite the electric field, whereas the positive nucleus remains fixed in its local crystal configuration. The net result is illustrated in Fig. 5.1. An equilibrium will form between the electric field pulling the electrons and the nucleus apart and the attractive forces between the negative electrons and the positive nucleus drawing them back together. As a result, each atom becomes a small dipole because of a small displacement of the electron cloud around the atom. The stronger the field, the more the electron cloud will be displaced, and the stronger is each dipole. We expect the induced dipole moment **p** of an individual atom to be proportional to the electric field:

$$\mathbf{p} = \alpha \mathbf{E}, \tag{5.1}$$

© The Author(s), under exclusive license to Springer Nature Switzerland AG 2026

A. Malthe-Sørenssen, *Elementary Electromagnetism Using Python*, Undergraduate Texts in Physics, https://doi.org/10.1007/978-3-032-19876-1_5

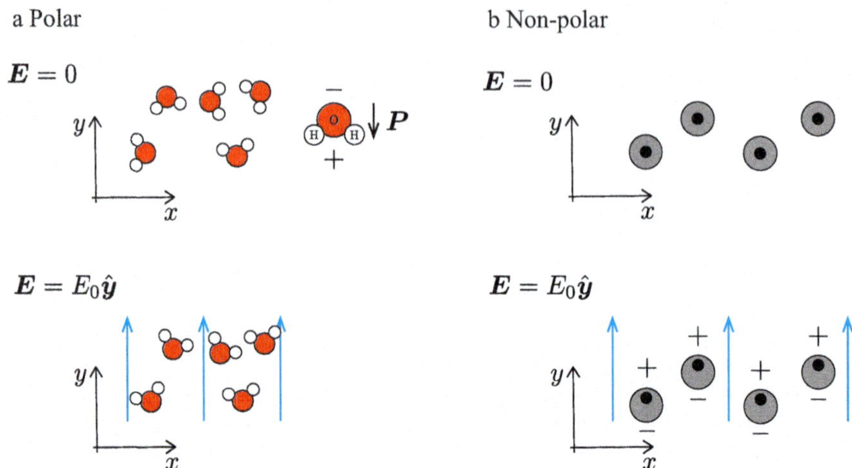

Fig. 5.1 Illustration of the polarization of: **a** a polar material and **b** a non-polar material

where α is called the *atomic polarizability*. The value of α depends on details in the structure of the atoms or the configuration they are in (in a molecule, a crystal, etc.). However, if the electric field is too strong, the atom will be pulled completely apart, ionizing it, such as in a lightning. The free ions and electrons then become a conducting material. (For systems consisting of molecules or crystals, the situation is a bit more complicated as they may polarize more easily in some directions than in other. This may lead to a more complicated relationship between the effective polarization vector **p** and the electric field described by a polarization tensor.)

Simple Model for Atomic Polarizability

As part of our modeling approach, let us see how we can make a simplified model for atomic polarizability. We model an atom as a nucleus with charge q and the surrounding electronic cloud as a uniform, spherical distribution of charge $-q$ with radius a. When the atom is put in an external electric field E_0, the electron cloud is displaced a distance $d \ll a$ relative to the nucleus. In equilibrium, the force on the electron cloud from the external field must be equal to the force from the nucleus on the electron cloud, which again is equal to the force from the electron cloud on the nucleus. What is the electric field inside the electron cloud? We recall that the electric field at a distance r from the center inside a uniform, spherical charge distribution is $E_r(r) = qr/(4\pi\epsilon_0 a^3)$. In equilibrium, the distance between the nucleus and the center of the cloud is d, so that the force on the nucleus is $F = qqd/(4\pi\epsilon_0 a^3)$. This must be equal to the force from the field, qE_0, so that $qE_0 = qqd/(4\pi\epsilon_0 a^3)$. We recall that the dipole moment for two charges q and $-q$ at a distance d is $p = qd$.

We therefore find that $p = qd = E_0 4\pi\epsilon_0 a^3 = \alpha E_0$. For this model, we see that the polarization (i.e. the dipole moment) is proportional to the applied field. It turns out that this model is within a factor four of the correct result for many simple atoms.

Polar Dielectric

In a polar dielectric the electrons are distributed relative to the positive charges so that molecules behave as individual dipoles. An example is water. Each water molecule is a small dipole with a dipole moment **p** that points from the oxygen atom towards the midpoint between the hydrogen atoms. If a polar dielectric is subject to an external electric field, there will be no net force on the dipoles, but the dipole molecules will tend to orient in the electric field with the positive part of the dipole pointing in the direction of the local field. Without an applied electric field, the dipoles will point in random directions, with no net effect, but with an applied electric field, the dipoles will tend to align with the field. The stronger the field, the stronger will be the alignment and the stronger the net dipole.

Let us address in detail what happens when a molecule such as a water molecule is placed in an electric field **E**. We model the water as a two-charge dipole with a charge q at a position $\mathbf{d}/2$ relative to the center of the molecule, in a direction toward the midpoint between the hydrogen atoms, and a charge $-q$ at a position $-\mathbf{d}/2$ relative to the center of the molecule, in a direction toward the oxygen atom. If the field is uniform, the net force from the field is $q\mathbf{E} - q\mathbf{E} = 0$. However, the torque around the center of the molecule will be

$$\tau = \mathbf{d}/2 \times q\mathbf{E} + (-\mathbf{d}/2) \times -q\mathbf{E} = q\mathbf{d} \times \mathbf{E}. \tag{5.2}$$

We replace $\mathbf{p} = q\mathbf{d}$ so that

$$\tau = \mathbf{p} \times \mathbf{E}. \tag{5.3}$$

This makes the molecule rotate so that the dipole moment will align with the electric field.

Polarization

The response of both a polar and a non-polar dielectric due to an external field is the formation of a set of dipoles that point more or less in the direction of the electric field. We say that the material becomes *polarized*. We describe the effect by the *polarization* **P**, which is the dipole moment per unit volume. The sum of the dipole moments of all the dipoles (with index i) in a small volume element dv is then:

$$\mathbf{P}\,\mathrm{d}v = \sum_{i \text{ in } \mathrm{d}v} \mathbf{p}_i = N_v\,\mathrm{d}v \frac{1}{N_v\,\mathrm{d}v} \sum_{i \text{ in } \mathrm{d}v} \mathbf{p}_i = N_v\,\mathrm{d}v \langle \mathbf{p} \rangle \;, \tag{5.4}$$

where $N_v\,\mathrm{d}v$ is the number of dipoles in the element $\mathrm{d}v$ and $\langle \mathbf{p} \rangle$ is the average dipole moment in the volume $\mathrm{d}v$.

The polarization vector depends on the total field. The mechanisms we have described for polarization implies that the polarization vector in a point $\mathbf{r}$ depends on the electric field in this point. We write this the following way: $\mathbf{P}(\mathbf{E})$. The polarization vector is a function of the (local) electric field. Notice that the polarization is a function of the *total* electric field – including the field set up by all the dipoles in the material. This may initially seem a bit counterintuitive. The polarization is a function of the total electric field, but the polarization will also affect the electric field, which again will affect the polarization. We often call such relationships, or equations, self-consistent solutions. Mathematically, it is not difficult to set up such a relation, and we also understand the physics of this process: The polarization is due to the electric field that affects the atoms/molecules, and this electric field must be the total electric field, including the field set up by the displacement of the molecules themselves.

Linear dielectrics. We saw that for a single atom the polarization was aligned with the electric field $\mathbf{p} = \alpha \mathbf{E}$ and for polar dielectrics we also expect the effective polarization to be aligned with the electric field. Materials that have this property also on the macroscopic scale are called *linear dielectrics*. We characterize a linear dielectric by the electric susceptibility χ_e:

$$\mathbf{P} = \chi_e \epsilon_0 \mathbf{E}. \tag{5.5}$$

The electric susceptibility is unitless. In vacuum $\chi_e = 0$ and in air $\chi_e \simeq 0$. Many materials are linear dielectrics, but not all. We previously mentioned that some molecules may have an anisotropic displacement of the charges when in an electric field, and similar effects may occur in a crystal, where the electron cloud may be more easily displaced in some directions. These effects may imply that the dielectric is anisotropic or non-linear. Most materials we discuss in this text are linear dielectrics.

Linear dielectric

In a linear dielectric the polarization, $\mathbf{P}$, in a point is proportional to the total electric field in that point, $\mathbf{E}$:

$$\mathbf{P} = \chi_e \epsilon_0 \mathbf{E}. \tag{5.6}$$

Here, $\mathbf{P}\,\mathrm{d}v$ is the net dipole moment in the small volume $\mathrm{d}v$ and χ_e is the electric susceptibility.

Test your understanding
(a) A sphere with radius a has a uniform polarization $\mathbf{P} = P_0\hat{\mathbf{z}}$. What is the total dipole moment $\mathbf{p}$ of this sphere? (b) A charge $+Q$ is held close to a cube with side L made from a dielectric material. In which direction is the net force on the cube from the charge? Attractive, repulsive or zero? (c) How would your answer in (a) change if the charge is negative, $-Q$?[1]

5.2 Bound Charges

We have now argued that the effect of an applied field on a dielectric material is a reorientation of permanent dipoles (for a polar dielectric), or a local displacement of the electron cloud to create dipoles (non-polar dielectric). While there are no free charges in the material, the small induced or aligned dipoles will lead to a distribution of charges in the dielectric material. We call these charges *bound charges*, because they are bound to the atoms or molecules and are not free to move around. Our plan is to relate the polarization $\mathbf{P}$ to a density of bound volume and surface charges. We can then use this distribution of charges and Coulomb's law or Gauss' law to calculate the electric field due to the bound charges, that is, the electric field due to the polarization.

Figure 5.2 illustrates the alignment of dipoles that make up the polarization $\mathbf{P}$ inside a dielectric material. There are no net charges inside the volume, only the bound charges. Each small dipole in itself does not contribute any net charge—it has a positive and a negative end, with equal and opposite charges. However, if we look at a part of the system, such as the part inside the volume in Fig. 5.2a, then the bounding surface intersects some of the dipoles so that one side of the dipole is inside the volume and one side is outside the volume. This will lead to sets of net charges on the inside (and outside) of the surface.

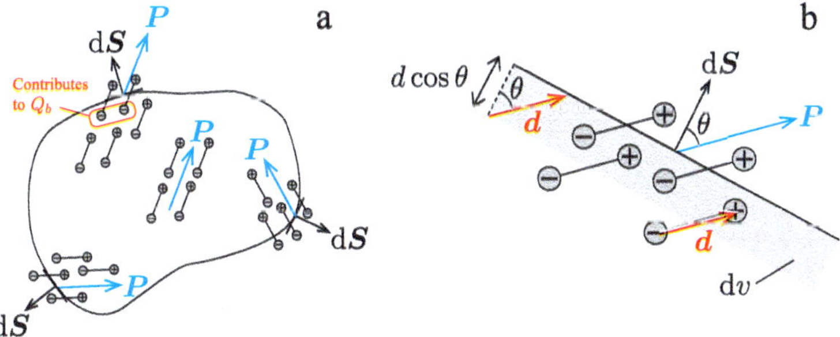

Fig. 5.2 Illustration of the bound charges inside a volume v. **a** a small volume element dv

[1] (a) $4\pi a^3\mathbf{P}/3$; (b) Attractive; (c) Attractive.

Bound charge within a volume. How much net bound charge is within the volume near the surface? We look at a small surface element $\mathrm{d}\mathbf{S}$ with surface normal $\hat{\mathbf{n}} = \mathrm{d}\mathbf{S}/|\mathrm{d}\mathbf{S}|$ as illustrated in Fig. 5.2b. In this point, the polarization vector is $\mathbf{P}$. We assume that the polarization is due to small dipoles with dipole moment $\mathbf{p} = Q\mathbf{d}$. We see that only dipoles that cross the surface contribute to the bound charge within the volume. Dipoles that are completely within the volume do not contribute to the net charge inside the volume since both the negative and the positive charges in the dipole are inside the volume and the *net* contribution is therefore zero. However, dipoles that cross the surface contribute to the net charge because only one part of the dipole is inside the volume. We see that all dipoles that are within a range $d\cos\theta = \mathbf{d}\cdot\hat{\mathbf{n}}$ cross the surface and hence contribute to the net bound charge inside the volume. In Fig. 5.2b, each dipole contributes a charge $-Q$ to the bound charge. If N_v is the number of dipoles per unit volume, then the number of dipoles in a volume $\mathrm{d}v = d\cos\theta\,\mathrm{d}S$ is $N_v\,\mathrm{d}v$. We can rewrite this as

$$N_v\,\mathrm{d}v = N_v d\cos\theta\,\mathrm{d}S = N_v\mathbf{d}\cdot\hat{\mathbf{n}}\,\mathrm{d}S = N_v\mathbf{d}\cdot\,\mathrm{d}\mathbf{S}, \tag{5.7}$$

The contributions from these dipoles to the bound charge $\mathrm{d}Q_b$ in the volume $\mathrm{d}v$ is then

$$\begin{aligned}\mathrm{d}Q_b &= (-Q)N_v\,\mathrm{d}v = -QN_v\mathbf{d}\cdot\,\mathrm{d}\mathbf{S} = -N_v(Q\mathbf{d})\cdot\,\mathrm{d}\mathbf{S}\\ &= -N_v\mathbf{p}\cdot\,\mathrm{d}\mathbf{S} = -\mathbf{P}\cdot\,\mathrm{d}\mathbf{S},\end{aligned} \tag{5.8}$$

where we have used that $p = Q\mathbf{d}$ and that $N_v\,\mathrm{d}v\mathbf{p} = \mathbf{P}\,\mathrm{d}v$ and therefore that $N_v\mathbf{p} = \mathbf{P}$. In order to find the net bound charge from the whole surface, we integrate over the surface:

$$Q_b = -\oint_S \mathbf{P}\cdot\,\mathrm{d}\mathbf{S}. \tag{5.9}$$

We call this the *bound charge* because this charge is bound to the material. It is not free to move. And it is a charge that appears due to the application of an external electric field. Without an electric field, there will be no bound charge. However, this bound charge is real in the sense that it also sets up an electric field.

Bound Volume Charge Density

In (5.9) we found the total bound charge Q_b in a volume v enclosed by the surface S. We can rewrite this in terms of the volume charge density of the bound charge $\rho_{v,b}$:

$$Q_b = -\int_S \mathbf{P}\cdot\,\mathrm{d}\mathbf{S} = \int_v \rho_{v,b}\,\mathrm{d}v. \tag{5.10}$$

We can apply the divergence theorem to this surface integral, getting

$$Q_b = -\int_v \nabla \cdot \mathbf{P}\, \mathrm{d}v = \int_v \rho_{v,b}\, \mathrm{d}v. \tag{5.11}$$

This is valid for any surface S enclosing a volume v, and therefore the arguments of the integrals must also be equal:

$$-\nabla \cdot \mathbf{P} = \rho_{v,b}. \tag{5.12}$$

This implies that in a region where the polarization $\mathbf{P}$ is uniform, the divergence is zero, and hence the volume density of bound charges is zero, $\rho_{v,b} = 0$.

Bound Surface Charge Density

What is the surface charge density on the interface of a dielectric, that is, on the boundary between a dielectric and vacuum? We look at a small volume Δv of height Δh and area ΔS as illustrated in Fig. 5.3. We assume that there is a surface charge density of bound charge, $\rho_{s,b}$, on the interface surface. The bound charge inside the volume Δv is therefore $\Delta Q_b = \rho_{s,b}\Delta S$. From (5.9) we know that the bound charge inside the volume is also given as the integral of $\mathbf{P}$ over the enclosing surface S:

$$\Delta Q_b = \rho_{s,b}\Delta S = -\oint_S \mathbf{P} \cdot \mathrm{d}\mathbf{S} \tag{5.13}$$

As $\Delta h \to 0$, the contributions to the integral from the side walls of the cylinder goes to zero. Vacuum is not dielectric and the polarization vector, $\mathbf{P}$, there is zero. The contribution to the integral from the top surface is therefore zero. We are therefore left with the contribution to the integral from the bottom surface inside the dielectric. This surface points into the dielectric away from the interface, thus the surface element is $\Delta\mathbf{S} = -\Delta S\hat{\mathbf{n}}$. The integral is therefore approximately:

$$\rho_{s,b}\Delta S = -\oint_S \mathbf{P} \cdot \mathrm{d}\mathbf{S} = -\mathbf{P} \cdot \Delta\mathbf{S} = \mathbf{P} \cdot \hat{\mathbf{n}}\, \Delta S. \tag{5.14}$$

The surface charge density of bound charge is therefore

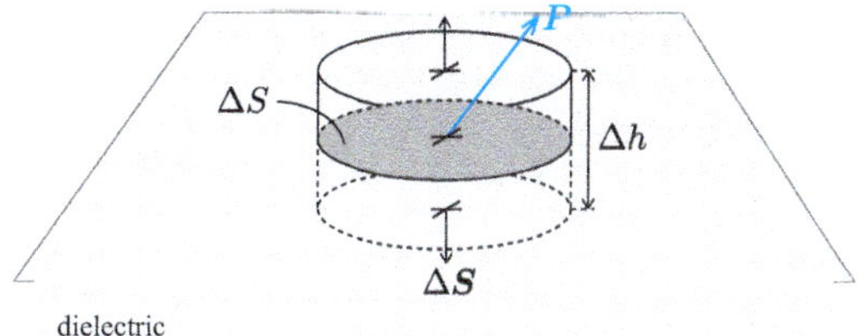

Fig. 5.3 Illustration of the bound charges inside a small volume $\mathrm{d}v$ across an interface between a dielectric (bottom) and vacuum (top)

$$\rho_{s,b} = \mathbf{P} \cdot \hat{\mathbf{n}} \tag{5.15}$$

Bound charge density

The *bound charge* enclosed by a closed surface S is

$$\Delta Q_b = -\oint_S \mathbf{P} \cdot d\mathbf{S}. \tag{5.16}$$

The *bound volume charge density* $\rho_{v,b}$ is

$$\rho_{v,b} = -\nabla \cdot \mathbf{P}. \tag{5.17}$$

The *bound surface charge density* $\rho_{s,b}$ at the surface of a dielectric material with surface normal $\hat{\mathbf{n}}$ is:

$$\rho_{s,b} = \mathbf{P} \cdot \hat{\mathbf{n}}. \tag{5.18}$$

Test your understanding

(a,b) The figure shows how the polarization $\mathbf{P}$ varies through a material. Is the bound volume charge density $\rho_{v,b}$ and the bound surface charge density at the top surface, $\rho_{s,b}$ positive, negative or zero? (c) The figure shows a dielectric body with polarization $\mathbf{P} = P_0\hat{\mathbf{z}}$. What is the bound surface charge density on the bottom surface? (d) The figure shows a dielectric sphere with polarization $\mathbf{P} = P_0\hat{\mathbf{z}}$. What is the bound surface charge density and the bound volume charge density of the sphere.[2]

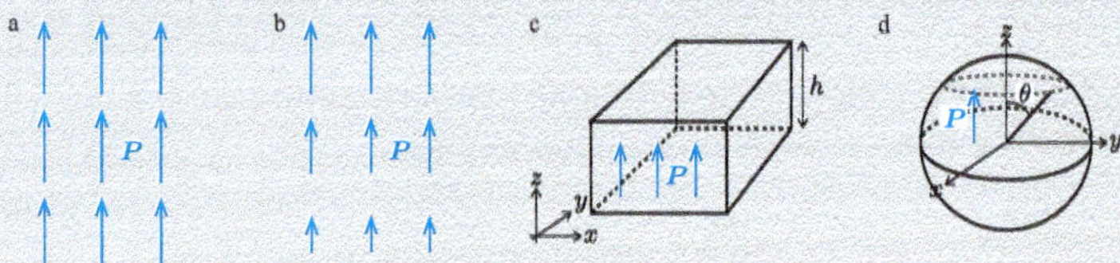

Example: Infinite Polarized Slab

An infinite dielectric slab of thickness d has a uniform polarization $\mathbf{P}$ *in the direction normal to the slab surface. Find the bound surface charge densities and the electric field in the slab.*

[2] (a) $\rho_{v,b} = 0$, $\rho_{s,b} > 0$; (b) $\rho_{v,b} < 0$, $\rho_{s,b} > 0$; (c) $\rho_{s,b} = -P_0$; (d) $\rho_{s,b} = -P_0 \cos\theta$; $\rho_{v,b} = 0$.

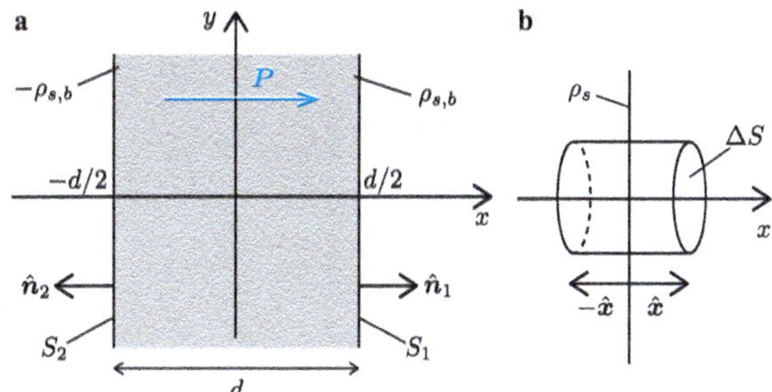

Fig. 5.4 Illustration of an infinite dielectric slab of thickness d with a uniform polarization $\mathbf{P}$

Specifying the problem. Figure 5.4a illustrates the geometry of the system. We choose the x-direction to be along $\mathbf{P}$, so that $\mathbf{P} = P\hat{\mathbf{x}}$.

Finding bound charges from the polarization. Because the polarization is uniform inside the slab, we know that $\rho_{v,b} = -\nabla \cdot \mathbf{P} = 0$ inside the slab. However, there may still be a bound surface charge density. For surface S_1, the surface normal is $\hat{\mathbf{n}}_1 = \hat{\mathbf{x}}$. We find the surface charge density from $\rho_{s,b} = \mathbf{P} \cdot \hat{\mathbf{n}}_1 = P$. Similarly, for surface S_2 we have $\hat{\mathbf{n}}_2 = -\hat{\mathbf{x}}$ and $\rho_{s,b} = \mathbf{P} \cdot \hat{\mathbf{n}}_2 = P\hat{\mathbf{x}} \cdot -\hat{\mathbf{x}} = -P$.

Applying Gauss' law. We can then use the surface charge densities to find the electric field. For a surface charge density ρ_s on an infinite surface, we recall that the electric field only has a component normal to the surface, $\mathbf{E} = E_x(x)\hat{\mathbf{x}}$, and that the field is mirror symmetric around the surface: $E_x(x) = -E_x(-x)$. We find the electric field from Gauss' law on a cylindrical Gauss surface as illustrated in Fig. 5.4b. Only the cylinder end surfaces ΔS contribute to the flux $\Phi = 2E_x \Delta S = \rho_s \Delta S/\epsilon_0$ and therefore $E_x = \rho_s/(2\epsilon_0)$.

Electric field from bound charges. We find the electric field from the superposition of the field from the two bound surface charge densities $\rho_{s,b}$ and $-\rho_{s,b}$. We find that for $x > d/2$, the net field is $E_x = \rho_s/(2\epsilon_0) - \rho_s/(2\epsilon_0) = 0$. For $-d/2 < x < d/2$, the net field is $E_x = -\rho_s/(2\epsilon_0) - \rho_s/(2\epsilon_0) = -\rho_s/\epsilon_0 = -P/\epsilon_0$. For $x < -d/2$, the two contributions cancel, $E_x = -\rho_s/(2\epsilon_0) + \rho_s/(2\epsilon_0) = 0$. The electric field (from the polarization $\mathbf{P}$) is therefore $\mathbf{E} = -(P/\epsilon_0)\hat{\mathbf{x}}$ inside the slab.

Method: Finding the electric field from polarization

When dielectric materials are subject to electric fields, they will become polarized with a polarization $\mathbf{P}$. We can find the density of bound charges from the polarization $\mathbf{P}$:

- The volume density of bound charge, $\rho_{v,b} = -\nabla \cdot \mathbf{P}$
- The surface density of bound charge at an interface between the dielectric and vacuum (or air) is $\rho_{s,b} = \mathbf{P} \cdot \hat{\mathbf{n}}$, where $\hat{\mathbf{n}}$ is the surface normal pointing from the dielectric into to vacuum/air.

We can then calculate the electric field $\mathbf{E}$ and the electric potential V from the charge densities by integrating Coulomb's law on the appropriate form as we have learned previously.

5.3 Generalized Gauss' Law

The polarization $\mathbf{P}$ depends on the electric field, where for linear dielectric media $\mathbf{P} = \chi_e \epsilon_0 \mathbf{E}$, and we can find the bound charges from the polarization. How can we put all of this in a common framework? We can do this by rewriting Gauss' law to include the bound charges due to polarization in addition to the free charges, and then use the same methods we developed previously to find the electric field.

Gauss' law states that

$$\oint_S \mathbf{E} \cdot d\mathbf{S} = \frac{Q_{\text{in}}}{\epsilon_0} = \frac{Q_f + Q_b}{\epsilon_0}. \tag{5.19}$$

where Q_{in} is the total charge inside the volume, which is the sum of the free charges and the bound charges. What are the free charges? We use the word *free charge* to specify the charges we discussed in previous chapters and contrast these charges from the *bound* charges that are due to polarization of a dielectric material. We would like to have a version of Gauss' law that only includes the free charges, which are the ones that we know where are without solving an equation for the polarization. We rewrite Gauss' law by replacing the bound charges Q_b with the integral over the polarization:

$$Q_b = -\oint_S \mathbf{P} \cdot d\mathbf{S}. \tag{5.20}$$

getting

$$\epsilon_0 \oint_S \mathbf{E} \cdot d\mathbf{S} = Q_f - \oint_S \mathbf{P} \cdot d\mathbf{S}. \tag{5.21}$$

We rewrite so that only the free charge is on the right-hand side:

$$\oint_S (\epsilon_0 \mathbf{E} + \mathbf{P}) \cdot d\mathbf{S} = Q_f. \tag{5.22}$$

Displacement field

We introduce the *displacement field* or simply the $\mathbf{D}$-field:

$$\mathbf{D} = \epsilon_0 \mathbf{E} + \mathbf{P}. \tag{5.23}$$

Table 5.1 Values for the relative permittivity, ϵ_r, and dielectric strength, E_c, for various materials

Material	ϵ_r	E_c(MV/m)	Material	ϵ_r	E_c(MV/m)
Vacuum	1		Water	81	
Air	1.0005	3	Porcelain	6	12
Paper	3–5	16	Rubber (neoprene)	6.7	12
Wood	2–5	10	Mica	7	150
Polystyrene	2.5	24	Strontium titanate	300	8
Oil	4	12	Barium titanate	1200	100
Glass (pyrex)	5	14			

Gauss' law in a dielectric material then becomes:

$$\oint_S \mathbf{D} \cdot d\mathbf{S} = Q_{\text{free in } S}, \tag{5.24}$$

where $Q_{\text{free in } S}$ is the free (non-bound) charge inside the surface S.

Linearly polarizable media. For a linearly polarizable media $\mathbf{P} = \epsilon_0 \chi_e \mathbf{E}$. For such a medium we get

$$\mathbf{D} = \epsilon_0 \mathbf{E} + \mathbf{P} = \epsilon_0 \mathbf{E} + \epsilon_0 \chi_e \mathbf{E} = \epsilon_0 (1 + \chi_e) \mathbf{E} = \epsilon_r \epsilon_0 \mathbf{E} = \epsilon \mathbf{E}, \tag{5.25}$$

where $\epsilon_r = (1 + \chi_e)$ is called the *relative permittivity* and $\epsilon = \epsilon_r \epsilon_0$ is called the *absolute permittivity* or simply the permittivity.

In general, a material is *linearly polarizable* if the relation between **P** and **E** is linear. If the material is also *isotropic*, the relation does not depend on the direction of **E** and χ_e and ϵ_r are scalars. If χ_e does not depend on the position in space we say that the medium is *homogeneous*.

Effect of dielectric on the electric field. A distribution of free charges set up an electric field **E** in vacuum. If we replace the vacuum with a dielectric material with relative permittivity ϵ_r, what happens to the electric field inside the dielectric? We notice that **D** only depends on the free charges. The **D**-field is therefore the same in both situations. However, the electric field will be $\mathbf{E} = \mathbf{D}/(\epsilon_r \epsilon_0)$ in the dielectric. This is a factor $1/\epsilon_r$ smaller than in vacuum.

Typical values of ϵ_r. Typical examples of linear, isotropic media are water, air, gases, and glass. Examples are shown in Table 5.1. (Crystals are often anisotropic). You can find the relative permittivity of many materials in physical tables.

Test your understanding
(a) A set of small charges have been inserted into a dielectric rod, resulting in a spatial charge density ρ. In order to find $\mathbf{D}$ in this rod, how will you treat the charge density ρ? As free charges; as bound charges, neither, ρ is the sum of free and bound charges. (b) If you place a dielectric material in an external field $\mathbf{E}_e$, a field $\mathbf{E}_p$ will be set up by the bound charges. The superposition of these fields is $\mathbf{E}_t = \mathbf{E}_e + \mathbf{E}_p$. Which of these three fields are used in the formula $\mathbf{D} = \epsilon_0 \mathbf{E} + \mathbf{P}$?[3]

Gauss' Law on Differential Form

Gauss' law can be rewritten on differential form using the divergence theorem:

$$\oint_S \mathbf{D} \cdot \mathrm{d}\mathbf{S} = \int_v \nabla \cdot \mathbf{D} \, \mathrm{d}v = Q_{in} = \int_v \rho_{\text{free}} \, \mathrm{d}v \tag{5.26}$$

This gives us:

Gauss' law on differential form

Gauss' generalized law on differential form is

$$\nabla \cdot \mathbf{D} = \rho_{\text{free}}, \tag{5.27}$$

where it is common just to use ρ for the free charges.

There is no Coulomb's law for D. Notice that even if Gauss' law for the displacement field $\mathbf{D}$ looks very similar to Gauss' law for the electric field—you simply have to replace the total charge ρ with the free charges ρ_{free}—it is in general not possible to find the displacement field by integrating the free charge density in space, as we did for the electric field. This is because the divergence alone is not sufficient to determine the field: you also need to know the curl of the field. For the electric field, the curl is zero, but the curl of the displacement field is not always zero. The curl of the displacement field is

$$\nabla \times \mathbf{D} = \epsilon_0 \left(\nabla \times \mathbf{E}\right) + \left(\nabla \times \mathbf{P}\right) = \nabla \times \mathbf{P}. \tag{5.28}$$

and the curl of $\mathbf{P}$ is not always zero. This means that in general there does not exist any scalar potential for $\mathbf{D}$.

[3] (a) As free charges; (b) $\mathbf{E}_t$.

Method: Using Gauss' law for linear dielectrics

Updated method for applying Gauss' law to linear dielectric materials:

- Find a set of surfaces that enclose a volume such that $\mathbf{D} \cdot \hat{\mathbf{n}}$ is constant on each such surface element. (It may be zero on some of the surfaces—zero is a constant!) For linear dielectrics you may use your physics knowledge of the symmetries of **E**, because $\mathbf{D} = \epsilon \mathbf{E}$.
- This often requires that you find a simplified description of the field in a chosen coordinate system, such as $\mathbf{D} = D_r(r)\hat{\mathbf{r}}$.
- Find the flux integral.
- Use Gauss' law to find the displacement field as a function of charge and position.
- Find the electric field from $\mathbf{E} = \mathbf{D}/\epsilon$.
- Notice that the surface does not have to enclose all the charges—it is allowed and indeed often necessary to chose a surface that contains only some of the charges. However, the electric field must be a constant on the surfaces you have chosen.

Dielectric Breakdown

Materials are dielectric only up to a given field strength. From the models we proposed above, it is clear that if the electric field becomes too large, the electron cloud will be pulled completely away from the positive charges in the nucleus, ionizing the material. This leads to the formation of free charges, and the material becomes a conductor. In many cases this gives a channel of ionized electrons which is often seen as a lightning. For air this happens for fields that are larger than $E \geq 3\ 10^6$V/m. This limit is called the *dielectric strength*, E_c. We will look at dielectric breakdown in detail later when we address lightning strikes. You can find E_c for some materials in Table 5.1.

Example: Point Charge in a Dielectric Medium

Find the electric field from a point charge Q in the origin in a linear, isotropic dielectric material with permittivity ϵ. Find the bound surface charge density on the outer interface of a spherical dielectric of radius a and dielectric constant ϵ.

Applying Gauss' law. We apply the same approach as we did for Gauss' law for a single charge. Because $\mathbf{D} = \epsilon \mathbf{E}$, we can assume that the **D**-field has the same symmetries as the **E**-field. We use Gauss' law on a sphere with radius r:

$$\oint_S \mathbf{D} \cdot d\mathbf{S} = D4\pi r^2 = Q, \tag{5.29}$$

where Q is the free charge. This gives

$$\mathbf{E} = \frac{\mathbf{D}}{\epsilon} = \frac{Q}{4\pi \epsilon r^2}\hat{\mathbf{r}}. \tag{5.30}$$

Discussing variations of the problem. What if the dielectric medium only stretches out a distance a? In that case, the electric field is given by the expression above for $r < a$. When $r > a$ the field is the same as for a point charge in vacuum, that is,

$$\mathbf{E} = \begin{cases} \frac{Q}{4\pi\epsilon r^2}\hat{\mathbf{r}} & r < a \\ \frac{Q}{4\pi\epsilon_o r^2}\hat{\mathbf{r}} & r > a \end{cases} \tag{5.31}$$

Bound charges at an interface. What happens at the boundary between the two regions? We expect that there will be surface charges on the spherical surface at $r = a$. For $r < a$, the electric field is $\mathbf{E} = Q/(4\pi\epsilon r^2)\hat{\mathbf{r}}$. The polarization is related to $\mathbf{E}$ through $\epsilon_0 \mathbf{E} + \mathbf{P} = \epsilon_0\epsilon_r \mathbf{E}$, which gives $\mathbf{P} = \epsilon_0(\epsilon_r - 1)\mathbf{E}$. We can therefore find the surface charge density when $r \to a$ as

$$\rho_s = \mathbf{P} \cdot \hat{\mathbf{n}} = \epsilon_0(\epsilon_r - 1)E = \epsilon_0(\epsilon_r - 1)\frac{Q}{4\pi\epsilon_0\epsilon_r a^2} = \frac{(\epsilon_r - 1)}{\epsilon_r}\frac{Q}{4\pi a^2}. \tag{5.32}$$

Method: Find the bound charges in a linear dielectric

If we have found $\mathbf{D}$, we can also find both the electric field $\mathbf{E} = \mathbf{D}/\epsilon$ and the polarization $\mathbf{P} = (\epsilon - \epsilon_0)\mathbf{E}$. We can then use this to find the bound volume charge density from $\rho_{v,b} = -\nabla \cdot \mathbf{P}$ and the bound surface charge densities from $\rho_{s,b} = \mathbf{P} \cdot \hat{\mathbf{n}}$, where $\hat{\mathbf{n}}$ is the normal vector on the interfacial surfaces.

Example: Coaxial Cable Filled with a Dielectric Medium

Find the electric field inside and outside a coaxial cable consisting of an inner cylinder with radius a and charge density ρ, and an outer cylindrical shell with radii b and c and charge density $-\rho$. The region between the inner and outer shells are filled with a dielectric with permittivity ϵ.

Draw the system and choose approach. The system is illustrated in Fig. 5.5. We assume that material from $0 < r < a$ has permittivity ϵ_0, that the material from

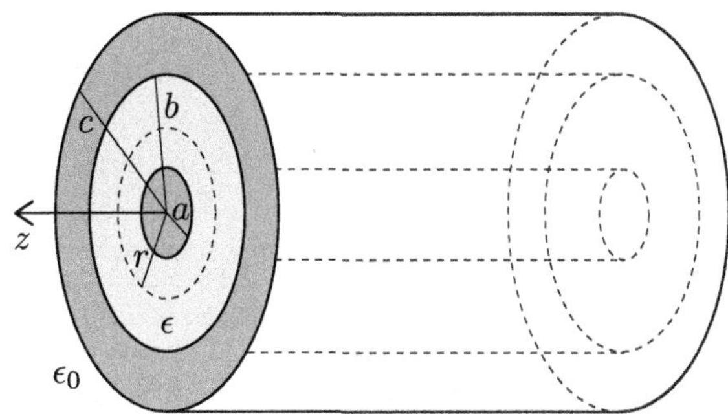

Fig. 5.5 Illustration of a coaxial cable

$a < r < b$ has permittivity ϵ, and that the material from $b < r$ has permittivity ϵ_0. We plan to use Gauss' law on integral form.

Symmetry. We notice that the system has cylindrical symmetry. For Gauss' law, we therefore plan to use a cylinder surface with radius r and length L as the Gauss surface. Due to the symmetry, we assume that the **E**-field, and therefore also the **D**-field, only have radial components. The flux through the top and bottom surfaces of a cylinder of length L will therefore be zero, since the field will be normal to the surface normal here. We notice that $\mathbf{D} \cdot \mathrm{d}\mathbf{S} = D_r(r)\,\mathrm{d}S$. We can therefore pull this factor outside the surface integral when we apply Gauss' law:

$$\oint_S \mathbf{D} \cdot \mathrm{d}\mathbf{S} = 2\pi r L D_r = Q \tag{5.33}$$

We address the different regions separately:

When $r < a$, the charge density is ρ so that the charge inside the cylindrical Gauss surface of radius r is $Q = \pi r^2 L\rho$, where $\pi r^2 L$ is the volume of the cylinder. We therefore find

$$2\pi r L D_r = \pi r^2 L\rho \Rightarrow D_r = \frac{1}{2}\rho r. \tag{5.34}$$

We use that $\mathbf{E} = \mathbf{D}/\epsilon_0$, giving

$$E_r = \frac{1}{2\epsilon_0}\rho r \tag{5.35}$$

When $a < r < b$, the charge density is zero, and the charge inside the Gauss surface is the charge inside the inner cylinder with radius a: $Q = \pi a^2 L\rho$. In this region, the dielectric constant is ϵ:

$$2\pi r L D_r = \pi a^2 L\rho \Rightarrow D_r = \frac{\rho a^2}{2r} \Rightarrow E_r = \frac{\rho a^2}{2\epsilon r}\,. \tag{5.36}$$

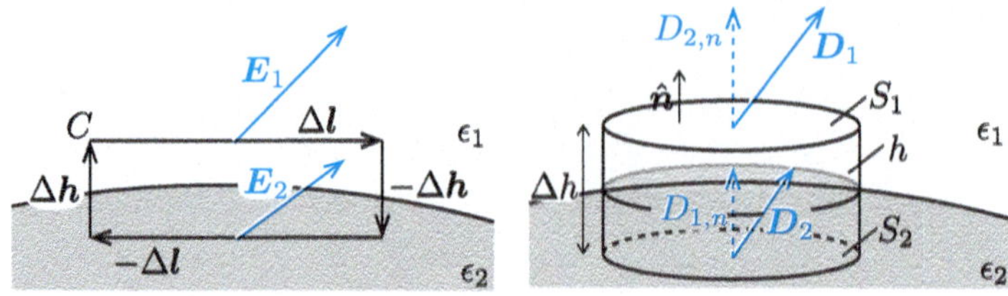

Fig. 5.6 Illustration of the boundary conditions for the **E** and **D** fields

When $b < r < c$, the charge inside is the Gauss surface is the sum of the charge inside the inner cylinder plus the part of the outer cylinder that is inside the Gauss surface at r:

$$Q = \pi a^2 L\rho + (r^2 - b^2)(-\rho)L. \tag{5.37}$$

The **D**-field is therefore

$$2\pi r L D_r = \pi a^2 L\rho - \pi r^2 \rho L + \pi b^2 \rho L \Rightarrow D_r = \frac{\rho}{2r}\left(a^2 + b^2 - r^2\right). \tag{5.38}$$

and similarly for the electric field:

$$E_r = \frac{\rho}{2\epsilon_0 r}\left(a^2 + b^2 - r^2\right) \tag{5.39}$$

When $c < r$, the charge inside the Gauss surface is the sum of the charges in the inner and the outer cylinder: $Q = \pi a^2 L\rho - \pi(c^2 - b^2)L\rho$. If this is non-zero, there will be a field outside the cylinder, otherwise the field is zero:

$$D_r = \frac{\rho}{2r}\left(a^2 + b^2 - c^2\right),\ E_r = \frac{\rho}{2\epsilon_0 r}\left(a^2 + b^2 - c^2\right). \tag{5.40}$$

What value of ϵ to use. Notice that the value of ϵ that is used for the conversion between **D** and **E** is the local value of the dielectric constant.

5.4 Boundary Conditions for E and D

What happens on the interface between two (dielectric) media, such as the interface between a plastic and air, or between water and your skin? If we know the electric field **E** and the displacement field **D** on one side of the interface, what are they on the other side? The situation is illustrated in Fig. 5.6.

Tangential boundary condition. We use that the line integral of **E** over a closed loop is zero to relate the field immediately outside the interface to field immediately inside the interface. We construct an integration loop C with a very small width Δh

consisting of two line segments normal to the interface, $\Delta\mathbf{h} = \Delta h\hat{\mathbf{n}}$ and $-\Delta\mathbf{h} = -\Delta h\hat{\mathbf{n}}$, and two tangential line segments $\Delta\mathbf{l}$ and $-\Delta\mathbf{l}$, as illustrated in Fig. 5.6. In the limit when the width Δh goes to zero, the only contributions to the integral are from the two tangential pieces, $\Delta\mathbf{l}$ and $-\Delta\mathbf{l}$:

$$\oint_C \mathbf{E}\cdot d\mathbf{l} \simeq \mathbf{E}\cdot\Delta\mathbf{h} + \mathbf{E}_1\cdot\Delta\mathbf{l} - \mathbf{E}\cdot\Delta\mathbf{h} - \cdot\mathbf{E}_2\cdot\Delta\mathbf{l} \rightarrow_{\Delta h\rightarrow 0} \mathbf{E}_1\cdot\Delta\mathbf{l} - \cdot\mathbf{E}_2\cdot\Delta\mathbf{l}. \tag{5.41}$$

The line integral of **E** around a closed loop is always zero. The presence of a surface charge density ρ_s on the surface does not change this. Therefore

$$\oint_C \mathbf{E}\cdot d\mathbf{l} = \mathbf{E}_1\cdot\Delta\mathbf{l} - \mathbf{E}_2\cdot\Delta\mathbf{l} = 0 \tag{5.42}$$

Since $\Delta\mathbf{l}$ is tangential to the surface, this means that the tangential components of the electric fields are equal:

$$E_{1t} = E_{2t}. \tag{5.43}$$

Normal boundary condition. We use Gauss' law to relate the normal component of the field on each side of the interface to the free charge on the interface, ρ_s. We construct an Gauss surface in the form of a cylinder S of height Δh and base area ΔS. We apply Gauss' law for dielectrics, getting

$$\oint_S \mathbf{D}\cdot d\mathbf{S} = \mathbf{D}_1\cdot\hat{\mathbf{n}}\Delta S + \mathbf{D}_2\cdot(-\hat{\mathbf{n}})\Delta S + \int_h \mathbf{D}\cdot d\mathbf{S} = q = \rho_s\Delta S. \tag{5.44}$$

The surface S consists of the top and bottom surfaces of the cylinder, S_1 and S_2, and the curved cylinder surface, h. As the height Δh of the cylinder approaches zero, the contribution to the integral from the curved cylinder surface h becomes negligible. Here, we have introduced $\hat{\mathbf{n}}$ as a normal vector to the top and bottom surfaces. We divide by ΔS on both sides and conclude that the normal components of **D** are related to the surface charge density of *free charges*, ρ_s, on the interface:

$$D_{1n} - D_{2n} = \rho_s. \tag{5.45}$$

Two dielectric media. If both media are dielectric and there are no free surface charges, then the normal components of **D** do not charge across the surface: $D_{1n} = D_{2n}$. However, the normal components of the electric field will typically change across an interface between two different dielectric materials, due to the bound charges at the interface. For example, for a point charge in a dielectric sphere of radius a, there is no change in D_n across the interface at a, but there will be a change in E_n.

Boundary conditions for the electric field

At an interface between a dielectric material 1 and a dielectric material 2 with a surface normal $\hat{\mathbf{n}}$ pointing from material 2 to material 1, we have the following boundary conditions:

$$E_{1,t} = E_{2,t}, \tag{5.46}$$

and

$$\mathbf{D}_1 \cdot \hat{\mathbf{n}} - \mathbf{D}_2 \cdot \hat{\mathbf{n}} = \rho_s. \tag{5.47}$$

Example: Cavities in a Dielectric Material

A linear dielectric has an electric field $\mathbf{E} = E_0\hat{\mathbf{z}}$ *and a displacement field* $\mathbf{D} = \epsilon\mathbf{E}$. *There are no free charges. The dielectric has either a thin cylindrical hole or a wide, short cylinderical hole directed with their axes along the field. What are the fields inside the holes (cavities) in the two cases?*

Specification of problem. We plan to use the boundary conditions of the **E** and **D** fields at the boundaries. We assume that the holes are small and do not perturb the **E** or **D**-fields inside the dielectric. The relevant boundary for the long, thin cylinder is the boundary along the long side, because the top boundary is very small and only affects the field close to the top and the bottom. Most of the cavity is only close to the side boundaries. Similarly, the relevant boundary for the wide, short cylinder is on the top and bottom of the cylinder, because the side boundary is very small, and only contributes to the field close to the sides. Most of the cavity is only close to the top and bottom surfaces.

The long, thin cylinder. The electric field inside the dielectric is directed along the long, thin cylinder. This means that the electric field only has a tangential component inside the dielectric material outside the cavity. Because the tangential component of the electric field is the same across the boundary, the tangential component immediately inside the cylinder must also be tangential and the same. The **D**-field is parallel to the **E**-field in the dielectric material. The normal component of the **D**-field, the component normal to the side surface of the long cylinder, is therefore also zero. Because there are no free charges, the normal component of the **D**-field is also zero immediately inside the cylinder.

The wide, short cylinder. The normal vector to the interface at the top (and bottom) of the wide, short cylinder points in the direction of the **E**-field and the **D**-field. The electric field does therefore not have a tangential component inside the dielectric, and therefore also no tangential component inside the cavity. The normal component of the **D**-field across the boundary is $D_{1n} - D_{2n} = \rho_s$, but there are no free charges, so $\rho_s = 0$, and $D_{1n} = D_{2n}$ across the boundary. The **D**-field inside the cylinder is therefore the same as the **D**-field in the dielectric. We also find that $D_{1,n} = \epsilon_0 E_1 = \epsilon E_0 = D_{2,n}$ and therefore $E_1 = E_0(\epsilon/\epsilon_0)$. If the dielectric is water and the hole is a dilute gas, then the field inside the hole is about 80 times that of the field in the water.

Summary

The **polarization vector P** is the dipole moment per unit volume

$$\mathbf{P} = \frac{1}{v} \sum_{i \text{ in } v} \mathbf{p}_i .$$

The polarization depends on the **total electric field**. In a **linear dielectric**:

$$\mathbf{P} = \chi_e \epsilon_0 \mathbf{E},$$

where **E** is the local electric field. The constant χ_e is called the electrical susceptibility.

The **bound charge** Q_b inside a closed surface S is

$$Q_b = - \oint_S \mathbf{P} \cdot \mathrm{d}\mathbf{S}.$$

The **volume bound charge density** $\rho_{v,b}$ is

$$\rho_{v,b} = -\nabla \cdot \mathbf{P},$$

and thc **surface bound charge density** $\rho_{s,b}$ is

$$\rho_{s,b} = \mathbf{P} \cdot \hat{\mathbf{n}},$$

where $\hat{\mathbf{n}}$ is the surface normal.

The **displacement field D** is defined as

$$\mathbf{D} = \epsilon_0 \mathbf{E} + \mathbf{P}$$

Gauss' law in a dielectric material is

$$\oint_S \mathbf{D} \cdot \mathrm{d}\mathbf{S} = Q_{\text{free in } S},$$

where the charge $Q_{\text{free in } S}$ is the free (not bound) charge inside the surface S.

In **linearly polarizable media** $\mathbf{D} = \epsilon\mathbf{E}$, where $\epsilon = \epsilon_r\epsilon_0$ is called the absolute permittivity.

Gauss' law on differential form becomes $\nabla \cdot \mathbf{D} = \rho_{v,\text{free}}$.

Boundary conditions for the electric field. The tangential components of the electric field is continuous across a boundary between two different dielectric materials, $E_{1,t} = E_{2,t}$. The normal component of the displacement field is related to the free surface charge density ρ_s across a boundary between two different dielectric materials, $D_{1,n} - D_{2,n} = \rho_s$.

Exercises

Discussion Exercises

5.1 Temperature dependence. Water is polarized when subject to an electric field. Do you think the polarization increases or decreases if you increase the temperature? Explain your reasoning.

5.2 Continuous change. Does the scalar potential and the electric field always change continuous across boundaries? Argue that they change continuously or find examples where they do not.

5.3 Continuity of fields. A charge q lies in the origin and is enclosed by a spherically shaped dielectrical material of radius a, permittivity ϵ with its center in the origin. Are the electrical fields (**E** and **D**) continuous across the interface from the sphere to the vacuum outside? Explain your reasoning.

Tutorials

5.4 Electric field through water. Two infinite planes with charge densities ρ and $-\rho$ are parallel with a distance d from each other.
(a) What is the electric field between the plates in vacuum?
(b) If the region between the planes instead is filled with water (with $\epsilon_r = 81$), what is now the electric field between the planes?
(c) Explain with words why the presence of water makes the electric field much smaller.
(d) Above and below the planes, there is no water. Does the electric field here change when the water is added?

5.5 Dielectric plate in uniform field. We will now build intuition and skills by addressing a particular situation in detail and develop a model for the charge distribution, the polarization, the electric field and the displacement field in the system. We start with a system consisting of empty space with a uniform electric field $\mathbf{E}_0 = E_0\hat{\mathbf{y}}$.

(a) What kind of charge distribution may be the cause of such a field? What assumptions would you then need to make about the system?

We then place an infinitely long linearly dielectric plate of thickness d and dielectric constant ϵ in the xz-plane. We will now try to find the electric field, the distribution of bound charges, and the polarization everywhere.

(b) Draw the plane into a coordinate system and specify where the top and the bottom of the plane are in the coordinate system.

(c) Make a sketch of the behavior of bound dipoles when the dielectric is placed into the uniform electric field. What are the consequences for the net bound charge in the system? Explain the sign of the bound charges on the interfaces between the plate and the vacuum.

(d) Your friend Q states that this problem is simple to solve. The polarization inside the plate is simply $\mathbf{P} = \chi_e \epsilon_0 \mathbf{E}_0$. Do you agree with him? Explain your reasoning.

Approach 1: Starting from the bound charges. We will now solve the problem in two different ways. In approach 1, we will assume a given surface density of bound charges, use this to find the electric field due to the charges, then find the polarization, and finally find a self-consistent solution so that the polarization due to the total field is consistent with the charge density related to the polarization. In method 2, we will use the displacement field approach to solve the same problem.

(e) Assume that there is a bound surface charge $\rho_{s,b}$ on the top surface and a bound charge $-\rho_{s,b}$ on the bottom surface. What is the electric field due to these charge densities? What is the total electric field, $\mathbf{E}_T$?

(f) What is the polarization, **P**, for this total electric field?

(g) Find the charge density on the top surface from the polarization and use this to find an equation for $\rho_{s,b}$. Solve this equation and find expressions for **P** and $\mathbf{E}_T$.

Approach 2: Using the displacement field. While approach 1 provides us with insight into the physics of the problem, the methods is rather cumbersome. Let us now instead use the displacement field to solve the same problem.

(h) What is the displacement field, **D**, outside the dielectric?

(i) What are the boundary conditions at the interface between the dielectric and the vacuum outside?

(j) Show that $E_T = (\epsilon_0/\epsilon)E_0$.

Exercises

5.6 Torque on dipole. A dipole with two charges $-Q$ and Q a distance d apart is in a field $\mathbf{E} = E_0\hat{\mathbf{y}}$.

(a) What orientations of the dipole gives zero torque around its center?

(b) What orientations give the maximum torque?

(c) Does it matter where the center of the dipole is located?

5.7 Bound charges. A dielectric sphere with radius a has a uniform polarization $\mathbf{P} = P\hat{\mathbf{z}}$.

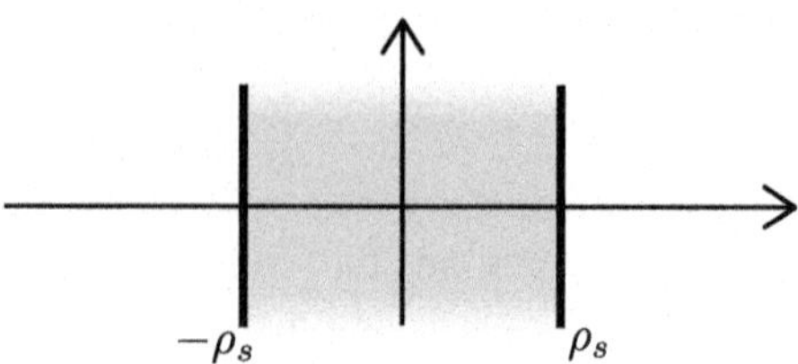

Fig. 5.7 Infinite dielectric slab

(a) What is the bound volume charge density inside the sphere?
(b) Sketch the bound surface charge density $\rho_{s,b}(\theta)$ on the surface of the sphere as a function of the angle θ in the xz-plane.
(c) What is the net bound charge on the whole sphere?

5.8 Charge in dielectric. A charge q is placed in the origin in (1) vacuum or (2) an infinite, isotropic, linear dielectric.
(a) Is the displacement field at a distance a from the charge larger, equal, or smaller in case (1) than in case (2)?
b) Is the electric field at a distance a from the charge larger, equal, or smaller in case (1) than in case (2)?
c) Is the polarization at a distance a from the charge larger, equal, or smaller in case (1) than in case (2)?

5.9 Charge in hole. A charge q is placed in the origin in the center of a spherical empty hole (vacuum) or radius a surrounded by an infinite, isotropic, linear dielectric.
(a) What is the electric field as a function of r, the distance to the charge, in the hole?
(b) What is the displacement field as a function of r, the distance to the charge, in the hole?
(c) What is the displacement field as a function of r outside the hole? Sketch the displacement field from $r < a$ to $r > a$.
(d) What is the electric field as a function of r outside the hole? Sketch the electric field from $r < a$ to $r > a$.

5.10 Continuity. Figure 5.7 illustrates a system consisting of an infinite slab of a dielectric with dielectric constant ϵ in vacuum. There are surface charge densities ρ_s and $-\rho_s$ on the two surfaces.
(a) Sketch the free and bound surface charge densities in the system.
(b) Sketch the **D**-field in the system.
(c) Sketch the **E**-field in the system.

5.11 A simple model for polarization. We develop a simplified model for polarization and the effective field in a dielectric material.

(a) An infinite, thin plane with surface charge density ρ is placed in the xy-plane. Find the electric field as function of z.

We will now introduce a simplified model for a dielectric material in the form of a plate in the xy-plane with thickness h. The plate consists of positive and negative charges which we model as a positive charge density ρ and a negative charge density $-\rho$ with the same shape as the plate. Initially, the two charge distributions are on top of each other, so that the total charge in each point is zero. If there is an electric field inside the plate, then each of the two charge distributions will be slightly displaced, so that the two distributions are shifted a distance Δz relative to each other: The positive charge distribution is moved a distance $\Delta z/2$ in the direction of the field and the negative charge distribution is moved a distance $-\Delta z/2$.

(b) When there is an electric field $\mathbf{E} = E\hat{\mathbf{z}}$ in the plate, the positive charge distribution and the negative charge distributions will be displaced and no longer overlap perfectly. Make a sketch of the system. What is the charge density on the top and bottom of the plate in this case? We call these *induced* surface charge densities.

(c) What is the electric field, E_i, set up by these induced surface charge densities? (E_i is the field from the induced charges, not the total electric field).

(d) Now, we assume that the displacement Δz of the charge densities is proportional to the total local electric field, $\Delta z = \chi E_{tot}$, where χ is a constant. (This is similar to the assumption we made in our model for polarization of atoms). Find an expression for the induced field inside the plate, E_i, as a function of the total electric field, E_{tot}, inside the plate.

(e) We place the plate in an external field $\mathbf{E}_0 = E_0\hat{\mathbf{z}}$. What is the total electric field $\mathbf{E}_{tot}$ inside the plate?

(f) Now, assume that a plate with dielectric constant ϵ is placed in a uniform electric field $\mathbf{E}_0 = E_0\hat{\mathbf{z}}$. Use the boundary condition relations to show that the electric field inside the plate is $\mathbf{E} = (\epsilon_0/\epsilon)\mathbf{E}_0$. Use this to relate ρ_s and χ to ϵ. Check that this expression is reasonable.

Homework

5.12 Polarized cube. A dielectric cube with sides of length a and dielectric constant ϵ is placed in the first octant with a corner in the origin. There is vacuum outside the cube.

(a) Assume that the cube has a uniform polarization $\mathbf{P} = P_0\hat{\mathbf{x}}$. What is the density of bound charges inside the cube? And on the sides?

(b) Assume that the cube has a polarization $\mathbf{P}(x, y, z) = P_0 xy/a^2\hat{\mathbf{x}}$. What is the distribution of bound charges in and on the cube now?

5.13 A cell membrane. In this exercise we will develop a simple model for a cell membrane. We assume that the cell membrane is shaped as a spherical shell with inner radius a and outer radius b. The membrane consists of a dielectric material with dielectric constant ϵ. There is a ion pump through the cell membrane, so that on the inner surface (with radius a) there is a charge $+Q$, while on the outer surface (with radius b), there is a charge $-Q$.

(a) What symmetry does the electric field have for this system?
(b) What is the electric field as a function of r, the distance to the center of the cell?
(c) What is the potential difference between the outer and the inner surfaces?
(d) For a typical nerve cell we can assume that the potential difference is 70mV, the inner radius is $a = 1\mu$m, and the thickness of the membrane is $b - a = 6$nm, and $\epsilon = 81\epsilon_0$. How large charge is there on the cell surface? How many ions with a charge $-e$ does this correspond to?

5.14 Long cylinder. An infinitely long cylinder of radius a and dielectric constant ϵ has a uniform volume charge density ρ and is placed in air.
(a) Find the electric field inside and outside the cylinder.
(b) Find the bound charge on the cylinder.

5.15 Hollow sphere. A hollow spherical shell of inner radius a, outer radius b and dielectric constant ϵ surrounds a uniformly charged sphere of radius a.
(a) Find the displacement field **D**.
(b) Find the electric field **E**.
(c) Find the bound volume charge density.

Chapter 6
Laplace Equation

In these cases, it can be useful to reformulate the problem in a different form—as a differential equation for the electric potential. We have found two sets of differential equations for the electric field: Gauss' law on differential form, $\nabla \cdot \mathbf{E} = \rho/\epsilon_0$; and that the curl of the electric field is zero, $\nabla \times \mathbf{E} = 0$. This gives us a set of partial differential equations that can be solved, but it may again be practically challenging. However, if we introduce the potential V so that $\mathbf{E} = -\nabla V$, these equations can be reformulated in a simpler form because $\nabla \cdot \mathbf{E} = -\nabla^2 V = \rho/\epsilon_0$ and $\nabla \times \nabla V = 0$ because the curl of a gradient is always zero. This equation is called *Poisson's equation*. Together with appropriate boundary conditions, solving Poisson's equations gives us the potential which again gives us the electric field, which in turn allows us to find the distribution of charges. Solving Poisson's equation is therefore equivalent to finding the potential or electric field from an integral over the charges. The approach provides us with new, powerful tools to study electrostatic systems, but it requires new mathematical and numerical methods.

In this chapter we will demonstrate how we can solve problems in electrostatics by solving Poisson's equation, we will demonstrate fundamental mathematical properties of the equations such as the uniqueness of solutions given a set of boundary conditions, and we will show how Poisson's equation can be used to model complex electrostatic systems.

6.1 Laplace's and Poisson's Equations

Motivational Example

Figure 6.1 illustrates a lightning striking from a cloud onto the ground in Grand Canyon. How would we model such a system using the physics we have learned so far? We would need to simplify the system: We could first simplify it to a sketch

© The Author(s), under exclusive license to Springer Nature Switzerland AG 2026

A. Malthe-Sørenssen, *Elementary Electromagnetism Using Python*, Undergraduate Texts in Physics, https://doi.org/10.1007/978-3-032-19876-1_6

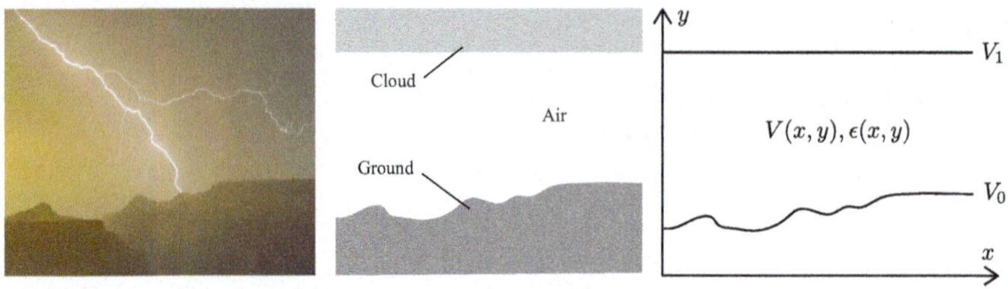

Fig. 6.1 Real image of a lightning strike, simplified drawing of the geometry, and model system. (Image by NPS Photo/M. Quinn)

of a cloud and the ground. We simplify it further to a physics problem by assuming that the cloud has a potential V_1 relative to the potential V_0 of the ground. What then happens during the lightning strike?

We would need to know some additional physics for this: The effect of dielectric breakdown. When the electric field across a small, dielectric region exceeds a maximum value, the charges (electrons) are no longer bound, but are ripped from their bound positions, and become free. We call this *dielectric breakdown*. The result is the formation of a plasma and motion of charges, which we will discuss further down. However, in order to find out if and where a dielectric breakdown would occur, we need to find the electric field in space, for example, by finding the electric potential, and then finding the field from the potential. How can we find the electric potential everywhere in space, if we only know the potential in the regions that correspond to the cloud and the ground? We need to develop new methods for this!

Finding the Electric Potential

We know how to find the potential from a given set of charges, Q_i, by summing their individual contributions in the sum $V(\mathbf{r}) = \sum_i Q_i/(4\pi\epsilon_0 R_i)$. However, if we do not know where the charges are, but only what the potential is in specific points in space, then we need a different approach. We know that the electric field can be found from Gauss' law on differential form:

$$\nabla \cdot \mathbf{D} = \nabla \cdot \epsilon \mathbf{E} = \rho \,, \tag{6.1}$$

where we have used that $\mathbf{D} = \epsilon \mathbf{E}$. If we assume that ϵ is uniform, we can put it outside the differential (∇) operator, getting

$$\epsilon \nabla \cdot \mathbf{E} = \rho \;\Rightarrow\; \nabla \cdot \mathbf{E} = \frac{\rho}{\epsilon}. \tag{6.2}$$

We then insert that $\mathbf{E} = -\nabla V$, getting:

$$\nabla \cdot \mathbf{E} = -\nabla^2 V = \frac{\rho}{\epsilon}. \tag{6.3}$$

We call the ∇^2 operator the *Laplace operator*. This equation is called *Poisson's equation* and in regions with no free charges, $\rho = 0$, we get *Laplace's equation*:

Poisson's and Laplace's equations

Poisson's equation is valid in regions with a charge density ρ and a uniform dielectric constant:

$$\nabla^2 V = \frac{\partial^2 V}{\partial x^2} + \frac{\partial^2 V}{\partial y^2} + \frac{\partial^2 V}{\partial z^2} = -\frac{\rho}{\epsilon}. \tag{6.4}$$

In the special case when $\rho = 0$, we get *Laplace's equation*:

$$\nabla^2 V = \frac{\partial^2 V}{\partial x^2} + \frac{\partial^2 V}{\partial y^2} + \frac{\partial^2 V}{\partial z^2} = 0. \tag{6.5}$$

Notice that if ρ is zero everywhere, that is, there are no charges anywhere, there will be no electric fields. Laplace's equation is useful when there may be plenty of charges elsewhere or at the boundaries of the system—and these charges set up electric fields—but there are no free charges inside the system where we want to solve Laplace's equation to find the electric potential and the electric field.

In addition to these equations, to find V, we need to specify the *boundary conditions* for the problem. To find the potential in a region v, we may for example need the values for the potential on a surface S enclosing v. For example, to find the electric potential in the air in Fig. 6.1, we need to know the values of the potential V at the boundaries, that is, at the surface of the ground and the surface of the clouds. This is indeed why it is called *boundary conditions*. We will later come back to what types of boundary conditions are necessary for the solution to be unique, and what different types of boundary conditions we may have for Poisson's or Laplace's equation.

Test your understanding

Can these expressions be solutions to Laplace's or Poisson's equation on the interval $-1 < x < 1$? If they are a solution to Poisson's equation, what is $\rho(x)$? (a) $V(x) = 0$; (b) $V(x) = 2\sin x$; (c) $V(x) = 2x - 3$.[1]

[1] (a) Yes/Yes; (b) No/Yes ($\rho(x) = \epsilon 2 \sin x$; (c) Yes/Yes.

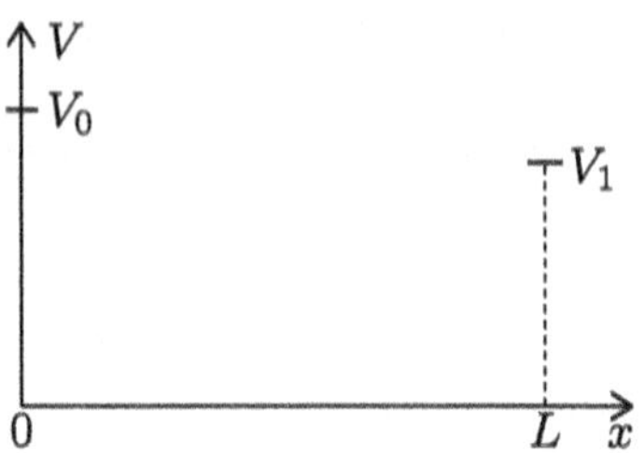

Fig. 6.2 Illustration of 1d problem

Example: Laplace's Equation in One Dimension

For a system of length L with $V(x = 0) = V_0$ and $V(x = L) = V_1$. What is the electric potential in the region $0 < x < L$, when there are no charges in this region and the material is a homogeneous dielectric with a constant ϵ.

Draw the system. We start by making a drawing of the system in Fig. 6.2 which includes the boundary conditions.

Approach. Our plan is to solve the problem by solving Laplace's equation, because there are no charges in the region of interest. In one dimension, Laplace's equation simplifies to:

$$\nabla^2 V = \frac{\partial^2 V}{\partial x^2} + \frac{\partial^2 V}{\partial y^2} + \frac{\partial^2 V}{\partial z^2} = \frac{\partial^2 V}{\partial x^2} = 0 \tag{6.6}$$

because the partial derivatives with respect to y and z are zero for a one-dimensional system. (One dimension in this context simply means that we assume that the solution only depends on x and not on y or z). We can find the solution to this equation by integrating twice with respect to x, getting an expression on the form $V(x) = A + Bx$, where A and B are constants that need to be determined by the boundary conditions. Since there are two constants, we need two boundary conditions to specify them. (This means that if we only knew that $V(0) = V_0$, we would not be able to find both constants). In this case, we see that the boundary conditions give that $V(0) = A = V_0$ and $V(L) = V_0 + BL = V_1$, and therefore $B = (V_1 - V_0)/L$. The full solution is therefore $V(x) = V_0 + x(V_1 - V_0)/L$, which correspond to a line from V_0 at $x = 0$ to V_1 at $x = L$.

Extending the problem. *What is the potential if the charge distribution is not zero, but instead has the form $\rho(x) = \rho_0 \sin(\omega x)$, where $\omega = 2\pi/L$? The potential is still $V(0) = V_0$ and $V(L) = V_1$ at the boundaries.*

Approach. In this case, we need to find the potential from Poisson's equation, because charge density is not zero. In one dimension Poisson's equation simplifies to:

$$\frac{\partial^2 V}{\partial x^2} = -\frac{\rho}{\epsilon} = -\frac{\rho_0}{\epsilon} \sin(\omega x), \tag{6.7}$$

where we have inserted the expression for $\rho(x)$. If you remember how to solve such differential equations, you are all set. Otherwise, it is for now enough to know that there exists a unique solution to this differential equation with its boundary condition. This means that if we find *a* solution, it must be *the* solution. Here, we simply try a general form that we from experience expect to work: $V(x) = B\cos(\omega x) + C\sin(\omega x) + Dx + E$. (Do not worry if you do not know where this solution came from. It is not important for the following, and you will build the experience as your work more examples and exercises). We insert this equation in Poisson's equation, to show that it indeed is a solution to the equation:

$$\frac{\partial^2 V}{\partial x^2} = -B\omega^2\cos(\omega x) - C\omega^2\sin(\omega x) = -\frac{\rho_0}{\epsilon}\sin(\omega x), \tag{6.8}$$

where we see that $B = 0$ and $-C\omega^2 = -\rho_0/\epsilon$, that is, $C = \rho_0/(\omega^2\epsilon)$. The solution therefore has the form $V(x) = \rho_0/(\omega^2\epsilon)\sin(\omega x) + Dx + E$. The two constants D and E must be determined from the boundary conditions. The boundary condition at $x = 0$ gives $V(0) = E = V_0$. The boundary condition at $x = L$ gives $V(L) = DL + E = DL + V_0 = V_1$ and $D = (V_1 - V_0)/L$. The full solution is therefore $V(x) = \rho_0/(\omega^2\epsilon)\sin(\omega x) + (V_1 - V_0)x/L + V_0$.

Test your understanding

A function $V(x, y) = -V_0x^2\cos y$. (a) What is $\partial^2 V/\partial x^2$ in the points $(0, 0)$, $(0, \pi/2)$? (b) What is $\partial^2 V/\partial y^2$ in the points $(0, 0)$, $(\pi/2, 0)$?[2]

Poisson's Equation Versus the Equations of Motion

You already know a differential equation that may look similar to Poisson's equation. When you apply Newton's laws and solve the equations of motion for a time-varying external force $F(t) = mf(t)$:

$$\frac{d^2x}{dt^2} = f(t),\ x(t_0) = x_0,\ \frac{dx}{dt}(t_0) = v_0. \tag{6.9}$$

First, we notice that this equation looks similar to Poisson's equation in one dimension:

$$\frac{d^2V}{dx^2} = -\rho(x)/\epsilon,\ V(a) = V_a,\ V(b) = V_b. \tag{6.10}$$

[2] (a) $-2V_0$, 0; (b) 0, $V_0\pi^2/4$.

You may therefore be tempted to use some of the techniques you already know to solve one-dimensional equations of motion. However, there are important differences. One difference may be in the boundary conditions. For equations of motion we usually have an *initial value problem.* We know x and its derivative at a given time: $x(t_0) = x_0$ and $x'(t_0) = v_0$. For Poisson's equation we usually have a *boundary value problem*, where we know the value of $V(x)$ for two different values of x, e.g. $V(a) = V_a$ and $V(b) = V_b$. This becomes particularly important for numerical solution methods. For an initial value problem, we can simply integrate forward in time using a numerical, iterative scheme. However, for a boundary value problem, we cannot do this, since we need to ensure that we end up at a particular value $V(b) = V_b$. Thus, we must use other numerical methods. Finally, Poisson's equation is generalized to a partial differential equation in higher dimensions, and Newton's equations do not have a corresponding generalization. We can therefore use some intuition from our previous experience, but we also need to build new intuition about partial differential equations and their solution methods.

Why is Laplace's Equation So Common?

You will meet Laplace's equation repeatedly in physics and other disciplines. Indeed, it is such a beautifully simple equation, $\nabla^2 V = 0$, which makes it mathematically interesting. But the equation is also of fundamental importance in physics. One situation where Laplace's equation appear are in variations of the diffusion equation. The time development of a concentration field, $c(\mathbf{r}; t)$, or a temperature field, $T(\mathbf{r}; t)$ is given by the diffusion equation

$$\frac{\partial c}{\partial t} = D\nabla^2 c \text{ or } \frac{\partial T}{\partial t} = \alpha \nabla^2 T. \tag{6.11}$$

where D is called the diffusivity and α the thermal diffusivity. These equations are called the *diffusion equation* and the *heat equation*. In the *stationary state*, that is when the time derivative is zero, these and other transport equations reduce to Laplace's equation. We will later see that this is also the case in electrostatics.

6.2 Boundary Conditions

The solution of Laplace's or Poisson's equations depends on the boundary conditions. Let us examine how the boundary conditions determine the solutions, what properties the solutions have, and what types of boundary conditions can be imposed.

Existence and Uniqueness

From mathematics we know that there exists a solution and that the solution is unique for a given set of boundary condition. More precisely:

Existence and uniqueness

For Laplace's equation defined in a region v enclosed by a smooth surface S, Lax-Milgram's Lemma (Evans 2010) ensures that

- there *exists a solution* $V(\mathbf{r})$
- the *solution is unique* given a set of values $V = V_S$ at the boundary surface S

Properties of the Solution

A solution $V(\mathbf{r})$ to Laplace's equation in a region v enclosed by the surface S has several important properties:

- A function $V(\mathbf{r})$, which is a solution to Laplace's equation, does not have any extrema inside the region v. All the extrema are on the boundary S.
- The value of V at a point $\mathbf{r}$ is the average value of V over a spherical surface of radius R around $\mathbf{r}$.

We provide brief proofs of these properties at the end of the chapter.

Different Types of Boundary Conditions

Boundary conditions can specify the values of the function on the boundary, called Dirichlet boundary conditions, or the derivative of the function on the boundary, called Neumann boundary conditions.

Dirichlet boundary conditions. The typical boundary condition for Laplace's and Poisson's equations is to specify the value of the potential at the boundary. This type of boundary condition is called a *Dirichlet boundary condition*. Boundaries can be either the external boundaries or internal boundaries. Figure 6.3 illustrates boundary conditions in one and two dimensions. Dirichlet boundary condition here correspond to specifying the value of the potential at specific positions on the boundaries, $\mathbf{r}_i$: $V(\mathbf{r}_i) = V_i$

Neumann boundary conditions. Another type of boundary condition is to specify the derivative of V, $\partial V/\partial n$, in a given direction n, typically normal to the boundary.

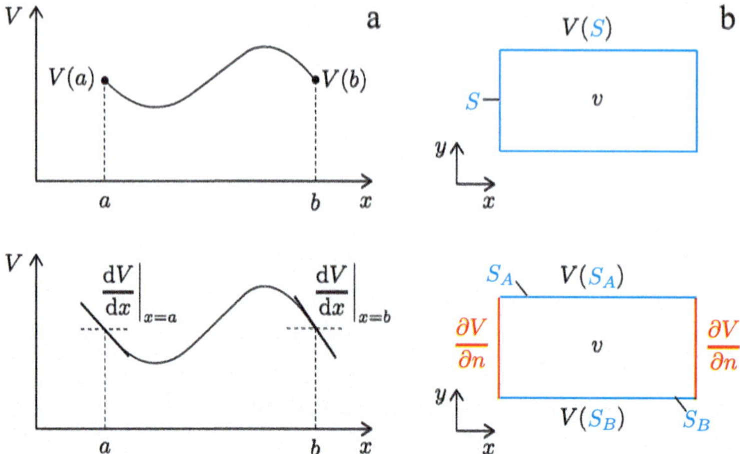

Fig. 6.3 Boundary conditions in **a, b** one dimension and in **c, d** two dimensions. **a** and **c** show Dirichlet boundary conditions where the values at the boundary are specified. **b** and **d** illustrate Neumann boundary conditions where the derivatives at the boundaries are given. **d** shows mixed boundary conditions

This type of boundary condition is called a *Neumann boundary condition*. In the one-dimensional case, Neumann boundary conditions corresponds to specifying $\partial V/\partial x$ at the boundaries, which corresponds to specifying the electric field at the boundaries, because $E_x = -\partial V/\partial x$. The same interpretation is valid in higher dimensions, where we interpret $\partial V/\partial n$ as the gradient in the direction along n, which again is related to the electric field. A common type of Neumann boundary conditions is that the derivative is zero at the boundaries, that is, that the electric field is zero in the direction normal to the boundary.

Test your understanding

(a) If a region v does not contain any charges. What can we say about $V(\mathbf{r})$ in this region? (b) If a region v which does not contain any charges is enclosed by a surface S and the potential is V_0 everywhere on the surface S, what can we say about $V(\mathbf{r})$ in this region?[3]

∇^2 *in Various Coordinate Systems*

When we apply Laplace's and Poisson's equation to find the electric potential, we choose a version of the equations that follows the symmetry of the boundary condi-

[3] (a) V satisfies $\nabla^2 V = 0$; (b) $V = V_0$ in v.

tions. If a system has boundaries that follow a cylindrical symmetry, then we expect the solution also to have cylindrical symmetry. We therefore provide the Laplace operator, $\nabla^2 V$, in the three most common coordinate systems:

The Laplace operator

Cartesian coordinates:

$$\nabla^2 V = \frac{\partial^2 V}{\partial x^2} + \frac{\partial^2 V}{\partial y^2} + \frac{\partial^2 V}{\partial z^2}.$$

Cylindrical coordinates:

$$\nabla^2 V = \frac{1}{r}\frac{\partial}{\partial r}\left(r\frac{\partial V}{\partial r}\right) + \frac{1}{r^2}\frac{\partial^2 V}{\partial \phi^2} + \frac{\partial^2 V}{\partial z^2}.$$

Spherical coordinates:

$$\nabla^2 V = \frac{1}{r^2}\frac{\partial}{\partial r}\left(r^2\frac{\partial V}{\partial r}\right) + \frac{1}{r^2 \sin\theta}\frac{\partial}{\partial \theta}\left(\sin\theta\frac{\partial V}{\partial \theta}\right) + \frac{1}{r^2 \sin^2\theta}\frac{\partial^2 V}{\partial \phi^2}.$$

Test your understanding

Is $V(r) = C/r$, where C is a constant, a possible solution of Laplace's equation in spherical coordinates?[4]

Example: Cylindrical System

A cylindrical system consists of two concentric cylindrical surfaces with radius a and b. Assume that the potential at $r = a$ is V_a and at $r = b$ is V_b. There are no free charges for $a < r < b$. Find the potential for $a < r < b$.

Approach. We assume that the cylinders are infinitely long and place them along the z-axis. We recognize that the system has cylindrical symmetry, and we expect $V = V(r)$, so that there is no ϕ or z dependence. Because there are no free charges between the cylinder surfaces, the system must obey Laplace's equation in this region.

[4] Yes.

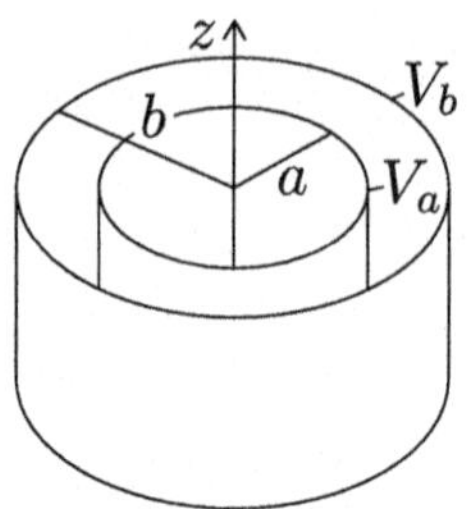

Fig. 6.4 System of two cylindrical surfaces. The potential is V_a at the inner surface and V_b at the outer surface

System sketch. Figure 6.4 provides a sketch of the system.

Solution. We use Laplace's equation in cylindrical coordinates, but only include the r-derivative since the potential cannot depend on ϕ or z:

$$\nabla^2 V = \frac{1}{r}\frac{\partial}{\partial r}\left(r\frac{\partial V}{\partial r}\right) = 0 \tag{6.12}$$

For $r > a > 0$, we multiply by r, getting

$$\frac{\partial}{\partial r}\left(r\frac{\partial V}{\partial r}\right) = 0 \tag{6.13}$$

This means that what is inside the parenthesis must be a constant, that is,

$$r\frac{\partial V}{\partial r} = A \Rightarrow \frac{\partial V}{\partial r} = \frac{A}{r} \tag{6.14}$$

This means that $V = A\ln(r/a) + B$, where we have decided to measure r in units of a. We determine the constants A and B from the boundary conditions $V_a = A\ln(a/a) + B$ and $V_b = A\ln(b/a) + B$. We see that $V_a = A\ln 1 + B = 0 + B = B$. We insert this value of B into the condition at $r = b$: $V_b = A\ln(b/a) + V_a$, which gives $A = (V_b - V_a)/\ln(b/a)$. This gives us

$$V(r) = V_a + (V_b - V_a)\frac{\ln(r/a)}{\ln(b/a)}. \tag{6.15}$$

We check that this function indeed satisfies the boundary conditions: $V(a) = V_a$ and $V(b) = V_a + (V_b - V_a) = V_b$.

6.3 Numerical Solutions: Finite Difference Methods

Finding exact solutions to Poisson's equations in two or three dimensions often require cumbersome mathematical methods. Fortunately, there are numerous numeri-

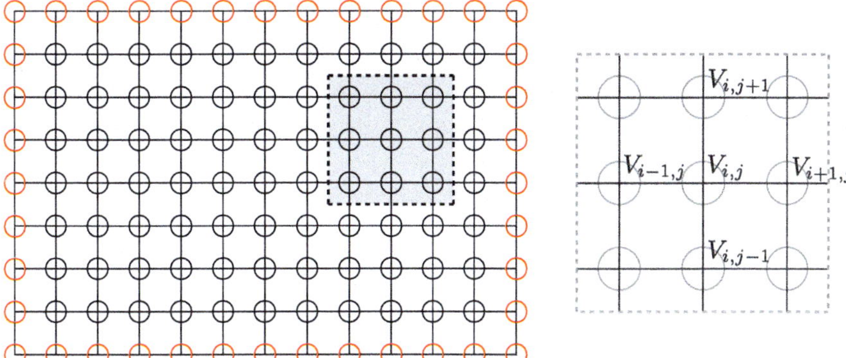

Fig. 6.5 Illustration of the discretized grid

cal methods to solve Poisson's equation spanning from simple methods, such as Finite Difference method, to advanced methods such as finite element methods. Here, we will focus on a very simple, but inefficient, solution method that also provides insight into the physics of the problem.

Discrete Lattice

We introduce a method to solve Laplace's equation, but the same approach may be used to solve Poisson's equation. We would like to solve the equation

$$\nabla^2 V = 0 \tag{6.16}$$

with a given set of boundary conditions. We will solve this on a *discrete lattice*. That is, we will find the solution at lattice positions $(x_i, y_i) = (i\Delta x, j\Delta y)$ as illustrated in Fig. 6.5.

Discretizing the Equation on a Square/Cubic Lattice

We can discretize the equation on this lattice by first finding the discrete derivative and then taking the derivative once more. This is usually done on what is called a staggered grid: We find the derivative at in-between positions $x_i + \Delta x/2$ and $x_i - \Delta x/2$, and then use this to find the second derivative.

The derivative at $x_i + \Delta x/2$ is approximately:

$$\frac{\partial V}{\partial x}(x_i + \Delta x/2) \simeq \frac{V(x_i + \Delta x, y_i) - V(x_i, y_i)}{\Delta x}, \tag{6.17}$$

and similarly

$$\frac{\partial V}{\partial x}(x_i - \Delta x/2, y_i) \simeq \frac{V(x_i, y_i) - V(x_i - \Delta x, y_i)}{\Delta x}, \tag{6.18}$$

The second derivative in (x_i, y_i) is then:

$$\frac{\partial^2 V}{\partial x^2} \simeq \frac{\frac{\partial V}{\partial x}(x_i + \Delta x/2, y_i) - \frac{\partial V}{\partial x}(x_i - \Delta x/2, y_i)}{\Delta x}, \tag{6.19}$$

and similarly for the y-direction:

$$\frac{\partial^2 V}{\partial y^2} \simeq \frac{\frac{\partial V}{\partial y}(x_i, y_i + \Delta y/2) - \frac{\partial V}{\partial y}(x_i, y_i - \Delta y/2)}{\Delta y}, \tag{6.20}$$

We can simplify the notation by introducing $V(i\Delta x, j\Delta x) = V_{i,j}$ when $\Delta x = \Delta y$. We then get that

$$\frac{\partial^2 V}{\partial x^2} + \frac{\partial^2 V}{\partial y^2} = \frac{1}{\Delta x^2}\left(V_{i+1,j} + V_{i-1,j} + V_{i,j+1} + V_{i,j-1} - 4V_{i,j}\right) = 0 \,. \tag{6.21}$$

We can solve for $V_{i,j}$, getting

$$V_{i,j} = \frac{1}{4}\left(V_{i+1,j} + V_{i-1,j} + V_{i,j+1} + V_{i,j-1}\right). \tag{6.22}$$

Notice that this looks like a local average. The solution is such that the value in each point is the average of the values at the nearest neighbors around it. We therefore expect the solution to be very smooth.

This is a linear system of equations, which must be supplemented by the boundary conditions. Here, we will specify the boundary conditions by Dirichlet boundary conditions through a function $B(x, y)$ on a boundary S. The boundary conditions are that $V(x, y) = B(x, y)$ for all points (x, y) on the boundary S.

Test your understanding

The following are discrete values $V_{i,j}$. Draw them onto a two-dimensional lattice. Are these possible solutions to the discrete Laplace's equation? (a) $V_{1,2} = +1$, $V_{2,2} = 0$, $V_{3,2} = 0$, $V_{2,1} = -1$, $V_{2,3} = -1$; (b) $V_{1,2} = +1$, $V_{2,2} = 2$, $V_{3,2} = 3$, $V_{2,1} = 1$, $V_{2,3} = 3$; (c) $V_{1,2} = +1$, $V_{2,2} = 2$, $V_{3,2} = 1$, $V_{2,1} = 1$, $V_{2,3} = 1$; (d) Does it matter if you exchange any of the values that are not in the center?[5]

[5] (a) No; (b) Yes; (c) No; (d) No.

Solving the Discretized Equations Using Jacobi Iterations

There are various numerical schemes to solve these sets of equations. First, we will solve this system of equations using a simple, iterative scheme (which is an inefficient method to solve the system of equations, but the method is easy to understand) called Jacobi iterations. We see that in the *final solution* for each point (x_i, y_i), the potential is the average of the surrounding potentials. Our strategy is therefore the following: If we iterate this average many times, we will slowly smooth things out and converge towards a solution. This suggests the following method:

1. Initialize $V_{i,j}$.
2. On the boundary we set $V_{i,j} = B_{i,j}$, that is, equal to the boundary condition $B(x, y)$.
3. For each point in the interior of the volume v, calculate new potentials as the average of the surrounding potentials. Repeat until it converges (sufficiently well).

Numerical Implementation

First, we make a function to solve the equation using Jacobi iterations. We start by importing necessary libraries, including `numba`, which is a library that is used to accelerate loops in functions.

```
import numpy as np
import matplotlib.pyplot as plt
from numba import jit
```

Second, we write a function using the algorithm above:

```
@jit(cache=True)
def solvepoisson(b,nrep):
    # b = boundary conditions, = NaN on interior of region
    # nrep = number of iterations
    # returns potential on the same grid as b
    V = np.copy(b)
    for i in range(len(V.flat)):
        if (np.isnan(b.flat[i])):
            V.flat[i] = 0.0
    Vnew = np.copy(V) # See comment in text
    Lx, Ly = b.shape[0], b.shape[1]
    for n in range(nrep):
        for ix in range(Lx):
            for iy in range(Ly):
                if (np.isnan(b[ix,iy])):
                    Vnew[ix,iy] = (V[ix+1,iy]+V[ix-1,iy]+\
                                   V[ix,iy+1]+V[ix,iy-1])/4
        V, Vnew = Vnew, V # Swap pointers to arrays
    return V
```

Why do we introduce the new array Vnew and not only update V in each step? If we did this, we would gradually change V as we were moving through the loops, and we would no longer use the values in the previous step to calculate the next step, instead we would use some of the values from the old step and some values from the new step. This would become a mess and would not correspond to the algorithm presented above. Also notice the use of nan in the array b, which represents the boundary conditions. In the positions where b contains a value, this is the boundary condition in this point, whereas in the positions where b does not contains any numerical value, where it is set to nan, there are no boundary conditions. (This method is improved below to also work for internal boundary conditions. Here, it only works for boundary conditions at the external boundaries, and all the values at the external boundaries must be set).

Defining Boundary Conditions

We are now ready to use this function to find the potential for a given set of boundary conditions. Let us start with a square system that spans from $x_A = -a$ to $x_B = +a$ and from $y_A = -a$ to $y_B = a$, where $a = 1$cm with N cells in each direction. A cell (i, j) corresponds to the position (x_i, y_j) where $x_i = x_A + i\,(x_B - x_A)/(N-1)$ and $y_j = y_A + j\,(y_B - y_A)/(N-1)$. Notice that we divide by $(N-1)$ because we start counting at $i = 0$ and stop counting at $i = N-1$, so that the last element is at $x_{N-1} = x_B$. Now, let us assume that we have Dirichlet boundary conditions with $B(x, x_A) = 1$ and $B(x, x_B) = B(x_A, y) = B(x_B, y) = 0$. Now, to set these boundary conditions, we must set values at the corresponding positions in the b-matrix. First, we set all values in the b-matrix to be nan, then we set the boundary values at the outer boundaries of the b-matrix. We plot the resulting b-matrix in Fig. 6.6 to check our boundary conditions.

```
N = 40
a = 1.0 # meters
x_A, x_B, y_A, y_B = -a,a,-a,a
x = np.linspace(x_A,x_B,N)
y = np.linspace(y_A,y_B,N)
# Set up matrix of b-values
b = np.zeros((N,N),float)
b[:] = float("nan")
b[0,:] = 0.0
b[N-1,:] = 0.0
b[:,0] = 1.0
b[:,N-1] = 0.0
# Visualize results
rx,ry = np.meshgrid(x,y,indexing="ij")
plt.pcolormesh(rx,ry,b), plt.colorbar()
```

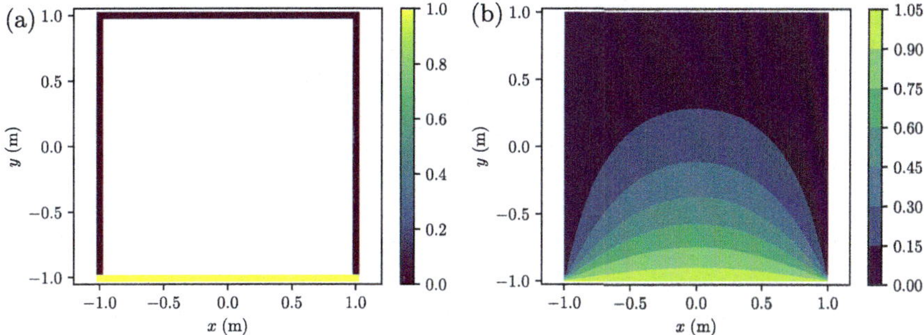

Fig. 6.6 **a** Illustration of the boundary conditions $B(x, y)$ shown with a color-scale. The colors indicate where $B(x, y)$ is defined. White interior positions show where `b = nan`. **b** Plot of the resulting potential $V(x, y)$

Finding the Potential

We apply Jacobi's method using the `solvepoisson`-function to find the potential $V(x, y)$ with the given boundary conditions $B(x, y)$ and plot the resulting $V(x, y)$ in Fig. 6.6. Here, we choose `nrep=2000` iterations.

```
nrep = 2000
V = solvepoisson(b,nrep)
plt.contourf(rx,ry,V), plt.colorbar()
```

We get an even better impression of the potential with a surface plot, as shown in Fig. 6.7.

```
from mpl_toolkits.mplot3d import Axes3D
from matplotlib import cm
fig = plt.figure()
ax = fig.add_subplot(111, projection="3d")
ax.plot_surface(rx,ry,V,cmap=cm.viridis)
ax.view_init(20, 60), ax.set_xlabel("$x$")
ax.set_ylabel("$y$"), ax.set_zlabel("$V$")
plt.show()
```

Finding the Electric Field

When we have found the potential, we calculate the electric field using $\mathbf{E} = -\nabla V$. The electric field in the x-direction is

$$E_x = -\frac{\partial V}{\partial x} \simeq -\frac{V(x_i + \Delta x, y_j) - V(x_i, y_j)}{\Delta x} = -\frac{V_{i+1,j} - V_{i,j}}{\Delta x}, \tag{6.23}$$

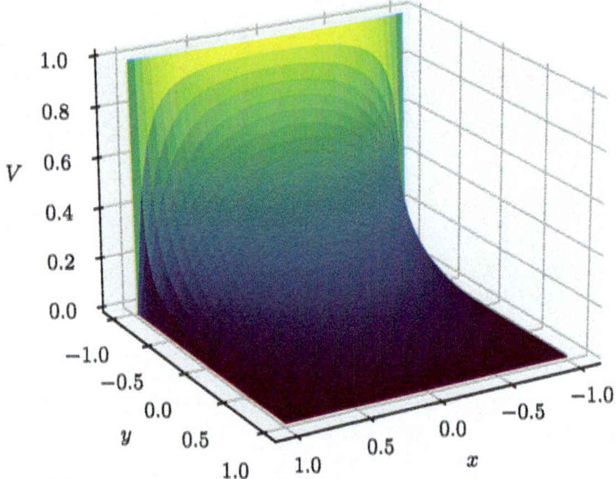

Fig. 6.7 Surface plot of the potential $V(x, y)$

and similarly in the other direction. We can use the `gradient`-function in Python to calculate the gradient of $V_{i,j}$: $(V_{i+1,j} - V_{i,j}, V_{i,j+1} - V_{i,j})$. However, to find the electric field with correct units, we need to divide by Δx, which we find from $\Delta x = x_{i+1} - x_i$:

```
deltax = x[1]-x[0]
Ex,Ey = np.gradient(-V/deltax)
plt.contourf(rx,ry,V,levels=15), plt.colorbar()
plt.quiver(rx,ry,Ex,Ey), plt.axis("equal")
```

The resulting plot is shown in Fig. 6.8. To get a better impression of the functional form of $\mathbf{E}(x, y)$ we plot $E_y(x, y)$ along a line $x = 0$. We realize that $x = 0$ is in the middle of the matrix, that is, at an index `i = int(N/2)`. We use `:` to get all the values for all the `j`-indicies. The resulting plot is shown in Fig. 6.8.

```
plt.plot(y,Ey[int(N/2),:],"-r")
plt.xlabel("$y$ (m)")
plt.ylabel("$E_y$ (V/m)")
```

General Solution Method

The approach provided above only works when we specify the boundary condition on all the external boundaries. However, in many cases we only want to specify the borders on some objects in the system, such as on the boundaries of conductors that we use to construct the system from. These may be shaped like two boxes, two thin lines, or two cylindrical objects. But instead of specifying what the potential should be at the external boundaries, we may want to model that the boundaries are far away by saying that the potential is flattening out, without specifying its value. Flattening out means that the derivative of the potential approaches zero. Therefore, we would like

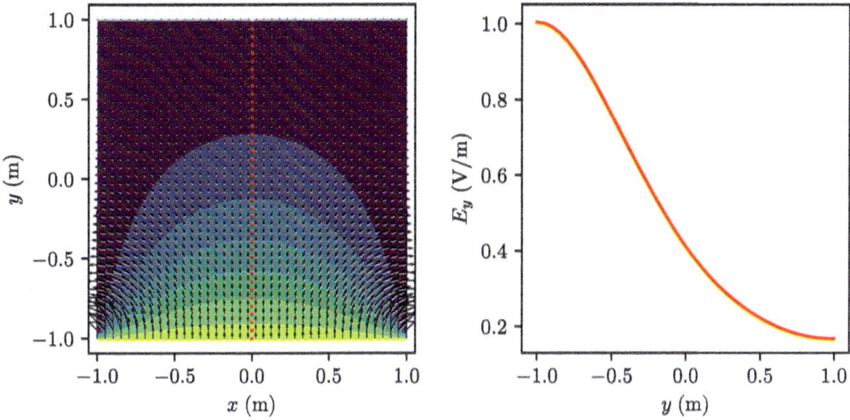

Fig. 6.8 **a** Plot of $V(x, y)$ and $\mathbf{E}(x, y)$. The red lines indicate the cross-section shown in **b**. **b** Plot of $E_y(x = 0, y)$

to have an approach that has Dirichlet boundary conditions on some closed regions and then Neumann boundary conditions on the external boundaries. We rewrite the program to explicitly consider what happens at the external boundaries. It turns out that we can include Neumann boundary conditions at a boundary point $V_{0,j}$ simply by not including the element $V_{-1,j}$ in the averaging sum in each iteration step. (We prove this in an exercise). The rewritten and more general solution function becomes:

```
@jit
def solvepoissonneumann2d(b,nrep):
    # b = boundary conditions
    # nrep = number of iterations
    # returns potentials
    V = np.copy(b)
    for i in range(len(V.flat)):
        if (np.isnan(b.flat[i])):
            V.flat[i] = 0.0
    Vnew = np.copy(V)
    Lx = b.shape[0]
    Ly = b.shape[1]
    for n in range(nrep):
        for ix in range(Lx):
            for iy in range(Ly):
                ncount = 0.0
                pot = 0.0
                if (np.isnan(b[ix,iy])):
                    if (ix>0):
                        ncount = ncount + 1.0
                        pot = pot + V[ix-1,iy]
                    if (ix<Lx-1):
                        ncount = ncount + 1.0
                        pot = pot + V[ix+1,iy]
```

```
                if (iy>0):
                    ncount = ncount + 1.0
                    pot = pot + V[ix,iy-1]
                if (iy<Ly-1):
                    ncount = ncount + 1.0
                    pot = pot + V[ix,iy+1]
                Vnew[ix,iy] = pot/ncount
            else:
                Vnew[ix,iy]=V[ix,iy]
        V, Vnew = Vnew, V # Swap pointers to arrays
    return V
```

The `if (np.isnan(b[ix,iy])):` condition ensures that it is only where `b` is equal to `nan` that new values are calculated in each step. In positions where `b` is not equal to `nan`, the boundary conditions values are used instead. This approach allows us to address different types of geometries. Such as this case with two finite planes with given potentials. The resulting plots are shown in Fig. 6.9. This method is robust and can be used to model many types of systems in detail.

```
N = 100
a = 1.0 # meters
x_A, x_B, y_A, y_B = -a,a,-a,a
x = np.linspace(x_A,x_B,N)
y = np.linspace(y_A,y_B,N)
rx,ry = np.meshgrid(x,y,indexing="ij")
# Set up matrix of b-values
b = np.zeros((N,N),float)
b[:] = float("nan")
L14 = int(N*0.25)
L34 = int(N*0.75)
b[int(N*0.25),L14:L34] = -1.0
b[int(N*0.75),L14:L34] = 1.0
# Visualize results
nrep = 100000
V = solvepoissonneumann2d(b,nrep)
dx = x[1]-x[0]
Ex,Ey = np.gradient(-V)
plt.contourf(rx,ry,V)
plt.quiver(rx,ry,Ex,Ey)
plt.colorbar(), plt.axis("equal")
```

6.4 Numerical Solutions: Implicit Methods

Above we introduced the method of Jacobi iterations to find an approximative solution to Laplace's equation for the two-dimensional boundary value problem. This is just one of many methods to solve Laplace's equation in general and the discretized version of Laplace's equation in particular. There are also other strategies such as

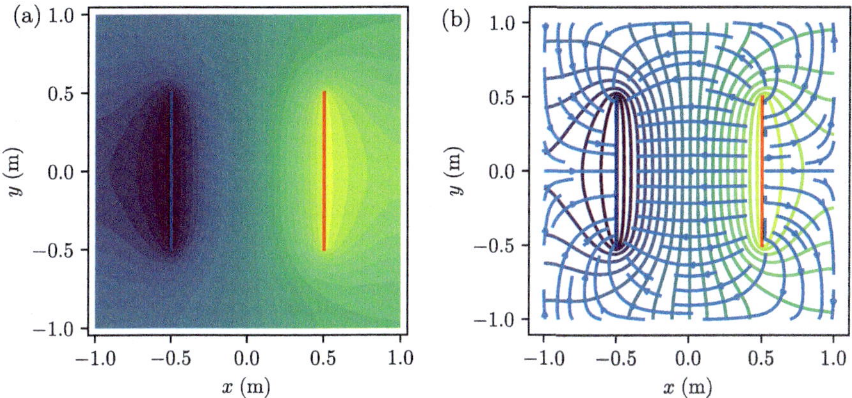

Fig. 6.9 Plot of $V(x, y)$ (**a**) and $\mathbf{E}(x, y)$ (**b**)

finite element methods and machine learning methods. Here, we will generalize the solution methods for the discretized equations and discuss their properties.

We found that in two dimensions Laplace's equation is discretized to:

$$\frac{\partial^2 V}{\partial x^2} + \frac{\partial^2 V}{\partial y^2} = V_{i+1,j} + V_{i-1,j} + V_{i,j+1} + V_{i,j-1} - 4V_{i,j} = 0, \tag{6.24}$$

where boundary conditions can be in the form of specified values of $V(x, y)$ on a boundary S. Finding a *solution* to this discretized version of Laplace's equation means to find values $V_{i,j}$ that satisfies both (6.24) and the boundary conditions. The method of Jacobi iterations provides an *approximative* solution. We did not find the values $V_{i,j}$ that exactly satisfied (6.24), but values that were sufficiently near the exact solution. Let us be more precise by following the approach introduced in Ida (2015), where we look at a small lattice to enumerate the problem exactly.

Description of the System

We address the electric potential $V(x, y)$ on a quadratic system of length $L \times L$ with boundary conditions $V = 0$ on the lines $x = 0$, $x = L$ and $y = L$ and $V = V_0$ on the line $y = 0$. The system is discretized using a $N \times N$ elements square lattice with spacing $\Delta x = L/(N-1)$ so that $V_{i,j} = V(i\Delta x, j\Delta x)$, where i and j run from 0 to $N-1$. This system is illustrated in Fig. 6.10 with the detailed enumeration for $N = 6$. For this system, there are 16 internal points, where we need to find the values for $V_{i,j}$ from (6.24), and there are 20 boundary values that are given.

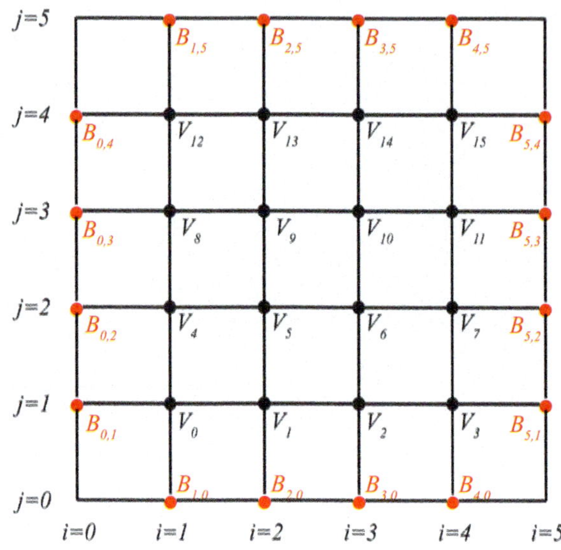

Fig. 6.10 Illustration of an 6×6 system of points for the solution of $V(x, y)$ on the lattice $V_{i,j}$. Boundary nodes are illustrated in red and internal nodes are illustrated in black

Equations for the Unknown Values of $V_{i,j}$

Each value of $V_{i,j}$ are determined from (6.24). For points that do not have any boundary values as its nearest neighbors, such as for (2, 3), the equation is

$$V_{3,3} + V_{1,3} + V_{2,4} + V_{2,2} - 4V_{2,3} = 0, \tag{6.25}$$

where all the values $V_{i,j}$ are unknowns. However, for points that are near the boundary, such as for (1, 3), the equation still has the same form:

$$V_{2,3} + V_{0,3} + V_{1,4} + V_{1,2} - 4V_{1,3} = 0, \tag{6.26}$$

but we replace $V_{0,3}$ with its boundary value $B_{0,3}$, which is a given constant:

$$V_{2,3} + B_{0,3} + V_{1,4} + V_{1,2} - 4V_{1,3} = 0, \tag{6.27}$$

We get one such linear equation for each of the 16 internal points: 16 equations in total and 16 unknown values of $V_{i,j}$. We therefore have a set of linear equations that we need to solve to find the solution. These equations can be solved using an approximative method, such as Jacobi iterations, or with an exact method, such as Gaussian elimination.

Enumeration

However, if we are to write all 16 linear equations, it is better to introduce a new enumeration scheme for the points, where we use a single index n for each point (i, j). Here, we use the scheme: $n = (j - 1)(N - 2) + (i - 1)$ so that $n = 0, 1, \ldots, 15$. This scheme is illustrated in Fig. 6.10. We also change the sign of the equation to simplify the notation. At $(2, 3)$ we get:

$$V_{3,3} + V_{1,3} + V_{2,4} + V_{2,2} - 4V_{2,3} = 0 \rightarrow -V_{10} - V_8 - V_{13} - V_5 + 4V_9 = 0. \tag{6.28}$$

and at $(1, 3)$ the equation becomes:

$$V_{2,3} + B_{0,3} + V_{1,4} + V_{1,2} - 4V_{1,3} = 0 \rightarrow -V_9 - B_{0,3} - V_{12} - V_4 + 4V_8 = 0, \tag{6.29}$$

where we notice that we need to keep track of the boundary value $B_{0,3}$. We can rewrite this on standard form for linear equations, where the constants are on the right hand side:

$$- V_9 - V_{12} - V_4 + 4V_8 = B_{0,3}, \tag{6.30}$$

System of Equations

We write down all the 16 equations for the internal points. On matrix form this gives us the following system of equations:

$$\begin{bmatrix}
4 & -1 & 0 & 0 & -1 & 0 & 0 & 0 & 0 & 0 & 0 & 0 & 0 & 0 & 0 & 0 \\
-1 & 4 & -1 & 0 & 0 & -1 & 0 & 0 & 0 & 0 & 0 & 0 & 0 & 0 & 0 & 0 \\
0 & -1 & 4 & -1 & 0 & 0 & -1 & 0 & 0 & 0 & 0 & 0 & 0 & 0 & 0 & 0 \\
0 & 0 & -1 & 4 & 0 & 0 & 0 & -1 & 0 & 0 & 0 & 0 & 0 & 0 & 0 & 0 \\
-1 & 0 & 0 & 0 & 4 & -1 & 0 & 0 & -1 & 0 & 0 & 0 & 0 & 0 & 0 & 0 \\
0 & -1 & 0 & 0 & -1 & 4 & -1 & 0 & 0 & -1 & 0 & 0 & 0 & 0 & 0 & 0 \\
0 & 0 & -1 & 0 & 0 & -1 & 4 & -1 & 0 & 0 & -1 & 0 & 0 & 0 & 0 & 0 \\
0 & 0 & 0 & -1 & 0 & 0 & -1 & 4 & 0 & 0 & 0 & -1 & 0 & 0 & 0 & 0 \\
0 & 0 & 0 & 0 & -1 & 0 & 0 & 0 & 4 & -1 & 0 & 0 & -1 & 0 & 0 & 0 \\
0 & 0 & 0 & 0 & 0 & -1 & 0 & 0 & -1 & 4 & 1 & 0 & 0 & -1 & 0 & 0 \\
0 & 0 & 0 & 0 & 0 & 0 & -1 & 0 & 0 & -1 & 4 & -1 & 0 & 0 & -1 & 0 \\
0 & 0 & 0 & 0 & 0 & 0 & 0 & -1 & 0 & 0 & -1 & 4 & 0 & 0 & 0 & -1 \\
0 & 0 & 0 & 0 & 0 & 0 & 0 & 0 & -1 & 0 & 0 & 0 & 4 & -1 & 0 & 0 \\
0 & 0 & 0 & 0 & 0 & 0 & 0 & 0 & 0 & -1 & 0 & 0 & -1 & 4 & -1 & 0 \\
0 & 0 & 0 & 0 & 0 & 0 & 0 & 0 & 0 & 0 & -1 & 0 & 0 & -1 & 4 & -1 \\
0 & 0 & 0 & 0 & 0 & 0 & 0 & 0 & 0 & 0 & 0 & -1 & 0 & 0 & -1 & 4
\end{bmatrix}
\begin{bmatrix} V_0 \\ V_1 \\ V_2 \\ V_3 \\ V_4 \\ V_5 \\ V_6 \\ V_7 \\ V_8 \\ V_9 \\ V_{10} \\ V_{11} \\ V_{12} \\ V_{13} \\ V_{14} \\ V_{15} \end{bmatrix}
=
\begin{bmatrix} B_{0,1} + B_{1,0} \\ B_{2,0} \\ B_{3,0} \\ B_{4,0} + B_{5,1} \\ B_{0,2} \\ 0 \\ 0 \\ B_{5,2} \\ B_{0,3} \\ 0 \\ 0 \\ B_{5,3} \\ B_{0,4} + B_{1,5} \\ B_{2,5} \\ B_{3,5} \\ B_{5,4} + B_{4,5} \end{bmatrix}$$

This is a matrix equation on the form $Ax = B$, where the unknown x contains the potentials, V. Notice how you can read out the form of the equation from this matrix. You see that there is a 4 on the diagonal (this would be 2 in one dimension and 6 in three dimensions), which corresponds to $4V_i$ in the equation. You also see that there is a -1 for each of the nearest neighbors: These are the -1 that are next to the 4 on the diagonal: $-V_{i+1,j}$ etc. These are replaced by zeros for some of the rows: for the nodes that are next to a boundary. Then there are -1's that are offset by two zeros. These correspond to the nodes that are on the row above or below the current row. The number of positions they are offset from the diagonal (with value 4) correspond to the number of nodes in a row. See for yourself that the values in the matrix makes sense when you compare with Fig. 6.10.

Implicit Numerical Solution

We now have a system of linear equations that we can solve with any solution method that is efficient and precise. There are many tools in Python to help you solve such systems of linear equations. Here, we will set up the equations and solve them with a simple tool.

Setting up the matrices. We now generate the matrix $A_{n,m}$ and the vector B_n by inserting the relevant values for each (i, j). We look through each point (i, j) and insert a 4 at the corresponding $A_{n,n}$. We then insert a -1 for each neighbor $(i + 1, j)$, $(i - 1, j)$, $(i, j + 1)$ and $(i, j - 1)$ at the corresponding position in A, unless they are part of the boundary, in which case we instead add the boundary value to the B-vector. We write a function to set up the system of equations, solve the system, and put the values back into an array that includes both the internal and the boundary values for V:

```
def poissonimplicit(b):
    N = b.shape[0]
    # Setting up system of equations
    LM = (N-2)*(N-2)
    A = np.zeros((LM,LM),float)
    B = np.zeros((LM),float)
    for j in range(1,N-1):
        for i in range(1,N-1):
            # 4Vi,j-Vi+1,j-Vi-1,j-Vi,j+1-Vi,j-1
            n = (j-1)*(N-2) + (i-1)
            A[n,n] = 4 # Diagonal
            if (i>1):    A[n,n-1] = -1
            else:        B[n] = B[n] + b[i-1,j]
            if (i<N-2): A[n,n+1] = -1
            else:        B[n] = B[n] + b[i+1,j]
            if (j>1):    A[n,n-(N-2)] = -1
            else:        B[n] = B[n] + b[i,j-1]
            if (j<N-2): A[n,n+(N-2)] = -1
            else:        B[n] = B[n] + b[i,j+1]
    x = np.linalg.solve(A,B)
    # Generate full V matrix
    V = np.zeros((N,N),float)
```

```
    for j in range(0,N):
        for i in range(0,N):
            if (i<1)or(i>=N-1)or(j<1)or(j>=N-1):
                V[i,j] = b[i,j]
            else:
                n = (j-1)*(N-2) + (i-1)
                V[i,j] = x[n]
    return V
```

We then set up the boundary condition in the array b and solve the system:

```
import numpy as np
import matplotlib.pyplot as plt
# Setting up the boundaries
N = 6
b = np.zeros((N,N),float)
b[:] = float("nan")
b[:,0] = 1.0
b[:,N-1] = 0.0
b[0,:] = 0.0
b[N-1,:] = 0.0
# Setup physical model
a = 1.0 # meters
x_A, x_B, y_A, y_B = -a,a,-a,a
x = np.linspace(x_A,x_B,N)
y = np.linspace(y_A,y_B,N)
rx,ry = np.meshgrid(x,y,indexing="ij")
# Solve equation
V = poissonimplicit(b)
# Visualize
plt.contourf(rx,ry,V), plt.axis("equal")
```

The resulting plot is shown in Fig. 6.11. Notice that this is an exact solution of the linear system of equations, which approximates Laplace's equation. It is not necessarily an exact solution to Laplace's equation!

6.5 Proofs and Notebooks

Proof that the Extremes of V are on the Boundaries

Let us assume that V is a solution to Laplace's equation in a region v and has a maximum in a point inside v. This implies that V will decrease away from this point, which again implies that the gradient ∇V must point in toward the point. The electric field $\mathbf{E} = -\nabla V$ must therefore point out from the point. Gauss' law then implies that there must be a positive free charge in this point. But this cannot be true for Laplace's equation, since $\rho = 0$ in the region v. Consequently, V does not have any local maxima inside v. We can apply a similar argument for a minimum in a point inside v.

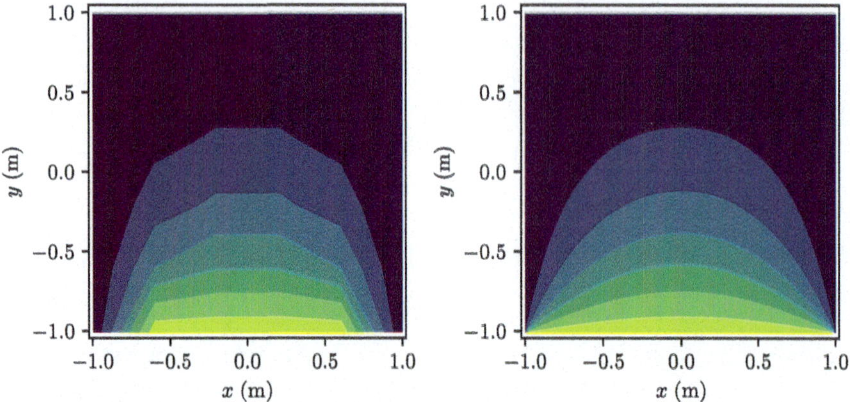

Fig. 6.11 Plot of $V(x, y)$ for the implicit solution of Laplace's equation for $N = 6$ (left) and $N = 100$ (right)

Proof that Laplace's Equation has a Unique Solution

We can use this to prove that Laplace's equation has a unique solution in υ given that the values of V on the boundaries are specified, $V(S)$. Let us assume that there are *two* solutions to Laplace's equation, V_1 and V_2, that both satisfy Laplace's equation in υ. We then define $V = V_1 - V_2$. If the two solutions V_1 and V_2 are identical at the boundaries, we notice that V must be zero at the boundaries. Because V is a solution to Laplace's equation the extreme must be on the boundary. This means that $V = 0$ in the whole volume υ, and we conclude that $V_1 = V_2$. The solution must therefore be unique.

Notebooks

You can find several relevant notebooks addressing elements from this chapter here: Solving Laplace's equation in inhomogeneous dielectric media; Polarized water bubbles; Machine-learning methods to solve Laplace's equation.

Summary

Poisson's equation can be used to find the electric potential in a volume with a given charge density ρ: $\nabla^2 V = -\rho/\epsilon$.

Laplace's equation can be used to find the electric potential in a volume with zero charge density: $\nabla^2 V = 0$.

Laplace's and Poisson's equations have a unique solution on a volume v if the boundary conditions on the surface enclosing v are given.

Dirichlet boundary conditions provide a given value for the potential at the boundaries. **Neumann boundary conditions** provide a given value for the gradient of the potential at the boundaries.

We have now established several methods to relate the charge density ρ, the electric field $\mathbf{E}$ and the electric potential V:

$$
\begin{aligned}
\rho_v \rightarrow \mathbf{E} &: \quad \mathbf{E} = \frac{1}{4\pi\epsilon_0}\int_v \frac{\rho_v \mathbf{R}}{R^3}\,\mathrm{d}v' \\
\mathbf{E} \rightarrow \rho_v &: \quad \nabla \cdot \mathbf{E} = \rho_v/\epsilon_0 \quad (\nabla \times \mathbf{E} = 0) \\
\rho_v \rightarrow V &: \quad V = \frac{1}{4\pi\epsilon_0}\int_v \frac{\rho_v}{R}\,\mathrm{d}v' \\
V \rightarrow \rho_v &: \quad \nabla^2 V = -\rho_v/\epsilon_0 \\
V \rightarrow \mathbf{E} &: \quad \mathbf{E} = -\nabla V \\
\mathbf{E} \rightarrow V &: \quad V = -\int_{\mathrm{r}}^{\mathrm{ref}} \mathbf{E} \cdot \mathrm{d}\mathbf{l}
\end{aligned}
$$

Exercises

Discussion Exercises

6.1 Laplace operator. In one dimension, a linear curve is a solution to Laplace equation. Are there any other solutions that are not linear? In two dimensions, a plane may also be a solution to the Laplace equation. Are there any other solutions and what would that depend on?

6.2 Discontinous potential. Is it possible to have a discontinuous potential? Provide examples or arguments for why it is not possible. (Hint: $\mathbf{E} = -\nabla V$).

Tutorials

6.3 Laplace operator in one dimension. We call ∇^2 the Laplace operator. In this exercise, we will address its properties. Figure 6.12 shows two differential possible potential $V(x)$.

(a) Where is the Laplace operator applied to the function, $V(x)$, $\mathrm{d}^2V/\mathrm{d}x^2$, negative, zero or positive in the two figures?

(b) Can the potential, $V(x)$, to the left in the figure be a solution to Laplace's equation?

(c) Can the potential, $V(x)$, to the left in the figure be a solution to Poisson's equation? In that case, what does the charge distribution look like?

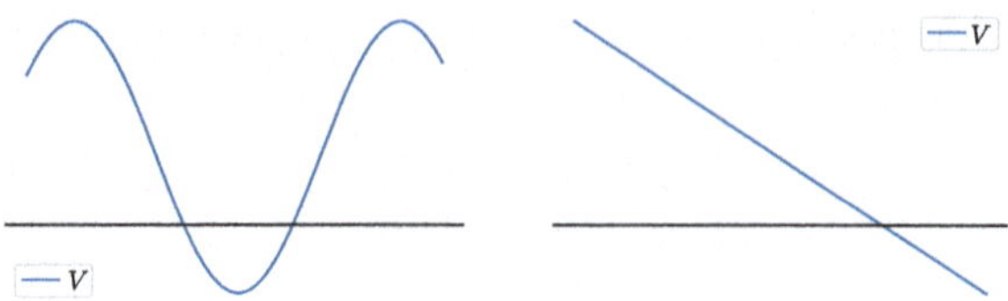

Fig. 6.12 Two possible potentials $V(x)$

(d) Can the potential, $V(x)$, to the right in the figure be a solution to Laplace's equation?

6.4 Boundary conditions in one dimension. You know that the potential is 1.5 V higher in $x = 0$ cm than in $x = 1$ cm. You can assume that there are no charges between or outside these two points. We want to find the electric potential along the x-axis using Laplace's equation.
(a) Make a sketch of the system. What do you know about the *boundary conditions*? What is the potential infinitely far away?
(b) Explain why we can use Laplace's equation to find the electric potential along the x-axis.
(c) Write down Laplace's equation and the boundary conditions you know. Find a *general solution* of the equation, that is, a solution to the equation without having used the boundary conditions yet.
(d) Find the *special solution* of Laplace's equation, that is, the solution that is consistent with the boundary conditions.
(e) Find the electric field.
(f) Do you recognize this electric field? What kind of physical system do you think this corresponds to in the real, three-dimensional world?

6.5 The Laplace operator in two and three dimensions. Figure 6.13 shows examples of potential surfaces $V(x, y)$.
(a) What is the sign of the Laplace operator applied to $V(x, y)$ in the points A, B, C, D, E, and F?
(b) The Laplace operator in two dimensions is $\nabla^2 V = \frac{\partial^2 V}{\partial x^2} + \frac{\partial^2 V}{\partial y^2}$. Which of these functions may satisfy Laplace's equation? $V_A(x, y) = Ax + By$, $V_B(x, y) = A \sin x + B \sin y$; $V_C(x, y) = A\, e^{-y} \sin x$

6.6 The Laplace-equation in cylindrical coordinates. In cylindrical coordinates the Laplace equation is given as

$$\nabla^2 V = \frac{1}{r}\frac{\partial}{\partial r}\left(r\frac{\partial V}{\partial r}\right) + \frac{1}{r^2}\frac{\partial^2 V}{\partial \phi^2} + \frac{\partial^2 V}{\partial z^2}$$

(a) Check that this equation has the correct units. (This is often a smart check to see if you have written the formula correctly. All parts must have the same unit.)

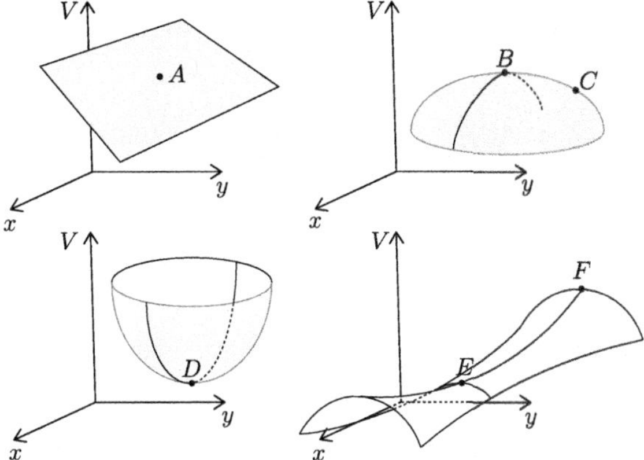

Fig. 6.13 Illustration of four different potential surfaces $V(x, y)$

(b) Show that the potential of an infinitely long line charge, $V(r) = A \ln r + B$, satisfies Laplace's equation.

6.7 The Laplace-equation and dimensionality. Your friend Q tells that she has found an error in the electromagnetic laws. She imagines that she places a small charge Q with a small radius a in the origin. The surface of the charge has a potential V_0. If she now solves Laplace's equation in one dimension, she find that the electric potential must decay or increase linearly away from the charge. If she solves Laplace equation in two dimensions the electric potential is a logarithm of the distance to the charge, $V(r) = A \ln r + B$. But she knows that the electric potential to a single charge should be $V(r) = \frac{Q}{4\pi\epsilon_0 r}$ from Coulomb's law.

Here, something is obviously wrong, she says. Maxwell's equations or Laplace's equation must be wrong. Help me solve this, so that we can get the Nobel prize together! What is wrong with the argument?

6.8 The discrete Laplace operator. The discrete Laplace operator is:

$$\bar{\nabla}^2 V = \frac{1}{\Delta x^2} \left(V(x + \Delta x) + V(x - \Delta x) - 2V(x) \right)$$

Figure 6.14 shows three potentials $V(x)$.
(a) What is the discrete Laplace operator in $x = 0$ in case A?
(b) What is the discrete Laplace operator in $x = 0$ in case B?
(c) What is the discrete Laplace operator in $x = 0$ in case C?

6.9 Manual Jacobi iteration. We solve Laplace's equation numerically in one dimension in the points $x_0 = 0$, $x_1 = 1$, $x_2 = 2$, $x_3 = 3$, $x_4 = 4$. The boundary conditions are $V(x_0) = 4$ and $V(x_4) = 0$.

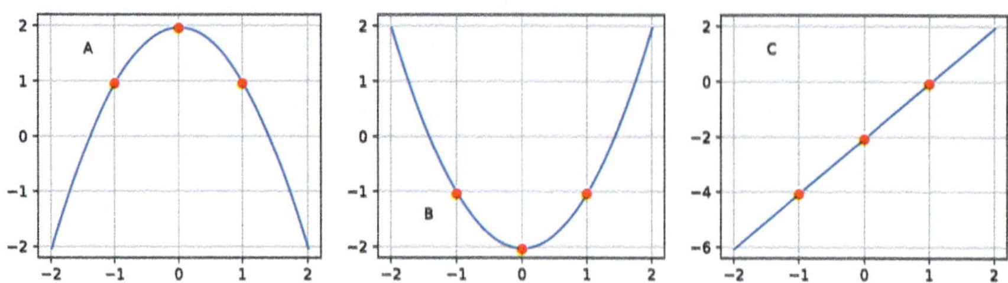

Fig. 6.14 Three potentials $V(x)$

(a) What initial values do you choose for V_i for $i = 0, \ldots, 4$?
(b) What are the V_i values after the first Jacobi-iteration?
(c) Do 5 Jacobi iterations, record and find V_i.
(d) What do you expect V_i to converge toward?

6.10 Numerical solution of Laplace's equation. We solve Laplace's equation numerical in two dimensions using the program `solvepoisson` introduced in this chapter. Copy the code from the example and test-run it to see that it reproduces the figures. Let us test perturbations to the code.
(a) What is the solution if the boundary conditions are $V = 0$ on all the outer boundaries?
(b) What is the solution if the boundary conditions are $V = 0$ on all the outer boundaries and $V = 1$ in a point in the middle?
(c) What is the solution if the boundary conditions are $V = 0$ on all the outer boundaries and $V = 1$ on a circle in the middle with radius of 0.1 of the width of the system?
(d) How can we find the electric field from the potential? Plot the electric field in the three cases above.
(e) How can you find the electric potential along a line along the x-axis? Find and plot the electric potential as a function of x for part (b) and (c) above. What functional forms do you think it is reasonable to compare with in these cases?
f) You friend Q says that he has invented a brilliant way to protect himself from electric field—by sitting inside a closed surface that has a constant electric potential. Is this correct? Use results you have found here to argue for or against this position. How do you think this can be realized? (We will address this in the next chapter).

Exercises

6.11 Laplace operator. Which of the following scalar potentials are or can be solutions to Laplace's equation and Poisson's equation?

(a) $V(x) = ax + b$
(b) $V(x) = A \sin \omega x$
(c) $V(x) = c$
(d) $V(x, y) = Ax + By$
(e) $V(r, \theta, \phi) = a/r + b$
(f) $V(x, y) = ax^2 y$

6.12 Spatial variation of potentials (Laplace equation). We will study a system in the xy-plane. At $x = 0$ and $x = L$ the scalar potential is $V_0 = 0$. The volume charge density between $x = 0$ and $x = L$ is zero, and the dielectric constant is ϵ.
(a) Make a sketch of the system. What quantities is it useful to include in the sketch?
(b) What is the scalar potential in the region between $x = 0$ and $x = L$? Be very precise in how you determine this and what assumptions you make. What is the electric field in this region?

We now introduce a new feature in the system: The potential at $x = L/2$ is $V_1 > V_0$. The volume charge density is still zero, and the dielectric constant is ϵ.
(c) Update your drawing to reflect this.
(d) Find the scalar potential $V(x, y)$ everywhere between $x = 0$ and $x = L$. What is the electric field?
(e) What kind of system could this situation represent?
f) What is the surface charge density on the surfaces at $x = 0$, and $x = L$. Discuss what happens at $x = L/2$.

6.13 Surface charges. Consider a sphere with radius a with a charge Q uniformly distributed on its surface.
(a) What is the electric field as a function of r? (Find the behavior for both $r < a$ and $r > a$)
(b) What is the scalar potential, $V(r)$?
(c) Does $V(r)$ satisfy Laplace equation? (Where does it/does it not satisfy the equation).
(d) Use your result for E to find the surface charge at $r = a$.

6.14 Laplace's equation in two dimensions. The Python programs provided in the text provides skeleton programs to solve Laplace's equation on a rectangular area with a given set of Dirichlet boundary conditions on the boundary S, $V(S)$.
(a) Apply the program to the test case where $V(x = 0, y) = 0$ and $V(x = L, y) = V_1$. How can you check that the results are correct? Find the exact solution and compare with the numerical results.
(b) Use the program to find V and $\mathbf{E}$ for the case when $V = 0$ for $x = 0$ and $x = L$, and $V = V_1$ for $y = 0$ and $y = L$. Discuss the potential and field you get. Does it correspond to your intuition?
(c) Use the program to explore a situation of choice. Predict what you will get first, and then run the program to test your intuition.

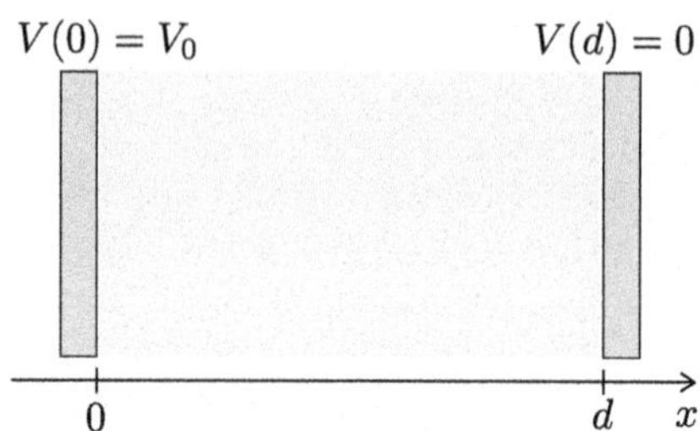

Fig. 6.15 System of two metal plates.

Homework

6.15 Laplace's equation for a long cylinder. In cylindrical coordinates, Laplace's equation is given as

$$\nabla^2 V = \frac{1}{r}\frac{\partial}{\partial r}\left(r\frac{\partial V}{\partial r}\right) + \frac{1}{r^2}\frac{\partial^2 V}{\partial \phi^2} + \frac{\partial^2 V}{\partial z^2} = 0.$$

In this exercise we will address an infinitely long cylindrical shell along the z-axis as a model for a cabel or the axon in a nerve cell.
(a) Explain why the potential V only can depend on r and not on ϕ or z.
(b) We assume that the potential is V_0 on the inside of the cylindrical shell at $r = a$ and $V_1 = 0$ on the outside of the cylindrical shell at $r = b$. There are no free charges in the area between $r = a$ and $r = b$. Find the electrical potential as a function of r by solving Laplace's equation.

6.16 Charge distribution between metal plates. Two plane metal plates with a spacing of $d = 1$ cm have a uniformly distributed volume charge density $\rho_v = -10^{-5}$ C/m^3 in the area between them. The medium between the plates have relative permittivity $\epsilon_r = 1$. One plate is grounded ($V = 0$) while to other plate is at a potential $V_0 = 10$ V. The system is illustrated in Fig. 6.15.
(a) Find the electric potential between the plates as a function of x when we assume that the plates have infinite extent. (Hint: Poisson's equation gives a second order differential equation in one variable. Solve this equation with boundary conditions $V(0) = V_0$ and $V(d) = 0$.)
(b) Find the electric field as a function of x.
(c) For which value of x does the potential have a minimum? Find V_{min}.
(d) Sketch the potential $V(x)$, and the x-component of the electric field as a function of x.

6.17 Poisson's equation in one dimension. Poisson's equation is

$$\nabla^2 V = -\frac{\rho}{\epsilon},$$

and is valid in this form when the permittivity does not vary over the area we solve the equation. We can solve Poisson's equation for example using finite differences by approximating the second derivative of the potential as:

$$\frac{\mathrm{d}^2 V_i}{\mathrm{d}x^2} \approx \frac{V_{i+1} - 2V_i + V_{i-1}}{\Delta x^2}$$

where $V_i = V(x_i)$, and we evaluate V in discrete points x_i.

We see from the form of the finite difference approximation that we cannot solve the equation by forward integration, as we do when we integrate the equations of motion using Forward Euler or Euler-Cromer's method. If we had started on the left side and tried to find V_{i+1} from V_i and V_{i-1}, we would not necessarily reached the correct boundary condition on the right hand side. Similarly, if we started on the right hand side, we would not have reached the correct boundary condition on the left hand side.

The solution is to use an *implicit* solver. We can do this by posing the complete set of equations as a matrix equation and solving this matrix equation by inverting the matrix,

$$A\mathbf{V} = \frac{\text{æ}}{\epsilon} \tag{6.31}$$

$$\mathbf{V} = A^{-1}\text{æ}. \tag{6.32}$$

where A is a matrix and $\mathbf{V}$ and **æ** are column vectors.

(a) Construct the matrix A so that $A\mathbf{V} = \frac{1}{\epsilon}\text{æ}$, where $\mathbf{V}$ and **æ** now are column vectors that contain the values of respectively V and ρ in discrete points. For now ignore what happens at the boundaries of the region of solution.

We need to make an additional consideration when it comes to the boundary values. When we multiply the matrix with V, the resulting boundary values we get for ρ will not reflect the elements outside the diagonal, as is the case for the elements inside ρ.

(b) Modify the first and the last element in ρ so that you can choose a specific potential value on the boundary of the region (Dirichlet boundary conditions).

(c) Write a program that takes the boundary conditions and the charge distribution as inputs, and return the potential. Test the program for the charge distribution from exercise 6.16.

6.18 Poisson's equation.

(a) Consider a cylinder with height $h = 1\text{m}$ and radius $r - 1\text{m}$. The inside of the cylinder is hollow, and the walls are very thin. Use Poisson's equation to find the electric potential on the walls of the cylinder. The bottom of the cylinder has electric potential $V = 10\text{V}$, and the top of the cylinder has zero potential. The cylinder is electrically neutral. (Hint: If you fold the cylinder out you get a square.)

(b) Find a difference equation that approximates the solution Poisson's equation, in one dimension. (Hint: Use the approximation $\mathrm{d}V_n/\mathrm{d}x \approx (V_{n+1} - V_n)/\Delta x$.)

(c) Imagine that we do not know the theoretical solution. We then need to find it numerically. Use the difference equation over to solve Poisson's equation numerically for the cylinder. (Hint: Notice that you only know the top and bottom conditions. You need to test for different initial conditions until you find a solution that fits.)(Hint: To test the initial conditions, you need to compare your result (with the tested initial conditions) with the potential at the bottom of the cylinder (that you know).)(Hint: A good testing range is $V_{n+1} \in [10\text{V}, 8\text{V}]$.)

6.19 Potential. In this exercise we will address a one-dimensional system with charge density $\rho(x)$. Assume that the electric potential is $V(0) = 0$ in the point $x = 0$ and $V(L) = V_0$ in the point $x = L$.
(a) Find the electric potential on the interval $0 < x < L$ when $\rho(x) = 0$.
(b) Find the electric potential on the interval $0 < x < L$ when $\rho(x) = \rho_0$, which is a constant.

6.20 Earnshaw's theorem. Earnshaw's theorem (in electrostatics) states that a charge Q cannot be kept in stable equilibrium by a set of charges Q_i. This means that it is not possible to make an electrostatic trap from charges alone. However, your friend Q has devised what he believes to be a charge trap. The system consists of eight equal positive charges, $+Q$, placed on the corners of a cube. This should keep a positive charge $+q$ inside the cube. Just look, he says, if a place a positive charge in the middle of the cube, all the forces cancel. Is he right, does his configuration provide a stable equilibrium for a positive charge? Is there any point inside the cube where a positive charge can be in a stable equilibrium? Can you provide an argument for Earnshaw's law from Laplace's equation? (Earnshaw's theorem is important for the design of fusion reactors, where we need to confine a hot plasma, that is, a hot fluid of charged particles. Earnshaw's theorem states that this plasma cannot be confined by electric fields alone.) (Hint: What does it mean that a point is a stable equilibrium? Use Laplace's equation to discuss the possibility of a minimum in the potential energy of the charge inside the box.)

6.21 Exact and numerical solution. We have addressed the boundary value problem for Laplace's equation where the potential is defined on the boundaries of an $a \times a$ square: $\nabla^2 V = 0$ and $V(x, 0) = 0$, $V(x, a) = 0$, $V(0, y) = V_0$ and $V(a, y) = 0$. For this problem we find a numerical solution. However, it is possible to find the exact solution for this problem in the case when we known that $V(x, y) \to 0$ when $x \to \infty$ instead of $V(a, y) = 0$. The solution is

$$V(x, y) = \frac{2V_0}{\pi} \tan^{-1} \frac{\sin(\pi y/a)}{\sinh(\pi x/a)}.$$

Compare the numerical solution to the exact solution by plotting $V(x, y)$ along a line $x = a/2$ and along a line $y = a/2$ for both the numerical and the exact solution.

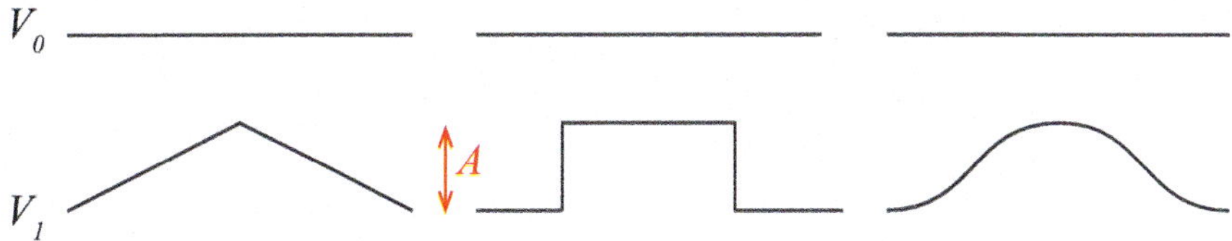

Fig. 6.16 Three surface structures

Modeling Projects

6.22 Dielectric breakdown in periodic geometries. You are building a small electrical component that has an irregular surface structure. Three such structures are shown in Fig. 6.16. You will apply a huge voltage across the gap—between surfaces A and B—and you worry that you may get dielectric breakdown. Your plan is to use the program that we have developed to find the potential:

```
import numpy as np
import numba
@numba.jit(cache=True)
def solvepoissonperiodic(b,nrep):
    # b = boundary conditions
    # nrep = number of iterations
    # returns potentials
    V = np.copy(b)
    for i in range(len(V.flat)):
        if (np.isnan(b.flat[i])):
            V.flat[i] = 0.0
    Vnew = np.copy(V)
    Lx,Ly = b.shape
    for n in range(nrep):
        for ix in range(Lx):
            for iy in range(Ly):
                etot = 0.0
                pot = 0.0
                if (np.isnan(b[ix,iy])):
                    if (ix>0):
                        etot = etot + 1
                        pot = pot + V[ix-1,iy]
                    if (ix<Lx-1):
                        etot = etot + 1
                        pot = pot + V[ix+1,iy]
                    iy1 = iy-1
                    if (iy1<0):
                        iy1 = Ly-1
                    etot = etot + 1
                    pot = pot + V[ix,iy1]
                    iy1 = iy+1
                    if (iy1>Ly-1):
                        iy1 = 0
```

```
                etot = etot + 1
                pot = pot + V[ix,iy1]

                Vnew[ix,iy] = pot/etot
            else:
                Vnew[ix,iy]=V[ix,iy]
        V, Vnew = Vnew, V # Swap pointers to arrays
    return V
```

to explore the electric field (the difference in potential between two neighboring points). We expect the dielectric material to break down first in a the point where the electric field is the largest. You can specify the structure in the b-array and then calculate the electric potential and field. For example, you can specify a cosine-shaped surface in the following way:

```
N = 200
a = 1.0 # meters
x_A, x_B, y_A, y_B = -a,a,-a,a
x = np.linspace(x_A,x_B,N)
y = np.linspace(y_A,y_B,N)
rx,ry = np.meshgrid(x,y,indexing="ij")
b = np.zeros((N,N),float)
b[:] = float("nan")
b[N-1,:]=-1.0
A = 30
P = 1.0
for iy in range(N):
    h = int(5+A - A*np.cos(iy/N*2*np.pi*P))
    b[0:h,iy] = 1.0
# Visualize the boundary conditions
plt.pcolormesh(rx,ry,b), plt.colorbar()
```

(a) Study the three suggested structures to see which one would be best for this purpose.
(b) Explore how the resolution (N) and discreteness of the simulation affects the results. (Remember to include the lattice spacing Δx when you calculate the electric field correctly: $E_{x,i,j} = (V_{i+1,j} - V_{i,j})/\Delta x$). Can we use this type of modelling to answer this type of question?

6.23 Modelling a 3D ion trap. In 1989, Wolfgang Paul and Hans Georg Dehmelt received a Nobel prize in physics for the creation of the *Paul trap*, although the actual trapping technique was developed by them back in the 1950s. The trapping technique allows us to trap and study properties of particles and atoms predicted by quantum mechanics. For example, at CERN physicists trap antiparticles using such a trap to measure their properties and compare them with theoretical predictions. Such traps can also be used to trap the fundamental computing units in quantum computers, known as *qubits*.

In this exercise, we'll look at how we can create such a trap using tools from electromagnetism. For our purpose we'll assume we're going to trap ions with a charge $q > 0$ and mass m. We'll assume that the trap is in a vacuum with no charge distribution within.

(a) We want a particle to remain in a small region by creating a spherical trap from a continuous charge distribution to create a generalized harmonic oscillator potential (similar to three decoupled springs):

$$V(x, y, z) = A(x^2 + y^2 + z^2),$$

where A is a constant. This would create a minimum in the potential energy in the origin. Unfortunately, nature is not that simple. Show that this potential cannot exist in a vacuum. (Hint: Use Laplace's equation.)
(b) To create a potential that can exist in a vacuum, we can modify the potential to be

$$V(x, y, z) = A(\alpha x^2 + \beta y^2 + \gamma z^2),$$

where $\alpha, \beta, \gamma \neq 0$ are real constants. Find constraints on these constants such that the potential can exist in a vacuum and show that from a suitable choice of constraints we can obtain the potential (there's more than one choice)

$$V(x, y, z) = A(x^2 + y^2 - 2z^2),$$

(c) Figure 6.17 shows a simplified experimental setup consisting of two finite paraboloid-shaped end-caps and a ring which will serve as electrodes. To trap positive ions, we want the end caps to be held at a static positive potential V_0 and the ring to be held at at static negative potential $-V_0$. Let the radius of the ring be r_0 and the end caps be placed a length z_0 from the center of the ring. Using cylindrical coordinates $x = r\cos\phi$, $y = r\sin\phi$ and $z = z$, the potential can be written as $V(r, z) = A(r^2 - 2z^2)$. Find the constant A and the relationship between r_0 and z_0 and use the results to show that the potential for the trap can be written as $V(r, z) = V_0(2z^2 - r^2)/r_0^2$.
(d) Show that this potential cannot trap the ion by using the second derivative test (Hessian determinant) or plot the surface.
So far we have found that a static electric field is *not* sufficient to trap charged particles in 3D. This is stated by *Earnshaw's theorem*. However, there is a way to fix this problem. Instead of working with a static potential, we replace the constant V_0 with the time dependent voltage $V_0\cos(\Omega t)$:

$$V(x, y, z, t) = \frac{V_0\cos\Omega t}{r_0^2}(2z^2 - x^2 - y^2),$$

(e) Using this potential, find the force on a particle with charge $q > 0$ and mass m. Solve the equations of motion numerically and plot the particle's trajectory in 3D. Are you able to trap the particle? Use $V_0 = 4000$ V, $\Omega = 100\pi$ Hz, $m \approx 5 \times 10^{-5}$ kg, and $q \approx 5 \times 10^{-5}$ C, which is the mass and charge of the cinnamon grain we will use for demonstration. A reasonable size for the trap is $z_0 \simeq 5$ mm.
(f) Instead we place N identical particles with charge $q > 0$ and mass m in the trap. Each particle, then, will experience a force from the field as well as from the

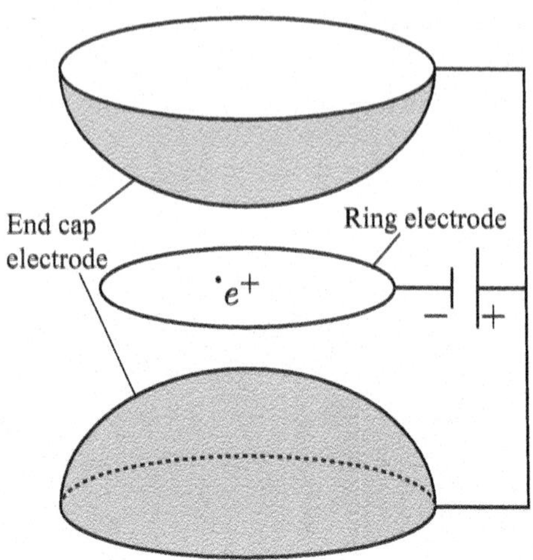

Fig. 6.17 A simple setup for the trap with (approximately) the potential $V(x, y, z)$. In reality the end caps must be paraboloids stretching out to infinity. Similarly, the ring must be a hyperboloid that stretches out to infinity in both directions. But inside the setup shown, $V(x, y, z)$ is a good approximation

electrostatic field formed by the charged particles themselves. Find the equations of motion for all the particles when you include the interaction between the particles and modify your code so it computes the position of all N particles. Compute particle trajectories for $N = 2$, $N = 5$ and $N = 15$ particles and plot their trajectories in 3D. Are you able to trap multiple particles?

References

Lawrence C. Evans. *Partial Differential Equations, Second Edition*. American Mathematical Society, 2010.

Nathan Ida. *Engineering Electromagnetics*. Springer, 2015.

Chapter 7
Ideal Conductors

7.1 Conductors

Conductors are materials that contain many charges—usually electrons—that can move around in the material. Metals are a class of materials that behave as conductors. Metals are found to the left in the periodic table (see Fig. 7.1) and the electrons in metals behave as a sea of electrons that can move around with very little resistance. This is essentially a quantum mechanical effect, but you can gain some insight into this from a classical model of the process: Metals have weakly bound electrons in the outer shells, which in metallic crystals are essentially free to move around as illustrated in Fig. 7.2. If the metal is not exposed to any electrical field, the electron cloud will surround the positive ions, making the system neutral with no macroscopic distribution of charges. Notice that the metal may be net neutral and still allow the electrons to move: the positive atoms remain, while the electrons move, allowing for a redistribution of charges within the system. However, not only metals are conductors, for example, plasmas and fluids with ions are also conductors. As you may know, while pure water is not a good conductor, water with dissolved salt can be a conductor, because the positive and negative salt ions will move if an electric field is applied.

Ideal Conductor

The charges in a conductor experience a form of resistance from the material. We call this effect electric resistance and will discuss it in detail later. In an *ideal conductor* we assume that the motion of the charges in response to an electric field is very rapid, so that it quickly reaches a state of equilibrium. What happens if we place an (ideal) conductor in an external electric field? This is illustrated in Fig. 7.3. Inside the circular conductor, there are charges that are free to move. Positive charges will move in the direction of the field (if they are free to move) and negative charges

© The Author(s), under exclusive license to Springer Nature Switzerland AG 2026
A. Malthe-Sørenssen, *Elementary Electromagnetism Using Python*, Undergraduate Texts in Physics, https://doi.org/10.1007/978-3-032-19876-1_7

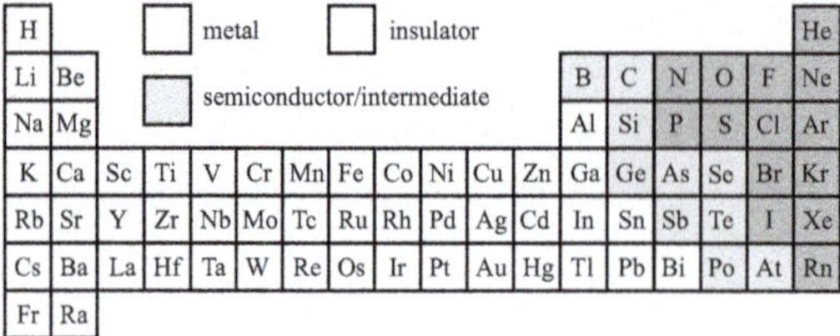

Fig. 7.1 Illustration of periodic table with metals and insulators

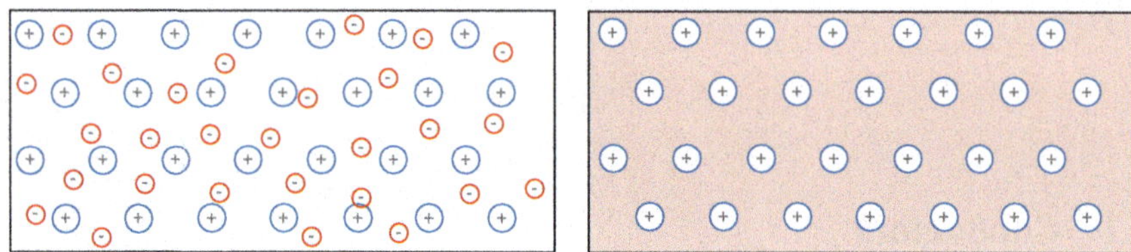

Fig. 7.2 Metallic atoms have electrons in the outer shell. In a crystal of metallic ions these electrons form a sea of conducting electrons that can move around with low resistivity illustrated by the red, smeared-out electrons on the right

will move in the direction opposite the electric field. (If only the electrons can move, the remaining atoms will be positively charged, and the effect will be the same as if both positive and negative charges can move). The negative charges will move in the direction of $-\mathbf{E}$ until they reach the end of the conductor—the surface. However, as the charges move, they will set up an electric field and change the total electric field in the system. Figure 7.3b illustrate the electric field set up by the charges on the surface of the conductor. The charges will build up at the surface until the electric field inside the conductor is zero: If the electric field is not zero, the charges will continue to move. Figure 7.3a shows the sum of the external field and the field set up by the surface charges in equilibrium. The electric field will then be zero inside the conductor. What are the consequences of this for an ideal conductor?

7.2 Properties of an Ideal Conductor in Equilibrium

When the ideal conductor is in equilibrium, that is, when all the charges have finished moving, the ideal conductor will therefore have the following properties which also are illustrated in Fig. 7.4.

Zero field inside a conductor. The electric field is zero inside an ideal conductor: $\mathbf{E} = 0$. If this was not the case, the electric field would move charges until they stop on the surface of the conductor. This process will continue, until field from these charges exactly cancels the external field, and the net field inside the conductor is zero.

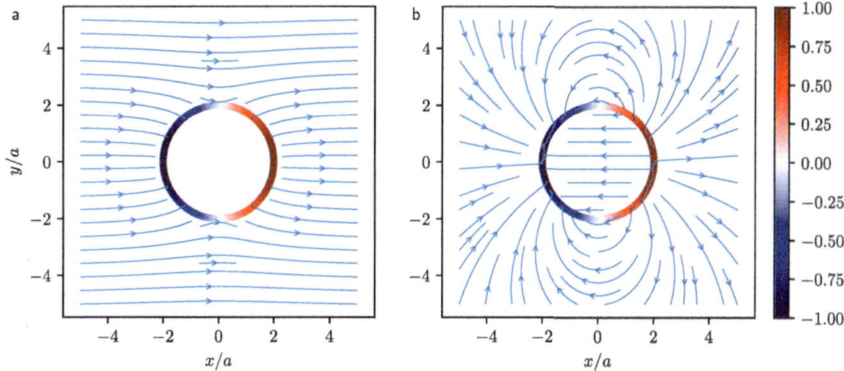

Fig. 7.3 The total electric field when a circular ideal conductor is placed in a uniform field. The field inside the conductor is zero. The colors show the surface charge density on the surface of the ideal conductor. The electric field from the surface charge density. These charges set up a uniform field inside the conductor, which cancels the external field in this region

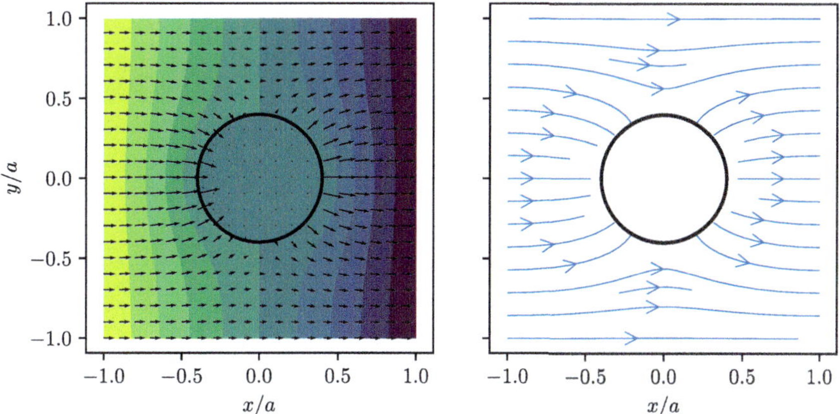

Fig. 7.4 Illustration of a conducting disk in a constant electric field showing that $\mathbf{E} = 0$ inside, the potential is constant inside the conductor, and $\mathbf{E}$ is normal to the surface

No free charges inside a conductor. There are no free charges inside an ideal conductor. We can see this from Gauss' law: Because $\mathbf{E} = 0$ inside the conductor, Gauss' law on differential form gives $\nabla \cdot \mathbf{E} = 0 = \rho/\epsilon_0$. All the charges must therefore be on the surface. The surface charge density is often called an *induced surface charge density*, since it often is the consequence of an external field.

Conductors are equipotential surfaces. The conductor must be an equipotential surface because the electric field is zero inside the conductor. We can see this by showing that the potential difference between two points A and B inside the conductor is zero: $V_{AB} = \int_A^B \mathbf{E} \cdot d\mathbf{l} = \int_A^B \mathbf{0} \cdot d\mathbf{l} = 0$, because $\mathbf{E} = \mathbf{0}$ inside the conductor.

There is no tangential field immediately outside a conductor. The tangential electric field is zero immediately outside the conductor. This is a consequence of the boundary conditions for the electric field and that the electric field is zero inside the conductor. The boundary condition across the boundary from inside to outside the conductor is $E_{2t} = E_{1t}$. Because E_t is zero inside the conductor, E_t must also be zero immediately outside the conductor.

The electric field immediately outside a conductor is $E_n = \rho_s/\epsilon$. The electric field immediately outside the conductor is normal to the surface of the conductor and is given as $E_n = \rho_s/\epsilon$. Because the tangential component is zero, there can only be a normal component of the field immediately outside the conductor. We can find the magnitude of this field from the boundary conditions, which states that $D_{2n} - D_{1n} = \rho_s$. On the inside the electric field and therefore also the displacement field is zero, $D_{1n} = 0$. We therefore get that $D_{2n} = \epsilon_0 E_n = \rho_s$ and $E_n = \rho_s/\epsilon_0$ when the outside is vacuum. If the conductor is a dielectric, we need to divide by ϵ instead. In general, we therefore have that immediately outside the conductor $\mathbf{E} \cdot \hat{\mathbf{n}} = \rho_s/\epsilon$. We can use this to find the electric field given that we know the surface charge density, and the opposite: We can find the surface charge density from the electric field.

Properties of an ideal conductor

- $\mathbf{E} = 0$ inside a conductor.
- $\rho = 0$ inside a conductor.
- A conductor is an equipotential surface.
- $E_t = 0$ immediately outside a conductor.
- $E_n = \mathbf{E} \cdot \hat{\mathbf{n}} = \rho_s/\epsilon$

Test your understanding

(a) A massive spherical ideal conductor has a net charge $Q > 0$. Where is the electric potential the largest? (b) A neutral sphere of copper has a spherical cavity in its center. A charge $+Q$ is placed in the center of the cavity (and the center of the sphere). What is the total charge on the outside of the copper sphere? (c) Two isolated copper spheres have a net charge $+Q$ each. If sphere A is bigger than sphere B, which sphere has a higher voltage?[1]

Example: A Conducting Plate in an External Field

In this example, we will demonstrate how we use the properties of a conductor to determine the charge distribution on the conductor in an electric field. An ideal conductor formed as a plate with area $A = a^2$ *and thickness* h *is placed in the*

[1] (a) Constant in the conductor; (b) $+Q$; (c) B.

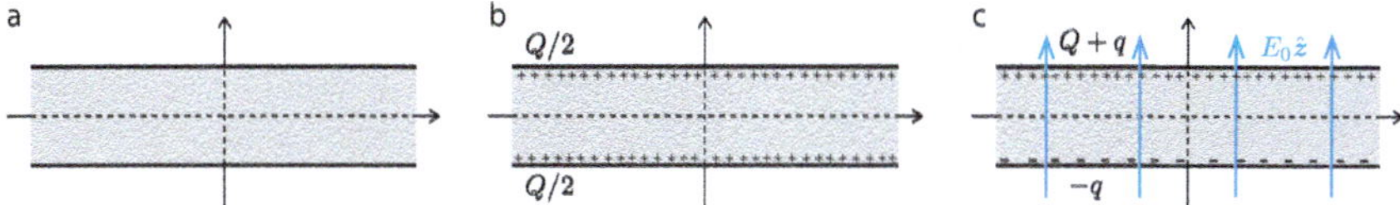

Fig. 7.5 **a** A conducting plate. **b** A charge $+Q$ is distributed on the top and bottom surfaces. **c** In an external field $E_0\hat{\mathbf{z}}$, the charge Q and charges $+q$ and $-q$ are distributed so that the net field inside the conductor is zero

xy-plane. You can assume that $a \gg h$ so that you can ignore the effects of the boundaries. The conductor has a total charge $+Q$. How are the charges distributed in the plate? If an electric field $\mathbf{E} = E_0\hat{\mathbf{z}}$ is applied to the system, how are the charges now distributed?

Approach. Our plan is to use the properties of an ideal conductor to infer the charge distribution. We will see how we can distribute the charges, so that the fields they set up ensure that the electric field inside the conductor is zero. We start by sketching the system in Fig. 7.5a, where we have drawn a cross-section of the system in the xz-plane.

Solution–part 1. A useful strategy is to start by assuming that the charges are uniformly distributed inside the conductor. This is not a stable situation. The positive charges making up the charge $+Q$ will repel each other. As a result, the charges will move out toward the boundaries. Since the plate is thin compared to its extent, we will assume that the charges only are on the top and the bottom surfaces in the figure. We realize that the charge density must be the same on both the top and the bottom surface, so that the electric field from the charges cancel inside the conductor. Therefore, there will be an equal charge $Q/2$ on the top surface and on the bottom surface, uniformly distributed on the surfaces as shown in Fig. 7.5b.

Solution–part 2. Then an electric field $\mathbf{E} = E_0\hat{\mathbf{z}}$ is applied. Again, we start by assuming that the charges are inside the conductor. They would then move in the direction of the electric field until they stop at the upper surface. Now, there are three possibilities. (i) The charge Q sets up an electric field that exactly matches the electric field $E_0\hat{\mathbf{z}}$. In this case, the system is in equilibrium. (ii) The charge Q on the top surface sets up an electric field that is larger than the applied electric field $E_0\hat{\mathbf{z}}$. The system is then still not in equilibrium. Some charges q will then move from the top surface to the bottom surface, so that the resulting net electric field inside the conductor is zero and the system is in equilibrium with a charge $Q - q$ on the top surface and a charge $+q$ on the bottom surface. (iii) The charge Q sets up an electric field that is smaller than the applied electric field $E_0\hat{\mathbf{z}}$. The system is then still not in equilibrium. Some charges q will then move from the bottom surface to the top surface, until the net electric field inside the conductor is zero. The system is then in equilibrium with charge $Q + q$ on the top surface and a charge $-q$ on the bottom surface as shown in Fig. 7.5c.

Discussion. Notice that charges will move around in the conductor until the system is in equilibrium and there is zero net electric field inside the conductor. However, this must be done in a way that does not change the total charge of the system.

Example: A Spherical Conductor

A spherical ideal conductor with radius a has a total charge q. Find the electrical potential and field everywhere.

Approach. To solve this problem, we will use key properties of ideal conductors. We know that the electric field inside the conductor is zero and that all the charges are on the surface. There are several ways to approach this problem.

1. We could use the spherical symmetry to assume that the charge is uniformly distributed on the surface. We have then mapped the problem onto a known problem: the case of a charged, spherical shell with radius a. For this system we know that the electric field outside the shell is the same as for a point charge in the origin (the center of the sphere), that is, $\mathbf{E} = E_r \hat{\mathbf{r}}$, $E_r = q/(4\pi\epsilon_0 r^2)$, and that the potential also will be the potential from a single point charge in the origin, $V(r) = q/(4\pi\epsilon_0 r)$.
2. Alternatively, we could use that the surface is an equipotential surface and then solve Laplace's equation.

7.3 Laplace's Equation in Systems with Conductors

Conductors and ideal conductors are common parts of electromagnetic models, systems and circuits. In such systems, we often place a conductor in contact with a given potential. What does this mean? In practice, it may imply that we place it in contact with a battery with a given potential difference between its two poles. In our models, it means that the conductor has a given value of the potential, V. Because the conductor is an equipotential surface, it means that the whole conductor will be at the same potential. This means that systems with conductors can be addressed using Laplace's or Poisson's equation, where the conductors are Dirichlet boundary conditions with a given value of the potential. In this case, we do not need to know the amount of charge present at the conductor, instead we know the potential. The distribution of charges can be calculated afterwards, when we have found the potential and the electric field. Thus, we can propose a new method to find the electric potential and fields in electrostatic problems:

Method for problems with conductors

1. Determine what parts of the system are conductors and what their potentials are. These are the boundary conditions.
2. Solve Laplace's or Poisson's equation with the conductors as boundary conditions.
3. Calculate the electric field from the electric potential, $\mathbf{E} = -\nabla V$.
4. Find the surface charge density from the electric field: $\rho_s = \epsilon_0 \mathbf{E} \cdot \hat{\mathbf{n}}$.

Test your understanding
A massive copper cube has a cubical cavity at its center. There is a charge $+Q$ outside the copper cube. (a) What is the electric field inside the cavity if there are no charges there? (b) We place a charge $+Q$ in the center of the cubical cavity. If the field around the charge (but still inside the cavity) the same as the field from a charge in vacuum?[2]

Example: Field from Two Finite Plates

Two conductors shaped as plates of height h and width d are placed at a distance w from each other. One conductor has the potential V_0 and the other the potential $-V_0$. What is the electric field and surface charge density in this system?

Approach. We plan to solve this problem numerically in the xy-plane by placing the plates parallel to the y-axis. We place the conductors in a system of size $a \times a$. We place the conductor at $+V_0$ from $x = w/2$ to $x = w/2 + d$ and from $y = -h/2$ to $y = +h/2$, and the other conductor at $-V_0$ from $x = -w/2 - d$ to $x = -w/2$ and from $y = -h/2$ to $y = +h/2$. Because we cannot model an infinite system, we may implement the outer boundary condition either by setting the potential to be 0 at the outer boundary or by setting the potential to have zero derivative at the out boundary. Here, we choose the latter.

Setting up boundary conditions. There are no free charges, so we need to solve Laplace's equation: $\nabla^2 V = 0$ on a square of size $a \times a$ from $(-a/2, -a/2)$ to $(a/2, a/2)$ with a resolution of $N = 200$ grid cells in each direction. We set up the system to use our previously developed function to solve Laplace's equation with Neumann boundary conditions. First, we set up the boundary conditions by creating two rectangles. For this we need a function `x2i` that converts from system coordinates to grid indexes on the $N \times N$ grid. A grid point i is at position $x_i = -a/2 + i(a/N)$, which gives $i = (x_i + a/2) * (N/a)$. Based on this, we write

[2] (a) Zero; (b) No, the inner surface of the cubic cavity will also be charged, so the system does not have spherical symmetry.

a short function `rectangleregion` that fills all grid points in a rectangular region with a given potential value. We put this together:

```
# Define rectangular boundaries
N = 200 # Resolution
a = 2.0 # Size of box in meters
Vb = 1.0 # Voltage of box, volt
h, d, w = 0.5*a, 0.05*a, 0.35*a
b = np.zeros((N,N))
b[:] = float("nan")
def x2i(x,L,N):
    return int(N*(x+L/2)/L)
def rectangleregion(b,V,x0,x1,y0,y1,L,N):
    for ix in range(x2i(x0,L,N),x2i(x1,L,N)):
        for iy in range(x2i(y0,L,N),x2i(y1,L,N)):
            b[ix,iy] = V
    return
x0, x1 = -w/2-d, -w/2
y0, y1 = -h/2, h/2
rectangleregion(b,-Vb,x0,x1,y0,y1,a,N)
x0, x1 = w/2, w/2 + d
y0, y1 = -h/2, h/2
rectangleregion(b,Vb,x0,x1,y0,y1,a,N)
# Visualize
x, y = np.linspace(-a/2,a/2,N), np.linspace(-a/2,a/2,N)
rx,ry = np.meshgrid(x,y,indexing="ij")
plt.contourf(rx,ry,b)
```

Solving Laplace's equation. We solve Laplace's equation and plot the results using:

```
V = solvepoissonneumann2d(b,100000)
plt.contourf(rx,ry,V,10)
Ex,Ey = np.gradient(-V)
plt.quiver(rx,ry,Ex,Ey)
```

The resulting plot is shown in Fig. 7.6a.

Surface charges. We know that the surface charge ρ_s is related to the normal component of the field immediately outside the conductor: $\rho_s = \epsilon_0 \mathbf{E} \cdot \hat{\mathbf{n}}$. We can use this to estimate the surface charge density in this system. For every point (i, j) on the conductors, we look for neighboring points that are not on the conductor, find the normal vector, and calculate the local surface charge density. This is implemented in the following program:

```
rhos = np.zeros((N,N))
for ix in range(N):
    for iy in range(N):
        if (not(np.isnan(b[ix,iy]))): # Boundary site
            drho = 0.0 # drho = En*nhat in each direction
            if (np.isnan(b[ix-1,iy])): # Left boundary
                drho = drho + Ex[ix,iy]*(-1)
            if (np.isnan(b[ix+1,iy])): # Right boundary
                drho = drho + Ex[ix,iy]*(+1)
```

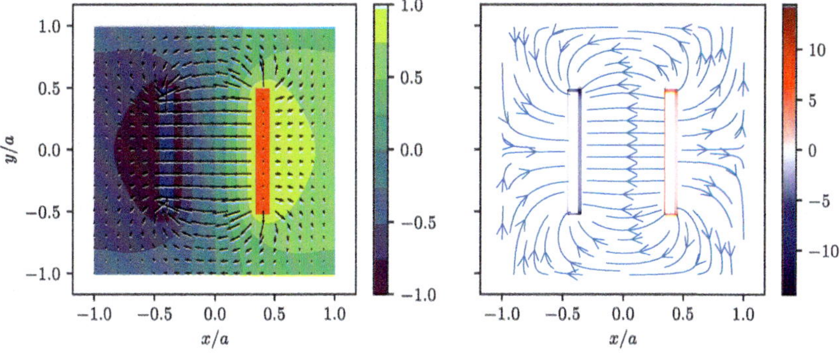

Fig. 7.6 **a** Plot of the electric potential and the electric field for a two-plate system. **b** Plot of the electric field and the charge distribution in a two-plate system

```
            if (np.isnan(b[ix,iy-1])): # Bottom boundary
                drho = drho + Ey[ix,iy]*(-1)
            if (np.isnan(b[ix,iy+1])): # Top boundary
                drho = drho + Ey[ix,iy]*(+1)
            rhos[ix,iy] = drho
```

which returns the matrix `rhos`, of the same size as `b` and `V`, with the surface charge densities. (We need to multiply with ϵ_0 to make it the correct dimensions). The results are shown in Fig. 7.6b. You may wonder why we use the electric field in the lattice position `ix,iy` for this calculation. Is not the field zero inside the conductor? Yes, but when we calculate the field using `gradient` in Python the boundaries are smeared out a bit, so that both sides of the boundaries are set to the electric field at the boundary.

Dimensionality of Laplace's equation. Why did we call this a system of plates, when they look like thin boxes in all the figures? This is a good question. When we solve Laplace's equation in the xy-plane as we have done here, we have really assumed that there is no variation in the potential in the z-direction, and therefore that the $\partial^2 V/\partial z^2$ term is zero. This means that the system does not change along the z-axis. A line or a rectangle in the xy-plane, therefore extends infinitely in the z-direction and is really a plane or an infinitely long rectangular prism

7.4 Imaging Methods and Mirror Charge

Another classical method used to solve complex problems with conductors is the use of mirror charges. We demonstrate this method through an example: A point charge Q is located a distance h above a grounded, conducting plane with potential $V_0 = 0$ as illustrated in Fig. 7.7. How can we find the surface charge distribution, $\rho_s(x, y)$, in the conductor plane?

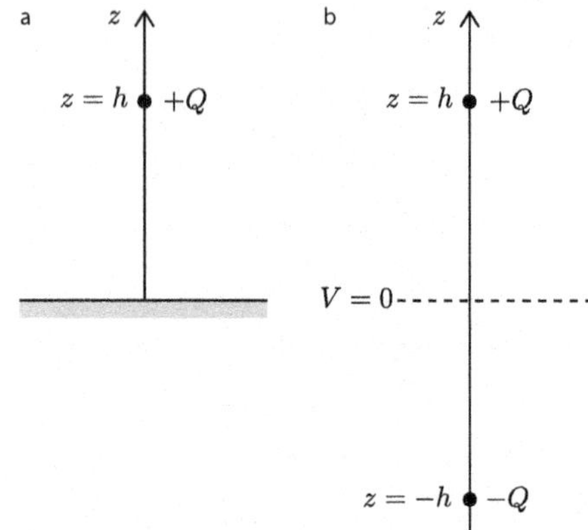

Fig. 7.7 **a** A point charge $+Q$ at $z = h$ above a conducting plane. **b** We remove the plane and replace it with a mirror charge $-Q$ at $z = -h$

The trick of mirror charges. We know that immediately outside the conductor, the electric field in the normal direction is $E_n(z = 0) = \rho_s/\epsilon$, and $E_t(z = 0) = 0$. The electric field in in the plane at $z = 0$ is a superposition of the field from the point charge Q and the charges in the plane. Because the plane is a conductor, it is an equipotential surface. The electric potential is therefore the same everywhere on the plane.

Now, we will introduce a *trick*. We can use the uniqueness property of the potential field. If we can find an electric potential that satisfies the boundary condition $V = 0$ on the plane, we know that this must be a unique solution to the problem. We therefore compare the two cases (a) and (b) illustrated in Fig. 7.7. Case (a) corresponds to the charge Q and the conducting plane, whereas case (b) corresponds to the charge Q and a charge $-Q$ in a position mirrored across the conducting plane. For both these systems the potential is zero at $z = 0$. In the part $z > 0$ we can solve Laplace's equation to find the electric potential with the given boundary condition at $z = 0$ and the charge Q. This would give us a solution for V for $z > 0$. However, we also see that the system with the charge Q and the mirror charge $-Q$ satisfies this boundary condition at $z = 0$. The electric potential from the sum of the charge Q and the charge $-Q$ must therefore be a solution to Laplace's equation for $z > 0$. However, because the solution is unique, this is *the* solution to Laplace's equation for $z > 0$. We have therefore found the potential using the method of mirror charges. We can then find the electric field from the electric potential, or we can simply find the electric field from the charge and the mirror charge, which very much simplifies the problem.

Simplified solution at $x = 0$, $y = 0$. We can apply the mirror charge method to find the electric field in any position $z \geq 0$. At $x = 0$, $y = 0$, $z = 0$ we find that

$$\mathbf{E} = -\frac{Q}{4\pi\epsilon h^2}\hat{z} - \frac{Q}{4\pi\epsilon h^2}\hat{z} = -2\frac{Q}{4\pi\epsilon h^2}\hat{z} \tag{7.1}$$

General solution in the plane. We can find the field in a position $(x, y, 0)$ by realizing that the field only has a component along the z-axis due to symmetry. Thus, the field from $+Q$ is

$$\mathbf{E}_{+} = -\frac{Q}{4\pi\epsilon} \frac{h}{(x^2 + y^2 + h^2)^{3/2}} \hat{\mathbf{z}} , \tag{7.2}$$

and similarly for $-Q$, so that the total field is

$$\mathbf{E} = -\frac{2Q}{4\pi\epsilon} \frac{h}{(x^2 + y^2 + h^2)^{3/2}} \hat{\mathbf{z}} . \tag{7.3}$$

For $z > 0$ this is therefore the electric field from the charge and the conductor. The charge distribution on the surface of the conductor is given by $\rho_s = \epsilon \mathbf{E} \cdot \hat{\mathbf{n}}$, where $\hat{\mathbf{n}} = \hat{\mathbf{z}}$:

$$\rho_s = -\frac{2Q\epsilon}{4\pi\epsilon} \frac{h}{(x^2 + y^2 + h^2)^{3/2}} , \tag{7.4}$$

Summary

Ideal conductors have a sea of charges that immediately move in response to an electric field, until the total electric field inside the conductor is zero.

There are no free charges inside an ideal conductor.

An ideal conductor is an **equipotential surface**.

There is no tangential field immediately outside an ideal conductor.

The electric field immediately outside an ideal conductor is $E_n = \rho_s/\epsilon$.

We can use the property that the ideal conductor is an equipotential surface as a boundary condition to Laplace's equation. This allows us to find the electrical potential and therefore the electrical field around conductors.

Exercises

Discussion Exercises

7.1 Copper between charges. We place two charges Q_A and Q_B a distance $2L$ from each other. Then we place a neutral copper sphere at the midpoint between Q_A and Q_B. Sketch the distribution of charges in the copper sphere if (i) both charges are positive, (ii) both charges are negative, (iii) the charges have opposite signs. What is the electric field and the potential inside the sphere in these situations?

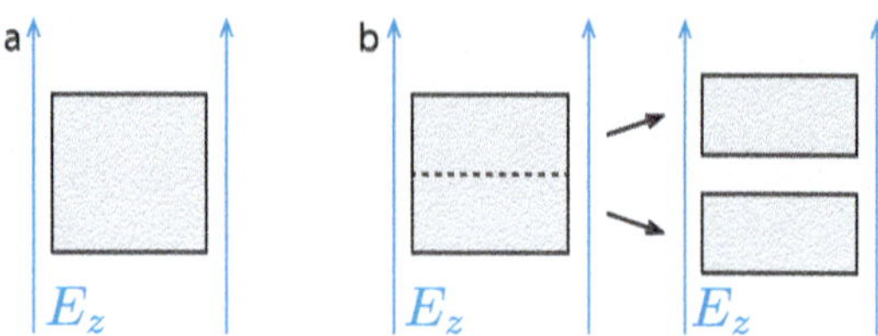

Fig. 7.8 A cube in an electric field

7.2 Where are the charges. "The free charges in a conductor are always on the external surface of the conductor". Is this statement always true? Sketch and explain.

7.3 Holes in a conductor. If you have a hole inside a conductor it acts as a Faraday cage by shielding the inside from the electric field on the outside. What happens if you have two holes inside a conductor. Would they still act as Faraday cages?

Tutorials

7.4 Conductors and electric field. A cube is made of an ideal conductor is placed in an electric field $\mathbf{E} = E_z\hat{\mathbf{z}}$ as shown in Fig. 7.8a.
(a) What is the electric field inside the cube?
(b) Sketch how the charge is distributed on the cube.
(c) Is the electric field around the cube homogeneous?
(d) While the cube is inside the electric field, your friend Q quickly cuts the cube in two pieces as shown in Fig. 7.8b. You now have two half-cubes in the homogeneous electric field. What is now the electric field inside the upper half cube?
(e) What is the charge distributions on the two cubes? (Hint: Check the the field is zero where it is supposed to with the charge distribution you have chosen.)
(f) We now turn off the external field. What is the charge distribution on the two half cubes now?

7.5 A plate in between. Figure 7.9a shows an ideal conducting plate with charge $+Q$ placed between two ideal conducting plates.
(a) Sketch the charge distribution on all the plates, when the plates is placed as shown in the figure. (Hint: Is the field zero inside *all* the conductors?)
(b) Sketch the electric field in the direction normal to the plates.
(c) In Fig. 7.9b an ideal conducting plate with charge $+Q$ is placed above to neutral conducting plates. Sketch the charge distribution on all the plates after the plate has been put in place.
(d) Sketch the electric field in the direction normal to the plates.

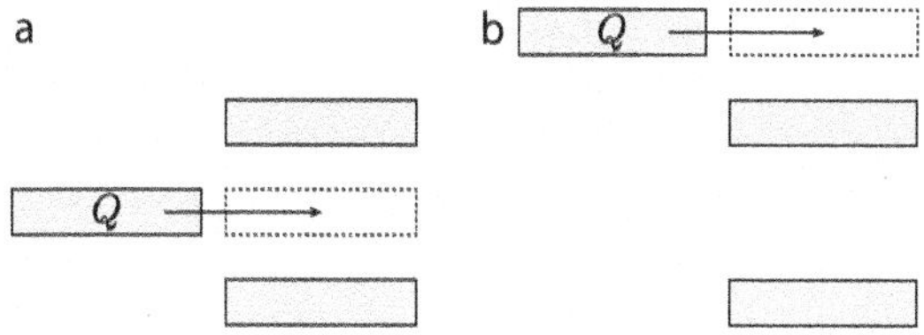

Fig. 7.9 Illustration of a conducting plate with charge $+Q$ in between two neutral conducting plates

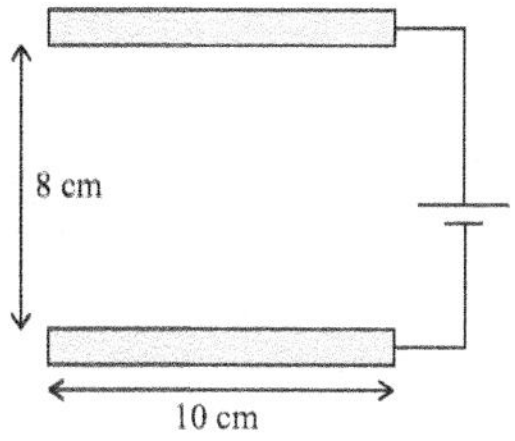

Fig. 7.10 Two plates connected by a battery

7.6 Secret briefcase. Your friend Q has invented a smart briefcase which consists of a spherical shell of copper. Q says that he can smuggle any charge he wants inside this briefcase, and you would not be able to figure out what is inside it, because the shell acts as a Faraday cage. Can Q really use this invention to smuggle charges into his exam without being caught by the E-field-detector at the entrance to the exam hall?

7.7 Millions of electrons. You connect a 1.5 V AA battery between two identical steel plates that are $10 \times 10 \times 1$ cm large and that are placed 8cm from each other as shown in Fig. 7.10. Then you cut the cables.
(a) Sketch the charge distribution on the two plates.
(b) What is the electric field in between the plates?
(c) You put one plate in your briefcase. How many free electrons did you then bring? You can assume that you pick up the plate with neutral and perfectly isolating gloves and place it in an isolated and neutral room in your briefcase.

7.8 Coaxial cable. A coaxial cable consists of a long conducting cylinder of radius a which is inside a conducting cylindrical shell of finite thickness with inner radius b and outer radius c. The two conductors are placed with the same axis and the space inbetween them is filled with a plastic, keep the cylinder and the cylindrical shell separated by a small distance. We will here assume that the cable is very long and neglect all end effects near the ends of the cable.
(a) Assume that the inner cylinder has a charge $+Q$ and the outer conductor has a charge $-Q$. Sketch how the charges are distributed on the inner cylinder and the outer cylinder shell. How do you know that this is the correct distribution and that there cannot be other possible distributions?
(b) In what regions of r, the distance from the cylinder axis, does the electric potential vary and where is it constant. Discuss the cases for $0 < r < a, a < r < b, b < r < c$ and $c < r$.

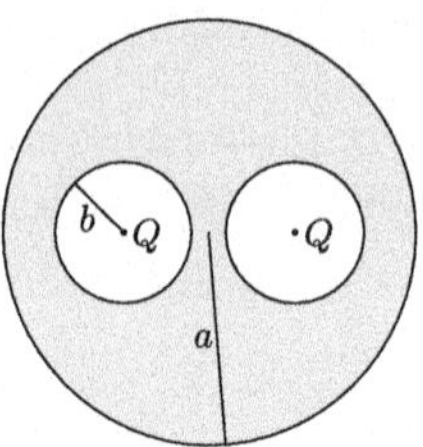

Fig. 7.11 A spherical conductor with two holes

(c) Assume instead that the inner conductor has a charge $+Q$ on it and the outer conductor is neutral. Sketch how the charges are distributed on the inner cylinder and the outer cylinder shell. How do you know that this is the correct distribution and that there cannot be other possible distributions?
(d) Assume instead that the outer conductor has a charge $+Q$ on it and the inner conductor is neutral. Sketch how the charges are distributed on the inner cylinder and the outer cylinder shell. How do you know that this is the correct distribution and that there cannot be other possible distributions?
(e) What is the potential difference between the inner and the outer conductor for this case (a charge $+Q$ on the outer conductor and a neutral inner conductor).

7.9 Shielding. A large sphere of radius a is made out of a conducting material and contains two spherical holes as shown in Fig. 7.11. Each hole has a charge $+Q$ placed in its center.
(a) Make a sketch of the surface charge for each of the three surfaces. Explain your answer.
(b) Find the surface charge densities ρ_a and ρ_b.
(c) Make a sketch of the electric field everywhere in the system. Is the electric field zero anywhere?
(d) If an external charge q was brought near the large sphere, how would the surface charge densities in the system change at the outer surface of the sphere and on the surfaces of the holes?

Homework

7.10 Faraday cage. You have a large conductor and carve out a small cavity in the conductor. A charge Q is brought near the conductor. (Such a system is called a Faraday cage). The are no charges inside the cavity.
(a) If the electric potential on the outer surface of the conductory is $V = 0$, what is the electric potential inside the cavity?
(b) What is the electric field inside the cavity in the conductor?

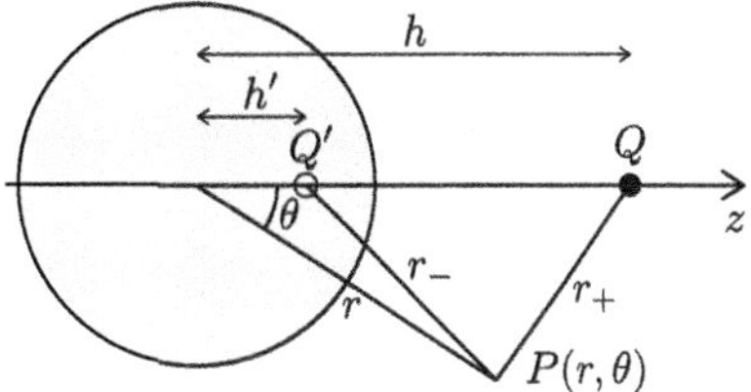

Fig. 7.12 Illustration of a charge Q above an conducting sphere of radius a

(c) This type of construct is called a Faraday cage. Can you provide examples of systems that may act like Faraday cages that you meet in your daily life?

7.11 Point charge near conducting sphere. A point charge Q is placed at a distance h to a grounded conduting sphere of radius a as shown in Fig. 7.12. (Grounded means that the potential is zero). The potential on the surface of the conducting sphere can be found by replacing the grounded sphere by an imaginary charge $Q' = -mQ$ at a distance h' from the center of the sphere as illustrated in the figure.
(a) Show that the potential in a point $P(r, \theta)$ relative to the center of the conducting sphere can be written as

$$V(r, \theta) = \frac{Q}{4\pi\epsilon_0}\left(\frac{1}{r_+ - \frac{m}{r_-}}\right),$$

where $r_+ = (r^2 + h^2 - 2hr\cos\theta)^{1/2}$ and $r_- = (r^2 + h'^2 - 2h'r\cos\theta)^{1/2}$. (Hint: Use the cosine rule of lengths in a triangle.)
(b) Show that for the potential on the surface of the conducting sphere to be zero, the imaginary charge Q' must satisfy the relations $hh' = a^2$ and $m = a/h$.
(c) Show that when Q is close to the sphere, that is, when $h \simeq a$, the value of m approaches 1 and the result corresponds to that of a point charge above a conducting plane.
(d) The surface charge distribution on the conducting sphere is

$$\rho(\theta) = -\frac{mQ}{4\pi a^2}\frac{1 - m^2}{(1 + m^2 - 2m\cos\theta)^{3/2}}.$$

Show that the total induced charge is $Q' = -mQ$. (Hint: Use that $\hat{\mathbf{n}}\,\mathrm{d}s = 2\pi a^2 \sin\theta\,\mathrm{d}\theta\,\hat{\mathbf{r}}$.)

7.12 Point charge above conducting plane. A point charge Q is in a height h above an infinitely large, conducting plane as shown in Fig. 7.13.
(a) Find the potential V and the electric field E_z along the z-axis for $0 < z < h$. (Hint: Use the mirror charge method.)
(b) Find the induced surface charge distribution $\rho_s(r)$ on the surface of the charge, where r is the distance to the z-axis.

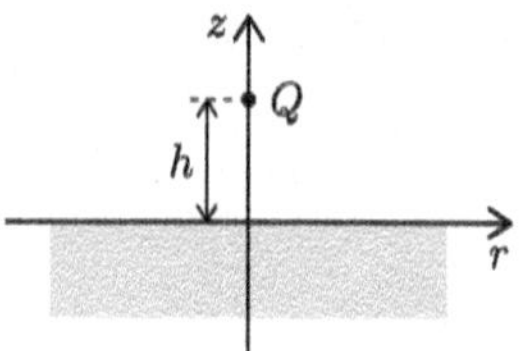

Fig. 7.13 Illustration of a charge Q above an infinite, conducting plane

(c) Show that the total induced charge on the surface is $-Q$.
(d) Sketch the electric field lines everywhere.
(e) Find the force on the point charge from the plane.

7.13 A lightning rod. In this exercise we will address if a lightning prefers to strike a pole or a rounded hilll. Lightnings occur due to dielectric breakdown of air, which is due to a high electric field. Here, we will make a model for a lightning rod that can address both a thin pole and a round object, calculate the electric field around this rod, and find how the field depends on the geometry of the rod. We model the rod as a sphere, which represents the tip of the rod or the roundness of a hill. The sphere has a radius a and we assume that it is an ideal conductor. We assume that the sphere has potential V_0. A cloud far above has the potential V_1.
(a) Make a sketch of the system.
(b) First, we assume that the clouds are far away, so we can find the electric field around a sphere or radius a with potential V_0, and that the potential is $V_1 = 0$ infinitely far away. Show that the solution to Laplace's equation in spherical coordinates for this system is $V(r) = \frac{V_0 a}{r}$.
(c) Find the electric field in spherical coordinates.
(d) Lighting will occur where there is dielectric breakdown, that is, where the magnitude of the electric field exceeds a critical value, E_c. Where will dielectric breakdown occur in this system? What does this tell you about the likelyhood of a lightning striking a rounded hill versus a pole?

7.14 A lightning rod in two dimensions. In this exercise we will address dielectic breakdown in a two-dimensional system and compare exact analytical results from a round conductor with numerical results from the round tip of a pole. We will model the tip of a lightning rod as a conducting cylinder of radius a in the xy-plane and infinite extent in the z-direction. This is not the same as a spherical shaped conductor, but allows us to compare with two-dimensional numerical calculations. The potential of the cylinder is $V(a) = V_0$ and the potential far away at a distance b is $V(b) = V_1 = 0$. Assume that the are no free charges in the system.
(a) Show that Laplace's equation in cylindrical coordinates for the system has the solution $V(r) = V_0 \ln(r/b)/\ln(a/b)$.
(b) Find the electric field in cylindrical coordinates.

(c) We model the same system numerically, with a cylinder at a height h above the ground, where the cylinder or radius a and the ground has potential $V_0 = 1V$ and the clouds above are modeled as a flat plate with potential $V_1 = 0$ at a distance H from the ground. Write a program to find the electric potential and the electric field in this system using a combination of Dirichlet and Neumann boundary conditions.
(d) Compare the numerical and the analytical solution close to the top of the cylinder.

Chapter 8
Capacitance

8.1 Capacitance

Motivational Example

Capacitance is a concept that we will introduce to describe the ability for a set of conductors to store electrical charge, and a *capacitor* is the physical realization of this system. You may think of a capacitor as an object and capacitance as the property of that object. (You may already have such an intuition about resistance and a resistor when it comes to electrical resistivity. This analogy is, as we will see, quite useful).

Figure 8.1 illustrates two (ideal) parallel plate conductors that are separated from each other by a distance d. We call such a configuration a *parallel plate capacitor*. If the system starts as neutral and we move a charge Q from the bottom plate to the top plate, the bottom plate will have a charge $-Q$ and the top plate a charge Q. The electric field between the plates will point from the $+Q$ plate to the $-Q$ plate and the magnitude of the field will be $E = Q/(A\epsilon)$,[1] where A is the area of the plate. The difference in electrical potential for this uniform electric field is $V = Ed = Qd/(A\epsilon)$. We define the *capacitance*, C, of this system as

$$C = \frac{Q}{V} = \frac{QA\epsilon}{Qd} = \frac{A\epsilon}{d}. \tag{8.1}$$

The capacitance depends only on the geometry of the system: the distance d between the plates, the area A of the plates, and the dielectric constant of the material they are embedded in.

---

[1] We have calculated this several times before. We apply Gauss' law to each of the plates. For the top plate we use a cylindrical Gauss surface with an axis normal to the plate and area S. The field will be symmetric around the plate and only depend on the distance to the plate. Gauss' law therefore gives that $2SE = \rho_s S/\epsilon$ and $E = \rho/(2\epsilon)$, where $\rho = Q/A$ is uniformly distributed on the plate. (The charges are only on the surfaces of the plate but we look at a volume that goes through the plate). The total electric field is therefore $E = 2\rho/(2\epsilon) = Q/(A\epsilon)$.

© The Author(s), under exclusive license to Springer Nature Switzerland AG 2026

A. Malthe-Sørenssen, *Elementary Electromagnetism Using Python*, Undergraduate Texts in Physics, https://doi.org/10.1007/978-3-032-19876-1_8

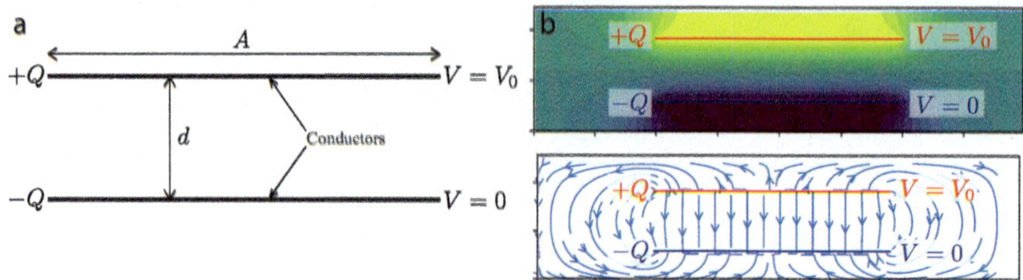

Fig. 8.1 Illustration of parallel plate capacitor. **a** The capacitor consisting of two plates of area A separated by a distance d. There is a potential difference V_0 between the plates generated by moving a charge Q from the bottom plate to the top plate so that the top plate has a charge Q and the bottom plate has a charge $-Q$. **b** The electrical potential V and electric field $\mathbf{E}$ around a finite-sized parallel-plate capacitor

Capacitors and Capacitance

The capacitance C defined as $C = Q/V$ is a property of the geometry of a system of conductors. This not only valid for two parallel plates, but for any configuration, because Coulomb's law for the electric field as well as for the electric potential is proportional to the charge: $V = \sum_i Q_i/(4\pi\epsilon_0 R_i)$: If all the charges are doubled, the field and the potential are doubled as well. This implies that the ratio of the charge and the potential is a constant, independent of the charge.

What does the capacitance tell about this system? As is hinted in the name, the capacitance describes a system's capacity to carry charge at a given potential difference. We see this from $C = Q/V$ which gives us $Q = CV$. The amount of charge the system can store at a given potential difference V between two conductors is proportional to C. Thus, if you were to carry as much charge as possible in a briefcase with a 9V battery to hold it, you would want to have a system with as large a capacitance as possible.

We call a system with a capacitance C a *capacitor*. A capacitor is a common component in electric circuits with a useful function as we will address later. Figure 8.2 shows examples of capacitors. Capacitance is also a useful aspect of a system that we may want to include in a model of that system. For example, a cell membrane as illustrated in Fig. 8.2 may be considered to act as a capacitor and have a given capacitance. Here, we will develop methods to calculate the capacitance for simple and complex geometries and for systems that consists of combinations of several capacitors. Later, we will see how we use capacitance and capacitors as elements in our models of real electromagnetic systems and in our analysis of electric circuits.

Capacitance

The **capacitance** of a system of two conductors with a potential difference V is

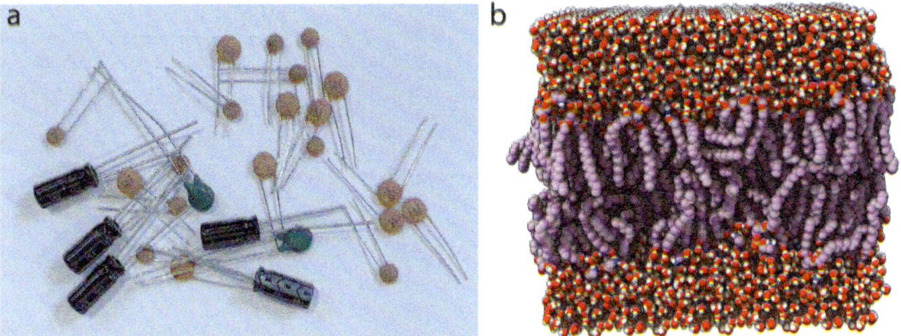

Fig. 8.2 Examples of capacitors. **a** Pictures of capacitors used in electric circuits. **b** A cell membrane consists of a set of lipid molecules called a lipid bilayer, that separate the inside and outside of the cell. There is little charge transport across the lipid bilayer, and therefore we may consider the cell membrane a capacitor. The typical thickness of a nerve cell membrane is 8–10 nm and the voltage difference, called the membrane potential, is typically 30–90 mV. (The inside is negatively charged)

$$C = Q/V, \tag{8.2}$$

where Q is the charge moved from one conductor to another to get the potential difference V. Thus, one conductor has the charge $+Q$ and the other conductor has the charge $-Q$ and the resulting potential difference between the two conducutors is V.

The unit of capacitance is F (farad), which is C/V.

(This may open for some confusion, because we use C for capacitance and C for coulomb, V for volt and V for electric potential. However, notice the difference in fonts between the unit V, which uses the normal font, and V, which uses the mathematical font.)

Applications of Capacitors

Capacitors have many practical applications and are important parts of realistic models of electromagnetic systems. For example, capacitors are often used to measure distances. A parallel-plate capacitor can be made very sensitive to small changes in the distance d between the two plates, $C = A\epsilon/d$, and this can be used to precisely measure nano-scale distances. A parallel-plate capacitor can also be very sensitive to small particles entering the area between the plates, effectively changing the dielectric properties of this area and thereby changing the capacitance. Many different types of sensors are based on these types of principles.

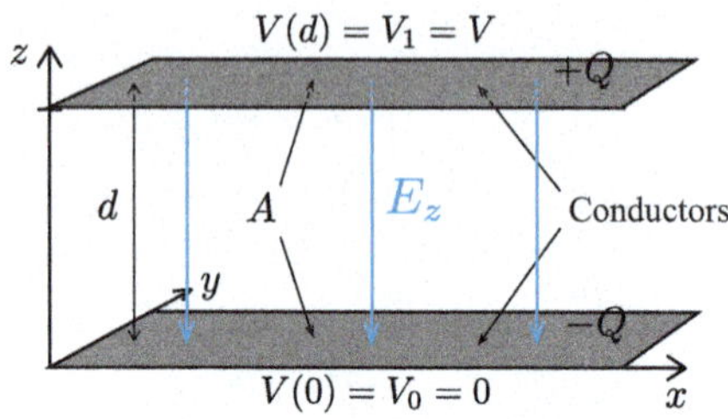

Fig. 8.3 Illustration of a parallel plate capacitor consisting of two plates of area A separated by a distance d. The top plate has a charge $+Q$ and the bottom plate a charge $-Q$ and the potentials are illustrated

Test your understanding

You have built a capacitor with capacitance C consisting of two parallel plates of area A and distance d. You have connected the capacitor to a battery with voltage V. (a) What happens to the capacitance C if you double the voltage on the battery? (b) What are the charges on the positive and negative sides if you double V? (c) You disconnect the voltage from the capacitor. What is the capacitance C of the capacitor now? [2]

Example: Parallel Plate Capacitor

Find the capacitance of the parallel plate capacitor illustrated in Fig. 8.3.

Approach. We may follow one of two approaches: (1) First assume a charge distribution of $+Q$ on one conductor and $-Q$ on the other, use this to find the electric field $\mathbf{E}$, use the electric field to find the potential difference $V = -\int_0^1 \mathbf{E} \cdot d\mathbf{l}$, and then find the capacitance from $C = Q/V$; Or (2) solve Laplace's equation to find the electric potential, $V(\mathbf{r})$, use the potential to find the electric field, $\mathbf{E} = -\nabla V$, and then relate the electric field to the charge, for example, using that $E_n = \mathbf{E} \cdot \hat{\mathbf{n}} = \rho_s/\epsilon_0$ on the surface of the conductors. We will pursue both approaches to demonstrate their use.

Method 1: From charge to field to potential.

Find the field. We can find the electric field using Gauss' law as we have done in previous chapters. We do not repeat this here, but simply use the result that the field from one plate is $|E_z| = Q/(2A\epsilon_0)$, pointing away from the plate if Q is positive. The total field from both plates is therefore $E_z = -Q/(A\epsilon_0)$, which points in the negative z-direction.

Find the potential difference. The potential difference between the top and the bottom plate is the line integral of the electric field from the top plate to the bottom plate:

[2] (a) Nothing; (b) $+2Q$ and $-2Q$; (c) The same.

$$V = V(d) - V(0) = \int_d^0 \mathbf{E} \cdot \mathrm{d}\mathbf{l} = \int_d^0 E_z \,\mathrm{d}z = -\frac{Q}{A\epsilon_0}(-d) = \frac{Qd}{A\epsilon_0}. \tag{8.3}$$

We check that the signs are reasonable: V_1 must be larger than V_0 because the electric field points from V_1 to V_0, therefore $V = V_1 - V_0$ must be positive, which it indeed is. The signs are therefore ok.

Find the capacitance. Finally, we find the capacitance from:

$$C = \frac{Q}{V} = \frac{Q}{Qd/(A\epsilon_0)} = \frac{A\epsilon_0}{d}. \tag{8.4}$$

Method 2: From potential to field to charge.

Find the potential using Laplace's equation. We assume that the distance d is small compared to the extent of the plates, so that the potential only depends on z and not on x and y. We want to solve $\nabla^2 V = \mathrm{d}^2 V/\mathrm{d}z^2 = 0$ for $0 < z < d$, where $V(0) = 0$ and $V(d) = V$. We recognize that the solution must be on the form $V(z) = c_1 z + c_2$. The boundary conditions determine the constants c_1 and c_2: $V(0) = 0$ gives $c_2 = 0$ and $V(d) = c_1 d = V$ gives $c_1 = V/d$. Therefore $V(z) = Vz/d$. (We notice that in method 1 we only found $V(z = d)$ and not the general form $V(z)$.)

Find the electric field. We find that $\mathbf{E} = -\nabla V = -(V/d)\hat{\mathbf{z}}$.

Find the surface charge density and the charge Q. The surface charge density on the bottom side of top plate must be given by $\mathbf{E} \cdot \hat{\mathbf{n}} = \rho_s/\epsilon_0$, where $\mathbf{E} = -(V/d)\hat{\mathbf{z}}$ and $\hat{\mathbf{n}} = -\hat{\mathbf{z}}$, so that $\rho_s/\epsilon_0 = (V/d)$. The total charge on the bottom side of the top plate is then $Q = \rho_s A = V\epsilon_0 A/d$. Similarly, the total charge on the top side of the bottom plate is $-V\epsilon_0 A/d$. Notice that there is no charge on the top side of the top plate or the bottom side of the bottom plate, because the electric field is zero outside the plates on these sides.

Find the capacitance. We find the capacitance as:

$$C = \frac{Q}{V} = \frac{(V/d)A\epsilon_0}{V} = \frac{A\epsilon_0}{d}. \tag{8.5}$$

Interpretation. First, we notice that we can increase the capacitance by decreasing the spacing d, or by increasing the area A or the dielectric constant. What may be the disadvantage of decreasing the distance d between the plates? We see that for a given V, the electric field is $E_z = (V/d)$. Decreasing d while keeping V constant will increase the charge stored in the capacitor, but also increase the electric field. If d becomes too small, we may have dielectric breakdown, leading to the failure of the capacitor and charges leaking from one conductor to another.

How would the results be modified if we instead had a *dielectric material* with dielectric constant ϵ between the plates? The only change in the calculations we did above would be to change ϵ_0 with ϵ, increasing the capacitance.

Example: Effect of Dielectric Materials

We charge the capacitor illustrated in Fig. 8.3 with a battery V_0 so that it holds a total charge Q_0, disconnect the conductors from the battery, and then introduce a dielectric material with dielectric constant ϵ into the space between the two plates? What are the new charges, potential difference and electric field in the system?

Approach. We use the result for the capacitance $C = A\epsilon/d$ and address the different steps in the process in detail, by first charging the capacitor in vacuum, then disconnecting it, and finally inserting the dielectric. We realize that the charges cannot change when the conductors are disconnected, because the charges do not have anywhere to go.

Solution. We charge the capacitor by applying a voltage V_0. In vacuum, the capacitance is $C_0 = A\epsilon_0/d$. The charge on the capacitor is then $Q_0 = C_0 V_0$. The electric field inside the capacitor with vacuum is then $E_0 = V_0/d$. We then decouple the capacitor, so that it retains its charge Q_0, but we no longer keep the potential difference by applying an external voltage. The charge on the capacitor then remains Q_0.

Then, we insert the dielectric into the gap. The charge remains the same, $Q_1 = Q_0$, but the capacitance of the new system is $C_1 = \epsilon A/d$, where $C_0 = (A\epsilon_0/d)$ so that $C_1 = (\epsilon/\epsilon_0)C_0 = \epsilon_r C_0$. This means that the potential also changes to:

$$V_1 = \frac{Q_1}{C_1} = \frac{Q_0}{C_1} = \frac{C_0 V_0}{\epsilon_r C_0} = \frac{V_0}{\epsilon_r}. \tag{8.6}$$

The electric field is then reduced to $E = E_0/\epsilon_r$ as expected. This is due to the polarization of the medium, that sets up a field that acts in the opposite direction, hence reducing the effective field.

Discussion. Notice that if we performed the experiment *in a different way*, by not disconnecting the battery while we insert the dielectric. We would then apply the voltage V_0 to the system with the dielectric. In this case, the charge, Q_2, in the system would be different: $Q_2 = C_1 V_0$, because the battery would supply more charge to keep the voltage constant. However, the electric field would be the same as in vacuum, $E_2 = V_2/d = V_0/d = E_0$.

Test your understanding

A capacitor consists of two parallel quadratic conducting plates with sides L, thickness h and separation d. The upper plate has a charge $+Q$ and the lower has a charge $-Q$. (a) How are the charges distributed in the plates? (b) If we double the charge on each plate from Q to $2Q$, how does the capacitance change? (c) If we double the length L and halve the separation d, what is the change in capacitance in the system?[3]

[3] (a) Lower side of upper plate and upper side of lower plate; (b) No change; (c) 8 times larger.

Calculating the Capacitance

From these examples we can extract a *general method* that we can use to calculate the capacitance of a system:

Method: Finding the capacitance of a system

Method 1: From charge to field to potential.

1. We assume that one conductor has a charge $+Q$ and that the other conductor has a charge $-Q$.
2. We calculate the electric fields due to these charges using e.g. Gauss' law or Coulomb's law while utilizing possible symmetries in the system and find the total electric field $\mathbf{E}$ using the superposition principle.
3. We calculate the potential difference between the two conductors, $V = -\int_0^1 \mathbf{E} \cdot d\mathbf{l}$, using the total electric field we found.
4. We find the capacitance from $C = Q/V$, where we insert our value for V and the value for Q we started with.

Method 2: From potential to field to charge.

1. We solve Laplace's equation to find the potential $V(\mathbf{r})$ for the system. Boundary conditions are given by assuming that one conductor has a potential V_1 and the other a potential V_0. Usually, we choose $V_0 = 0$.
2. We find the electric field $\mathbf{E} = -\nabla V$.
3. We find the surface charge density on the conductors from $\rho_s = \epsilon \mathbf{E} \cdot \hat{\mathbf{n}}$. We sum/integrate the surface charge density to find the total charge Q on one of the conductors. (The other conductor should then have a charge $-Q$).
4. We find the capacitance from $C = Q/V$, where we insert our value for Q and the potential difference $V = V_1 - V_0$ we started with.

8.2 Combining Several Capacitors

The methods we have introduced allow us to calculate the capacitance of a single capacitor consisting of two conductors. Capacitors are common elements in electric circuits and common features in models we build of electromagnetic systems. However, in many cases we want to combine several capacitors with individual capacitances C_i into one combined component with capacitance C. For example, we may connect several capacitor components together, in parallel, in series or in a combination. In this case, we can derive expressions for the combined capacitance if we can assume that the capacitors are independent components that do not interact. What do we mean by not interacting? This means that the electric field from one component does not interact with another component, for example, leading to a displacement of the charge distribution. For a set of parallel plate capacitors this is usually the case, because the electric field outside of the space between the capacitors is generally very small. (We assumed it to be zero in the case of a thin and large capacitor).

Circuit Diagram of Capacitor

First, let us introduce a compact way to describe a system of several (non-interacting) capacitors. We do this in the form of a *circuit diagram*. A circuit diagram consists of a set of thin lines that represent ideal conductors connecting various parts of the components. In general, we do not care about the shape of these lines: We draw them as simple as possible often using straight lines, because the way we draw them has no impact on our system. (This will change when we discuss magnetic induction later). Because the lines represent ideal conductors, the electric potential is the same everywhere along the line: The lines connect components at the same potential. Figure 8.4 illustrates an example diagram. We draw a capacitor as two parallel plates. Notice that the potential on each side of the capacitor will not be the same, but will depend on the charge Q on the capacitor. The voltage drop, the difference in potential across the component, is $V = Q/C$, where Q is the charge on the capacitor. Usually, in circuits, we will have voltage sources that provide a given potential difference and not charge sources, so that we will know the potential difference V and calculate the charge $Q = CV$.

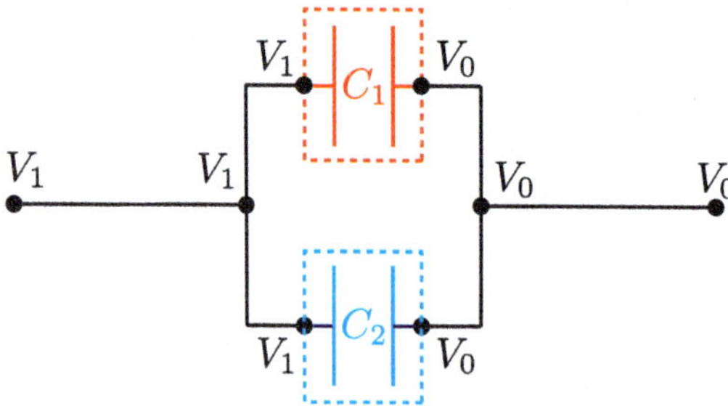

Fig. 8.4 Illustration of circuit diagram of two capacitors in parallel. The electric potential at various points (filled circles) along the circuit are shown. The top capacitor is shown in red and the bottom in blue. However, usually components are drawn in the same color as the connecting wires

When we draw individual components in a circuit diagram, we implicitly mean that they do not interact: The electric (or later magnetic) field inside a component does not interact with the fields inside other components.

Parallel Coupling of Capacitors

First, we address the case of n capacitors, $C_1, C_2, \ldots, C_n$, connected in parallel as illustrated in Fig. 8.5. The goal is to find the value C for a single capacitor to replace all the capacitors so that the effective capacitance is the same.

We notice that all the n capacitors have the same potential on the top, V_1, and on the bottom, V_0. This is indeed how we interpret the solid line drawn between the components. The charge on each capacitor is Q_i as indicated in the figure, where $C_i = Q_i/V$, where $V = V_1 - V_0$ is the same for all the capacitors.

All the conductors on the top side are connected so that they are at the same potential. It is like they are one large conductor. Similarly for the bottom side. The total charge on the top side is $Q = \sum_i Q_i$ and the total charge on the bottom side is $-Q = \sum_i -Q_i$. The total capacitance of the whole system is then

$$C = \frac{Q}{V} = \frac{\sum_i Q_i}{V} = \sum_i \frac{Q_i}{V} = \sum_i C_i. \tag{8.7}$$

This proves that there is a simple additive law for (independent) capacitors in parallel, that is, for capacitors whose *fields do not interact*.

$$C = C_1 + C_2 + C_3 + \ldots + C_n = \sum_i C_i. \tag{8.8}$$

We can therefore replace these n capacitors with a single capacitor with the capacitance C.

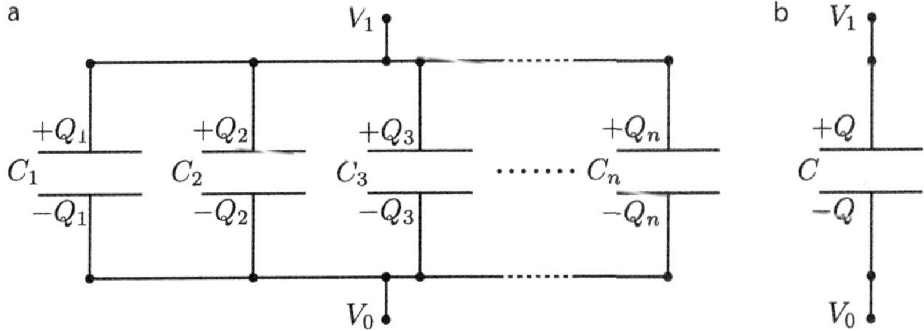

Fig. 8.5 **a** Illustration of circuit diagram of n capacitors in parallel. **b** The n capacitors are replaced by a single capacitor with capacitance C

Fig. 8.6 **a** Illustration of circuit diagram of n capacitors in series. **b** The n capacitors are replaced by a single capacitor with capacitance C

Series Coupling of Capacitors

Figure 8.6 illustrates a series coupling of n capacitors with capacitances $C_1, C_2, \ldots, C_n$. We want to replace them by a single capacitor with capacitance C. What should the value of C be? (We assume that the electric field from one capacitor does not affect another capacitor).

We notice from the figure that the right part of capacitor 1 is connected to the left side of capacitor 2. This means that the total charge on these two capacitors must be conserved: What is taken from the right side of capacitor 1 is added to the left side of capacitor 2, and so on. Therefore, all the capacitors will have the same charge Q.

What about the voltage differences across the capacitors? We have denoted the voltage on the left V_1 and on the right V_0, (which we without loss of generality can set to zero, $V_0 = 0$). We denote the voltage to the left of capacitor i, V_i, as shown in the figure. The voltage difference across the first capacitor with capacitance C_1 is then $\Delta V_1 = V_1 - V_2$. In addition, we know that $C_1 = Q/(\Delta V_1)$ so that $\Delta V_1 = Q/C_1$. Similarly, for the second capacitor, C_2, we have that $\Delta V_2 = V_2 - V_3 = Q/C_2$.

We sum the voltage differences for all the capacitors, noticing that subsequent pairs of potentials (e.g. $-V_2 + V_2$) sum to zero:

$$\begin{aligned}\sum_i \Delta V_i &= V_1 - V_2 + V_2 - V_3 + \ldots + V_n - V_0 = V_1 - V_0 \\ &= \frac{Q}{C_1} + \frac{Q}{C_2} + \ldots + \frac{Q}{C_n} = Q\sum_i \frac{1}{C_i} = Q\frac{1}{C}.\end{aligned} \tag{8.9}$$

The total effective capacitance is therefore:

$$\frac{1}{C} = \frac{1}{C_1} + \frac{1}{C_2} + \ldots + \frac{1}{C_n} = \sum_i \frac{1}{C_i}\,. \tag{8.10}$$

Test your understanding
(a) Two capacitors C_1 and C_2 are connected in parallel. Will the resulting combined capacitance always be larger or smaller than each of C_1 and C_2? (b) Two capacitors C_1 and C_2 are connected in series. Will the resulting capacitance always be larger or smaller than each of C_1 and C_2?[4]

[4] (a) Larger than both C_1 and C_2; (b) Smaller than both C_1 and C_2.

8.3 Energy Stored in an Electric Field

What is the energy stored in the electric field? What does this question even mean? For a capacitor, we can calculate how much energy is needed to build up a given charge difference, Q, between the two conductors. There will be an electric field and a potential difference associated with this charge difference. We could therefore say that since there is an energy needed to create this charge difference, there is also an energy needed to create the electric field. We expect to get this energy back if we discharge the capacitors. Our plan is to start to study a simple capacitor, introduce an energy associated with the electric field of this capacitor, and generalize this to build a general description of the energy associated with an electric field.

Energy of a Parallel Plate Capacitor

Figure 8.7 illustrates a part of the charging process of a parallel-plate capacitor. We start from a state where both plates have the same potential and no charge. We charge the capacitor in increments by moving a small amount of charge $\mathrm{d}q$ from one conductor to another. The figure illustrates a step in this process where the charge on the top plate is q, and the potential difference is $V(q) = q/C$. For this situation, the work $\mathrm{d}W$ needed to move a charge $\mathrm{d}q$ from one conductor to another is $\mathrm{d}W = V(q)\,\mathrm{d}q = (q/C)\,\mathrm{d}q$.

The total work done when the capacitor is fully charged with a charge Q, is the integral of the work during each of these steps from $q = 0$ to $q = Q$.

$$W = \int \mathrm{d}W = \int_0^Q \frac{q}{C}\,\mathrm{d}q = \frac{1}{2}\frac{Q^2}{C}. \tag{8.11}$$

We interpret this work as the potential energy, U, stored in the capacitor. We insert $Q = CV$ and get:

$$U = \frac{1}{2}\frac{Q^2}{C} = \frac{1}{2}QV = \frac{1}{2}CV^2. \tag{8.12}$$

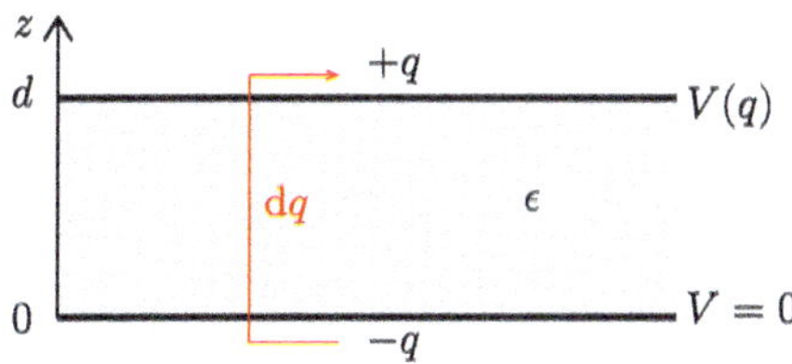

Fig. 8.7 Illustration of a parallel-plate capacitor when the charge is $+q$ on the top surface, $-q$ on the bottom surface, and the potential is $V(q)$. A small charge $\mathrm{d}q$ is transferred from the bottom surface to the top surface

If we connect a motor to the capacitor, and the motor is driven by the discharge of the charge, then the total amount of work done by the motor will be U. Although we here argued for the behavior of a parallel plate capacitor, our argument is general and applies to any capacitor with capacitance C. We can therefore generalize our result:

Energy of a capacitor

The potential energy stored in a capacitor C with a charge Q and/or a potential V is:

$$U = \frac{1}{2}QV = \frac{1}{2}CV^2 = \frac{1}{2}\frac{Q^2}{C}. \tag{8.13}$$

Test your understanding
(a) Two plate capacitors have the same area A and the same charge Q, but the gap of capacitor B is double the gap of capacitor A. What capacitor has stored the most energy? (b) Two plate capacitors have the same area A and gap d, but capacitor B has double the amount of charge as capacitor A. What capacitor has stored the most energy?[5]

Energy of a Charge Distribution

The energy stored in a capacitor corresponds to the work done to create the charge distribution. We can generalize this by finding the work needed to create a charge distribution of N charges Q_i at positions $\mathbf{r}_i$ by adding one charge at a time to the system. We place the first charge, Q_1, in position $\mathbf{r}_1$. This requires no work. Then we bring the second charge, Q_2, in from infinity to its position at $\mathbf{r}_2$. The work required to do this is $W_2 = Q_2 V_1(\mathbf{r}_2)$, where $V_1(\mathbf{r}_2)$ is the electric potential from charge Q_1. This work is the energy, W_2, needed to create the charge distribution. Notice that this work only depends on the distance between the two charges:

$$W_2 = Q_2 \frac{Q_1}{4\pi\epsilon_0 |\mathbf{r}_2 - \mathbf{r}_1|}. \tag{8.14}$$

Then we bring in the third charge. The work, W_3, needed to create a system of three charges is the work done when we bring in the new charge to position $\mathbf{r}_3$ *plus* the work, W_2, done to create the system of two charges:

[5] (a) $U_B > U_A$; (b) $U_B > U_A$.

$$
\begin{aligned}
W_3 &= Q_3 \frac{Q_1}{4\pi\epsilon_0 |\mathbf{r}_3 - \mathbf{r}_1|} + Q_3 \frac{Q_2}{4\pi\epsilon_0 |\mathbf{r}_3 - \mathbf{r}_2|} + W_2 \\
&= Q_3 \frac{Q_1}{4\pi\epsilon_0 |\mathbf{r}_3 - \mathbf{r}_1|} + Q_3 \frac{Q_2}{4\pi\epsilon_0 |\mathbf{r}_3 - \mathbf{r}_2|} + Q_2 \frac{Q_1}{4\pi\epsilon_0 |\mathbf{r}_2 - \mathbf{r}_1|} .
\end{aligned}
\tag{8.15}
$$

We see that this energy is the sum of potential energies for all pairs of charges, where we call the potential energy of a pair of charges the potential energy of one charge in the field set up by the other charge or vice versa. (Because $|\mathbf{r}_i - \mathbf{r}_j| = |\mathbf{r}_j - \mathbf{r}_i|$). This repeats itself each time we add a new charge. The total energy therefore becomes the sum of the potential energies for each pair of charges. We might be tempted to write this as:

$$
W_N = \sum_i \sum_j \frac{Q_i Q_j}{4\pi\epsilon_0 |\mathbf{r}_i - \mathbf{r}_j|} . \tag{8.16}
$$

However, there are two problems with this expression. First, we should not include the potential energy of a charge i with itself. Therefore i should not be equal to j in this sum. Second, if we do this sum, we will add both a term for the potential energy of a charge i in the field set up by charge j *and* the potential energy of a charge j in the field set up by charge i. The two potential energies are identical, and we should only add one potential energy for each such interaction. In the sum we have counted each such interaction twice, and we must therefore divide by two:

$$
W_N = \frac{1}{2} \sum_i \sum_{j \neq i} \frac{Q_i Q_j}{4\pi\epsilon_0 |\mathbf{r}_i - \mathbf{r}_j|} . \tag{8.17}
$$

We rewrite this as

$$
W_N = \frac{1}{2} \sum_i Q_i \sum_{j \neq i} \frac{Q_j}{4\pi\epsilon_0 |\mathbf{r}_i - \mathbf{r}_j|} , \tag{8.18}
$$

and notice that the sum over j is simply the potential at the point $\mathbf{r}_i$ due to the other $N - 1$ charges at positions $\mathbf{r}_j$. We can therefore write the energy of the system as:

$$
W = \frac{1}{2} \sum_i Q_i V(\mathbf{r}_i) . \tag{8.19}
$$

where $V(\mathbf{r}_i)$ is interpreted as the potential at a point $\mathbf{r}_i$ due to all the other charges in the system. This is the amount of energy needed to create the distribution of discrete charges. We can easily generalize this to the case of a continuous charge distribution $\rho(\mathbf{r})$, for which the energy in a volume v is:

$$
W = \frac{1}{2} \int_v \rho(\mathbf{r}') V(\mathbf{r}') \, \mathrm{d}v' . \tag{8.20}
$$

Energy Density

We interpret the total energy needed to create a charge distribution to be the energy of the charge distribution. However, the electric field is set up by the charges. We can therefore interpret this energy as the energy needed to create the corresponding electric field. This motivates the introduction of an energy density. For the parallel plate capacitor, there is only an electric field between the plates (for an infinite system, that is), and the energy U is

$$U = \frac{1}{2}CV^2. \tag{8.21}$$

Here, the electric field is uniform, $E = V/d$, in the region between the two plates and the capacitance C is $C = \epsilon A/d$, where A is the area of the plate and d is the spacing. We insert $V = Ed$ and $C = \epsilon A/d$ into (8.21) and find that

$$U = \frac{1}{2}CV^2 = \frac{1}{2}\frac{\epsilon A}{d}(Ed)^2 = \frac{1}{2}\epsilon E^2 Ad, \tag{8.22}$$

where we recognize Ad as the volume that the electric field occupies inside the capacitor. The *energy density*, which is the energy per unit volume:

$$u_e = \frac{U}{v} = \frac{U}{Ad} = \frac{1}{2}\epsilon E^2. \tag{8.23}$$

For the parallel plate capacitor, we then get that the total energy is $U = u_e\, Ad$. While we have derived this result for a parallel-plate capacitor, the result is general for any linear, isotropic medium.

Energy density of the electric field

The energy density $u_e(\mathbf{r})$ of the electric field for a linear, isotropic medium is defined as:

$$u_e = \frac{1}{2}\epsilon E^2. \tag{8.24}$$

Notice that $\mathbf{E}(\mathbf{r})$ does not have to be uniform and that the energy density (in general) varies in space. The energy U in a volume v is the integral of the energy density over the volume:

$$U = \int_v u_e(\mathbf{r}')\, dv'. \tag{8.25}$$

This result can be further generalized to any linear medium, including anisotropic media, by introducing $\mathbf{D} = \epsilon \mathbf{E}$:

$$u_e = \frac{1}{2} \mathbf{D} \cdot \mathbf{E}. \tag{8.26}$$

Example: Spherical Capacitor

Figure 8.8 illustrates a spherical capacitor consisting of an inner conducting sphere of radius a and an outer conducting spherical shell of inner radius b and outer radius c. Find the capacitance of this system and the energy stored of this system.

Approach. We can use both method 1 and method 2 to solve this problem. Let us start by using method 1 where we assume a given charge Q on the capacitor, find the electric field, from the electric field find the potential, and then find the capacitance as $C = Q/V$.

Distributing charges. We start by distributing the charges $+Q$ on the inner shell and $-Q$ on the outer shell. How is the charge distributed? On the inner shell, we know that there cannot be any free charges inside the conductor. Therefore, all the charges must be on the outer surface of the sphere. What about the outer shell? Also here, there cannot be any charges inside the conductor, so all the charges must be on the surfaces of the conductor. We also know that the electric field must be zero inside the outer conductor shell. How must the charges be organized to ensure this? We expect the charge distribution and the electric field to have spherical symmetry. We can use Gauss' law on a spherical Gauss surface inside the outer conductor. The electric field on this surface must be zero, and therefore the net charge inside the Gauss surface must be zero. The net charge on the surface at a is $+Q$. The net charge on the surface at b must therefore be $-Q$ to ensure the net charge inside the Gauss surface to be zero. The $-Q$ charge on the outer shell is therefore on the inner surface at b.

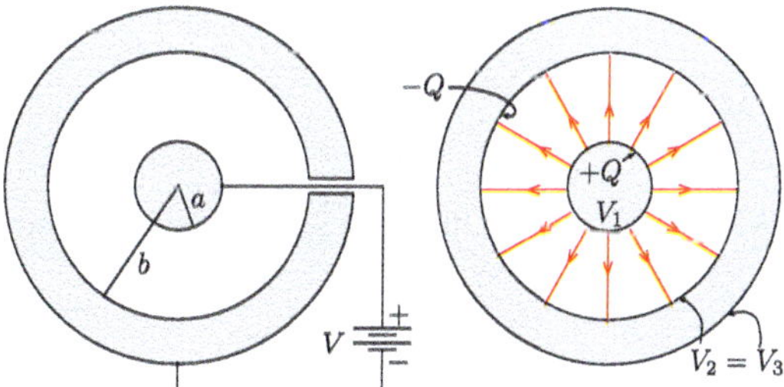

Fig. 8.8 Illustration of a spherical capacitor. (The gap in the drawing is to illustrate how we can connect the inner sphere to a voltage. Assume the outer spherical shell to be without any holes.)

Finding the electric field. We can then find the electric field in the region $a < r < b$ by using Gauss' law with a spherical Gauss surface of radius r. Gauss' law gives that $4\pi r^2 E_r = Q/\epsilon$, and therefore $E_r = Q/(4\pi\epsilon r^2)$.

Finding the potential difference. Then we find the potential difference between the inner and outer conductors. We first find the potential for all $a < r < b$:

$$V(r) - V(a) = \int_r^a \mathbf{E} \cdot \mathrm{d}\mathbf{l} = \frac{Q}{4\pi\epsilon} \int_r^a \frac{\mathrm{d}r}{r^2} = -\frac{Q}{4\pi\epsilon}\left(\frac{1}{a} - \frac{1}{r}\right). \tag{8.27}$$

The potential is highest at $r = a$ and decays towards $r = b$, as expected when there is a positive charge at $r = a$. We find the potential difference ΔV as $\Delta V = V(a) - V(b)$, which is

$$\Delta V = -(V(b) - V(a)) = \frac{Q}{4\pi\epsilon}\left(\frac{1}{a} - \frac{1}{b}\right). \tag{8.28}$$

Calculating the capacitance. The capacitance is therefore

$$C = \frac{Q}{\Delta V} = 4\pi\epsilon \frac{1}{\frac{1}{a} - \frac{1}{b}} = 4\pi\epsilon \frac{ab}{b - a}. \tag{8.29}$$

This result can also be simplified to the case when there is only a single sphere of radius a in a large grounded cage. In this case, $b \to \infty$ and $C = 4\pi\epsilon a$.

Energy stored in spherical capacitor. We now have an expression for both the electric field, $\mathbf{E}(\mathbf{r})$, the electric potential $V(\mathbf{r})$ and the capacitance. How can we use this to find the energy stored in the capacitor?

Approach. Our plan is to use the energy density, $u = \frac{1}{2}\epsilon \mathbf{E} \cdot \mathbf{E}$, and integrate this over the whole volume of the capacitor.

Solution. In the region $a < r < b$ the electric field is

$$E(r) = \frac{Q}{4\pi\epsilon} \frac{1}{r^2}, \tag{8.30}$$

so that the energy density is

$$u_e = \frac{1}{2}\epsilon E^2 = \frac{Q^2}{32\pi^2 \epsilon r^4}. \tag{8.31}$$

What is the energy, $\mathrm{d}U$, in a shell between r and $r + \mathrm{d}r$?

$$\mathrm{d}U = u_e \,\mathrm{d}v = u_e(4\pi r^2 \,\mathrm{d}r) = \frac{1}{2}\epsilon \frac{Q^2}{(4\pi\epsilon)^2} \frac{1}{r^2} 4\pi r^2 \,\mathrm{d}r = \frac{Q^2}{8\pi\epsilon} \frac{\mathrm{d}r}{r^2}. \tag{8.32}$$

The energy for the whole region is the integral from $r = a$ to $r = b$:

$$U = \frac{Q^2}{8\pi\epsilon}\int_a^b \frac{\mathrm{d}r}{r^2} = \frac{Q^2}{8\pi\epsilon}\frac{b-a}{ab}. \tag{8.33}$$

We can now use that $C = 4\pi\epsilon ab/(b-a)$, getting:

$$U = \frac{1}{2}\frac{Q^2}{C}, \tag{8.34}$$

as it should be!

Summary

The **capacitance** C for a set of two conductors is defined as $C = Q/V$ when the two conductors have the charges $+Q$ and $-Q$ and their potential difference is V.

The **capacitance** C **of a parallel plate capacitor** with area A, spacing d and dielectric constant ϵ is $C = A\epsilon/d$.

We find the capacitance of a system using two main methods. In **method 1**, we assume that the conductors have charges $+Q$ and $-Q$, calculate the electric field, from the field we find the potential difference V, and then we find $C = Q/V$. In **method 2**, we assume that the two conductors are equipotential surfaces with a potential difference V and solve Laplace's equation for the potential between the conductors, we find the electric field from the potential, and the surface charge density from the electric field, sum the surface charge density to find the charge Q and $-Q$ and calculate $C = Q/V$.

We can **combine capacitors** $C_1, C_2, \ldots, C_n$, to one effective capacitor C by combining capacitors *in parallel* getting $C = \sum_i C_i$ or *in series* getting $1/C = \sum_i 1/C_i$.

The **energy stored** in a capacitor is $U = Q^2/(2C)$ or $U = CV^2/2$. The **energy density** of an electric field is $u_e = (1/2)\epsilon E^2$ or $u_e = (1/2)\mathbf{D}\cdot\mathbf{E}$. The total energy of a volume is $U = \int_v u_e\,\mathrm{d}v$.

Exercises

Discussion Exercises

8.1 Dielectric materials. How can you explain with words why the capacitance of a parallel plate capacitor increases if you place a dielectric material with higher dielectric constant in the spacing between the plate.

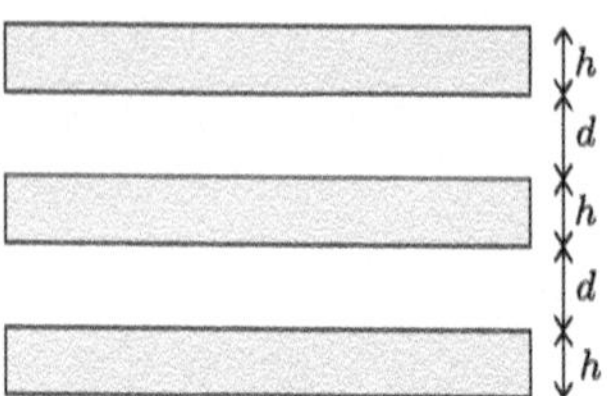

Fig. 8.9 Illustration of a three-layer capacitor

8.2 Capacitance in pieces. You have a parallel plate capacitor with sides L and a spacing of d. You cut it into four identical pieces. What is the capacitance of one of the four pieces compared to the original capacitor?

8.3 The shape of a capacitor. "A capacitor always consists of two identical conductors that have been translated relative to each other". Is this true? Explain.

Tutorials

8.4 Cylinder. You have built a capacitor formed as a conducting cylinder with a radius a enclosed in a conducting cylindrical shell with inner radius b.
(a) Assume that a charge Q is on the inner cylinder. How will the charge be distributed?
(b) Assume that there is a charge $-Q$ on the outer shell. How will this charge be distributed?
(c) You may use that the electric field from a cylindrical charge Q on a cylinder of radius a and length L is $E_r(r) = (Q/L)/(2\pi\epsilon_0 r)$. What is the potential difference between the two conductors.
(d) What is the capacitance of the system?

8.5 Double trouble. You have built a capacitor which consists of two planar plates, each with area A. The are placed at a distance d from each other. You explain to your friend Q that the capacitor stores charge.

That was smart, Q says. But if I double the charge stored from Q to $2Q$, then the electric field will also double. In addition, I need to transfer twice the amount of charge from one plate to the other to charge the capacitor. The work I must do to charge the capacitor is therefore not only double, but four times as large. Because the work is related to the potential energy and therefore the electric potential, this means that the voltage also must be four times larger. I therefore need to have four times the voltage difference to have double the charge on the capacitor.

Is Q's argument correct?

8.6 Sandwich. We build a smart capacitor consisting of three parallel, conducting plates with thickness h, spacing d, and area A as in Fig. 8.9. You may assume that the area A is so large that you can disregards edge effects and assume that the plates are infinitely large.

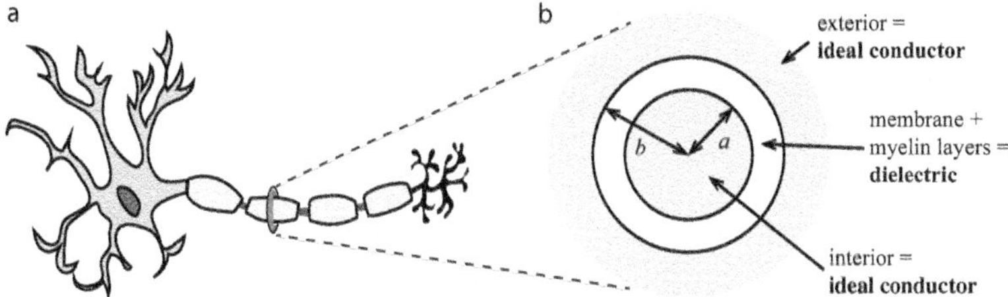

Fig. 8.10 Illustration of a nerve cell. **a** Physical system, **b** Model system showing a cross-section through the axon

(a) What is the electric field from an infinite plane in the xy-plane with surface charge density $\rho_S = Q/A$? (It is sufficient to remember or find the result).
(b) We assume that the middle conductor has a charge $-Q$ and the each of the two other conductors have a charge $Q/2$. How is the charge distributed on the conductors? (Hint: Carefully consider how the charges must be placed and what principles you apply to reach this conclusion.)
(c) What is the electric field normal to the conductors, $E_z(z)$? You may use the results from part (a).
(d) What is the potential difference between the positive and the negative conductor?
(e) What is the capacitance of the system?
(f) How do you interpret this result?

8.7 Capacitance of the axon. A nerve cell is illustrated in Fig. 8.10. The cell has a long axon, which is approximately cylindrical in shape. A cross section of the (cylindrical) axon shows that the cell consists of a center, a thin outer membrane and a myelin sheath outside the membrane. We will assume that the interior and the exterior of the cell can be considered as good conductors. We will model this system as a cylindrical capacitor: We model the inner cylindrical part of the cell as an ideal conductor ($r < a$); we model the cell membrane and the myelin sheath as a cylindrical shell with dielectric constant ϵ ($a < r < b$); and we model the exterior as an ideal conductor ($b < r$).

In this exercise, we will find the capacitance of this system by (i) making a drawing of the system, (ii) drawing the expected charge distributions, (iii) finding the electric field for this charge distribution, (iv) finding the potential by integrating the electric field, and (v) finding the capacitance by relating the charges to the potential difference.

We start from an initially neutral system. We then apply a potential difference V across the dielectric shell. In this process, a charge Q is transferred from the outer shell to the inner shell. Afterwards, the exterior of the dielectric shell is at $V = 0$ and the inner part is at V.
(a) Make a sketch of the distribution of charges on a cross section of the cylinder. What principles do you use when you make this drawing? Is the surface charge density the same on the inner and outer surfaces?
(b) Find the electric field everywhere in terms of Q. (Hint: Choose a Gaussian surface and use Gauss' law.)

(c) Find the potential $V(r = a)$ at the inner cylinder, when the potential $V(r = b) = 0$ at the outer cylinder.
(d) Find the capacitance of the cylindrical shell.

Exercises

8.8 Capacitance. What is the definition of capacitance? Explain the symbols you use.

8.9 Parallel plates. Two thin metal plates are placed in parallel in isolating brackets at a distance d. You can assume that the plates are so large and so close to each other that you can ignore edge effects and that all the charge are on the internal surfaces of the plates (on the side facing the other plate). The surface charge density on one plate is $+\rho_s$ and on the other it is $-\rho_s$. How does the following quantities change (if they do change) when the two plates are moved closer to each other?
(a) How does the surface charge density on each plate change?
(b) How does the magnitude of the electric field between the plates change?
(c) How does the absolute value of the potential difference between the plates change?
(d) How does the capacitance to the two plates change?

8.10 Capacitance.
(a) You have two capacitors of different capacitance C_1 and C_2. You wish to construct a circuit with the highest possible total capacitance using these two capacitors. Should you wire them in parallel or in series?
(b) What is the largest and the smallest capacitance you can get from connecting four 2.0 μF capacitors?
(c) A capacitor has a capacitance of 5.0 μF when connected to a 9.0-V battery. How much energy is stored in the capacitor?

Homework

8.11 Coaxial cable. A coaxial cable consists of an inner cylindrical conductor with radius a and outer cylindrical shell conductor with inner radius b. A dielectric with permittivity ϵ is placed between the conductors. The inner conductor has a potential V_0 while the outer conductor is grounded (has potential 0).
(a) Find the **E**-field between the conductors and the capacitance per unit length C'.
(b) Calculate the numerical value for C' when $\epsilon = 3\epsilon_0$ and $\frac{b}{a} = 7$.
(c) Find how much electric energy is stored per unit length of the cable.
(d) What is the net force acting on the outer conductor from the inner conductor?

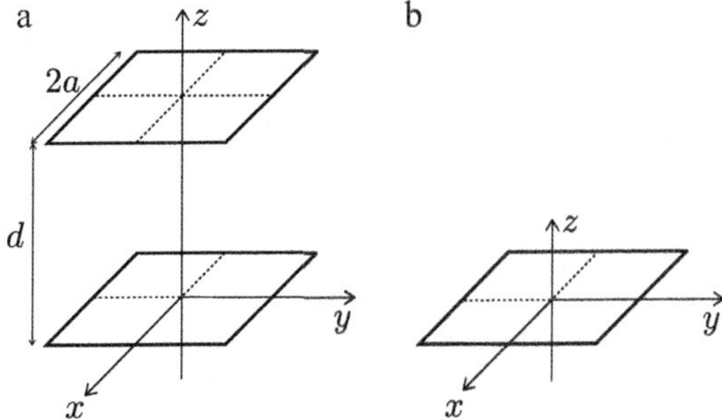

Fig. 8.11 A system of two square conductors forming a capacitor

8.12 Capacitor of two spheres.
(a) Assume that you have a single metallic sphere with a charge $+Q$. How will the charge be distributed in the sphere? (You can assume that the sphere is an ideal conductor).
(b) We will now study a system of two metallic spheres. Both spheres have a radius a and they are placed with a distance d between the centers of the spheres. You can assume that $d > 2a$. One sphere has a charge $+Q$ and the other a charge $-Q$. Assume that the charges are uniformly distributed on each sphere. What is the potential difference between the spheres?
(c) What is the capacitance of the system?
(d) We have assumed that the charge is uniformly distributed on each sphere. Is this a reasonable assumption. Do you think that this assumption has made the capacitance too big or too small compared to the actual situation?

8.13 Capacitor from two circuits. In this exercise we will study a capacitor consisting of two conductors. Each conductor is a square formed from four straight wires of length $2a$. The two conductors are placed at a distance of $d = 2a$ from each other as shown in Fig. 8.11a. We assume that we can model each square as consisting of four linear charge distributions with a uniform line charge density.

Single square First, we address a single square as shown in Fig. 8.11b. Assume that the square consists of four line charges with uniform line charge density so that the whole square has a total charge Q.
(a) What is the electric field in the origin? What can be said about the electric potential without calculating it?
(b) Show that the electric potential along the z-axis for the square with charge Q is $V(z) = \frac{Q}{4\pi\epsilon_0 a}\text{arcsinh}\frac{a}{\sqrt{z^2+a^2}}$. (Hint: Address a single line and then use the superposition principle. You may use that the $\int \frac{\mathrm{d}x}{\sqrt{C^2+x^2}} = \text{arcsinh}\frac{x}{C}$.)
(c) Write a program to find the electric potential for the square. Visualize the potential and the electric field in the yz-plane.
(d) Check your results by comparing with the analytical expression. (Hint: Visualize the analytical and the numerical expression $V(z)$ in the same plot.)

Two squares We will now address a system of two squares, one with charge $+Q$ in $z = 0$ and one with a charge $-Q$ in $z = 2a$ so that they are at a distance d from each other.

(e) We want to find an approximation to the potential difference between the twos squares to estimate the capacitance of the system. As a first approach we want to use the expression we found for the potential along the z-axis. We assume that the contribution from the positive square to the potential difference can be approximated as the difference in potential between the points $(0, 0, 0)$ and $(0, 0, 2a)$. What is the total potential difference between the two squares using this approximation? What is the capacitance of the system using this approximation? (Hint: You can calculate it for one of the squares and use the superposition principle.)

(f) You are not satisfied with this approximation. You would like to use your numerical model to find a better value, but still use the same approach. Instead of using the potential from the positive square in a point on the z-axis, you wish to find the potential from the positive square in a point at the middle of one of the negatively charged lines. Then you multiply with two to find the total potential difference. What is the potential difference with this approximation? And what is the capacitance? (You can assume that the diameter of the line charge is $h = a/20$).

(g) Plot the potential from the positive square along one of the lines of the square. What do you see? What does this tell you? Explain how you would find the capacitance of the system without getting into this problem.

Chapter 9
Current and Resistance

9.1 Current

Figure 9.1 illustrates a cross-section through a conductor. Let us assume that charges are moving along the conductor so that a charge Δq passes through the cross-sectional area A in a short time interval Δt. We define the *current*, I, through the cross-sectional area A as the instantaneous rate at which positive charges are passing through the surface: $I = \Delta q/\Delta t = \mathrm{d}q/\mathrm{d}t$.

Definition of current

The **current** I through a surface S is defined as the instantaneous rate at which positive charges are passing through the surface:

$$I = \frac{\mathrm{d}q}{\mathrm{d}t} \tag{9.1}$$

where $\mathrm{d}q$ is the amount of charge passing through the surface S in an infinitesimal time $\mathrm{d}t$. Current is measured in coulomb per second (C/s), which is called **ampere** in the SI system. Ampere is considered a basic unit in the SI system, such as kg, meters or seconds.

Test your understanding
A charge Q that flows through a circular cross-section with radius a during the time t gives a current I_1. If the charge $2Q$ flows through a circular cross-section with radius $2a$ in the time t, what is the current I_2 relative to I_1?[1]

[1] $I_2 = 2I_1$.

© The Author(s), under exclusive license to Springer Nature Switzerland AG 2026

A. Malthe-Sørenssen, *Elementary Electromagnetism Using Python*, Undergraduate Texts in Physics, https://doi.org/10.1007/978-3-032-19876-1_9

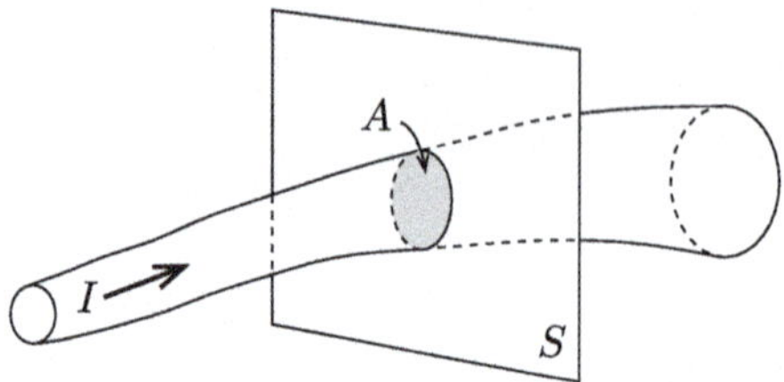

Fig. 9.1 Illustration of a conductor of cross-section A passing through a surface S. An amount of charge Δq is passing through the surface along the conductor in a short time Δt.

Microscopic Picture

How can we relate the concept of current to the microscopic picture of charges and electric fields we have introduced so far? What happens to a particle with charge q, such as an electron, inside a real conductor? We often call such a particle a charge carrier. Figure 9.2a illustrates a real conductor such as a metal. Such a system consists of a sea of electrons that can move and a stationary lattice of positive charges, ions, that are vibrating, but otherwise not moving far from their positions in the lattice. If we apply an electric field $\mathbf{E}$ to this system, a small charge q, such as an electron, will be subject to a force $\mathbf{F} = q\mathbf{E}$. In Fig. 9.2b we illustrate the resulting forces on the red electrons.

In Fig. 9.2c we have illustrated the motion of an electron as it bounces across the system under the influence of both the force from the external field and the random forces resulting from collisions with the ions making up the lattice. The result is a motion of many small random steps, but with an overall drift in the direction of the force from the electric field. After a time Δt the electron has moved a distance $\Delta\mathbf{r}$. This gives rise to an *average* or *drift* velocity, where all the small fluctuations have been averaged out, $\mathbf{v}_d = \Delta\mathbf{r}/\Delta t$.

We can connect this picture with our definition of current by summing up the contributions from all the charge carriers in a small volume and finding how much charge moves through a surface $\mathrm{d}\mathbf{S}$ per unit time. We describe the charge carriers by N, the number of charge carriers per unit volume. The number of charge carriers in a small volume $\mathrm{d}v$ is $N\mathrm{d}v$. We assume that each of these charge carriers have a

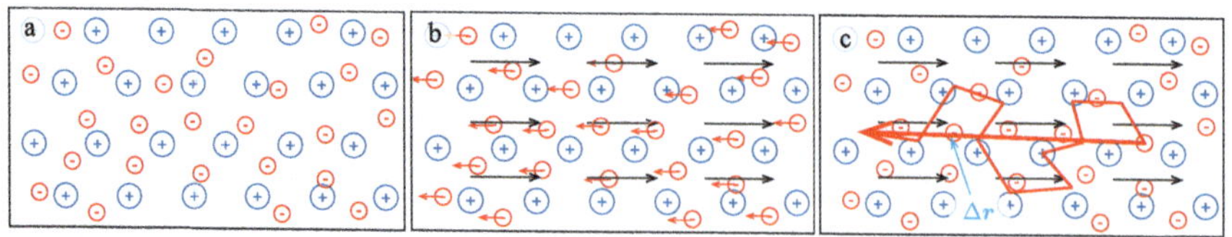

Fig. 9.2 **a** Illustration of a metal with a sea of freely moving electrons and a vibrating lattice of positive ions. **b** An electric field is applied, indicated by black arrows, leading to force on all the electron illustrated by red arrow. **c** The motion of an electron is due to both the force from the external electric field and forces from collisions (interactions) with the lattice ions, resulting in a directed random walk

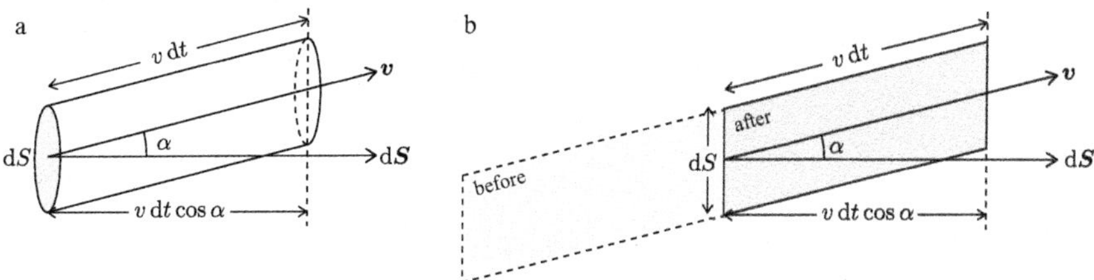

Fig. 9.3 Illustration of the volume $\mathrm{d}v = v\,\mathrm{d}t\,\mathrm{d}S\,\cos\alpha$ moving through a surface $\mathrm{d}\mathbf{S}$ in a time interval $\mathrm{d}t$

charge q and move with a drift velocity $\mathbf{v}_d$. In a small time interval $\mathrm{d}t$, each charge moves a distance $\mathbf{v}_d\,\mathrm{d}t$. Figure 9.3 illustrates the situation. All the charges that are in a volume $\mathrm{d}v = v_d\,\mathrm{d}t\cos\alpha\,\mathrm{d}S$ will have moved through the surface $\mathrm{d}S$ in the time $\mathrm{d}t$, where α is the angle between the surface normal to $\mathrm{d}\mathbf{S}$ and the velocity $\mathbf{v}_d$ so that $\mathrm{d}\mathbf{S}\cdot\mathbf{v}_d = \mathrm{d}S v_d\cos\alpha$ and $\mathrm{d}v = \mathbf{v}_d\,\mathrm{d}t\cdot\mathrm{d}\mathbf{S}$.

The amount of charge in the volume $\mathrm{d}v$ is then $\mathrm{d}q = Nq\,\mathrm{d}v$. The amount of charge flowing through the surface $\mathrm{d}\mathbf{S}$ in the time interval $\mathrm{d}t$ is $Nq\mathbf{v}_d\,\mathrm{d}t\cdot\mathrm{d}\mathbf{S}$. This is the $\mathrm{d}q$ in the definition of the current $\mathrm{d}I$ through the surface $\mathrm{d}\mathbf{S}$:

$$\mathrm{d}I = \frac{\mathrm{d}q}{\mathrm{d}t} = \underbrace{Nq\mathbf{v}_d}_{\mathbf{J}}\cdot\mathrm{d}\mathbf{S} = \mathbf{J}\cdot\mathrm{d}\mathbf{S}, \tag{9.2}$$

where we introduce the term current density, $\mathbf{J}$, so that

$$\mathrm{d}I = \mathbf{J}\cdot\mathrm{d}\mathbf{S}. \tag{9.3}$$

The total current I through (an open) surface S is therefore the sum of all these elements $\mathrm{d}I$ integrated (summed) over the whole surface:

$$I = \int_S \mathbf{J}\cdot\mathrm{d}\mathbf{S}. \tag{9.4}$$

Current density

We define the current density $\mathbf{J}$ so that the current $\mathrm{d}I$ through a surface element $\mathrm{d}\mathbf{S}$ is

$$\mathrm{d}I = \mathbf{J}\cdot\mathrm{d}\mathbf{S}. \tag{9.5}$$

We can relate the current density to the volume density of charge carriers N, their charge q and their drift velocity $\mathbf{v}_d$ through

$$\mathbf{J} = Nq\mathbf{v}_d. \tag{9.6}$$

If there are more than one type of charge carrier, the current density is the sum of the contributions from all the charge carriers, $\mathbf{J} = \sum_i N_i q_i \mathbf{v}_{d,i}$.

Test your understanding
(a) Positive ions are flowing out through a channel through a cell membrane, and negative ions are flowing in. The charge of the ions has the same magnitude. The densities of the ions are the same and they have the same velocity. Is there a net current through the membrane?
(b) A current I flows through a wire of length L and square cross-section with sides a. If the current is uniformly distributed across the cross-section, what is the current density J?[2]

9.2 Ohm's Law on Microscopic Form

Intuitively, the drift velocity $\mathbf{v}_d$ and therefore $\mathbf{J}$, must depend on the electric field, $\mathbf{E}$. Here, we introduce a microscopic theory for the motion of charge carriers and use this to derive Ohm's law: $\mathbf{J} = \sigma \mathbf{E}$, where σ is called the conductivity of the material.

Our goal is to relate the drift velocity $\mathbf{v}_d$ to a classical microscopic picture of the processes in a metal by making a *microscopic model* of the process. (Notice that the movement of electrons in a conductor may be better described using quantum mechanics. This is therefore a simplified model). We will assume that the electrons move around like in a gas and collide with the vibrating ions in the crystal. Between each collision, we will assume that the electrons are only affected by the force from the electric field. This means that the velocity $\mathbf{v}$ a time t after a collision, and before a new collision, is:

$$\mathbf{v} = \mathbf{v}_0 + (q/m)\mathbf{E}t , \tag{9.7}$$

where m is the mass of the electron, $q = -e$ is the charge of the electron, and $\mathbf{v}_0$ is the velocity after the previous collision. Let us assume that time between each collision is τ. We call this the *mean free time*. How far does the charge get between each collision? The displacement between two collisions is $\Delta\mathbf{r}$ (which is the integral of $\mathbf{v}$ over a time τ):

$$\Delta\mathbf{r} = \int_0^\tau (\mathbf{v}_0 + (q/m)\mathbf{E}t)\, dt = \left(\mathbf{v}_0 + \frac{1}{2}\frac{q}{m}\mathbf{E}\tau\right)\tau . \tag{9.8}$$

Now, we take the average over many collisions. (We use the symbols $\langle a \rangle$ to indicate a time average). The initial velocities, $\mathbf{v}_0$ are in random directions, thus their average is zero. Averaged over many collisions, we find:

[2] (a) Yes, there is a net current; (b) $J = I/a^2$.

$$\langle \Delta \mathbf{r} \rangle = \frac{1}{2}\frac{q}{m}\mathbf{E}\tau^2, \tag{9.9}$$

and the average velocity, the *drift velocity*, is:

$$\mathbf{v}_d = \langle \mathbf{v} \rangle = \frac{\langle \Delta \mathbf{r} \rangle}{\tau} = \frac{q\tau}{2m}\mathbf{E}. \tag{9.10}$$

The current density is therefore:

$$\mathbf{J} = Nq\mathbf{v}_d = \frac{Nq^2\tau}{2m}\mathbf{E} = \sigma\mathbf{E}. \tag{9.11}$$

This relation is general for many types of materials, and is called *Ohm's law*:

$$\mathbf{J} = \sigma\mathbf{E} \tag{9.12}$$

Ohm's law (on local form)

Many real conductors obey Ohm's law on local form or Ohm's microscopic law:

$$\mathbf{J} = \sigma\mathbf{E}, \tag{9.13}$$

where σ is called the *conductivity*, $\mathbf{J}$ is the current density and $\mathbf{E}$ is the electric field. It tells us how well a material conducts (electrical) current. For an insulator $\sigma = 0$, while for a superconductor or an ideal conductor $\sigma = \infty$.

Notice, that the *resistivity* ρ is $\rho = 1/\sigma$, which is what is often found in tables. The conductivity σ depends strongly on the temperature in a material. (In statistical physics you will learn about theories for the temperature dependence of σ based on quantum mechanics.)

Test your understanding

(a) What is the direction of the current density in point 1 and 2 in the figure? (b) What is the direction of the electric field in points 3 and 4? (c) A current I flows through the system from left to right. Rank the current densities and the electric fields in A and B.?[3]

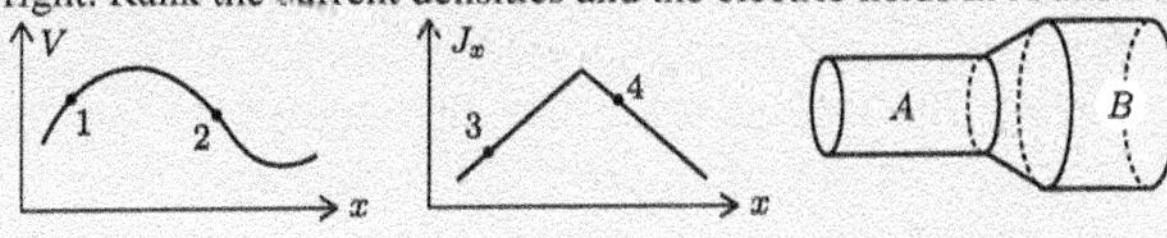

[3] (a) 1: $J_x > 0$, 2: $J_x > 0$; (b) 3: $E_x < 0$, 4: $E_x > 0$; (c) $J_A > J_B$, $E_A > E_B$.

Table 9.1 Examples of resistivities ρ for conductors (C), semi-conductors (S), and insulators (I) at room temperature. (Data from Handbook of Chemistry and Physics, 102nd ed.)

Material	ρ (ohm m)	Material	ρ (ohm m)
Silver (C)	1.59×10^{-8}	Salt water (S)	4.4×10^{-2}
Copper (C)	1.68×10^{-8}	Diamond (S)	2.7
Gold (C)	2.21×10^{-8}	Silicon (S)	2.5×10^{3}
Aluminum (C)	2.65×10^{-8}	Water (I)	2.5×10^{5}
Iron (C)	9.61×10^{-8}	Wood (I)	$10^{8} - 10^{11}$
Graphene (C)	1×10^{-8}	Glass (I)	$10^{10} - 10^{14}$

Conductors and Resistors

You may now be confused because we earlier said that the electric field had to be zero inside an ideal conductor. Now we have to modify this picture. For good conductors such as metals, the electric field needed to drive an electric current is so small, that we will usually assume that the voltage drops along a real conductor is negligible. However, non-conductors, such as semi-conductors and insulators, are made of poorly conducting materials, where there is a non-negligible field needed to drive a current and where we cannot ignore the potential drop along the insulator. We will usually model systems as combinations of *conductors*, which are ideal conductors with no potential drop, and *resistors*, which is our model of the parts of a system that has a non-negligible resistivity. We usually group materials into *conductors*, *semi-conductors*, and *insulators*. Table 9.1 provides the resistivities for various common materials.

9.3 Resistance

We apply this theory to find the current through a wire with a given conductivity σ, constant cross-sectional area S and length L as illustrated in Fig. 9.4. One side is at potential V_0 and the other at $V_1 = 0$. The wire has the same shape and cross-sectional area all across.

To find the current, we need to find the current density $\mathbf{J}$ and integrate across the cross-section to find the current. How can we find the current density and relate it to the potential difference V? Our plan is to relate the current density to the electric field through Ohm's law, $\mathbf{J} = \sigma \mathbf{E}$ and then use this to relate the current to the potential difference. We can do this either by first finding the electric field and then use this to find the current density, or by first finding the current density using symmetry consideration and use this to find the electric field. We will here demonstrate both these methods.

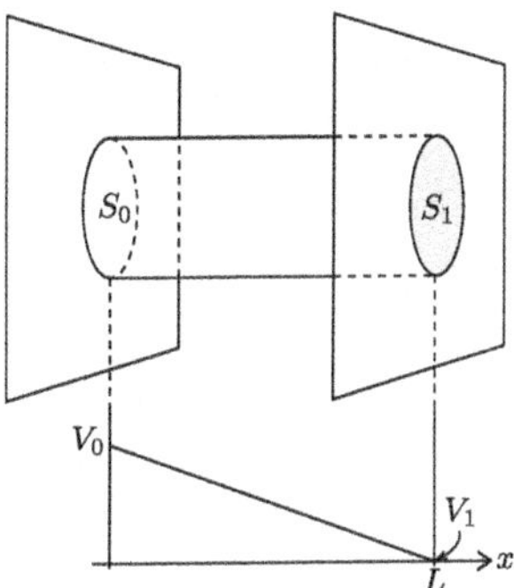

Fig. 9.4 A wire with conductivity σ, a constant cross-sectional area S and length L is subject to a potential difference $V_0 - V_1$

Method: From Electric Potential to Current Density

In this approach, we find the electric potential, use this to find the electric field, and then use the electric field to find the current density. But how can we find the electric potential? We assume that there are no free charges inside the wire. This is an approximation, because there may be charges moving along the wire, but we assume these are few and do not impact the electric field significantly. We may then solve Laplace's equation for the system. We orient the system along the x-axis. The potential is uniformly equal to V_0 on the surface S_0 at $x = 0$ and similarly equal to $V_1 = 0$ on the surface S_1 at $x = L$. A solution to Laplace's equation for this system is $V(x) = V_0 - xV_0/L$. (We will study this in more detail below). The electric field is therefore $E_x = -\mathrm{d}V/\mathrm{d}x = V_0/L$. We can then apply Ohm's law to find the current density $J_x = \sigma E_x = \sigma V_0/L$, which is also uniform in space in the wire, and zero outside the wire because there are no charge carriers there and therefore no current density. The current across the area S_1 is then

$$I = \int_{S_1} \mathbf{J} \cdot \mathrm{d}S\hat{\mathbf{x}} = \int_{S_1} J_x \, \mathrm{d}S = \sigma(V_0/L) \int_{S_1} \mathrm{d}S = \sigma(V_0/L)S_1. \tag{9.14}$$

We can rewrite this to find the potential difference, $V = V_0$, in terms of the current:

$$V_0 = \underbrace{\frac{L}{\sigma S_1}}_{=R} I \tag{9.15}$$

The voltage (difference) $\Delta V = V_0 - V_1$ is therefore proportional to the current I. It is common to simple denote this voltage difference or voltage drop V instead of ΔV. We call the constant of proportionality the *resistance* R. We have now recovered Ohm's law for a resistor:

Ohm's law for a resistor

The voltage drop V across a resistor is given as

$$V = RI, \tag{9.16}$$

where R is a property of the resistor called the **resistance**. The resistance for a specific resistor can be calculated or measured experimentally. Resistance is measured in ohm which is defined as volt/ampere. This is the law you may recall as Ohm's law. To differentiate it from $\mathbf{J} = \sigma\mathbf{E}$, we will call $V = RI$ Ohm's law and $\mathbf{J} = \sigma\mathbf{E}$ Ohm's law on local or microscopic form.

Method: From Current Density to Electric Field

An alternative approach to find the resistance R for a resistor is to assume a particular symmetry for the current density and then use this to find an expression for the electric field and then find the potential difference from the electric field.

For this particular problem we realize that the system is the identical at any cross-section along the component. We therefore expect the current density $\mathbf{J}$ also to be uniform along the x-axis. Similarly, the conductor is the same at any point in the cross section, so we expect the current density also to be uniform across the whole cross section. There are no currents normal to the boundaries, as this would mean that current is leaking from the conductor out into the vacuum. This combined with the requirement that the current density is uniform, means that the current density is $\mathbf{J} = J_x\hat{\mathbf{x}}$ everywhere inside the conductor and zero outside the conductor.

Ohm's law on local form can be used to find the electric field: $\mathbf{E} = \mathbf{J}/\sigma = (J_x/\sigma)\hat{\mathbf{x}}$. We can then find the potential difference $V = V_0 - V_1$ from the left (0) to the right (1) across the length L by

$$V = V_0 - V_1 = \int_0^1 \mathbf{E} \cdot \mathrm{d}\mathbf{l} = \int_0^L \frac{J_x}{\sigma} dx = \frac{J_x}{\sigma} L. \tag{9.17}$$

which gives $J_x = V\sigma/L$. We can also relate the current density to the current by

$$I = \int_S \mathbf{J} \cdot \mathrm{d}S\hat{\mathbf{x}} = \int_S J_x \mathrm{d}S = J_x \int_S \mathrm{d}S = J_x S. \tag{9.18}$$

where the integral is across any surface S along the wire. This means that $J_x = I/S$. We equate these two expressions for the current density, getting:

$$\frac{V\sigma}{L} = \frac{I}{S} \Rightarrow V = \frac{L}{\sigma S} I = RI. \tag{9.19}$$

We have therefore found the same result as above, but with a different starting point. The method we have used here is very similar to the method we used to find the capacitance for a capacitor.

Method: Calculating the Resistance

Finding the resistance of a resistor from first principles in electromagnetism is a process you should understand and a skill that you should learn. From these examples we see the pattern of a general method for calculating the resistance of a resistor—following an approach which is very similar to what we developed for a capacitor. (We use the numbering from the capacitance method, which is the opposite of the order in which we solve the problem above).

Method: Finding the resistance of a system

Method 1: From current density to field to potential:

1. We assume that the current density, $\mathbf{J}$, has a particular symmetry or form inside the conductors based on considerations of the physics and geometry of the system.
2. We calculate the current, I, through a cross-section, S, of the conductor using $I = \int_S \mathbf{J} \cdot d\mathbf{S}$.
3. We calculate the electric field $\mathbf{E}$ using Ohm's law on local form: $\mathbf{E} = \mathbf{J}/\sigma$.
4. We calculate the potential difference between the two ends of the conductor, $V = \int_0^1 \mathbf{E} \cdot d\mathbf{l}$, using the electric field we found.
5. We find the resistance from $R = V/I$, where we insert our value for V and the value for I we found from the current density.

Method 2: From potential to field to current:

1. We solve Laplace's equation to find the potential $V(\mathbf{r})$ for the system. Boundary conditions are given by assuming that there is no flow normal to the conductors.
2. We find the electric field $\mathbf{E} = -\nabla V$.
3. We find the current density from Ohm's law on local form $\mathbf{J} = \sigma\mathbf{E}$. We sum/integrate the current density to find the current through a cross-section S through the conductor: $I = \int_S \mathbf{J} \cdot d\mathbf{S}$.
4. We find the resistance from $R = V/I$, where we insert our value for I and the potential difference $V = V_0 - V_1$ we started with.

We will mostly use method 1 in this text. Method 2 is generally more difficult, as you need to solve Laplace's equation with boundary conditions that can be complex for complex geometries.

Test your understanding
The figure shows a resistor with length L and cross-sectional area A. When a current I is passing through the system from left (V_1) to right ($V_0 = 0$), the current density is J_1, the electric field is E_1, the potential is V_1 and the resistance is R_1. We create a new system by putting together two blocks as in (a). We send a current through this joint system from left to right. (a) In **b** which properties (J_2, E_2, V_2, R_2) change and which are the same? Find the new values. (b) In **c** which properties (J_2, E_2, V_2, R_2) change and which are the same? Find the new values.[4]

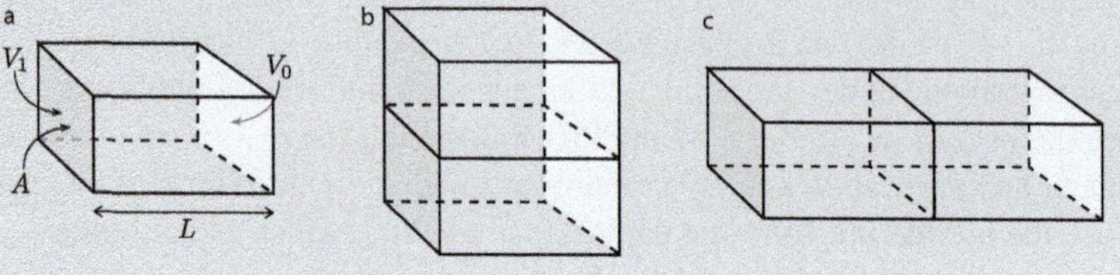

Example: Spherical Resistor

What is the resistance of a system consisting of a spherical shell resistor? The situation is illustrated in Fig. 9.5.

Approach. Our plan is to use method 1 to address this problem. We assume that a constant, stationary current I runs from the inner to the outer side of the spherical shell.

Solution. Because the system has a spherical symmetry, we will assume that the current density also has spherical symmetry so that $\mathbf{J} = J_r\hat{\mathbf{r}}$. (We do not expect any currents in the Φ or θ directions). The amount of charge per unit time $I = \mathrm{d}q/\mathrm{d}t$, through a spherical shell of radius r, where $a < r < b$, must be the same for all r, otherwise charge would disappear. We calculate the current from the current density:

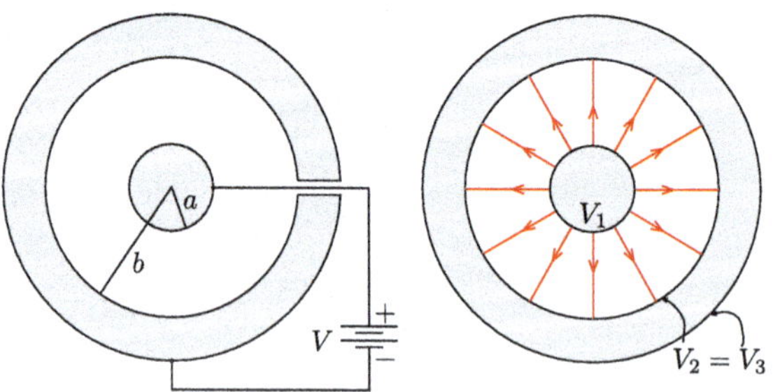

Fig. 9.5 Illustration of a cross-section through a resistor shaped as a spherical shell with inner radius a and outer radius b. A potential difference V is applied across the shell

[4] (a) $J_2 = J_{1/2}$, $E_2 = E_{1/2}$, $V_2 = V_{1/2}$, $R_2 = R_{1/2}$; (b) $J_2 = J_1$, $E_2 = E_1$, $V_2 = 2V_1$, $R_2 = 2R_1$.

$$I = \int_S \mathbf{J} \cdot d\mathbf{S} = 4\pi r^2 J_r. \tag{9.20}$$

and the current density is

$$\mathbf{J} = \frac{I}{4\pi r^2}\hat{\mathbf{r}} \text{ for } a < r < b \tag{9.21}$$

The electric field is found from Ohm's law on local form: $\mathbf{J} = \sigma\mathbf{E}$ and therefore $\mathbf{E} = \mathbf{J}/\sigma$

$$\mathbf{E} = \frac{I}{4\pi\sigma r^2}\hat{\mathbf{r}}. \tag{9.22}$$

We find the potential difference, V, from the electric field, which is the potential difference between the shell and the sphere, $V(b) = 0$ and $V(a) = V$. Hence, we find

$$V = V(a) = \int_a^{\text{ref}} \mathbf{E} \cdot d\mathbf{r} = \int_a^b \frac{I}{4\pi\sigma r^2} dr = -\frac{I}{4\pi\sigma}\left(\frac{1}{b} - \frac{1}{a}\right). \tag{9.23}$$

Now, we are ready to relate V to I, finding that

$$R = \frac{V}{I} = \frac{1}{4\pi\sigma}\left(\frac{1}{a} - \frac{1}{b}\right). \tag{9.24}$$

(Notice how this reminds us of what we did for the spherical capacitor. Indeed, this is a general result, that there is a relation between capacitance and resistance of a system. We can see this by looking at the current through a spherical surface, S, $I = \int_S \mathbf{J} \cdot d\mathbf{S} = \int_S \sigma\mathbf{E} \cdot d\mathbf{S} = (\sigma Q/\epsilon)$, thus $I/V = (\sigma/\epsilon)(Q/V)$, where $G = I/V$ is the conductance and C is the capacitance! This is a nice duality relationship.)

Using Laplace's Equation to Find the Field

How can we solve Laplace's equation as needed for method 2? We revisit the system in Fig. 9.4: A resistor with a constant cross-sectional area S of length L. We assume that the cross-section is a circle of radius a, $S = \pi a^2$, so that the resistor is a cylinder of radius a and length L. To apply method 2 to this system we need to solve Laplace's equation within the cylinder.

What are the boundary conditions for Laplace's equation? The potential is uniform $V_0 = 0$ and V_1 at both ends of the cylinder. What about the outer surface of the cylinder? We know that there cannot be any charges flowing (leaking) from the resistor into the vacuum outside it. This means that the current density cannot have a component normal to the outer surface, that is, $\mathbf{J} \cdot \hat{\mathbf{n}} = 0$. Using Ohm's law on local form we therefore see that $\mathbf{E} \cdot \hat{\mathbf{n}} = (1/\sigma)\mathbf{J} \cdot \hat{\mathbf{n}} = 0$, which again means that

$\partial V/\partial n = 0$ on the outer surface of the cylinder. This means that either V or its derivative is specified on all the surfaces. The potential V is therefore uniquely determined.

Consequently, we only need to find *a* solution and it will be *the* solution. For this problem, it is easy to guess the solution. If the cylinder is oriented along the x-axis, the potential $V = V_1 x/L$ satisfies the boundary conditions. We can then find the electric field:

$$\mathbf{E} = -\nabla V = -\frac{V_1}{L}\hat{\mathbf{x}}, \tag{9.25}$$

which is uniform. Notice that we were able to solve this problem because we could assume that all the current is in the conducting material only (inside the resistor). It would be much more difficult to solve Laplace's equation without the boundary condition $\partial V/\partial n = 0$ on the surface.

9.4 Energy Loss

As the electrons (or any charge carrier) moves along the conductor due to the electric field, they collide with vibrating lattice ions in the underlying crystal as illustrated in Fig. 9.2. This will transfer energy from the electrons to the lattice vibrations, increasing the thermal energy of the metal (and thus the temperature of the metal). But how much energy is dissipated?

Let us see what happens in a small volume unit $\mathrm{d}v$ with $N\,\mathrm{d}v$ charge carriers, where N is the charge carrier volume density. During a small time interval $\mathrm{d}t$, the charges in this volume will move a length $\mathrm{d}\mathbf{r} = \mathbf{v}_d\,\mathrm{d}t$. The electric field acts on the electrons. The force on each electron is therefore $\mathbf{F} = q\mathbf{E}$ (where $q = -e$ for an electron). The work, $\mathrm{d}W_1$, done by the electric field on one charge q over the time interval $\mathrm{d}t$ is

$$\mathrm{d}W_1 = q\mathbf{E}\cdot\mathrm{d}\mathbf{r} = q\mathbf{E}\cdot\mathbf{v}_d\,\mathrm{d}t, \tag{9.26}$$

and the work, $\mathrm{d}W$, done on all $N\,\mathrm{d}v$ charges, each with a charge q, is therefore:

$$\mathrm{d}W = N\,\mathrm{d}v\,\mathrm{d}W_1 = N\,\mathrm{d}v\,q\mathbf{E}\cdot\mathbf{v}_d\,\mathrm{d}t = \underbrace{Nq\mathbf{v}_d}_{=\mathbf{J}}\cdot\mathbf{E}\,\mathrm{d}v\,\mathrm{d}t, \tag{9.27}$$

The power loss, P, is $P = \mathrm{d}W/\mathrm{d}t = \mathbf{J}\cdot\mathbf{E}\,\mathrm{d}v$ and the power loss per unit volume, p_J, is therefore:

$$p_J = \frac{\mathrm{d}P}{\mathrm{d}v} = \mathbf{J}\cdot\mathbf{E} \tag{9.28}$$

We call this the power loss, because this is the work the electric field does to move the electrons. This work is therefore converted into thermal energy, and thus "lost" from the electric system. We often call this the dissipated energy, and you can feel its effect by an increase in temperature in the system.

Power loss

The power loss per unit volume, p_J, is

$$p_J = \mathbf{J} \cdot \mathbf{E}. \tag{9.29}$$

We find the total power loss, P, for a system (such as a resistor) by integrating the power loss per unit volume:

$$P = \int_v p_J \, \mathrm{d}v = \int_v \mathbf{J} \cdot \mathbf{E} \, \mathrm{d}v. \tag{9.30}$$

For a resistor, we can relate the work $\mathrm{d}W$ done on the charge $\mathrm{d}Q = qN\,\mathrm{d}v$ to the electric potential: The work done on a charge $\mathrm{d}Q$ as it is moved from one side of the resistor with potential V_1 to the other side with potential V_0 is $\mathrm{d}W = \mathrm{d}Q(V_1 - V_0) = \mathrm{d}QV$, where V is the potential drop across the resistor. For a constant V and I, the stored electric energy will be unchanged, so this energy is dissipated (converted to heat). The power loss is therefore:

$$P = \frac{\mathrm{d}W}{\mathrm{d}t} = \frac{\mathrm{d}QV}{\mathrm{d}t} = V\frac{\mathrm{d}Q}{\mathrm{d}t} = VI. \tag{9.31}$$

Power loss for a component

The power loss for a resistor with a current I and potential drop V is

$$P = VI. \tag{9.32}$$

We often only call this the *power*, the *power loss* or the *power dissipated* in the component.

This is called the **Joule heating law** . For a resistor with resistance R, we know that $V = RI$ so that $P = RI^2$ or $P = V^2/R$.

Test your understanding
(a) Two resistors R_1 and R_2 ($R_1 > R_2$) are connected in parallel. Through which of the resistors is the power loss the greatest? (b) Two resistors R_1 and R_2 ($R_1 > R_2$) are connected in series. Through which of the resistors is the power loss the greatest? (c) Two cylindrical resistors of length L are made of a material with conductivity σ. One has a radius a and the other $2a$. A current I goes through each resistor. Which resistor has the largest power loss per volume, p_J? (d) And in which resistor is the power loss the largest?[5]

[5] (a) R_2; (b) R_1; (c) a; (d) a.

Example: Power in a Spherical Resistor

What is the power loss in the spherical resistor illustrated in Fig. 9.5?

Approach. We compare the two methods: We find the power by either integrating the power loss per unit volume across the resistor or by using the expression for the power loss, P.

Solution. The power loss per unit volume is $p_J = \mathbf{E} \cdot \mathbf{J}$. We found that $\mathbf{J} = I/(4\pi r^2)\hat{\mathbf{r}}$ and that $\mathbf{E} = \mathbf{J}/\sigma$. Then

$$p_J = \mathbf{J} \cdot \mathbf{E} = \frac{1}{\sigma}\left(\frac{I}{4\pi r^2}\right)^2 . \tag{9.33}$$

We find the total power dissipated in the component by integrating over the volume from a to b, where a volume element is $\mathrm{d}v = 4\pi r^2 \, \mathrm{d}r$. The integral is:

$$\begin{aligned} P &= \int_v p_J \, \mathrm{d}v = \int_a^b \frac{1}{\sigma}\left(\frac{I}{4\pi r^2}\right)^2 4\pi r^2 \, \mathrm{d}r \\ &= \frac{I^2}{4\pi\sigma}\int_a^b \frac{\mathrm{d}r}{r^2} = \frac{I^2}{4\pi\sigma}\left(\frac{1}{a} - \frac{1}{b}\right) = I^2 R , \end{aligned} \tag{9.34}$$

where we have used the result $R = 1/(4\pi\sigma)((1/a) - (1/b))$ we found previously. This means that we have demonstrated that the result we get is the same as using $P = I^2 R$ directly.

9.5 Conservation of Charge

We know that the net charge is conserved. This is a fundamental principle. Charges can arise, as when an electron is emitted from a neutral atom, creating a negative electron ($-e$) and a positive ion ($+e$), but the net change in charge is zero. Similarly, a positron with charge $+e$ and an electron with charge $-e$ may combine form neutral photons (light), but the net charge remains zero. What consequences do the conservation of charge have for the flow of current? It means that charges can move, in the form of current, but they do not appear or disappear.

Let us formulate this conservation law, *the conservation of charge*, mathematically and apply it to conservation of currents. A volume v is enclosed by the surface S as illustrated in Fig. 9.6a. The charge Q inside the volume is the integral of the charge density, ρ:

$$Q = \int_v \rho \, \mathrm{d}v , \tag{9.35}$$

Fig. 9.6 **a** Illustration of a volume v enclosed by the surface S. **b** Illustration of a the currents I_1, I_2 and I_3 flowing in a conductor through the closed surface S

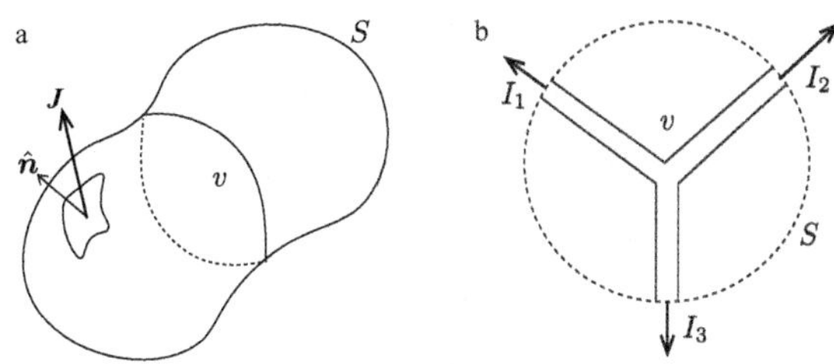

and the current out of the surface S is the integral of the current density over the surface S:

$$I_S = \oint_S \mathbf{J} \cdot \mathrm{d}\mathbf{S}. \tag{9.36}$$

In a small time interval $\mathrm{d}t$ the net change in charge inside the surface, $\mathrm{d}Q$ must correspond to the charge flowing through the surface in this time interval, $I_S\,\mathrm{d}t$. The charge is reduced (the change is negative), if the flow out of the surface is positive:

$$-\,\mathrm{d}Q = I_S\,\mathrm{d}t \;\Rightarrow\; -\frac{\mathrm{d}Q}{\mathrm{d}t} = I_S. \tag{9.37}$$

As always we check the sign to ensure it is correct: We notice that the integral over the surface tells us the amount of charge flowing *out of the surface*, which must correspond to a reduction in the charge inside, $-\mathrm{d}Q$. This is a formulation of the principle of conservation of charge.

We can rewrite the conservation of charge in integral form, by using the expression for Q from (9.35):

$$\oint_S \mathbf{J} \cdot \mathrm{d}\mathbf{S} = -\frac{\mathrm{d}}{\mathrm{d}t}\int_v \rho\,\mathrm{d}v, \tag{9.38}$$

If we study systems where all the currents are stationary, that is they are independent of time, then there cannot be a current into a volume or out of a volume forever, because this would lead to an infinite build-up (or removal) of charge. Thus, for a stationary system we must have that:

$$\oint_S \mathbf{J} \cdot \mathrm{d}\mathbf{S} = 0. \tag{9.39}$$

Figure 9.6 illustrates a similar situation, but where all the currents are localized to a conductor such as a copper wire. In this case, the total current I_S out through the surface S is the sum of the currents $I_S = I_1 + I_2 + I_3$, where the currents I_1, I_2 and I_3 are positive when they are in the direction out of the surface. For a stationary system, for which $\mathrm{d}Q/\mathrm{d}t = 0$, conservation of charge implies that

$$\sum_i I_i = 0. \tag{9.40}$$

We call this law *Kirchhoff's current law*. It is only valid when there are no charge sources or sinks. If there is a capacitor inside the surface, it is possible to build up charge, and Kirchhoff's currents law would not be valid.

Conservation of charge

For a volume v enclosed by a surface S the net charge flowing out through the surface S must correspond to the reduction in charge inside the volume v:

$$\oint_S \mathbf{J} \cdot d\mathbf{S} = -\frac{d}{dt}\int_v \rho dv. \tag{9.41}$$

For a stationary system, where all the currents are constant and there is no build-up or removal of charges, we get **Kirchhoff's current law**:

$$\oint_S \mathbf{J} \cdot d\mathbf{S} = 0. \tag{9.42}$$

For an intersection of n conductors (wires), **Kirchhoff's current law** becomes:

$$\sum_{i=1}^{n} I_i = 0. \tag{9.43}$$

where each current I_i is the current *out of* the intersection (or *into* the intersection—the point is that they must all be either into or out of the intersection).

Test your understanding
The figure shows three possible systems where currents are flowing into and out from a point. The arrows indicate positive directions for the currents. What figures satisfies Kirchhoff's currents law?[6]

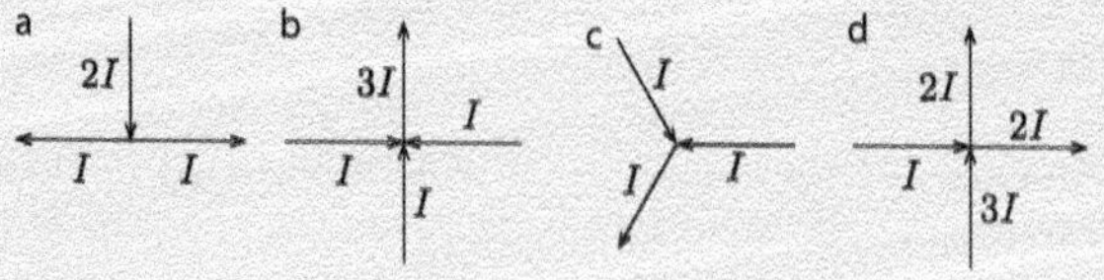

[6] (a) yes; (b) yes; (c) no; (d) yes.

Combining Resistors

How can we find the effective resistance of a resistor that consists of several parts? Let us demonstrate this through two examples: Two resistors in series and two resistors in parallel.

Resistors in series. Figure 9.7a illustrates two resistors A and B with resistances R_A and R_B placed after each other in *series*. They are connected to an ideal conductor on the left side and on the right side and they are joined by an ideal conductor as illustrated in the figure. What is the effective resistance of this combined system, that is, if we replaced this system with a single resistor, what would be the resistance R of this resistor?

The effective resistance R relates the voltage drop across the whole system, $V_2 - V_0 = V$, and the current, I, through the system: $R = V/I$. In a stationary state, when no charges are building up inside the system, the current through any cross-section must be the same. The current is therefore the same through both components A and B: $I_A = I_B = I$. The voltage drop across component A is $V_A = V_2 - V_1$ and the voltage drop across component B is $V_B = V_1 - V_0$. For each of the components we know that $V_A = R_A I_A = R_A I$ and $V_B = R_B I_B = R_B I$. We therefore see that the total voltage drop, V is:

$$V = V_2 - V_0 = V_A + V_B = R_A I + R_B I = \underbrace{(R_A + R_B)}_{=R} I = RI, \tag{9.44}$$

where $R = R_A + R_B$ is the effective resistance for the whole component. We can expand the same reasoning to a sequence of n resistors R_i with effective resistance $R = \sum_i R_i$.

Resistors in parallel. Figure 9.7b illustrates two resistors A and B with resistances R_A and R_B placed in parallel. The are connected to an ideal conductor on the left and the right side. What is the effective resistance R of this combined system?

Because both resistors are connected to an ideal conductor on the left and on the right side, the potential on the left side is the same for both resistors and the potential on the right side is the same for both resistors. The potential drops across the resistors, V_A and V_B, are therefore the same, $V_A = V_B = V$. What about the current? Due to the conservation of current, we know that the current I flowing through the whole system must be the sum of the currents in each of the resistors. (You can also argue for this using Kirchhoff's current law: The current flowing out of the left junction, that is the left side of the resistors, is I_A, I_B and $-I$, so that $-I + I_A + I_B = 0$ according to Kirchhoff's law, and $I = I_A + I_B$). The voltage drop across resistor A is then $V_A = V = I_A R_A$ and the voltage drop across resistor B is $V_B = V = I_B R_B$. The currents are therefore $I_A = V/R_A$ and $I_B = V/R_B$. The total current I is:

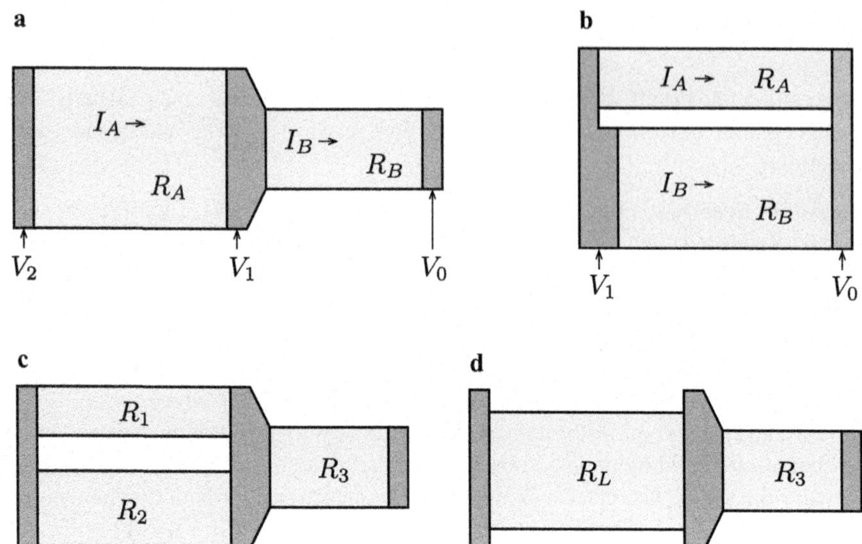

Fig. 9.7 Illustration of **a** a system of two resistors in series, **b** a system of two resistors in parallel, **c** a system of three resistors, **d** the same system but where the left two resistors in parallel have been combined to a new resistor with resistance R_L

$$I = I_A + I_B = \frac{V}{R_A} + \frac{V}{R_B} = \underbrace{\left(\frac{1}{R_A} + \frac{1}{R_B}\right)}_{1/R} V = \frac{V}{R}, \tag{9.45}$$

where $1/R = 1/R_A + 1/R_B$ is the effective resistance for the whole component. We can expand the same reasoning to a sequence of n resistors R_i with effective resistance $1/R = \sum_i 1/R_i$.

More advanced combinations. We can address more advanced combinations by grouping the system into parts that are connected in series or parallel. For example for the system illustrated in Fig. 9.7c, we can first combine the two resistors in parallel on the left into one resistor $R_L = (1/R_1 + 1/R_2)^{-1}$, as shown in Fig. 9.7d, and then combine this new resistor R_L with the other resistor R_3 in series to find the total resistance $R = R_L + R_3 = (1/R_1 + 1/R_2)^{-1} + R_3$ as illustrated in the figure. (Notice that it would *not* be correct to group resistors R_1 and R_3 into one resistor in series and then combine this with resistor R_2 in parallel.)

Combinations of resistors

Series coupling. The effective resistance R of n resistors R_i connected in series is

$$R = \sum_i R_i \quad \text{(Series coupling)} \tag{9.46}$$

Parallel coupling. The effective resistance R of n resistors R_i connected in parallel is

$$\frac{1}{R} = \sum_i \frac{1}{R_i} \quad \text{(Parallel coupling)} \tag{9.47}$$

Combinations. You can find the effective resistance of any combination of resistors in both series and parallel by grouping the resistors hierarchically into groups that are connected in series or parallel.

Test your understanding
You have a resistance R. (a) If you connect a resistance $r < R$ in parallel with R, will the resulting resistance be smaller or larger than R? (b) If you connect a resistance $r < R$ in series with R, will the resulting resistance be smaller or larger than R?[7]

Conservation of Charge on Differential Form

We can rewrite the principle of conservation of charge on differential form to make it more generally applicable and so that you may recognize its similarity to other conservation laws. For a volume v enclosed by the surface S, the conservation of charge within the volume is formulated as:

$$\oint_S \mathbf{J} \cdot \mathrm{d}\mathbf{S} = -\frac{\mathrm{d}}{\mathrm{d}t} \int_v \rho \, \mathrm{d}v. \tag{9.48}$$

We can reformulate the left side using the divergence theorem:

$$\oint_S \mathbf{J} \cdot \mathrm{d}\mathbf{S} = \int_v \nabla \cdot \mathbf{J} \, \mathrm{d}v. \tag{9.49}$$

We combine this to get:

$$\oint_S \mathbf{J} \cdot \mathrm{d}\mathbf{S} = \int_v \nabla \cdot \mathbf{J} \, \mathrm{d}v = -\frac{\mathrm{d}}{\mathrm{d}t} \int_v \rho \, \mathrm{d}v. \tag{9.50}$$

This is valid for any volume v, which means that the equality also must hold for the arguments inside the integrals:

---

[7] (a) smaller; (b) larger.

$$\nabla \cdot \mathbf{J} = -\frac{\partial \rho}{\partial t}. \tag{9.51}$$

And when ρ is a constant (in time), we get

$$\nabla \cdot \mathbf{J} = 0. \tag{9.52}$$

Continuity equation for charges

The **continuity equation** for charges, which formulates the principle of conservation of charge, is on differential form:

$$\nabla \cdot \mathbf{J} = -\frac{\partial \rho}{\partial t}. \tag{9.53}$$

Laplace Equation for Conducting Materials

Why can we use Laplace equation to find the potential inside a conductor? There are moving charges inside the conductor, so why can we assume that the charge density is zero?

For the case of non-ideal conductors, we can derive Laplace equation in a different way, by using the conservation of charges instead. If we address a stationary situation, then the conservation of charges states that

$$\nabla \cdot \mathbf{J} + \frac{\partial \rho}{\partial t} = 0. \tag{9.54}$$

The term $\partial \rho / \partial t$ must be zero for a stationary state, otherwise charges will build up indefinitely inside the system. This means that

$$\nabla \cdot \mathbf{J} = 0. \tag{9.55}$$

If we combine this with two other features: (i) that the current density is related to the electric field through Ohm's law: $\mathbf{J} = \sigma \mathbf{E}$, and (ii) that the electric field can be found from a potential $\mathbf{E} = -\nabla V$. We insert this in (9.55) and find:

$$\nabla \cdot \mathbf{J} = \nabla \cdot (\sigma (-\nabla V)) = -\sigma \nabla^2 V = 0 \quad \Rightarrow \quad \nabla^2 V = 0, \tag{9.56}$$

where we have assumed that the conductivity σ is uniform. We see that we recover Laplace equation for the electric potential. We did not assume that the charge density was zero, but instead assumed that the conductivity was uniform, that Ohm's law holds, and that the electric field could be found from a potential.

Summary

The **current** I passing through a surface S is the amount of charge $\mathrm{d}q$ per unit time $\mathrm{d}t$ passing through the surface: $I = \mathrm{d}q/\mathrm{d}t$. Current is measured in **ampere** which is a basic unit in the SI system.

We describe the motion of charges in space with the **current density** $\mathbf{J}(\mathbf{r})$. The current $\mathrm{d}I$ passing through a surface element $\mathrm{d}\mathbf{S}$ is $\mathrm{d}I = \mathbf{J} \cdot \mathrm{d}\mathbf{S}$, and the current I passing through a surface S is $I = \int_S \mathbf{J} \cdot \mathrm{d}\mathbf{S}$.

Most materials obey **Ohm's law** on microscopic form $\mathbf{J} = \sigma\mathbf{E}$, where σ is a (temperature dependent) material property called the conductivity. The resistivity is the inverse quantity, $\rho = 1/\sigma$.

The **resistance** R of a resistor (a non-ideal conductor or an insulator) is defined as $R = V/I$, where V is the voltage drop across the resistor when a current I passing through it.

Ohm's law for a resistor is that the voltage drop across the resistor with resistance R is $V = RI$, when a current I is passing through the resistor.

The **resistance** R **of a cylindrical resistor** with cross-sectional area A, length L and conductivity σ is $R = L/(A\sigma)$.

We find the **resistance** of a system using two main methods. In **method 1**, we assume that the current density $\mathbf{J}$ has a particular symmetry and use this to find the current I, the electric field $\mathbf{E}$, the potential difference V from the electric field, and then the resistance from $R = V/I$. In **method 2**, we find the potential $V(\mathbf{r})$ from e.g. Laplace's equation, the electric field from the potential, $\mathbf{E} = -\nabla V$, the current density from the electric field, $\mathbf{J} = \sigma\mathbf{E}$, the current from the current density, $I = \int_S \mathbf{J} \cdot \mathrm{d}\mathbf{S}$ and the resistance from $R = V/I$.

We can **combine resistors** $R_1, R_2, \ldots, R_n$, to one effective resistor R by combining resistors *in series* getting $R = \sum_i R_i$ or *in parallel* getting $1/R = \sum_i 1/R_i$.

The **power loss** due to resistance is $p_J = \mathbf{J} \cdot \mathbf{E}$. The power loss in a volume v is $P = \int_v p_J \mathrm{d}v$. The power loss for a resistor is $P = VI = RI^2 = V^2/R$.

The **conservation of charges** demand that $\int_S \mathbf{J} \cdot \mathrm{d}\mathbf{S} = -\mathrm{d}/\mathrm{d}t \int_v \rho \,\mathrm{d}v$ on integral form and $\nabla \cdot \mathbf{J} = -\partial\rho/\partial t$ on differential form, where ρ is the volume charge density.

In a **stationary system** with no build-up of charges $\nabla \cdot \mathbf{J} = 0$ and **Kirchhoff's current law** $\sum_i I_i = 0$ is valid for any junction.

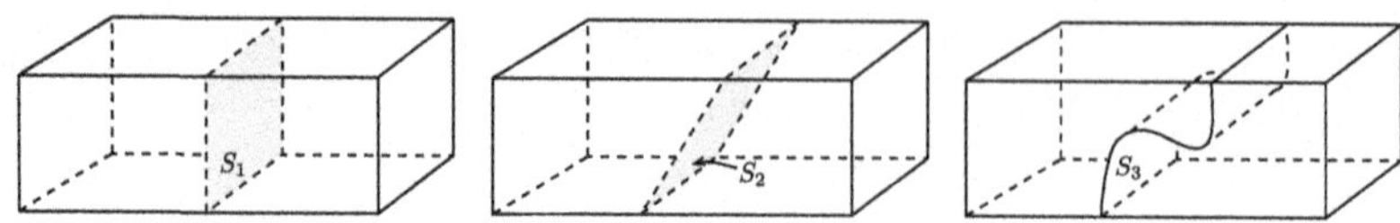

Fig. 9.8 Current along a rod

Exercises

Discussion Exercises

9.1 Current density and current. You have a long rod with quadratic cross-section and there is a current with homogeneous current density along the rod. Is the current in Fig. 9.8 through surface S_1 the same as through surface S_2?

9.2 Charge accumulation. Is the net current of charge out of a closed surface always zero? If not, can you find a counterexample?

9.3 Light bulb. If you buy a light bulb it is often specified how many watt the bulb consumes. If you buy a 10 W light bulb for your boat which runs on a 12 V battery, what would happen if you instead connected it to a 48V battery? (Assuming that the bulb does not break down). Do you think it would consume less or more power? (That is, be dimmer or brigher respectively).

9.4 Battery. A battery is assume to deliver a constant voltage, irrespective of the current. What happens if you connect the two poles of a battery with a wire? As you make the wire a better and better conductor, what would eventually determine the current through the battery as the wire becomes a superconductor with zero resistance?

Tutorials

9.5 Moving charges. Consider four situations:

1. An ion with a charge $+Q$ moves to the right
2. A neutral hydrogen atom (consisting and a proton with charge $+e$ and an electron with charge $-e$) moves to the right
3. Electrons are accelerated to the right where they hit a plate and stop
4. In an ionic solution large, positive ions are moving to the right, while equally many smaller, negative ions are moving to the left.

(a) In which of these situations is the net current zero?
(b) In which of these situations is there a net current to the right?

9.6 Different velocities. Figure 9.9 shows a conductor consisting of $N \times N$ channels that each has a length L and a cross section $a \times a$, where $N = 10^6$ and $a = 1$nm

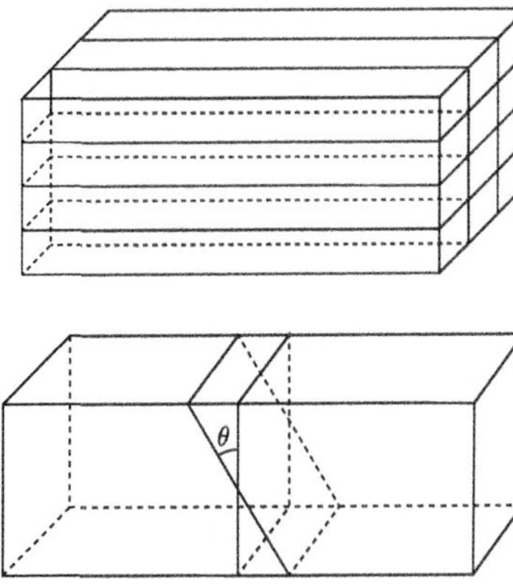

Fig. 9.9 A model conductor of $N \times N$ channels

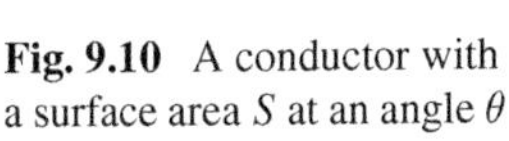

Fig. 9.10 A conductor with a surface area S at an angle θ

(a) Assume that you shoot an ion per second through each channel. The ion has the velocity v along the channel and the charge $Q = +e$. What is the current through each channel?
(b) What is the current through the whole conductor?
(c) What is the current density in the conductor?
(d) Assume instead that you all the time insert ions so that there at all times are M ions inside each channel and that they move with a velocity v along the channel. How many ions are inside a length b of a single channel? (We will now calculate using symbols and not numbers.)
(e) In this case, how many ions move out of the channel during a time interval Δt?
(f) What is the current through each channel?
(g) What is the current through the whole conductor?
(h) What is the current density in the conductor?

9.7 Current through inclined surface. A current I passes through a conductor with sides a and length L as shown in Fig. 9.10. Assume that the current denisty is uniform in the conductor.
(a) What is the current density in the conductor?
(b) We create a surface S which forms an angle θ with a cross-sectional surface normal to the x-axis as shown in the figure. Show that the current through S is the same for all angles $-\pi/2 < \theta < \pi/2$. (Hint: Use the angles between the normal vectors.)
(c) What is the current I if $\theta > \pi/2$?

9.8 Ohm's law for charges in a liquid. We will construct a simplified model for Ohm's law to guide our intuition. Assume that there are small particles of radius a and charge q in a liquid such as water. (Assume also that there is no charge transfer between the particle and the water, so that they keep the charge q.). We create an electric field E_x across the water.

(a) What forces will act on the charged particles?
(b) You may assume that the viscous force on a particle is $\mathbf{D} = -6\pi\eta a\mathbf{v}$, where η is a property of the liquid called the dynamic viscosity. What is the velocity of the particles after a very long time?
(c) How many particles N flow through an area A (which is normal to the velocity v) during a small time interval Δt? You can assume tha there are n particles per unit of volume in the liquid.
(d) What is the current density for this system?
(e) Here you can see what properties of this liquid with charges particles that contribute to the conductivity σ of the liquid. What properties of a conductor/insulator do you think contribute to the conductivity σ?

9.9 Quadratic cross section. An insulator with a quadratic cross-section a^2 and length L is made from a material with conductivity σ and is oriented along the x-axis. We will find an expression for the resistance of this component in two ways.

Method 1—we assume uniform current density. First, we assume that there is a current I in the x-direction. We also assume that the current density is uniform inside the component and zero outside.
(a) What is the current density inside the component?
(b) What is the electric field inside the component?
(c) What is the electric potential inside the component?
(d) What is the potential difference between the ends of the resistor?
(e) What is the resistance, R, of this component?

Method 2—We use Laplace's equation. Let us instead assume that each end of the resistor is connected to an ideal conductor. One end has the potential $V = V_0$, while the other end has the potential $V = V_1$.
(f) Solve Laplace's equation for the system and find $V(x)$.
(g) Find the electric field.
(h) Find the current density.
(i) Find the current I and the resistance R.

Properties of components and materials. We have now found that the resistance is $R = L/(A\sigma)$. Notice that the resistance R is a property of a thing, the component, while the conductivity σ is a material property.
(j) What happens with the resistance R of this component if we double all lengths in this system?

9.10 Resistance gymnastics. A resistor of cross-sectional area $A = a \times a$ and length L, which is made of a material with conductivity σ has the resistance $R = L/(\sigma a^2)$.
(a) We divide the resistor in the middle into two parts the are equally long, $L/2$, and that are connected in the middle. What is the resistance of this system of two resistors?
(b) We divide the resistor into small parts of length $\mathrm{d}x$. Show how you can find the total resistance of a resistor of length L through an integral.

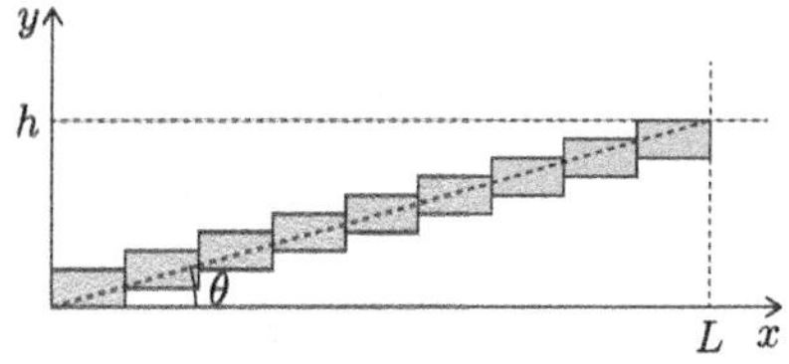

Fig. 9.11 Inclined resistor

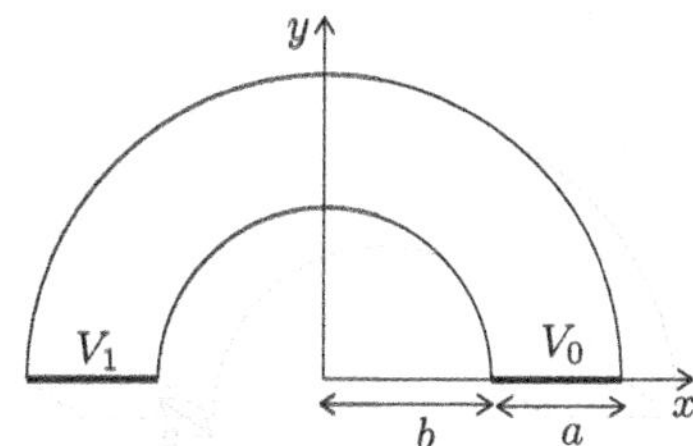

Fig. 9.12 A resistor is bent into a half circle. The black lines indicates ideal conductors with potentials V_0 and V_1

(c) The method of subdividing the system into small pieces and then summing up the contributions from each part through an integral can be applied in many situations. Let us address a resistor that consists of a sector θ of a cylinder shell with inner radius a, outer radius b and length h, where the current goes from the inner radius to the outer radius. What is the resistance $\mathrm{d}R$ of a thin slice from r to $r + \mathrm{d}r$?

(d) What is the resistance R of the whole system?

But this method does not always work. Figure 9.11 shows a resistor with cross section $a \times a$ and length L which is divided into many small pieces of length $\mathrm{d}x$, but where each piece is translated a bit along the y-axis, so that the resistor spans from $y = 0$ to $y = h$ and forms an angle θ with the x-axis.

(e) Why can we not use the same methods here and sum up all the contributions to find that the resistance is $R = L/(\sigma a^2)$?

(f) Can you find an expression for the resistance of this system by assuming that the volume is constant when the material is stretched so that the resistor spans from $x = 0$ to $x = L$ and from $y = 0$ to $y = h$?

Notice that this is the reason why we can *not* find the resistance of a conical resistor by slicing it into small cylinders and summing the contributions from all the cylinders (Fig. 9.12).

9.11 Bent resistor. We start from a resistor with cross section $a \times a$ and length L and bend it into a half circle of radius b as shown in Fig. 9.12. At each end we connect a good conductor so that there is a potential difference ΔV between the two ends.

(a) Sketch how you think that the electric field will be inside the conductor.

In this case, we cannot assume that the current density is uniform. We also do not really know what the electric field or the potential is. But we will be able to find a solution for the potential the satisfies the boundary conditions and Laplace's equation. Since the solution to Laplace's equation is unique, we know that this is the correct solution. We choose the potential so that it is V_0 on the right and V_1 on the left as in the figure.

(b) What symmetry will the electric potential have?
(c) What is Laplace's equation and the boundary conditions for this system?
(d) Find the electric potential $V(r, \phi)$.
(e) Find the electric field
(f) What is the current density?
(g) What is the current?
(h) And finally, what is the resistance?
(i) For this model, the electric field is not zero outside the conductor. (Can you argue why it cannot be zero immediately outside the conductor?) Why do we not include the current density outside the conductor in our calculations?

Exercises

9.12 Current density.
(a) For a homogeneous current density $\mathbf{J} = J_0\hat{x}$, what is the current through all the surfaces of a cube with corners at $(-1, -1, -1)$, $(-1, +1, -1)$, $(+1, -1, -1)$, $(+1, +1, -1)$, $(-1, -1, +1)$, $(-1, +1, +1)$, $(+1, -1, +1)$, and $(+1, +1, +1)$ all measured in units of a.
(b) A block with uniform charge density ρ is moving with a constant velocity $\mathbf{v} = v_0\hat{z}$. What is the current density $\mathbf{J}$ at a specific point in space inside the block (the point does not move along with the block). What would the current density $\mathbf{J}$ be if the point moved along with the block?

9.13 Cylindrical resistor. We will now look at charges that are leaking across the cylindrical membrane of a part of the axon of length L. We will assume that the myelin sheath is leaky (a conductor with conductivity σ) and that there is a radially-symmetric current density $\mathbf{J}$ leaking across it due to a potential difference. We will find the resistance R of a piece of the cylinder of length L. We assuming that the inner part of the cylidrical shell is connected to a potential V and that the outer part of the shell is connected to a potential $V = 0$.

We will find the resitance of this system by (i) making a drawing of the system, (ii) drawing the expected current density, (iii) finding the electric field for this current density, (iv) finding the potential by integrating the electric field, and (v) finding the resistance by relating the current to the potential difference. (Notice the similarity of this method and the method you used to find the capacitance!)

(a) Make a drawing of the cylindrical system. Sketch the current density.
(b) Given that there is a radially-symmetric current density, find the current density $\mathbf{J}$ expressed in terms of the total current I. (Hint: Integrate the current density along a cylindrical surface.)
(c) Find the electric field $\mathbf{E}$ by using the current-density version of Ohm's law.
(d) Use the electric field to relate the current I to the voltage difference between the inner $V(r = a)$ and outer $V(r = b) = 0$ surfaces of the axon.

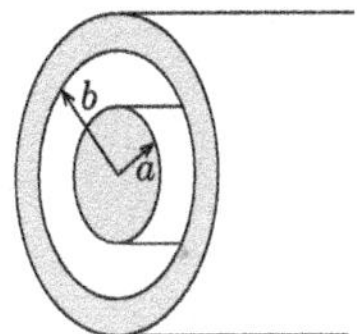

Fig. 9.13 Illustration of a coaxial cable

(e) Find the resistance of the cylinderical shell. Is this the resistance for flow along the axon?
(f) Interpret these expressions. Does the current you found actually depend on the length of the axon? Does the resistance?

9.14 Current of ions. Positive ions flow to the right through a liquid, negative ions flow to the left. The spatial density and velocity of both types of ions are identical. Is there a net current through the liquid?

Homework

9.15 Lightning. Lightning strikes in one end of a lightning rod of steel and induces a current of 30000 A which lasts for 65 μs. The lightning rod is a 1 m long and 2 cm in diameter, and the other end is connected to the ground through a 40 m copper wire with a diameter of 5 mm. The conductivity of steel and copper is respectively $\sigma_{\text{steel}} = 5.0 \cdot 10^6 \Omega^{-1}\text{m}^{-1}$ and $\sigma_{\text{copper}} = 5.8 \cdot 10^7 \Omega^{-1}\text{m}^{-1}$.
(a) Find the potential difference between the top of the lightning rod and the bottom of the copper wire as the current is passing through.
(b) Find the total energy dissipated in the lightning rod and the copper wire from the lightning strike.

9.16 Spherical current. We have a system consisting of two concentric spherical shells made of metal with radius a and b, where $a < b$. In the region between the two shells there is a weakly conducting material with conductivity σ.
(a) Assume that at a time $t = 0$ there is a charge $+Q$ on the inner shell and a charge $-Q$ on the outer shell. Find the current density as a function of position between the two shells, $\mathbf{J}(r)$.
(b) Find the current $I(t = 0)$ from the inner shell to the outer shell.
(c) Find the resistance of the resistor consisting of the material between the two shells.

9.17 Leaky coaxial cable. A 10 km long coaxial cable lies on the ocean floor. The internal radius is $a = 15$ mm, and the external is $b = 30$ mm as in Fig. 9.13. The cable itself is made of a superconductor, and the voltage carried by the cable is $V_0 = 520$ V.
(a) What is the current density $\mathbf{J}$ inside the cable?

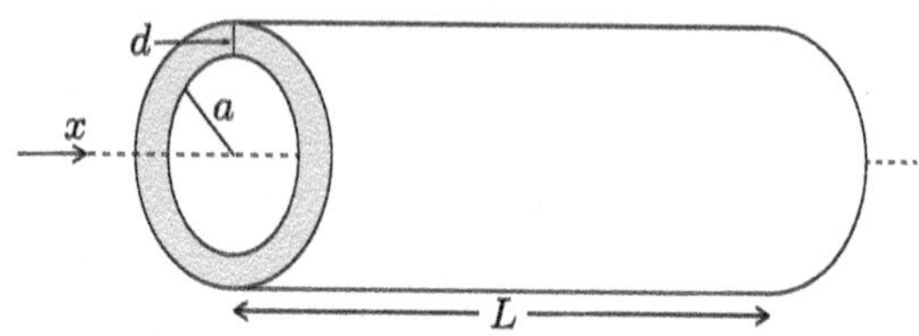

Fig. 9.14 A cylindrical resistor

(b) At one point the cable gets a leak and the entire space between the conductors are filled with seawater ($\sigma = 4\Omega^{-1}\text{m}^{-1}$).
How much power is lost to the seawater inside the cable?
(c) There is a generator connected to one of the ends of the cable. When the cable is filled with water, what is the resistance seen by this generator? Comment on this value in regards to the value of P_J.

9.18 A cylindrical component. In this exercise, we will study the behavior of a cylindrical nerve cell as illustrated in Fig. 9.14. We model a part of the cell as a cylindrical shell. The cylindrical shell has a length L. The inner radius is a. The thickness of the cylindrical shell is d. You can assume that L is much larger that a and d and that the system therefore has cylindrical symmetry.
First, we want to find the **capacitance** of the cylindrical shell. We assume that the regions inside and outside the cylindrical shell are ideal conductors. The cylindric shell is a dieletric material with dielectric constant ϵ.
(a) Assume that there is a charge Q on the inner surface of the cylindrical shell and a charge $-Q$ on the outer surface of the shell. Find the electric field everywhere in space.
(b) Find the scalar potential $V(r)$ as a function of the distance r to the axis of the cylindrical shell.
c) Show that the capacitance of the cylindrical shell is

$$C = \frac{2\pi\epsilon L}{\ln\left(1 + \frac{d}{a}\right)} . \tag{9.57}$$

Second, we want to find the **resistance** of the cylindrical shell. The cylindrical shell is a conductor with conductivity σ.
(d) Find the current density $\mathbf{J}$ in the cylindrical shel when a current I passes the shell from the inner to the outer surface.
(e) Find the electric field $\mathbf{E}$ in the cylindrical shell in this case.
(f) Find the scalar potential $V(r)$ in the cylindrical shell and use this to show that the resistance of the cylindrical shell is

$$R = \frac{\ln\left(1 + \frac{d}{a}\right)}{2\pi\sigma L} . \tag{9.58}$$

Fig. 9.15 Two cylinderical components

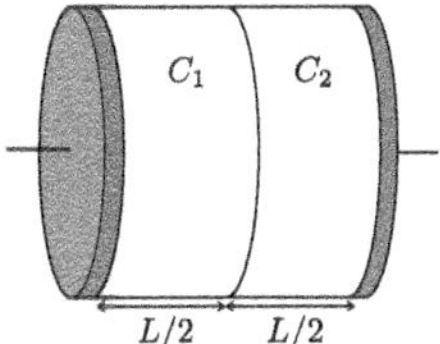

9.19 Two resistors. In this exercise we will study a cylindrical resistor of length L and radius a. The end surfaces of the cylinder are connected to conductors. The cylinder is divided into two cylindrical parts, each with a length $L/2$. One has a conductivity σ_1 and the other has a conductivity σ_2.
(a) Find the electric field everywhere inside the resistor when it is carrying a current I.
(b) Find the electric potential everywhere inside the resistor.
(c) Find the resistance R to the resistor.

9.20 Cylindric component. A cylindrical component consists of two thin cylindrical conductors of radius a, connected to two cylinders C_1 of radius a and length $L/2$ and C_2 of length $L/2$ and radius a as shown in Fig. 9.15.
(a) Assume that C_1 has permittivity ϵ_1 and C_2 has permittivity ϵ_2. Find the capacitance of the system consisting of the two cylinders.
(b) Assume that C_1 has conductivity σ_1 and the C_2 has conductivity σ_2. Find the resistance of the system consisting of the two cylinders.

9.21 Resistance of two cylinders.
(a) A resistor consists of a cylinder with radius r_0, length L and conductivity σ. Find the resistance of the cylinder in the direction of the cylinder axis.
(b) A resistor consists of two cylinders. The first has a radius a and the second has a radis $a/2$. Both have a length $L/2$ and conductivity σ. You can assume that the cylinders are connected at their ends with a thin layer of a very good conductor. Find the resistance of this system of two cylinders along the cylinder axis.
(c) A resistance consists of 3 cylinders, each with a length $L/3$ and conductivity σ. The first has a radius a, the second a radius $3a/4$ and the third a radius $2a/4 = a/2$. Find the resistance of this combined resistor along the axis of the cylinders.
(d) In the previous exercise, is the ordering of the cylinders important? If we continue with this subdivision into smaller and smaller pieces, you will get something that resembles a conical resistor. Do you think that this is a reasonable model for a conical resistor?
(e) What will the resistance be if we continue this subdivision many times, while the total system remains L long?

Chapter 10
Electric Circuits—Part 1

10.1 Electric Circuits

Figure 10.1 illustrates a system with two ideal conductors at potentials V_A and V_B. The left figure illustrates the resulting potential and electric field everywhere in space (found from Laplace's equation), when the system is in vacuum. We notice that the electric field points from V_A to V_B, but the field has a complicated shape. What happens if we place a conductor with a conductivity σ into this system as illustrated by the gray region in the right figure?

We solved this problem previously and found that the current density, $\mathbf{J} = \sigma \mathbf{E}$, is parallel with the conductor as illustrated in the figure. The current density cannot have any component normal to the conductor, as this would indicate that charges are leaking from the conductor out into the vacuum outside. Therefore, the electric field also cannot have any component normal to the conductor. (In the figure on the right we have shown the current density for clarity. The electric field will generally not be zero outside the conductor).

But why does the electric field change when the conductor is added? Immediately after the conductor is added to the system, we will expect the electric field to be as illustrated on the left. Consequently, charges will also move in directions normal to the conductor. However, these charges will not move outside the conductor but stop at the outer surface of the conductor. This will change the electric field inside the conductor, and this process will continue until the surface charges ensure that there is no electric field normal to the surface of the conductor. Then we will have reached the situation in the figure on the right. This process is very fast, and we can effectively think of it as instantaneous.

A similar process occurs in a curved conductor as in Fig. 10.2. When an electric potential difference is applied across the ends of this conductor, electric charges will pile up on the surface of the conductor until the electric field, and therefore also the current density, does not have a component normal to the surface of the conductor. The result is that the electric field will point along the conductor as illustrated in the

© The Author(s), under exclusive license to Springer Nature Switzerland AG 2026

A. Malthe-Sørenssen, *Elementary Electromagnetism Using Python*, Undergraduate Texts in Physics, https://doi.org/10.1007/978-3-032-19876-1_10

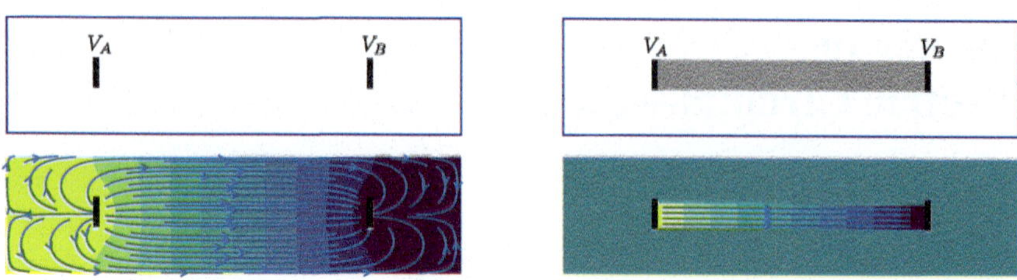

Fig. 10.1 (Left) Illustration of a system with two ideal conductors in vacuum at potentials $V_A > V_B$ and the resulting electric field. (Right) Illustration of a system with two ideal conductors at potentials $V_A > V_B$ and a real conductor with conductivity σ (in gray) and the resulting current density $\mathbf{J}$

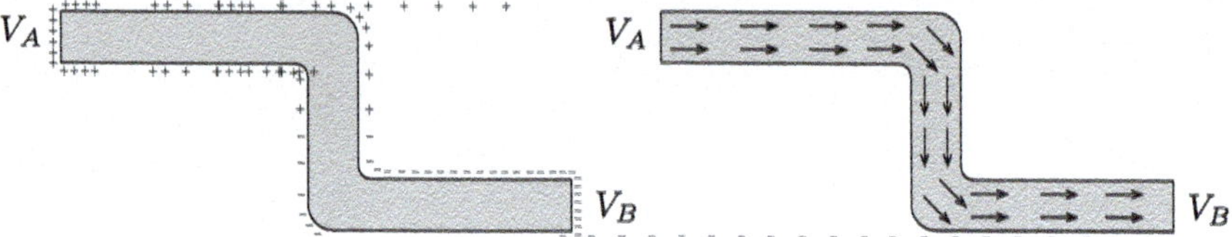

Fig. 10.2 Illustration of curved conductor. The surface charge density will organize in such a way that the electric field does not have a component normal to the surface of the conductor. The electric field will therefore point along the conductor

figure. Again, this process is effectively instantaneous. This means that we do not need to think about the shape of the conductor. The electric field and the current will follow whatever shape the conductor has. Typically, when we make a drawing of a system with a conducting wire we will not care too much about the shape of the wire but instead draw the wire in a way that is practical.

Circuit Diagrams

In Fig. 10.3 we have illustrated a system that consists of a good conductor, a wire made of a metal such as copper, and a bad conductor in the form of an insulator used to form a resistor. The top figure shows the real system, and the bottom shows the model drawing called a *circuit diagram*. A circuit diagram has several elements that are models of the real system, such as, in this case, ideal conductors drawn as lines and resistors drawn as rectangles. We realize that along the good conductor, the potential is essentially constant. This means that the shape or length of the ideal conductor does not matter. (At least not yet, it will matter later when we introduce inductance). We therefore often draw good conductors as straight lines. We need to remember that all points along these lines have the same potential. In the figure we have illustrated that the potential is the same in several places along the conductor illustrated with small circles.

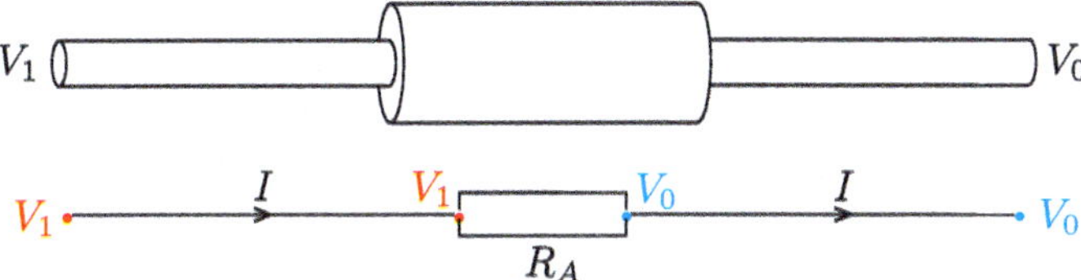

Fig. 10.3 (Top) Illustration of a system consisting of a good (ideal) conductor connected to a cylindrical insulator. (Bottom) Model in the form of a circuit diagram. Thin lines show ideal conductors. All points along a thin line have the same potential. Arrows indicate the direction of the currents

Drawing Resistors

The conductor consists of two parts: A good conductor (a copper wire) and a bad conductor (forming a resistor). We have drawn the good conductor as a thin line and we draw the bad conductor as a rectangle, which is still a conductor, but with much lower conductivity than the wire. Notice that the potential is not the same on each side of the resistor: There is a potential drop, $V = V_1 - V_0$, across the resistor. We characterize a resistor with a resistance R. For a resistor we know that the voltage drop is related to the current through Ohm's law: $V = RI$, where V is the voltage drop across the resistor, I the current through the resistor and R is the resistance.

We have now simplified the real system with an ideal system with two types of components: An ideal conductor drawn as a line and a resistor with resistances R respectively. We will often also simplify a real conductor, with a finite conductivity in this way, as a wire that is effectively an ideal conductor with zero potential drop connected in series with a resistor that represents the potential drop in the wire.

Charge Conservation and the Water Analogy

Where do the charges go in the diagram in Fig. 10.3? In this case positive charges are transported from high potential V_1 to the lower potential V_0. If this was the complete circuit, we would expect charges to build up at V_0 and be removed at V_1, gradually resulting in an increase in the potential V_0 and a decrease in potential at V_1 until they are equal and no more current would flow.

To keep the potential difference and to keep the current flowing, we insert an element in the circuit called a voltage source or a *battery*. (See the symbol used in the figure). We will not go into the detailed workings of a battery here but will simply assume that the potential of a charge is increased as it passes through the battery. This is illustrated in Fig. 10.4. Here, we have drawn the changes in the electric potential, which corresponds to the changes in the potential energy of a positive charge as it moves through the system. We may use a water analogy to guide our intuition about this system. When the charge moves along a wire, it does not change its potential. This

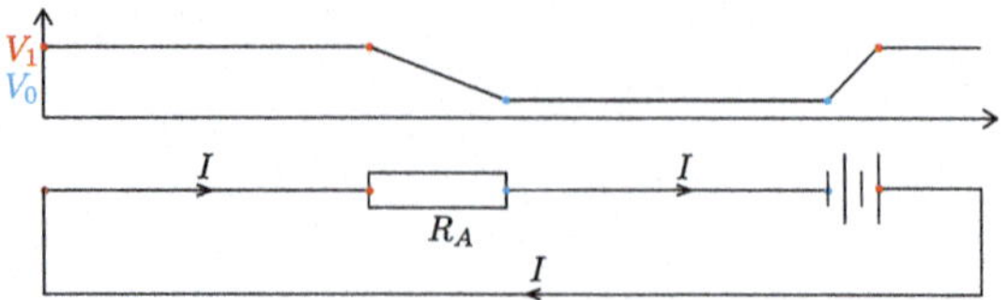

Fig. 10.4 Plot of the potential as a function of position along the top part of the circuit at the bottom. Notice the new symbol for the battery

is like a flat water channel. When the charge moves through a resistor, it decreases its potential. This is like a waterfall. When the charge moves through a battery, it increases its potential. This is like an elevator: It lifts the water up.

Circuits are loops. Usually, we draw circuits as loops with batteries as voltage sources. A battery provides a particular voltage difference, and this drives the current in the circuit. We draw circuits as loops, indicating that the current does not disappear, but moves along the various components in the circuit. For the simple diagram illustrated in Fig. 10.4 the current is the same in any cross section along the circuit.

Test your understanding
The figure shows a circuit diagram. (a) Sketch the potential between points a and b. (b) Sketch the current between points a and b.

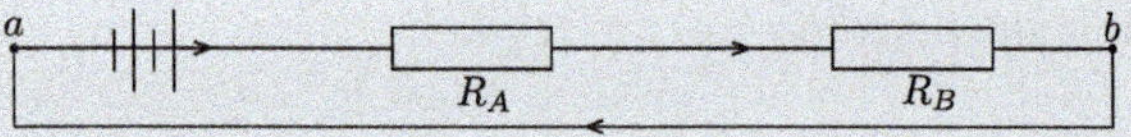

Loops and Kirchhoff's Voltage Law

How does the electric potential vary along a circuit? We recall that we previously found what we called Kirchhoff's voltage law:

$$\oint_C \mathbf{E} \cdot d\mathbf{l} = 0, \tag{10.1}$$

for any closed loop C. (We will modify this law later in electrodynamics). This integral is the change in electric potential around a loop. In Fig. 10.5a we illustrate a loop divided into segments and potential drops $\Delta V_i = V_{i+1} - V_i$ along the segments. Kirchhoff's voltage law for this circuit is then

$$\sum_{i=1}^{n} \Delta V_i = 0, \tag{10.2}$$

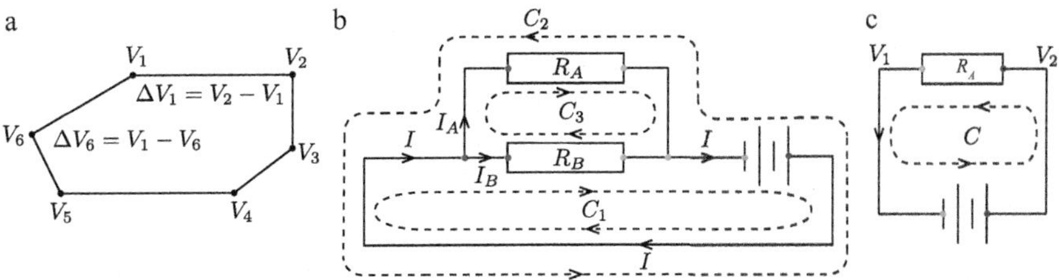

Fig. 10.5 **a** Illustration of a loop with vertices i, voltages V_i and voltage drops $\Delta V_i = V_{i+1} - V_i$. **b** Illustration of three loops C_1, C_2 and C_3 in a circuit. **c** Simple circuit with a single loop C

where $\Delta V_n = V_n - V_1$. Notice that for a resistor R we have a voltage *drop*, $V = RI$, whereas for a battery we have a voltage *gain*, e, which is a property of the battery.

Notice also that Kirchhoff's law is valid for *any closed circuit* or any *closed part of a circuit*. For example, it is valid along the circuits C_1, C_2 and C_3 illustrated in Fig. 10.5b. Notice that the orientation of the loop can be in the positive direction (C_1) or in the negative direction (C_2). You are free to choose what suits you best. However, for loop C_3 you must ensure that there is a voltage drop over the top resistor, but a voltage increase over the bottom resistor. We have voltage drops when we pass a resistor in the positive direction of the current and a voltage gain when we pass in the direction opposite the positive direction of the current. We therefore discuss the direction of the current in more detail below.

Kirchhoff's voltage law

Kirchhoff's voltage law states that the sum of potential drops around any closed circuit (loop) is zero:

$$\sum_i \Delta V_i = 0, \tag{10.3}$$

which follows from $\oint_C \mathbf{E} \cdot d\mathbf{l} = 0$ for any loop C. Notice that across a component there is a voltage drop in the positive direction of the current and a voltage gain in the direction opposite the current.

Example: Applying Kirchhoff's Voltage Law

Let us apply Kirchhoff's voltage law to the circuit in Fig. 10.5c. We choose the loop C shown in the figure. Along the loop there are two changes in voltage: Across the resistor and across the battery. For the resistor the voltage change is $V_{2 \to 1} = -RI$. Notice that this is a voltage drop. The potential V_1 is smaller than the potential at V_2.

For the battery the voltage change is $V_{1\to 2} = e$, which is the voltage delivered by the battery. Kirchhoff's voltage law gives:

$$\sum_i \Delta V_i = -RI + e = 0 \quad \Rightarrow \quad I = e/R. \tag{10.4}$$

We can therefore determine the current, given R and the voltage delivered by the battery.

Drawing Directions for the Currents

The way the conductors are connected provides additional information. In Fig. 10.5 the current I is flowing along the left conductor and through the battery. However, in the middle the conductors split. Intuitively, we understand that the current I is split into the currents I_A and I_B so that the current is conserved—just like current in a river or a garden hose.

We can analyze the junction using Kirchhoff's current law: $\sum_i I_i = 0$, but then we need to be careful with the direction of the currents and the signs used: All the currents I_i must either be into the junction or out of the junction. The positive directions are drawn with arrows in the figure. Kirchhoff's current law applied to the currents flowing out of the left junction is:

$$\sum_i I_i = (-I) + I_A + I_B = 0, \tag{10.5}$$

where I is the current flowing into the junction from the left and therefore $(-I)$ is the current flowing out of the junction to the left.

Kirchhoff's current law for circuits

For any junction in a circuit, Kirchhoff's current law applies:

$$\sum_i I_i = 0 \tag{10.6}$$

where all the currents I_i must flow either into the junction or out of the junction.

Test your understanding
The figure shows a circuit diagram with a battery e and two resistors R. (a) Write down Kirchhoff's voltage law for this circuit. (b) Find the current, I.[1]

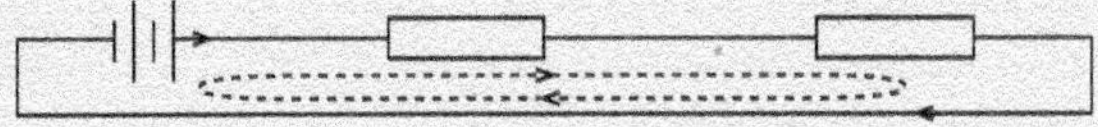

Example: Currents and Voltage Drops

Find the currents and voltage drops in the circuit in Fig. 10.5b when the battery has a voltage e.

Approach. We plan to combine Kirchhoff's voltage law and Kirchhoff's current law on all the closed circuits C_1, C_2 and C_3 to find equations for the currents, and from the currents find the voltage drops across the resistors.

Solution. We recall that the lines indicate ideal conductors. All points along a line have the same potential. The potential is illustrated by the color of the dots at the junctions. There are only two potentials V_1 and V_0 in this system with a voltage difference $V = V_1 - V_0 = e$, which is the voltage delivered by the battery.

Loop C_1. If we apply Kirchhoff's voltage law to the loop C_1 we see that the voltage drop across resistor R_B, $-I_B R_B$, and the voltage increase e across the battery sums to zero: $\sum_i \Delta V_i = -I_B R_B + e = 0$, and we find $I_B = e/R_B$.

Loop C_2. If we apply Kirchhoff's voltage law to the loop C_2, we see that the direction of the loop is opposite the direction of the currents, so that the voltage change over the battery in this circuit is $-e$ and the voltage change over resistor R_A is $R_A I_A$ so that $\sum_i \Delta V_i = -e + I_A R_A = 0$ and $I_A = e/R_A$.

Junctions and currents. We can then apply Kirchhoff's current law to any of the junctions. We use the currents flowing out of the left junction. The two currents I_A and I_B are flowing out of this junction and the current $-I$ is also flowing out of this junction (since I is flowing into the junction). Kirchhoff's current law therefore gives $\sum_i I_i = -I + I_A + I_B = 0$ and $I = I_A + I_B$, which indeed corresponds to our intuitive understanding: The current I flowing into the junction must equal the currents I_A and I_B flowing out of the junction to conserve currents.

Loop C_3. Instead of applying Kirchhoff's voltage law to loop C_2, we could also have applied Kirchhoff's law to the loop C_3: $\sum_i \Delta V_i = -I_A R_A + I_B R_B = 0$. Notice that the voltage drop across R_B is positive, because the direction of the loop is in the

[1] (a) $e - 2RI = 0$; (b) $I = e/2R$.

opposite direction to the direction of the current I_B. However, this gives us a relation between I_A and I_B, which must be combined with two other equations to find the three unknowns I, I_A, I_B.

Circuits and systems of linear equations. The system described in Fig. 10.5b has three unknowns, I, I_A and I_B. In general, we can think of circuits as setting up systems of linear equations. (Sometimes systems of linear differential equations, as we will soon see). We then need several linearly independent equations to find the unknowns, and we find these equations from Kirchhoff's laws and the voltage drops across resistors ($\Delta V = IR$) and other components. For the system in Fig. 10.5b we may get five equations: one for the application of Kirchhoff's voltage law for each of the three loops, and two from the application of Kirchhoff's current law for each of the two junctions. These equations will not be linearly independent: Kirchhoff's current law for the two junctions will be the same, and there are only two independent equations from the three loops. However, for more complicated circuits, it may be difficult to recognize which laws give independent equations. You should then instead use standard procedures from linear algebra. For example, you can find all the relevant equations and then reduce them to only the linearly independent ones using numerical or symbolic methods. (We demonstrate this in the exercises.)

Comparing with known results. For this problem we found that the total current was:

$$I = I_A + I_B = \frac{V}{R_A} + \frac{V}{R_B} = \left(\frac{1}{R_A} + \frac{1}{R_B}\right) V. \tag{10.7}$$

From this we also see that we can replace the two resistors in parallel with a single resistor with resistance $R = (1/R_A + 1/R_B)^{-1}$. This is the rule for combining resistors in parallel, which we found above. Sometimes we simplify problems by replacing complex resistors with simplified resistors, using the rules for combining resistors. This may provide you with better insight and oversight of the problem you are solving.

Example: Circuit Analysis

We can now use these tools to find the currents in the circuits in Fig. 10.6. For these systems, the battery voltage and the resistances of the resistors are given. This is a situation that corresponds to a typical physical situation, where we have a circuit and want to figure out what happens. What happens here means to find out what currents are flowing, the potential drops across the various components, and the power consumption of each component and the total circuit.

Circuit a. In circuit a there are three unknown currents, I, I_1 and I_2. Kirchhoff's current law for the currents flowing out of junction 1 gives us

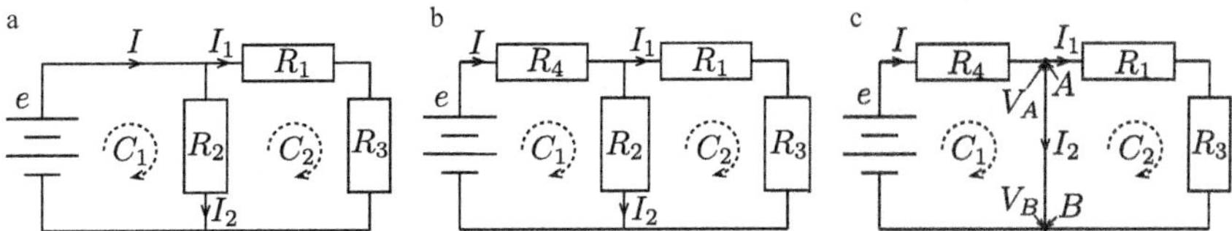

Fig. 10.6 Illustration of three circuits

$$\sum_i I_i = -I + I_1 + I_2 \tag{10.8}$$

We apply Kirchhoff's voltage law to loop C_1:

$$\sum_i \Delta V_i = e - R_2 I_2 = 0 \tag{10.9}$$

and similarly for loop C_2:

$$\sum_i \Delta V_i = -I_1 R_1 - I_1 R_3 + I_2 R_2 = 0, \tag{10.10}$$

This gives us three equations with three unknowns, which can be solved by linear algebra. Here, we see that $I = I_1 + I_2$ from (10.8), $I_2 = e/R_2$ from (10.9), and that

$$I_1 = \frac{R_2}{R_1 + R_3} I_2 = \frac{R_2}{R_1 + R_3} \frac{e}{R_2} = \frac{e}{R_1 + R_3}, \tag{10.11}$$

which we also could have seen by using a different loop through R_1, R_3 and the battery.

Circuit b. Circuit b can be solved using the same approach as for circuit a. Try to redo the arguments and find the resulting currents yourself.

Circuit c. What happens if we replace resistor R_2 with a wire? This corresponds to *short-circuiting* the circuit. Because we have introduced a wire without any resistance, we know that points A and B will have the same potential. This means that there will not be any current through resistors R_1 and R_3, because the current will go through the resistanceless wire from A to B instead. Beware such short circuits in your analysis as they may lead to problems.

Measuring Voltage and Current

We use a voltmeter to measure the potential difference between two points. An analogue voltmeter works by measuring the current through a coil in a magnetic

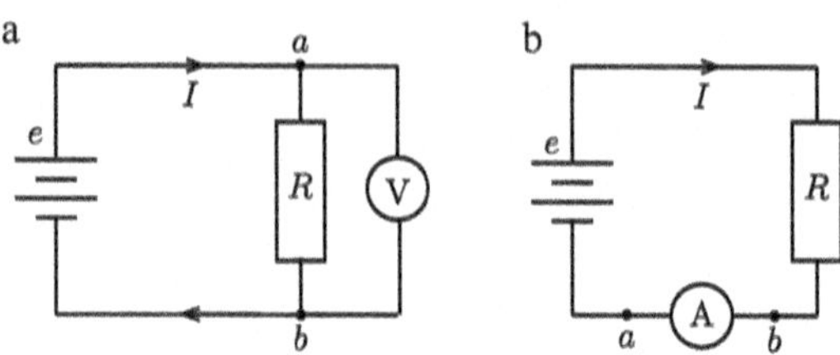

Fig. 10.7 Illustration of a circuit with **a** a voltmeter and **b** an amperemeter

field, where the effective deflection of the coil is proportional to the current, and the current is proportional to the voltage applied across the voltmeter. However, the internal resistance of the voltmeter is very high, so we typically assume that no current goes through the voltmeter. We illustrate a voltmeter in a circuit diagram with a V inside a circle as shown in Fig. 10.7. In the figure, the voltmeter measures the voltage difference between the points a and b. A voltmeter is applied in parallel to the component over which we wish the measure the potential difference.

Currents are measured by an amperemeter. They are built to have minimal impact on the current, that is, to have a very low internal resistance. We therefore typically assume that the potential drop across the amperemeter is negligible. Figure 10.7 shows that we draw an amperemeter as an A inside a circle. In the figure, the amperemeter measures the current flowing through points a and b. An amperemeter is applied in series with the wire through which we wish to measure the current.

Power

We know that the current is not "used up" throughout a circuit. Indeed, the current is conserved because charges are conserved. In the circuit in Fig. 10.5c the current is the same everywhere, but the voltage drops across the resistor. This means that there is a work done on a charge to move it from one side of the resistor to the other, and the resistors heats up as a result. The energy for this work is delivered by the battery. We recall that the power consumption in a resistor is

$$P = VI = RI^2 = \frac{V^2}{R}. \tag{10.12}$$

We could therefore say that energy is used up or converted from chemical energy in the battery to thermal energy, in the circuit. In physics we prefer not to use the word used up, in particular about energy, as energy conservation is one of the most fundamental laws of physics, and energy only passes from one form to another. We therefore say that energy is dissipated in the system. You will return to these questions when you study thermal and statistical physics.

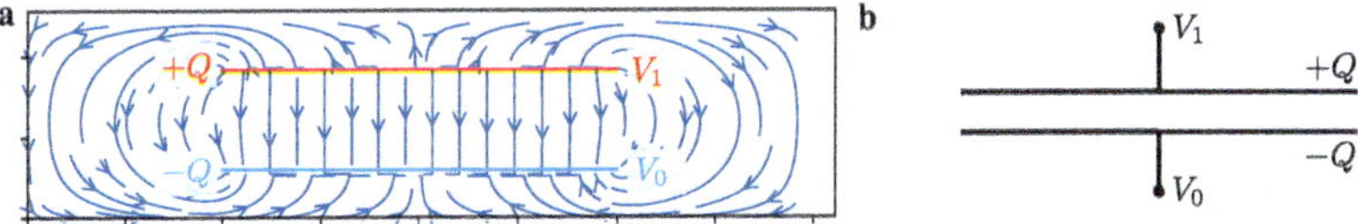

Fig. 10.8 **a** Illustration of a real capacitor with a streamplot of the electric field. **b** Model illustration of a capacitor used in a circuit diagram

10.2 Circuits with Capacitors

So far, we have only included resistors in our circuits, but we have also studied another component: the capacitor. How do we describe capacitors in circuits and how do circuits with capacitors behave?

A capacitor stores charge. We recall that a capacitor can function as a storage device for charge. A capacitor has a property, capacitance, C, which relates the voltage across the capacitor to the charge stored in the capacitor

$$C = \frac{Q}{V} . \tag{10.13}$$

A capacitor typically consists of two conductors (for example two parallel plates) as illustrated in Fig. 10.8. When a capacitor is charged with a charge Q it has a charge $+Q$ on one of the plates and a charge $-Q$ on the other plate. The two plates are not in contact, so no charge is flowing from one plate to another across the gap between the plates. When the capacitor is charged with a charge Q, the side with a charge $+Q$ has a potential V_1 and the side with charge $-Q$ has a potential V_0 so that the potential difference, $V = V_1 - V_0$ is $V = Q/C$.

Circuit with capacitor. In Fig. 10.8 we have illustrated the real system consisting of two conductors with an electric field between them on the left and a simplified illustration of the system on the right. In the model we draw the capacitor as two parallel plates independently of how the real capacitor looks like, just like we draw a resistor as a rectangle independently of the real shape of the resistor. This is the

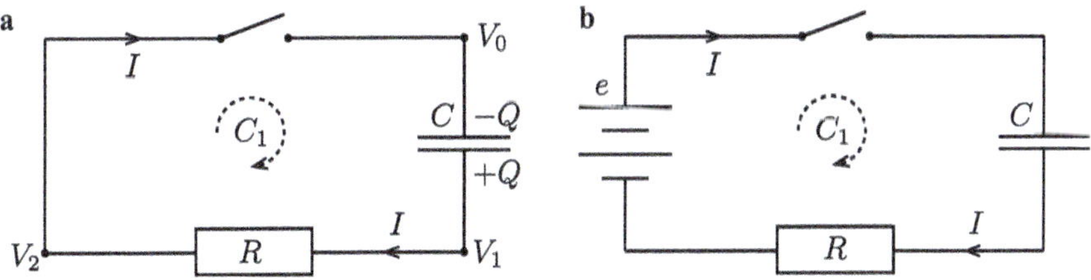

Fig. 10.9 **a** Illustration of circuit with capacitor C, resistor R and a switch. **b** Illustration of circuit with capacitor C, resistor R, battery e and a switch

symbol we use in a circuit diagram to indicate a capacitor. The capacitor then only has the properties described by the model, that the voltage across the capacitor is $V = Q/C$ given the charge Q or that the charge $Q = CV$ given the voltage V. The voltage is higher on the $+Q$ side, implying that there is a *voltage drop* of $V = Q/C$ from the $+Q$ side to the $-Q$ side across the capacitor.

Switch is open. Figure 10.9a illustrates a simple system consisting of a capacitor with capacitance C, a resistor with resistance R, a switch and a wire. Remember that all points along the wire have the same potential. When the switch is open, so that circuit is not connected, no current can flow in the wire. Let us assume that there is an initial charge Q on the capacitor. When the switch is open, there will not be any flow of current. Therefore, there will not be any potential drop across the resistor and $V_2 = V_1$. However, there is a potential drop across the capacitor of $V = Q/C$. The potential drop is from $+Q$ to $-Q$, so that $V_1 - V_0 = Q/C$. (This also means that there is also a potential difference across the switch. The left side of the switch is at $V_2 = V_1$ and the right side is at V_0). What happens when we close the switch and circuit becomes connected?

Switch is closed. When the switch is closed there is a wire from point 0 to 2 so these will have the same potential $V_2 = V_0$. There is a potential difference across the capacitor $V = V_1 - V_0 = (1/C)\,Q$, and there is a potential difference across the resistor $V_1 - V_0 = RI$. Kirchhoff's voltage law along the loop C_1 therefore gives

$$-RI + \frac{1}{C} Q = 0. \tag{10.14}$$

Notice that there is a *voltage drop* of $-IR$ over the resistor in the direction of the current, but that there is a *voltage gain* of $V = (1/C)\,Q$ over the capacitor because the voltage is higher on the $+Q$ side than on the $-Q$ side.

Current and charge in the circuit. But how are Q and I related? We recall that the current is the rate of change of the charge. There is a positive current I when the charge Q at the capacitor is flowing *out of* the capacitor *into* the wire. Therefore, $I = -\mathrm{d}Q/\mathrm{d}t$. We therefore get:

$$R\frac{\mathrm{d}Q}{\mathrm{d}t} + \frac{1}{C} Q = 0, \tag{10.15}$$

which can be rewritten as

$$\frac{\mathrm{d}Q}{\mathrm{d}t} = -\frac{1}{RC} Q. \tag{10.16}$$

Instead of a linear equation to find the current from the resistance, as we got for a circuit with only resistors, we now have a linear differential equation for the charge Q on the capacitor. When we have solved this equation, we can find the current by taking the time derivative $I = -\mathrm{d}Q/\mathrm{d}t$. To solve the differential equation, we also need the initial condition. We need to know what the charge Q of the capacitor is at a time $t = t_0$. Let us assume that the initial charge is $Q(t) = Q_0$ at $t = 0$. We recognize the differential equation, and that its solution is an exponential function of the form

$$Q(t) = Ae^{-t/(RC)} = Ae^{-t/\tau}, \tag{10.17}$$

where $\tau = RC$ is a characteristic time. We see that $Q(0) = Q_0$ gives that $A = Q_0$. You should convince yourself that this is a solution to the differential equation by inserting the solution in (10.16). The charge in the system is therefore $Q(t) = Q_0 e^{-t/\tau}$ and the current is $I = -\mathrm{d}Q/\mathrm{d}t = (Q_0/\tau)e^{-t/\tau}$.

Behavior after a long time. What happens after a long time? As time increases, the charge and the current will decay, effectively reaching zero after a long time. The characteristic time τ only depends on the properties of the circuit, the resistance R and the capacitance C, and not on the initial charge Q_0 on the capacitor.

Adding a battery. What happens if we add a battery with voltage e to the system as illustrated in Fig. 10.9b? We assume that the initial charge is $Q(t = 0) = 0$ and redo the calculation for a system with a battery. What is the charge and current in the system after an infinite time?

We apply Kirchhoff's voltage law to the loop C_1

$$e + V - IR = 0. \tag{10.18}$$

where $V = (1/C)Q$ and $I = -\mathrm{d}Q/\mathrm{d}t$ as above. We can now decide to solve the problem either in terms of $Q(t)$ or in terms of $V(t) = (1/C)Q(t)$. (If we solve for $V(t)$ we need to use that $Q(t) = CV(t)$ and that $I = -\mathrm{d}Q/\mathrm{d}t = -C\mathrm{d}V/\mathrm{d}t$). The differential equation for Q is:

$$e + \frac{1}{C}Q + \frac{\mathrm{d}Q}{\mathrm{d}t}R = 0 \tag{10.19}$$

and

$$\frac{\mathrm{d}Q}{\mathrm{d}t} = -\frac{e}{R} - \frac{1}{RC}Q \tag{10.20}$$

We solve this equation by first finding the homogeneous solution, when $e = 0$, which we found above, and then adding a constant to find the particular solution:

$$Q(t) = A + Be^{-t/\tau}, \tag{10.21}$$

where we know that $Q(0) = 0$, which gives $A = -B$ so that $Q(t) = A(1 - e^{-t/\tau})$. What are the initial conditions for the current? At $t = 0$ there is no charge on the capacitor, $Q(0) = 0$. From (10.20) we see that $\mathrm{d}Q/\mathrm{d}t = -e/R$ at $t = 0$. We use this initial condition for the derivative of $Q(t)$ from (10.21):

$$Q'(0) = \frac{A}{\tau}e^{-0/\tau} = -\frac{e}{R} \quad \Rightarrow \quad A = -\frac{e\tau}{R}. \tag{10.22}$$

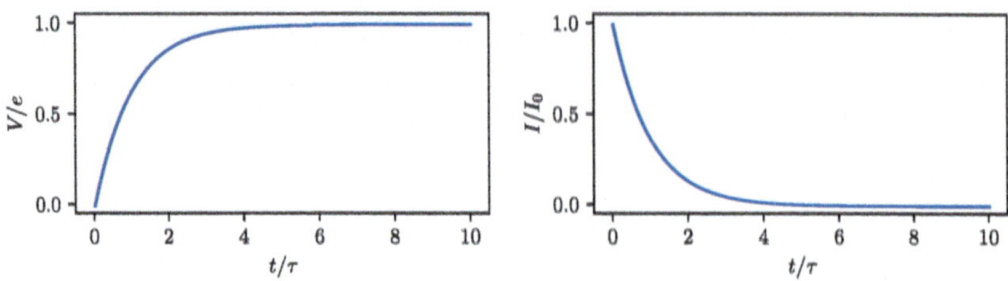

Fig. 10.10 Plot of the voltage $V(t)$ across the capacitor and the current I through the system as a function of time t for a system with a battery e, a capacitor C and a resistor R connected in series

We have therefore found both constants A and B in (10.21):

$$Q(t) = -\frac{e\tau}{R}\left(1 - e^{-t/\tau}\right). \tag{10.23}$$

Hmmm. Why did we get a negative charge on the capacitor? We need to look back on the figure. We notice that after a long time, we expect that the potential on the top of the battery and on the top of the capacitor must be the same, e, and on the bottom of both the battery and the capacitor the potential is zero. We have assumed that the charge is positive on the bottom of the capacitor and negative on the top, but our calculation shows that the result is the opposite. This is fine. You can always make such assumptions and then simply see what your calculations give you. Here, we found out that the charge is positive on the top and negative on the bottom.

If you like, you could instead express the result in terms of the potential drop V across the capacitor, $V = -(1/C)Q$:

$$V(t) = -\frac{1}{C}Q = e\left(1 - e^{-t/\tau}\right). \tag{10.24}$$

This shows that at $t = 0$, when the switch is closed, there is no charge and therefore no voltage drop across the capacitor. And after infinite time the potential drop across the capacitor is the same as the potential gain over the battery, since the current has then decayed to zero as illustrated in Fig. 10.10.

Test your understanding
The figure shows a circuit diagram. The potential in point A is $V_A = 0$. (a) Initially, there is no charge on the capacitor and the switch is open (as in the figure). What are the potentials V_B to V_E? (b) The switch is then closed. What are the potentials V_B to V_E when the system has reached a stationary state after a very long time?[2]

[2] (a) $V_B = V_C = V_D = e$, $V_E = 0$; (b) $V_B = e$, $V_C = V_D = V_E = 0$.

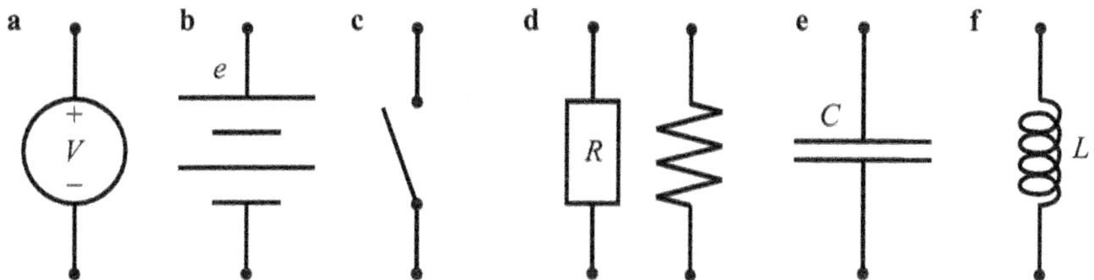

Fig. 10.11 Illustration of the various circuit symbols we have introduced. **a** Voltage source, **b** battery, **c** switch, **d** resistor, **e** capacitor, and **f** inductor (not introduced yet)

Components

We have now introduced several components to describe circuits as illustrated in Fig. 10.11.

- **Wire**: We draw wire or conductors as lines. We assume that the conductor is ideal and that all points along the line has the same potential.
- **Switch**: We introduce a switch that can break the circuit.
- **Resistors**: We draw resistors as small rectangles. (Sometimes a zig-zag line is also used as symbols for resistors, but we will not use this in this book). The voltage drop across a resistor in the direction of the current is $-IR$.
- **Capacitors**: We draw capacitors as two parallel lines perpendicular to the wire. The voltage drop across a capacitor is $-(1/C)Q$.
- **Battery**: A battery provides a constant potential gain e.
- **Voltage source**: A voltage source may provide a time-varying voltage gain $V(t)$.

These basic components combined with Kirchhoff's voltage and current law provides us with the tools to model and describe complex systems and solve their model behavior.

10.3 Circuits as Models

Circuits can be real systems that you build yourself or they can be simplified models of real systems so that we can address them using our knowledge of electromagnetics. For example, we know how to describe a real wire with a finite conductivity as illustrated in Fig. 10.12. However, if we want to include this wire in a description of a more complex system, we want to simplify its description. This can be done as illustrated in the figure as a system consisting of a wire of an ideal conductor, drawn as a line in the circuit diagram, connected in series with a resistor that represents the resistance of the whole wire. This is an example of a circuit model of a real system. The goal is that you should be able to simplify complex physical systems

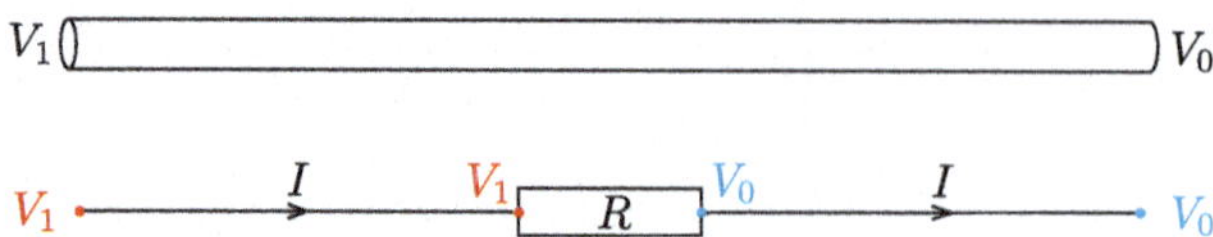

Fig. 10.12 Illustration of **a** a real wire with a finite conductivity and **b** the model of the system

to simpler circuit-based systems and then analyze them using your knowledge of electromagnetics. Let us demonstrate this by a few examples.

Example: Cell Membrane Potential

A cell membrane consists of a lipid bilayer with embedded ion channels and ion pumps. The lipid bilayer consists of a set of two layers of lipid molecules. Each lipid molecule is approximately linear, like a somewhat flexible rod. The lipid molecules are amphiphilic, which means that one end of the molecule is hydrophobic, and the other end is hydrophilic. In the outer lipid layer, the hydrophilic end points outwards, while in the inner lipid layer, the hydrophilic end points into the cell. The outside of the lipid bilayer is therefore polar, which means that the molecules form hydrogen bonds with the surrounding water molecules. The inner part of the lipid bilayer is non-polar. The lipid bilayer functions as a membrane, so that molecules that are on the outside of the layer cannot easily get through the layer. This allows cells to have different chemistry on the outside and inside of the cell membrane. See Fig. 10.13 for an illustration.

Cell membranes are not completely impermeable. They have various types of channels in the form of complex molecules, proteins, that allow some molecules or ions to flow through the cell membrane. *Ion pumps* use energy to pump ions from

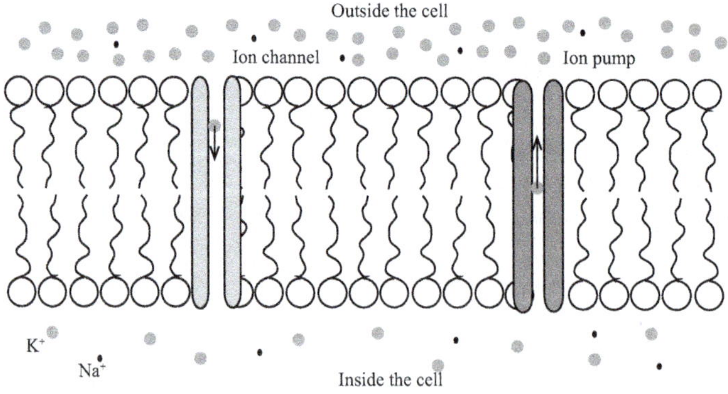

Fig. 10.13 Illustration of a cell membrane

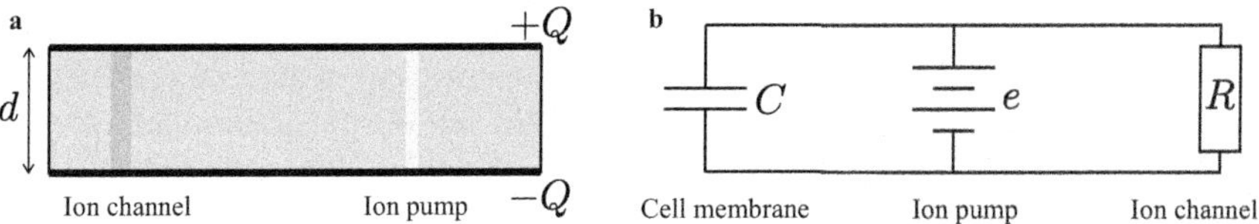

Fig. 10.14 Simplified models of a cell membrane. **a** Showing the cell membrane as a parallel-plate capacitor with ion pumps and ion channels. **b** Showing the corresponding circuit diagram

one side of the membrane to the other. For example, the sodium-potassium pump transports three sodium ions out of the cell and two potassium ions into the cell in each pumping cycle. Since both ions have a charge of $+e$, this means that the pump builds up a charge difference between the outside and the inside of the cell. Due to this pump, the outside of the cell will have positive charge, and the inside will have negative charge. There will therefore be an electric field from the outside of the cell to the inside of the cell. The inside of the cell will therefore have a lower electric potential than the outside. The ion pump uses chemical energy to perform work to transport charges across this potential. In addition, there are leaky ions channels that allows for example potassium ions to flow back into the cell. For nerve cells, the net effect of both the ion pumps and ion channels for several types of ions is a potential difference across the cell membrane of about -80 mV, which is called the membrane potential.

How can we create a simplified model of this system? Figure 10.14a contains a first iteration of a model. The cell membrane is illustrated as a dielectric medium of a given thickness and surface area. We represent this as a capacitor with capacitance $C = A\epsilon/d$. We assume that ion channels and ion pumps are small compared to the area and therefore do not affect the capacitance. But how should we represent the ion pump and ion channels in a model? The ion pump acts as a battery: It lifts charges up through a potential difference. We assume that all the ion pumps in total act as one battery with a voltage e. The ion channel acts as a leaky channel for ions: They will flow through the channel if there is a potential difference, but with some resistance. A potential difference is needed to drive the ions through the channel. We therefore model all the ion channels as one resistor with resistance R.

We have now developed a model of the electric current and electric charges across the cell membrane as illustrated in Fig. 10.14b. The model consists of a battery, a capacitor and a resistor in parallel. This is a simplified model of one aspect of the physics of a cell membrane, and we can use this model to address the electrical properties of the cell membrane. We can ask questions like, what happens if the ion pump stops pumping? What happens if we change the number of ion channels and therefore change the resistivity of the leaky channel? We can answer these questions by studying the circuit and describing its behavior.

We can also improve the details of this model. For example, we can assume that each ion pump has a voltage e_0 and that there is a given density N_P of ion pumps on the surface. The total number of ion pumps is then $N_P A$. The total effect of all of these pumps is like having $N_P A$ batteries in parallel. However, this does not change the potential difference of the battery, only the maximum power or current it can deliver, so that the total potential difference of the system still is $e = e_0$. Similarly, we can assume that there is a density N_C of ion channels. Each ion channel contributes with a resistivity R_0 and there are $N_C A$ such resistors in parallel. The total resistivity is therefore $1/R = N_C A/R_0$.

Example: Cable Equation

Let us extend the model for a cell to the axon, a long part of a nerve cell which transmits electric signals. The axon is approximately cylindrical in shape and consists of a cell interior, a cell membrane, and the extracellular system outside the cell as illustrated in Fig. 10.15. To make a model of this system, we divide the cylindrical axon into smaller cylindrical parts of length ΔL. The potential outside the cell is V_e. We assume that the outside of the cell is a relatively good conductor, so that the potential is the same, V_e, everywhere outside the cell. We model the behavior inside each such element by a single voltage V_i, in the middle of element i. The cell membrane acts as a capacitor with a capacitance C (which you know how to calculate for a cylindrical element), and there is a leaking current through the cell membrane through channels with an effective resistance R for each element. The capacitor C and the resistor R act in parallel for each such element.

In addition, an element is connected to the neighboring elements. Element i is connected to elements $i - 1$ and $i + 1$. There is a current going inside the cell between the elements with a resistance r. This makes up the model illustrated in Fig. 10.15. For simplicity we set $V_e = 0$.

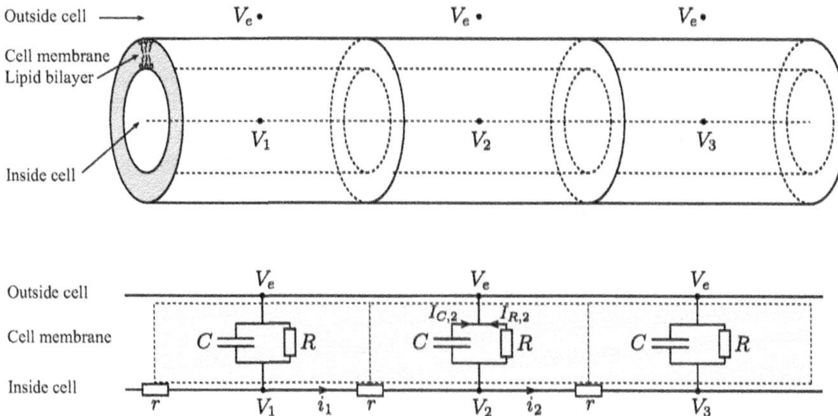

Fig. 10.15 Illustration of the physical system for the cable equation and the circuit model for the physical system

Derivation of the equations for the membrane potential. Let us find an equation for the behavior of element i in this system:

- We know that the current through the capacitor is related to the charge Q_i on the capacitor, $I_{C,i} = \mathrm{d}Q_i/\mathrm{d}t$. We can relate this to the voltage V_i across the capacitor, because $Q_i = CV_i$, getting $I_{C,i} = C\,\mathrm{d}V_i/\mathrm{d}t$. (The voltage drop across the capacitor is $V_i - V_e = V_i$ because $V_e = 0$.)
- We know that the current through the resistor R is related to the voltage drop, $I_{R,i} = (V_i - V_e)/R = V_i/R$.
- The current i_i through the resistor r is found through Ohm's law: $i_i = (V_i - V_{i+1})/r$.
- The current i_{i-1} through the resistor r is found through Ohm's law: $i_{i-1} = (V_{i-1} - V_i)/r$.

We apply Kirchhoff's current law in the point V_i:

$$i_{i-1} = i_i + \left(I_{C,i} + I_{R,i}\right), \tag{10.25}$$

which gives

$$\frac{V_{i-1} - V_i}{r} = \frac{V_i - V_{i+1}}{r} + \frac{V_i}{R} + C\frac{\mathrm{d}V_i}{\mathrm{d}t}. \tag{10.26}$$

We can rewrite this equation as:

$$C\frac{\mathrm{d}V_i}{\mathrm{d}t} + \frac{V_i}{R} = \frac{V_{i+1} - 2V_i + V_{i-1}}{r}. \tag{10.27}$$

Left boundary element. What about the left-most part of the system? We assume that $V_0(t)$ is a given voltage source, which we interpret as the signal sent out by the nerve cell. We therefore get:

$$C\frac{\mathrm{d}V_1}{\mathrm{d}t} + \frac{V_1}{R} = \frac{V_2 - 2V_1 + V_0(t)}{r}, \tag{10.28}$$

where $V_0(t)$ is a given function of time.

Right boundary element. What about the right-most part of the system? For the last element, V_N, there is no element to the right. This means that i_N is zero, since there is no current going to the right of the last element. We therefore get the equation $i_{N-1} = I_{C,N} + I_{R,N}$ and

$$\frac{V_{N-1} - V_N}{r} = \frac{V_N}{R} + C\frac{\mathrm{d}V_N}{\mathrm{d}t}. \tag{10.29}$$

We can rewrite this equation as:

$$C\frac{\mathrm{d}V_N}{\mathrm{d}t} + \frac{V_N}{R} = \frac{V_{N-1} - V_N}{r}. \tag{10.30}$$

Interpretation of the main equation. How can we interpret these equations? If we look at (10.27) we see that it is a differential equation in time. But how can we interpret the right-hand side of the equation? This looks like the discrete version of Laplace's equation. The index i refers to the spatial position of the various elements. Indeed, we can describe the length of an element at Δx so that $V_i(t) = V(i\Delta x, t) = V(x, t)$. We can then interpret (10.27) as:

$$C\frac{\partial V(x,t)}{\partial t} + \frac{V(x,t)}{R} = \frac{V(x+\Delta x,t) - 2\,V(x,t) + V(x-\Delta x,t)}{r}. \tag{10.31}$$

We recognize the right-hand side as the second derivative of $V(x, t)$ with respect to x:

$$\frac{\partial^2 V}{\partial x^2} \simeq \frac{V(x+\Delta x,t) - 2\,V(x,t) + V(x-\Delta x,t)}{\Delta x^2}. \tag{10.32}$$

We can therefore interpret (10.27) as:

$$C\frac{\partial V(x,t)}{\partial t} + \frac{V(x,t)}{R} = \frac{\Delta x^2}{r}\frac{\partial^2 V}{\partial x^2}. \tag{10.33}$$

This is similar to the diffusion equation, at least in the limit when $R \to \infty$. This equation is called the *Cable equation*.

Interpretation of the equation for element N. But how can we interpret the condition on the right-hand side of the system, given as (10.30)? In a stationary state, the time derivative is zero, and we can interpret the right-hand side of the equation as the spatial derivative of $V(x, t)$. In the stationary state, we therefore have that

$$\frac{\partial V(x,t)}{\partial x} = 0, \tag{10.34}$$

which corresponds to a Neumann boundary condition for $V(x)$ on the right-hand side.

Comparison with Laplaces's equation. We also notice that in the stationary state, that is when the time derivative in the system is zero, we see from (10.33) that we get:

$$\frac{V(x,t)}{R} = \frac{\Delta x^2}{r}\frac{\partial^2 V}{\partial x^2}. \tag{10.35}$$

In the case when $R \to \infty$ this reduces to:

$$\frac{\partial^2 V}{\partial x^2} = 0, \tag{10.36}$$

which is Laplace's equation.

Numerical scheme. The equations we found in (10.27) are discretized and it is simple to implement these equations in a simple Euler scheme to find the time development:

$$C\frac{\mathrm{d}V_i}{\mathrm{d}t} + \frac{V_i}{R} = \frac{V_{i+1} - 2V_i + V_{i-1}}{r}. \tag{10.37}$$

We start from a set of initial conditions, $V_i(t)$, and then find the solution at a time $t + \Delta t$ by using an explicit scheme where we approximate

$$\frac{\mathrm{d}V_i}{\mathrm{d}t} \simeq \frac{V_i(t+\Delta t) - V_i(t)}{\Delta t}. \tag{10.38}$$

This gives us the scheme:

$$V_i(t+\Delta t) \simeq V_i(t) - \frac{\Delta t}{RC}V_i(t) + \frac{\Delta t}{rC}\left(V_{i+1}(t) - 2V_i(t) + V_{i-1}(t)\right). \tag{10.39}$$

For the left-hand side, we recall that $V_0(t)$ is the external signal provided on the left-hand side. This is simply a given function in time, for example a square pulse, a sinusoidal function, or some other time-varying signal. For the right-hand side, we use the same time discretization to find that:

$$C\frac{\mathrm{d}V_N}{\mathrm{d}t} \simeq C\frac{V_N(t+\Delta t) - V_N(t)}{\Delta t} = -\frac{V_N}{R} + \frac{V_{N-1} - V_N}{r}, \tag{10.40}$$

which gives

$$V_N(t+\Delta t) = V_N(t) - \frac{\Delta t}{RC}V_N(t) + \frac{\Delta t}{rC}\left(V_{N-1} - V_N\right). \tag{10.41}$$

Python implementation. This numerical scheme can be almost directly implemented as it is into Python. We need two additional aspects: We need a function to specify the voltage source on the left side: `V0(t)`, and we need to specify the initial values for all the voltages, which corresponds to specifying the initial charges Q_i on the capacitors. Let us first assume that $V_i(0) = 0$ and that $V_0(t)$ is a step function which is V_0 for $t < t_0$ and 0 for $t \geq t_0$ so that $V_0(0) = V_0$. This corresponds to turning the voltage on the left side to V_0 at $t = 0$, keeping it constant at V_0, and then suddenly turning it to zero at t_0. We interpret this as a voltage pulse of height V_0 and of length t_0. We store the whole time evolution in the two-dimensional array `V[j,i]`, where `j` represents the time and `i` represents the physical position along the cell.

```
import numpy as np
import matplotlib.pyplot as plt
# Setting realistic paramters
C = 1e-10 #Capacitance
R = 1e11
r = 1e6
V0 = 100e-3
# Define voltage pulse step
```

```
def Vs(t,V0,t0):
    return V0*(t<t0)[0]
T = 0.1 # Simulation in seconds
t0 = T*0.1 # Pulse length
dt = 0.1*C*r
nsteps = int(T/dt) # Number of time steps
L = 100 # Number of capacitors
V = np.zeros((nsteps,L),float)
V[0,0] = V0 # Initial condition, t = 0
t = np.zeros((nsteps,1),float)
for j in range(0,nsteps-1):
    t[j+1] = t[j] + dt
    V[j+1,0] = Vs(t[j+1],V0,t0) # Left boundary condition (i=0)
    for i in range(1,L-1): # Internal elements
        V[j+1,i]=V[j,i]+(dt/C)*((V[j,i+1]-2*V[j,i]+V[j,i-1])/r-V[j,i]/R)
    # Right boundary condition (i=L-1)
    V[j+1,L-1] = V[j,L-1] + (dt/C)*((-V[j,L-1]+V[j,L-2])/r - V[j,L-1]/R)
# Plotting results
plt.subplot(2,1,1)
for i in range(0,L-1,10):
    plt.plot(t,V[:,i]/max(V[:,i]))
plt.subplot(2,1,2)
for j in range(0,nsteps-1,1000):
    plt.plot(V[j,:]/max(V[j,:]))
```

The resulting plots are shown in Fig. 10.16. In the top figure we show the voltage $V(x, t)$ as a function of time t at 8 different positions along the system. Notice that the value of V falls very rapidly as we study the system for larger values of x. This makes it difficult to interpret the results. We have therefore instead plotted $V(x, t)/V_{\max}$, where $V_{\max}$ is chosen so that when we plot $V(x, t)$ as a function of t for a given x it is the maximum of V over t for that given x, and similarly when we plot $V(x, t)$ as a function of x for a given t it is the maximum of V over x for that given t. This means that we can interpret the functional shapes, but we cannot compare the amplitudes of the curves. We clearly see how the initial signal, which is a square pulse for a short time t_0, is transmitted along the cable while being smeared out.

Exploring the model. You now have the basic tools needed to study the behavior of a signal traveling along a passive nerve cell. This provides you with a good basis to explore the behavior of this system and to extend the model to also include active components as you will find in real nerve cells and which ensure that the signal does not decay as rapidly as seen in the model presented here. This model can also be used to represent how signals decay in long electric signal cables, such as the telegraph cables laid across the Atlantic Ocean in the mid Nineteenth century. Indeed, these equations were used by Lord Kelvin for these purposes.

Stability of the numerical scheme. Notice that the numerical scheme we have used to solve this problem is an explicit scheme for solving the diffusion equation. In general, we know that this scheme, with $r \to \infty$, is numerically stable only if

$$\frac{\Delta t}{rC} \leq \frac{1}{2}. \tag{10.42}$$

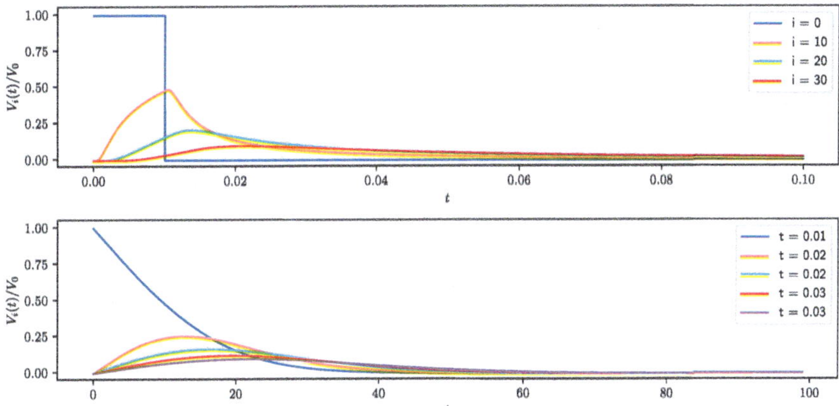

Fig. 10.16 Plots of $V(x, t)$ for the cable equation. Top figure is $V_i(t)$ as a function of t for various x-positions given by the index i. Bottom figure is $V_i(t)$ as a function of i for various t measured in seconds

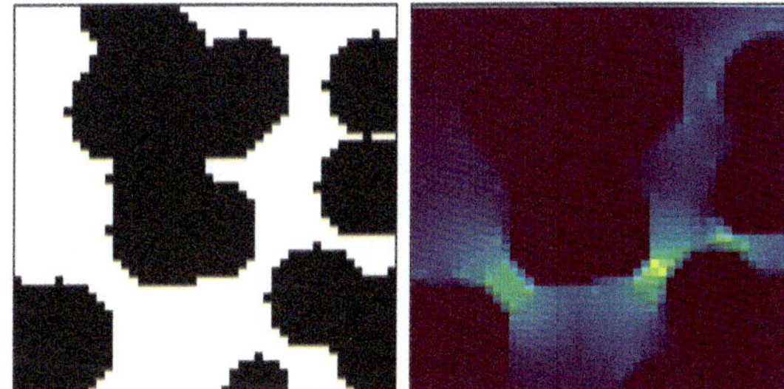

Fig. 10.17 A porous medium where white is the conductor and black is the hole and the resulting flow of current from left to right in the conductor

10.4 Current Distributions in Disordered Media

What would happen if we placed a porous material, that is a material with randomly placed holes, between two voltages? Figure 10.17 illustrates a material consisting of a conductor with finite conductivity (white) with holes (black) and the resulting flow of current from left to right. We now have the tools to model and study such systems by simplifying the problem to a circuit and then study the behavior of the circuit.

Model for Porous Medium

How can we model a two-dimensional random porous material as a circuit? Figure 10.18 illustrates a simplified model of a porous material. In the model, we have discretized the system into $N \times N$ voxels of size $d \times d$ each, which we call elements. Each element can be either present, with a conductivity σ, or absent, with

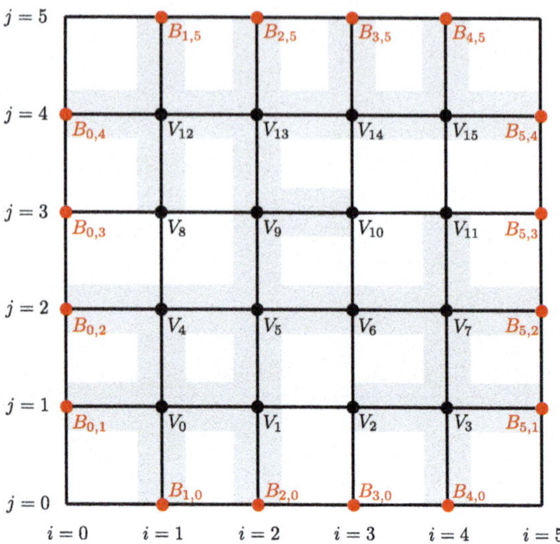

Fig. 10.18 Illustration of a 4 × 4 simulation system. Connections that are present are colored gray and connections that are absent are colored white

zero conductivity as illustrated in the figure. We want to model the voltages and current through this system, when we apply an external voltage V_A on the left side and V_B on the right side, where we set $V_B = 0$ for simplicity.

Figure 10.18 illustrates how we divide the system into a grid. At each grid point (i, j) there is a voltage $V_{i,j}$. We model the current from (i, j) to a neighboring grid point $(i + 1, j)$ using Ohm's law:

$$I_{(i,j);(i+1,j)} = \frac{1}{R_{(i,j);(i+1,j)}} \left(V_{i,j} - V_{i+1,j}\right) . \tag{10.43}$$

where $R_{(i,j);(i+1,j)}$ is the resistance for the connection between points (i, j) and $(i + 1, j)$. For simplicity, we instead introduce the term *conductance*, G:

$$G_{(i,j);(i+1,j)} = \frac{1}{R_{(i,j);(i+1,j)}} . \tag{10.44}$$

When we say that we describe porous system or a random material, in this model, this means that the conductances vary in space to reflect the porous system. Here, we will introduce a simplfied model for the conductance between the two elements at (i, j) and $(i + 1, j)$: The conductance may be either zero, corresponding to a missing connection, or G, corresponding to a present connection. We introduce a probability p for the connection between (i, j) and e.g. $(i + 1, j)$ to be present.

Applying Kirchhoff's Laws

The second step in our model is to realize that Kirchhoff's current law for each element means that:

$$\sum_n I_n = I_{(i,j);(i+1,j)} + I_{(i,j);(i-1,j)} + I_{(i,j);(i,j+1)} + I_{(i,j);(i,j-1)} = 0 \tag{10.45}$$

We combine this with Ohm's law from (10.43) and get:

$$\begin{aligned} &G_{(i,j);(i+1,j)}\left(V_{i,j} - V_{i+1,j}\right) + G_{(i,j);(i-1,j)}\left(V_{i,j} - V_{i-1,j}\right) + \\ &G_{(i,j);(i,j+1)}\left(V_{i,j} - V_{i,j+1}\right) + G_{(i,j);(i,j-1)}\left(V_{i,j} - V_{i,j-1}\right) = 0. \end{aligned} \tag{10.46}$$

This equation looks very much like the equations we found for voltage when we solved Laplace's equation. We will therefore try to solve this problem numerically using the same implicit scheme we developed for Laplace's equation.

Implicit Numerical Scheme

Let us first develop the numerical scheme for the small 4×4 system shown in Fig. 10.18. Each value of $V_{i,j}$ is determined from (10.46), where we must include both the values for the conductances and the values for the boundaries.

Internal element. Let us first look at an internal element $V_{2,3}$. We use the same enumeration scheme as we introduced previously, with a single index n for an element, where we form n as $n = (j-1)(N-2) + (i-1)$ so that $n = 0, 1, \ldots, 15$. Element (2, 3) therefore corresponds to index $n = 9$. We can therefore use Fig. 10.18 directly to read out the neighboring elements. In the figure, we have illustrated elements with zero conductance with a white and elements with finite conductance (G) with gray. This means that the four conductances leading out of element 9 are: $G_{9,10} = G$, $G_{9,8} = 0$, $G_{9,5} = G$, and $G_{9,13} = G$. The equation for V_9 is therefore:

$$-V_{10} - V_5 - V_{13} + 3V_9 = 0. \tag{10.47}$$

Boundary element. What happens to an element at the boundary? Let us look the element $n = 4$. The conductances leading out of element 4 are: $G_{4,5} = G$, $G_{4,(0,2)} = G$, $G_{4,8} = 0$, and $G_{4,0} = G$, where we have used the notation (0, 2) for the conductance to a boundary element, $B_{0,2}$. What is the equation for V_4?

$$G(V_4 - V_5) + G(V_4 - B_{0,2}) + 0(V_4 - V_8) + G(V_4 - V_0) = 0, \tag{10.48}$$

where we divide by G and reorder the equation to get:

$$-V_5 - V_0 + 3V_4 = B_{0,2}\,. \tag{10.49}$$

Dirichlet or Neumann. At the top boundary we can either use Dirichlet boundary conditions and provide values for each boundary element $B_{i,j}$ needed, or we can use Neumann boundary conditions, and instead assume that there is no current across the boundary, and that there is therefore no voltage difference across the boundary. Dirichlet boundary conditions for element $n = 13$ gives the equation:

$$-V_{12} - V_{14} - V_9 + 4V_{13} = B_{2,5}, \tag{10.50}$$

whereas Neumann boundary conditions for element $n = 13$ gives the equation:

$$-V_{12} - V_{14} - V_9 + 3V_{13} = 0, \tag{10.51}$$

which corresponds to assuming that $B_{2,5} = V_{13}$, which gives zero current into the boundary element $(2, 5)$. We will here use Neumann's boundary condition in the y-direction and Dirichlet boundary conditions in the x-direction so that there will be a current flowing from the left to the right.

System of equations. We write down all the 16 equations for the internal points. On matrix form this gives us the following system of equations using Neumann boundary conditions in the y-direction and Dirichlet boundary conditions in the x-direction:

$$\begin{bmatrix}
3 & -1 & 0 & 0 & -1 & 0 & 0 & 0 & 0 & 0 & 0 & 0 & 0 & 0 & 0 & 0 \\
-1 & 3 & -1 & 0 & 0 & -1 & 0 & 0 & 0 & 0 & 0 & 0 & 0 & 0 & 0 & 0 \\
0 & -1 & 3 & -1 & 0 & 0 & -1 & 0 & 0 & 0 & 0 & 0 & 0 & 0 & 0 & 0 \\
0 & 0 & -1 & 3 & 0 & 0 & 0 & -1 & 0 & 0 & 0 & 0 & 0 & 0 & 0 & 0 \\
-1 & 0 & 0 & 0 & 4 & -1 & 0 & 0 & -1 & 0 & 0 & 0 & 0 & 0 & 0 & 0 \\
0 & -1 & 0 & 0 & -1 & 4 & -1 & 0 & 0 & -1 & 0 & 0 & 0 & 0 & 0 & 0 \\
0 & 0 & -1 & 0 & 0 & -1 & 4 & -1 & 0 & 0 & -1 & 0 & 0 & 0 & 0 & 0 \\
0 & 0 & 0 & -1 & 0 & 0 & -1 & 4 & 0 & 0 & 0 & -1 & 0 & 0 & 0 & 0 \\
0 & 0 & 0 & 0 & -1 & 0 & 0 & 0 & 4 & -1 & 0 & 0 & -1 & 0 & 0 & 0 \\
0 & 0 & 0 & 0 & 0 & -1 & 0 & 0 & -1 & 4 & -1 & 0 & 0 & -1 & 0 & 0 \\
0 & 0 & 0 & 0 & 0 & 0 & -1 & 0 & 0 & -1 & 4 & -1 & 0 & 0 & -1 & 0 \\
0 & 0 & 0 & 0 & 0 & 0 & 0 & -1 & 0 & 0 & -1 & 4 & 0 & 0 & 0 & -1 \\
0 & 0 & 0 & 0 & 0 & 0 & 0 & 0 & -1 & 0 & 0 & 0 & 3 & -1 & 0 & 0 \\
0 & 0 & 0 & 0 & 0 & 0 & 0 & 0 & 0 & -1 & 0 & 0 & -1 & 3 & -1 & 0 \\
0 & 0 & 0 & 0 & 0 & 0 & 0 & 0 & 0 & 0 & -1 & 0 & 0 & -1 & 3 & -1 \\
0 & 0 & 0 & 0 & 0 & 0 & 0 & 0 & 0 & 0 & 0 & -1 & 0 & 0 & -1 & 3
\end{bmatrix}
\begin{bmatrix} V_0 \\ V_1 \\ V_2 \\ V_3 \\ V_4 \\ V_5 \\ V_6 \\ V_7 \\ V_8 \\ V_9 \\ V_{10} \\ V_{11} \\ V_{12} \\ V_{13} \\ V_{14} \\ V_{15} \end{bmatrix}
=
\begin{bmatrix} B_{0,1} \\ 0 \\ 0 \\ B_{5,1} \\ B_{0,2} \\ 0 \\ 0 \\ B_{5,2} \\ B_{0,3} \\ 0 \\ 0 \\ B_{5,3} \\ B_{0,4} \\ 0 \\ 0 \\ B_{5,4} \end{bmatrix}$$

This is a matrix equation on the form $Ax = B$ where the unknown x here contains the potentials, V_n.

Setting up the matrix. We now know how to set up the system of linear equations that we can solve with Python's linear equation solvers. However, first we need to set up the matrix $A_{n,m}$ and the vector B_n. We do this by looping through all elements (i, j), and then inserting all the relevant values for the equation used to determine $V_{i,j}$. For each neighbor (k, l) to (i, j) we insert the values from the equation $G_{(i,j),(k,l)}(V_{i,j} - V_{k,l}) = 0$, taking into consideration the value for the conductance and whether the element is a boundary element or an internal element. We write a function to set up the system equations, solve the system, and put the values back into an array of $V_{i,j}$ values as well as two arrays of the currents. This is done by first setting up three matrices to describe the porous medium: The matrix b describes the boundary conditions as previously. The matrices gx and gy describes the conductances in the x- and y-directions respectively.

```
# Generate random lattice
def makegxgy_random(N,p):
    # Setting up the boundaries
    b = np.zeros((N,N),float)
    b[:] = float("nan")
    #b[:,0] = 0.0
    #b[:,N-1] = 0.0
    b[0,:] = 0.0
    b[N-1,:] = 1.0
    # Setting up random conductances
    gx = 1.0*(np.random.rand(N,N)<p)
    gy = 1.0*(np.random.rand(N,N)<p)
    return b,gx,gy
```

Second, we write a function to set up and solve the system of equations given the porous medium described by b, gx and gy:

```
def findvoltage_for_gxgy(b,gx,gy):
    # Setting up the boundaries
    Nx,Ny = gx.shape
    N = Nx
    # Setting up system of equations
    LM = (N-2)*(N-2)
    A = np.zeros((LM,LM),float)
    B = np.zeros((LM),float)
    for j in range(1,N-1):
        for i in range(1,N-1):
            n = (j-1)*(N-2) + (i-1)
            # x- direction
            G = gx[j,i-1]
            if (i>1):
                A[n,n] = A[n,n] + G # Diagonal
                A[n,n-1] = -G
            # x+ direction
            G = gx[j,i]
            if (i<N-2):
                A[n,n] = A[n,n] + G # Diagonal
                A[n,n+1] = -G
            # y- direction
            G = gy[j-1,i]
            A[n,n] = A[n,n] + G # Diagonal
            if (j>1):
                A[n,n-(N-2)] = -G
            else:
                B[n] = B[n] + G*b[j-1,i]
            # y+ direction
            G = gy[j,i]
            A[n,n] = A[n,n] + G # Diagonal
            if (j<N-2):
                A[n,n+(N-2)] = -G
            else:
```

```
                B[n] = B[n] + G*b[j+1,i]
    Ainv = np.linalg.pinv(A)
    x = np.dot(Ainv,B)
    # Generate full V matrix
    V = np.zeros((N,N),float)
    for j in range(0,N):
        for i in range(0,N):
            if (i<1)or(i>=N-1)or(j<1)or(j>=N-1):
                V[j,i] = b[j,i]
            else:
                n = (j-1)*(N-2) + (i-1)
                V[j,i] = x[n]
    return V
```

The code is run with the following command

```
b,gx,gy = makegxgy_random(40,0.55)
V = findvoltage_for_gxgy(b,gx,gy)
```

To visualize the results, we may visualize the potential as shown in Fig. 10.19. However, this does not provide as much insight into the physics, because the potential varies slowly. Instead, we would like to study the current distribution. For each element, we calculate the absolute value of the current flowing into the element and then we visualize this quantity. This is implemented in the following program:

```
def findabscurrent(V,gx,gy):
    Nx = V.shape[0]
    Ny = V.shape[1]
    Ia = np.zeros((Nx,Ny),float)
    for i in range(1,Nx-1):
        for j in range(1,Ny-1):
            # Find contributions to current from each direction
            # x- direction
            G = gx[j,i-1]
            dV = abs(G*(V[j,i]-V[j,i-1]))
            # x+ direction
            G = gx[j,i]
            dV = dV + abs(G*(V[j,i]-V[j,i+1]))
            # y- direction
            G = gy[j-1,i]
            dV = dV + abs(G*(V[j,i]-V[j-1,i]))
```

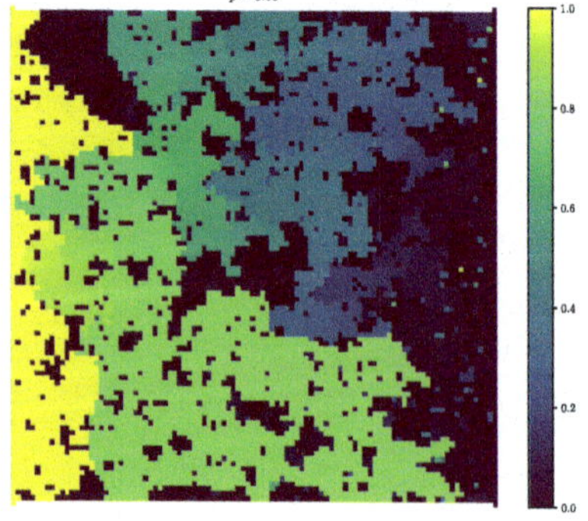

Fig. 10.19 Plot of $V(x, y)$ for a 100×100 simulation with $p = 0.50$. Current from high potential on the left side to low potential on the right side

```
            # y+ direction
            G = gy[j,i]
            dV = dV + abs(G*(V[j,i]-V[j+1,i]))
            Ia[j,i] = dV
    return Ia
```

and the results are plotted using:

```
Ia = findabscurrent(V,gx,gy)
plt.figure(figsize=(6,6))
ax1 = plt.subplot(1,1,1)
plt.imshow(Ia.T)
plt.gca().invert_yaxis()
ax1.set_aspect("equal", "box")
```

Notice that we either must transpose the resulting matrix before visualization, because Python is indexing a matrix `Ia[iy,ix]` with the y-coordinate-index as the first index. Alternatively, you can translate the system into physical coordinates and plot using `pcolormesh`. For a system from $(-a, -a)$ to (a, a), where a is a length with units meters, we can visualize the system using:

```
# Setup physical model
a = 1.0 # meters
Nx,Ny = Ia.shape
x_A, x_B, y_A, y_B = -a,a,-a,a
x = np.linspace(x_A,x_B,Nx)
y = np.linspace(y_A,y_B,Ny)
rx,ry = np.meshgrid(x,y,indexing="ij")
plt.figure(figsize=(8,8))
plt.pcolormesh(rx,ry,Ia)
plt.axis("equal"), plt.xlabel("$x/a$"), plt.ylabel("$y/a$")
```

Discussion of results. The resulting plot is shown in Fig. 10.20 for four different values of p. Here, we observe interesting properties of the system. First, we realize that there is a value for p below which there is no flow from one side to another. This value is called the percolation threshold for the system. For the system we have chosen for the random connections, which is called a bond percolation lattice (because it is the bonds, the connections between the elements, that are chosen at random), the percolation threshold for a square lattice is known to be for $p = 1/2$. If we choose p below $1/2$, there is typically no connecting path from one side to another, whereas if $p > 1/2$, there is a connecting path. Just at $p = 1/2$ the system behaves as a fractal, with a complex geometrical structure which is clear also from the way the current is distributed in the system in Fig. 10.20. You can read more about disordered systems and percolation theory in (1).

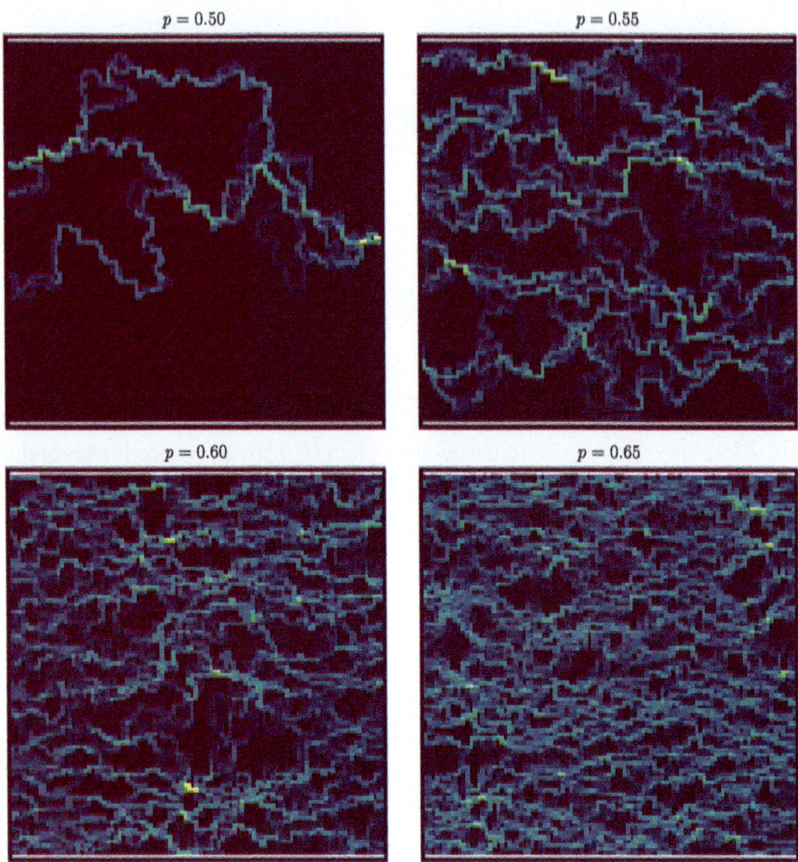

Fig. 10.20 Plot of $\sum_i |I_i|$ into each element for systems for various values of p. As p becomes larger, the current distribution becomes more homogeneous. For systems with $p < 1/2$ there are typically no connecting path from one side to another, and therefore no current

Relation to Laplace's Equation

If we set all the G-values in (10.46) to be the same and divide by G we get:

$$\left(V_{i,j} - V_{i+1,j}\right) + \left(V_{i,j} - V_{i-1,j}\right) + \left(V_{i,j} - V_{i,j+1}\right) + \left(V_{i,j} - V_{i,j-1}\right) = 0. \tag{10.52}$$

This is identical to the discrete version of Laplace's equation. How can we make sense of this? We know that Ohm's law on microscopic form is $\mathbf{J} = \sigma \mathbf{E}$ where $\mathbf{E} = -\nabla V$. In addition, we know that when there is no local build-up in charges, we have that $\nabla \cdot \mathbf{J} = 0$. This gives us that

$$\nabla \cdot \mathbf{J} = \nabla \cdot (-\sigma \nabla V) = -\sigma \nabla^2 V = 0. \tag{10.53}$$

This is indeed Laplace's equation.

Summary

The **electric field** point in the direction along a wire/conductor due to the immediate motion of charges to the surface of the conductor.

We analyze circuits using **Kirchhoff's laws for circuits**:

- **Kirchhoff's voltage law** states that the sum of voltage changes along a closed loop along a circuit is zero: $\sum_i \Delta V_i = 0$
- **Kirchhoff's current law** states that the sum of currents flowing into (or out of) a junction is zero: $\sum_i I_i = 0$.

The voltage drop or gain across various components are:

- Zero along a **wire**. The potential is the same along a wire
- $V = -IR$ for a **resistor** in the positive direction of the current
- $V = -(1/C)\,Q$ for a **capacitor** in the positive direction of the current
- $V = e$ for a battery.

We use circuits as models for complex electromagnetic systems, where we simplify the physical situation by combining the components in a circuit diagram and then analyze the circuit.

Exercises

Discussion Exercises

10.1 Current in a capacitor. Is the current on each side of a plate capacitor always the same? Explain why or why not.

10.2 Turning on the light. Why does an electric light bulb usually stop working when you turn the light on and not while it is on?

10.3 Lightning strike in a tree. A lightning strikes the top of a tree. How would you model this system as a circuit if you are interested in how the current passes through the air and the tree?

Tutorials

10.4 Current and eletric potential in a poor conductor. Figure 10.21 shows an non-ideal conductor shaped as a straight conductor (left) and a quarter of a cylindrical shell (right). There is a potential difference between the ends as shown in the figure.

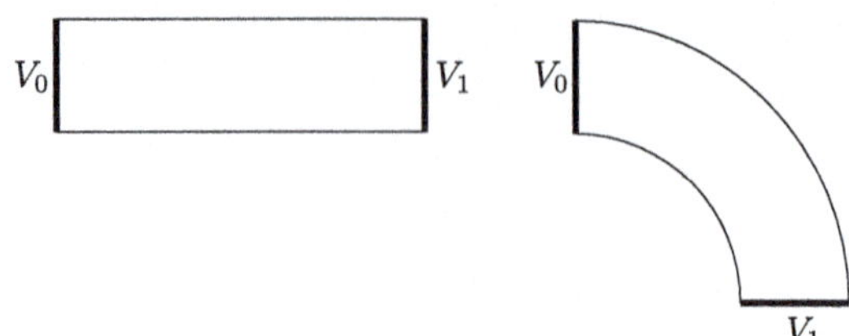

Fig. 10.21 A straight and a cylindrical conductor

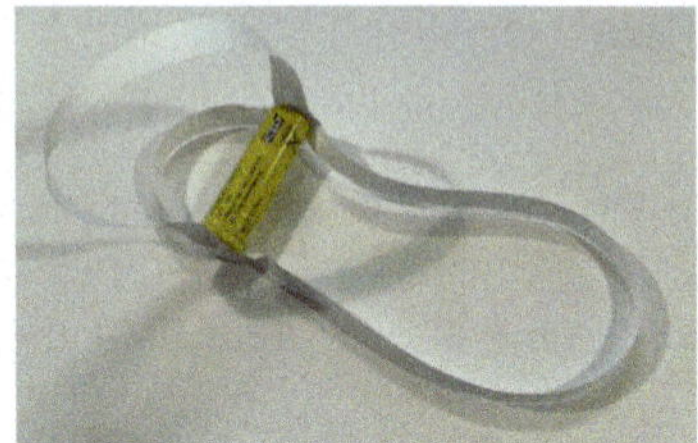

Fig. 10.22 Three ribbons connected to a battery

(a) Sketch the electric field inside the straight conductor.
(b) Sketch the charge distribution when there is a current in the straight conductor. (Hint: The charge distribution sets up the electric field.)
(c) We bend the straight conductor to the shape shown in the figure. What is now the shape of the electric field inside the conductor?
(d) What is now the charge distribution in the conductor?
(e) Does the electric field have to be zero outside the conductor?
(f) Is the electric field homogeneous inside the bent conductor? (Hint: Choose a circular path with constant radius from one end to the other. What is the integral of the electric field along this path? What does it tell you about the electric field?)
(g) How would you model this system as an electric circuit? Draw the circuit. (Hint: Change the system into a system with one of more circuit components connected with ideal conductors drawn as lines.)

10.5 Model system. You friend Q has made electrically conducting paper. He makes three ribbons that are 10cm, 20cm and 30cm long and connect them to a battery as shown in Fig. 10.22.
(a) Sketch this system as a circuit. (Hint: Remember that all points that are connected with lines have the same potential.)
(b) Q explains that his system is fine-tuned so that if one of the paper ribbons break, the current will increase in the other ribbons and the paper will start to burn. Is this explanation plausible?

10.6 Twin circuits. Figure 10.23 shows two circuits. Are the two circuits different or are they the same circuit drawn in two different ways? (Can you transform one into the other by stretching and moving the lines?)

10.7 Potential drop of resistor. Figure 10.24 shows a battery with a voltage U_0, which is connected to a cylindrical resistor R.

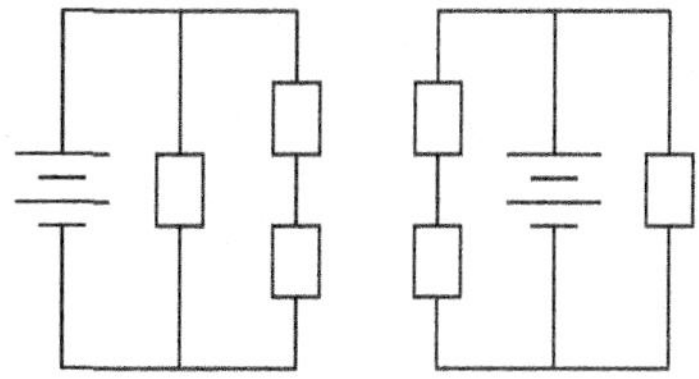

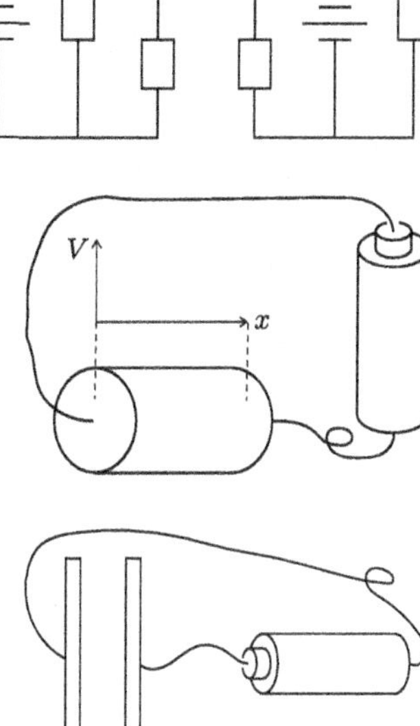

Fig. 10.23 Two circuits

Fig. 10.24 Battery and resistor

Fig. 10.25 A capacitor and a battery

(a) Sketch the electric field inside the resistor.
(b) Sketch the electric potential along the resistor.
(c) Draw a circuit diagram of this system.
(d) What is the voltage drop over the resistor along a path with a negative orientation. What have you assumed about the current when you gave your answer?
(e) Write down Kirchhoff's law for a path with a negative orientation (negative rotational direction from the right-hand rule).
(f) Write down Kirchhoff's law for a path with a positive orientation (positive rotational direction from the right-hand rule).
(g) What is the current in the resistor if the battery's potential is $U_0 = 1.5$ V and the resistance of the resistor is $R = 1$ k ohm.
(h) Assume that the current is directed in the other direction, that means that you change what you consider positive direction for the current. What is then the drop in potential across the resistor along a path that has a positive orientation (with respect to the right hand rule).
(i) Is assuming that the current goes in the opposite direction the same as turing the battery around?
(j) What is Kirchhoff's law for a path with a positive orientation?
(k) What is the current now?

10.8 Potential drop of capacitor. Figure 10.25 shows a capacitor consisting of two parallel plates of area $A = 1\ \text{cm}^2$ and spacing $d = 1$ mm. The capacitor is connected to a battery of $U_0 = 1.5$ V.

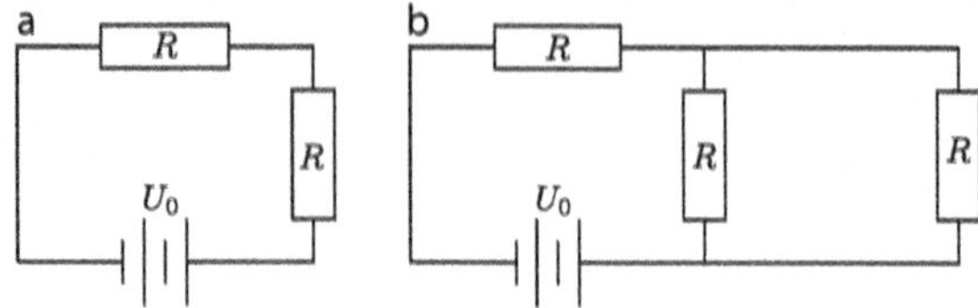

Fig. 10.26 **a** Circuit with two resistors and a battery. **b** Expanded circuit

(a) What is the potential on each side of the capacitor?
(b) Sketch the electric field in the capacitor.
(c) What is the charge on each side of the capacitor? Be careful in specifying on which side the charge is positive and negative.
(d) Draw a circuit diagram for this system.
(e) Assume that positive direction of the current is from the positive side of the battery, through the capacitor, and to negative side of the battery. What sign will the current have when you connect the discharged capacitor to the battery. Check that this is consistent with the sign of the charges you found above.

10.9 Kirchhoff's law. Figure 10.26 shows a circuit that contains a battery U_0 and two resistors R.
(a) What determins which way the current will flow in this circuit?
(b) Choose a positive direction for the current corresponing to the direction you think that the current will flow in. Write down Kirchhoff's law along a path in this direction.
(c) What is the current?
(d) Choose the opposite direction to be the positive direction for the current. Write down Kirchhoff's law for the same path with this choice of positive direction for the current.
(e) What is the current?
(f) We expand the circuit to include another resistor as shown in Fig. 10.26b. Define positive currents in the circuits. Where will you apply Kirchhoff's law of currents?
(g) Write down Kirchhoff's law of currents in one of the points you selected above.
(h) What is Kirchhoff's law of current in the other point?
(i) How many loops do you find in this circuit. Draw the loops into the figure.
(j) Write down Kirchhoff's law for all the loops you find.
(k) Find the currents.
(l) Redraw the circuit diagram in such as way that it is obvious that this is a resistance R connected in series with a resistor which consists of a parallel coupling of two resistors R.
(m) Did you use all the equations you found for Kirchhoff's voltage law and Kirchhoff's current law? Explain why not.

10.10 Equal currents. Your friend Q has made a circuit diagram (in Fig. 10.27) and to be sure, he has defined currents along all the lines. But maybe he has introduced too many symbols?
(a) What currents are the same as I_1?

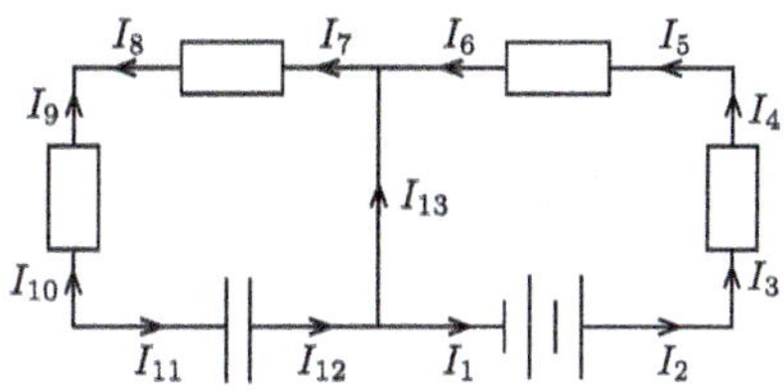

Fig. 10.27 Circuit diagram drawn by Q

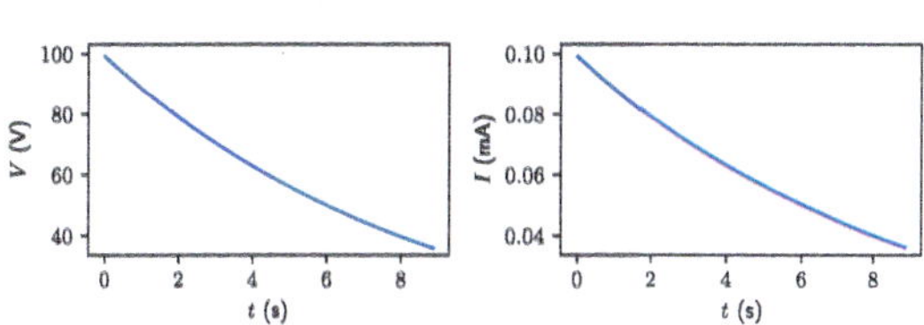

Fig. 10.28 Measured current and potential

(b) What currents are the same as I_{12}?
(c) What currents are the same as I_{13}?

10.11 Battery or capacitor. Your friend Q has invented a new battery. It is in a suitcase which is locked. By now you know Q and expect that the invention may have some downsides. Let us study the battery. Two wires are extending from the suitcase, one marked plus and one marked minus.
(a) You connect the wires to a resistor, an amperemeter and a voltmeter. Sketch how you would design this measurement circuit.
(b) How do you expect a battery to behave in what you can measure in the various devices?

When you use a resistor of $R = 100$ ohm the measurements show that both the potential and the current changes with time as shown in Fig. 10.28. Hmm. This is not what we expected. You ask Q what she used to construct this battery? "*Only two thin metal sheets of 1 square meter that I have folded ingeniously so that there is only 0.1mm between each fold. And then I have molded them into a material with a hundred times the permittivity of air. Genius, isn't it?*"
(c) We want to make a model for the current and the potential from the so called battery by assuming that it is a capacitor C. Draw the circuit.
(d) We assume that the plus wire corresponds to the positive side of the capacitor, where the charge is positive. Assume that positive direction for the current is from the positive side of the capacitor, through the resistor, and to the negative side of the capacitor. What is the relation between Q and I?
(e) We denote the potential across the capacitor as V. From $C = Q/V$ we see that the potential across the capacitor is $V = Q/C$. Use this and the expression you found for how I is related to Q of show that $\frac{dV}{dt} = -\frac{I}{C}$.
(f) Show that Kirchhoff's law gives that $I = V/R$ and insert this in the equation so that you get a differential equation for V.
(g) Show that the solution to this equation is $V(t) = V_0 e^{-t/\tau}$ where $\tau = RC$.
(h) What determines how fast this so-called battery will be discharged?

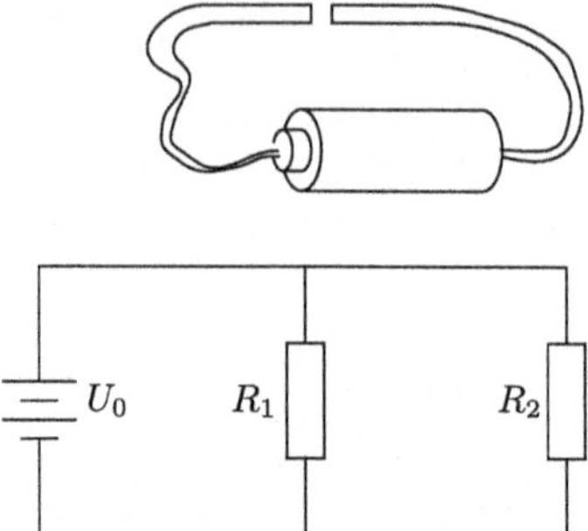

Fig. 10.29 Circuit with two cylindrical conductors, separated by a 1 mm layer of air

Fig. 10.30 Illustration of a simple circuit

(i) Does Q's statement about the system seem reasonable or do you suspect that she has included more components in the suitcase? (Hint: Check how long time the battery should use to discharge.)

10.12 Is the circuit closed. Your friend Q has made a circuit consisting of a battery, a switch, and two thick, cylindrical conductors with diameter 4mm as shown in Fig. 10.29. The conductors are not connected, but separated by a 1mm thin layer of air. Q turns the switch and shows you that for a short while there is a current through the circuit. *"Ha!"* says Q. *"This shows that all you have told me about current are wrong. A current may pass through air!"*
(a) What do you think happens in this system. Is is really correct that current passed through the air for a short moment?
(b) What happens if you gradually increase the voltage?

Homework

10.13 A simple circuit. Figure 10.30 illustrates a simple circuit driven by a battery with voltage U_0 and consisting of two resistors R_1 and R_2.
(a) Find the current I through the battery expressed in terms of U_0, R_1 and R_2.
(b) What is the power delivered by the battery?

10.14 From model to circuit. In this exercise we will address a system consisting of a battery of $e = 1.5$V, to very thin wires and a capacitor as illustrated in Fig. 10.31. The wires are both $L = 10$cm long, cylindrical wired with radius $a = 0.01$mm and are made of copper ($\rho = 1.68\ 10^{-8}$ ohm m). The capacitor consists of two parallel plates with area $A = (2\text{mm})^2$ at a distance $d = 1$mm and is filled with a material with $\epsilon = 6\epsilon_0$.
(a) How would you model this system as a circuit? Make a sketch of the circuit.
(b) You connect the battery at the time $t = 0$. Sketch the electric potential along the circuit immediately after you have connected the battery.
(c) What is the current in the circuit immediately after you have connected the battery?

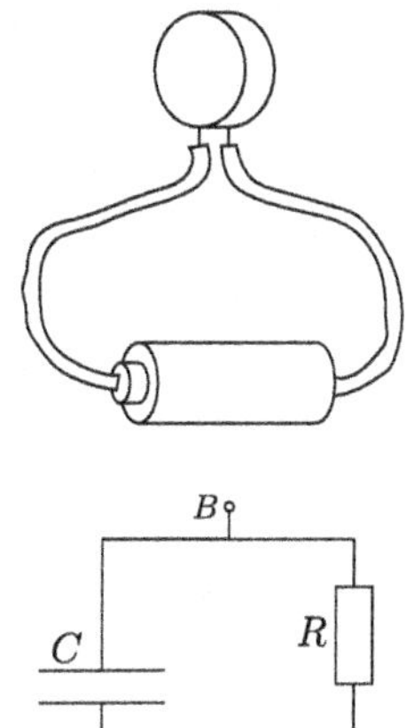

Fig. 10.31 A system consisting of a battery, two wires and a capacitor

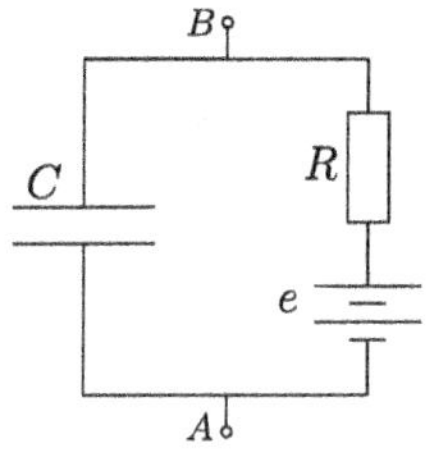

Fig. 10.32 A circuit model of a cell

(d) What is the voltage across the various components in the circuit after a very long time? And what is the current?
(e) Find a differential equation for the current in the circuit. What are the initial conditions for the equation? Solve the equation.

10.15 A circuit-model for a cell. We study a circuit as a model for a cell membrane. The circuit consists of a capacitor C, a resistor R, and a battery with an emf e connected as illustrated in Fig. 10.32. We measure the potential difference between points A and B, V_{AB}.
(a) At the time $t = 0$ the capacitor is without charge. What is the current I_R through the resistor at $t = 0$?
(b) What is the charge, Q_∞, at the capacitor when $t \to \infty$?
(c) Show that an equation to describe the charge, Q, on the capacitor can be written as

$$\frac{dQ}{dt} = \frac{1}{\tau}(Q_\infty - Q), \tag{10.54}$$

where $\tau = RC$.
(d) Write a program to find the voltage $V_{AB}(t)$ as a function of time for this system.

10.16 Wheatstone bridge. Figure 10.33 illustrates a Wheatstone bridge, which is used to measure resistance. In the circuit, R_3 is variable resistance, R_x is the unknown resistance, and R_1 and R_2 are given, known resistances. The voltmeter has infinite resistance.
(a) Find the voltage difference $V_a - V_b$ which is read from the voltmeter expressed in terms of the four resistances.

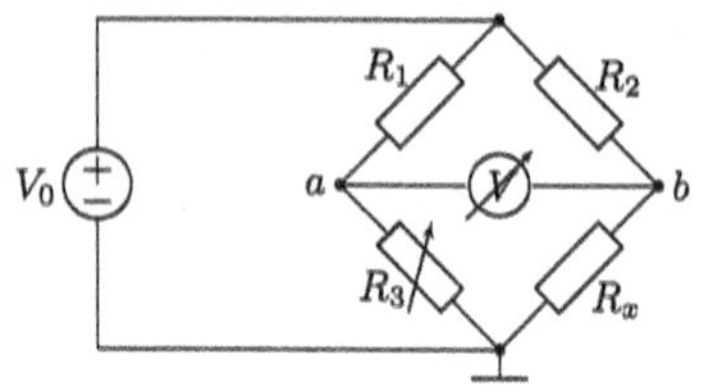

Fig. 10.33 Illustration of a Wheatstone bridge

(b) The variable resistance R_3 is adjusted until the measured voltage is zero. Find the unknown resistance R_x from the other resistances.

10.17 Lightning strike. In this problem we will develop a simple model for dielectric breakdown in a circuit. You may think of the system as a model for a lightning forming between clouds, or you may think of the model as an example of dielectric breakdown in a capacitor. We start from the system illustrated in Fig. 10.34a consisting of a battery with voltage V_0, a capacitor C and a resistor R_0.
(a) The circuit is open and has been open for a long time. Then at $t = 0$ the circuit becomes closed. What is the charge $Q(0)$ on the capacitor and the potential fall $V(0)$ over the capacitor immediately after the circuit is closed (at $t = 0$)?
(b) What is the potential fall V over the capacitor and the charge Q on the capacitor when there is no current flowing in the circuit? (A very long time after the battery has been connected to the circuit).
(c) Find the potential drop over the capacitor, $V(t)$.

We will model the dielectric breakdown of the capacitor by introducing a voltage-dependent resistor, R, in parallel to the capacitor as illustrated in Fig. 10.34b. The resistance depends on the voltage V across the resistor:

$$R(V) = \begin{cases} \infty & V < V_0/2 \\ x R_0 & V \geq V_0/2 \end{cases} . \tag{10.55}$$

where x is a small number, such as $1/1000$.
(d) Write a Python program to model the behavior of the system and plot the $V(t)$, the current $I_0(t)$ through resistor R_0 and the current $I(t)$ through the resistor R. Model the behavior from $t = 0$ when the circuit is closed.

10.18 Kirchhoff's circuit laws. In this exercise we are going to apply Kirchhoff's laws and solve problems using linear algebra.

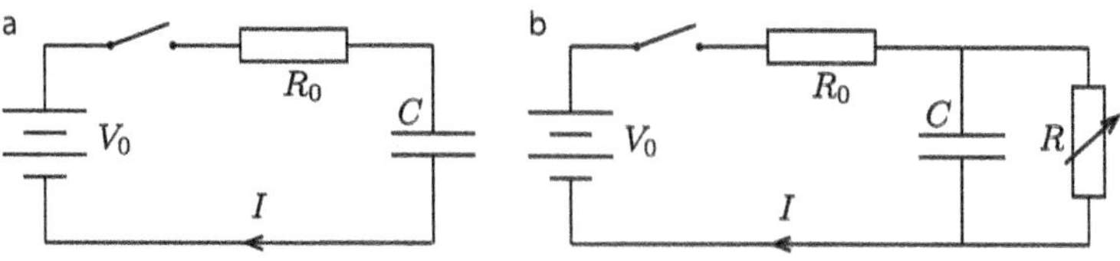

Fig. 10.34 Illustration of the RC circuit

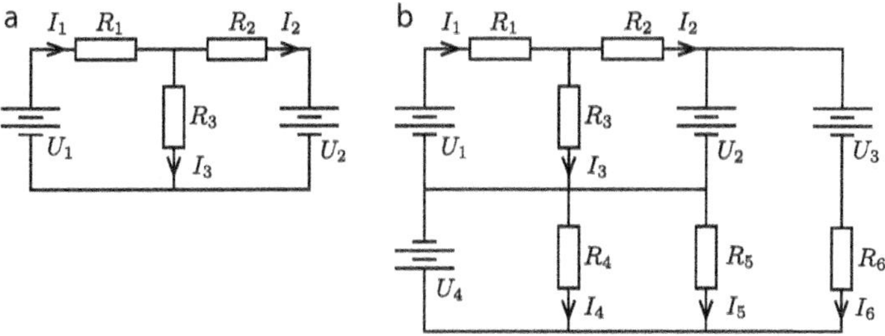

Fig. 10.35 Two complex circuits

(a) Figure 10.35a shows a circuit where $V_1 = 10\text{V}$, $V_2 = 5\text{V}$, $R_1 = 10\Omega$, $R_2 = 20\Omega$ and $R_3 = 30\Omega$. Using Kirchhoff's laws, find the equations needed to solve for the currents. Write it as a matrix equation and use Python to inverse the matrix and find the solution.

(b) Use the same approach for the circuit in Fig. 10.35b, where $R_4 = 40\Omega$, $R_5 = 50\Omega$, $R_6 = 60\Omega$, $V_3 = 12\text{V}$ and $V_4 = 24\text{V}$.

10.19 Circuit model of conducting sphere. A conducting sphere consists of an inner conducting sphere of radius a, and outer spherical shell with inner diameter b. (You may think of this as a simple model for a for a cell.) In the region between a and b, the material has a dielectric constant ϵ and a conductivity σ. A battery with a constant voltage difference V_0 is connected to the inner sphere and outer shell with a wire with resistance r.

(a) Make a drawing of the system.

(b) What is the resistance R of the sphere? (You may look up the result.)

(c) What is the capacitance C of the sphere? (You may look up the result.)

(d) Make a circuit model of the system. What components do you need? Should you add the components in series or in parallel? Provide an argument for you choice.

(e) Now, we have created a circuit model of the system. Let us argue qualitative how this model will behave. What is the current I through the battery immediately after you connect it?

(f) What is the current I through the battery after a very long time?

(g) Sketch the time development of the current if you connect the battery to the system at $t = 0$.

(h) We can also find the equations for this circuit and solve them. We introduce I as the current through the battery, I_R as the current through the resistance R and $I_C = \mathrm{d}Q_C/\mathrm{d}t$ as the current through the capacitor. How can you use Kirchhoff's current law in this circuit?

(i) Use Kirchhoff's voltage law on a circuit through the resistor R and the capacitor C. Show that if you combine this with Kirchhoff's current law you get:

$$R\frac{\mathrm{d}I_R}{\mathrm{d}t} = \frac{1}{C}(I - I_R)\,. \tag{10.56}$$

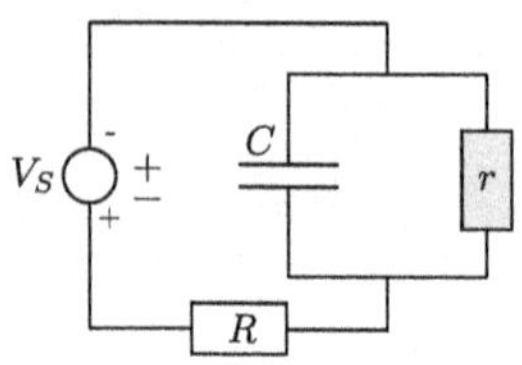

Fig. 10.36 A circuit model for an element in a cell membrane

(j) Now use Kirchhoff's voltage law for a path through the battery and both resistors to show that the differential equation for I_R is

$$RC\frac{dI_R}{dt} = \frac{V_0}{r} - \left(1 + \frac{R}{r}\right) I_R. \tag{10.57}$$

(k) Show that you can combine the two equations from the last to exercises to find an equation for I_R: $RC\frac{dI_R}{dt} = \frac{V_0}{r} - \left(1 + \frac{R}{r}\right) I_R$.
(l) We will not solve this equation here, but you can check that you get the expected limit in when $t \to \infty$.

Modeling Projects

10.20 Voltage gated ion channels. Figure 10.36 shows a simple model for a small element in a cell membrane. The resistor r represents an ion channel, but this channel is special: When the potential drop over the resistor is less than v, then the resistance is infinite, but when the voltage drop is larger than v, the resistance is r:

$$r = \begin{cases} r & , V_r > v \\ \infty & , V_r \le v \end{cases},$$

where V_r is the voltage drop across the resistor.
(a) Assume that the voltage V_s has been zero for a long time, so that the system is stationary. What is the voltage V_C over the capacitor and the current through the resistor R immediately after the voltage V_s is set to $V_s > 0$?
(b) First, assume that the voltage V_r over the resistor r is less than v. Show that the equation for the voltage V_C over the capacitor can be written as:

$$C\frac{dV_c}{dt} = \frac{V_s}{R} - \frac{V_C}{R}. \tag{10.58}$$

(c) Assume that the voltage V_r over the resistor r is larger than v. What is now the equation for the voltage V_C?
(d) Write a Python program to calculate the voltage V_C over the capacitor given the voltage source $V_s(t)$. You may assume that the voltage $V_C(0)$ has the value from part (a).

Reference

Percolation Theory Using Python, *Anders Malthe-Sorenssen*, **2024**, Springer

Chapter 11
Magnetic Field

11.1 The Magnetic Field

While you may have some intuition about electric fields, our intuition about magnetic fields is usually connected to either the Earth's magnetic field, which we and many animals use to navigate, or to permanent magnets as illustrated in Fig. 11.1. However, to build an understanding of permanent magnets, we first need to build an intuition about magnetic fields and how they arise due to currents. Then we will be ready to see how permanent magnets are due to permanent microscopic currents in a similar way to how polarized materials are due to small charge displacements.

While electrostatics is the theory of static charges, we will now address the behavior of stationary currents. We will see that there is an experimentally based law describing the interactions between currents and a moving charge, Biot-Savart's law, just as there was an experimentally based law for the interaction between one charge and a static charge, Coulomb's law.

From Coulomb's law to the magnetic force law

We had great success with introducing the electric field in electrostatics. Let us recall how we proceeded to build a theory for electrostatics. We started from the experimental observation, Coulomb's law, which states that the force from a charge Q_1 on a charge Q_2 is given as

$$\mathbf{F}_{\text{on 2 from 1}} = Q_2 \frac{Q_1}{4\pi\epsilon_0} \frac{\hat{\mathbf{R}}}{R^2} , \tag{11.1}$$

where the vector $\mathbf{R}$ is the vector from 1 to 2: $\mathbf{R} = \mathbf{r}_2 - \mathbf{r}_1$, where the position of charge Q_1 is $\mathbf{r}_1$ and the position of charge Q_2 is $\mathbf{r}_2$.

© The Author(s), under exclusive license to Springer Nature Switzerland AG 2026

A. Malthe-Sørenssen, *Elementary Electromagnetism Using Python*, Undergraduate Texts in Physics, https://doi.org/10.1007/978-3-032-19876-1_11

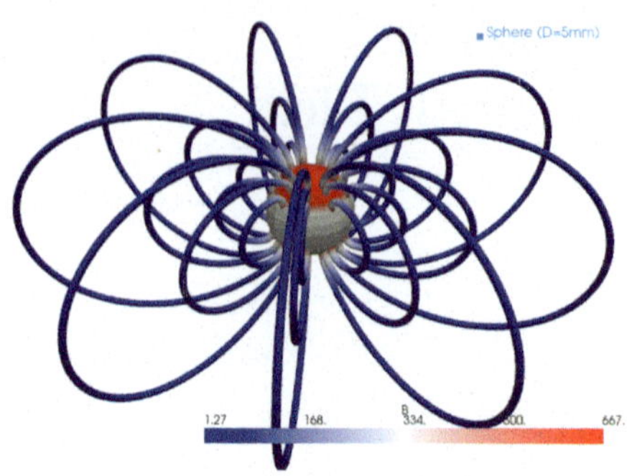

Fig. 11.1 Illustration of the magnetic field due to internal currents in a permanent magnet

Electric field. We realized that it was useful to introduce the electric field, **E**, from charge Q_1:

$$\mathbf{E} = \frac{Q_1}{4\pi\epsilon_0}\frac{\hat{\mathbf{R}}}{R^2} , \tag{11.2}$$

and then find the force on a charge Q_2 from

$$\mathbf{F}_{\text{on 2 from 1}} = Q_2\mathbf{E} , \tag{11.3}$$

where we understand that the electric field varies in space: $\mathbf{E} = \mathbf{E}(\mathbf{r})$. We also found that the superposition principle for forces gave us a superposition principle for the electric field: We found the electric field set up by many charges by summing or integrating the electric fields from each of charge:

$$\mathbf{E} = \sum_i \frac{Q_i}{4\pi\epsilon_0}\frac{\hat{\mathbf{R}}_i}{R_i^2} = \int_v \frac{\rho}{4\pi\epsilon_0}\frac{\hat{\mathbf{R}}}{R^2}\,\mathrm{d}v , . \tag{11.4}$$

This allowed us to divide the problem of finding the forces between two charges into two parts:

- first find the electric field set up by a distribution of charges
- then find the force acting on a specific charge at a specific point

We then went on to develop effective methods to calculate and discuss electric fields.

Forces between moving charges and electric currents. We will follow a similar method when introducing the magnetic field. We will first describe an experimental law for the interaction between *moving* charges and electric currents and use this to introduce the concept of a magnetic field created by electric currents and then address the force on a moving charge from a magnetic field.

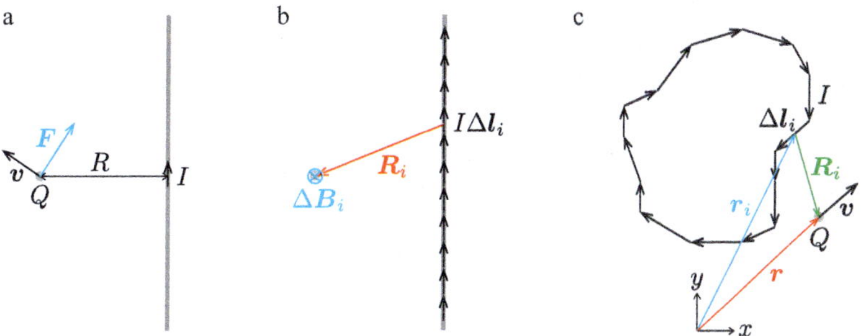

Fig. 11.2 **a** Illustration of the force on a moving charge from a very long wire carrying a stationary current I, **b** which is due to a sum of contributions from current elements $I\Delta\mathbf{l}_i$. **c** Illustration of a general current loop C consisting of current elements $I\Delta\mathbf{l}_i$ that combined forms a magnetic field $\mathbf{B}$ at the point of an observation charge Q

Figure 11.2a illustrates a system inspired by the original experiments of Biot and Savart in 1820. They found that the force on a moving charge close to a long wire with a current I is proportional to the current I, the charge Q, the velocity v, and decays as $1/R$, where R is the distance to the wire. Hmmm. This reminds us of the behavior of a long line charge, where we know that the electric field decays as $1/R$. We also recall that the $1/R$ dependence arose because we summed over all the charges along a long wire. Maybe something similar is at play for the force on a moving charge, but depending on the current instead of the static charge?

Indeed, detailed experimental studies have demonstrated that the force on a charge Q moving with a velocity $\mathbf{v}$ can be written as the sum of contributions from small *current elements* $I\Delta\mathbf{l}_i$:

$$\mathbf{F} = Q\mathbf{v} \times \left(\sum_i \frac{\mu_0}{4\pi} \frac{I\Delta\mathbf{l}_i \times \mathbf{R}_i}{R_i^3} \right) , \tag{11.5}$$

where the vector $\mathbf{R}_i$ is the vector from the current element $I\Delta\mathbf{l}_i$ to the moving charge Q as illustrated in Fig. 11.2b. If the particle with charge Q and velocity $\mathbf{v}$ is in position $\mathbf{r}$, and wire element $\Delta\mathbf{l}_i$ is at $\mathbf{r}_i$, then $\mathbf{R}_i = \mathbf{r} - \mathbf{r}_i$. The concept of the $\mathbf{R}$-vector is therefore the same as for charge distributions.

In a *stationary situation*, currents must go in loops. The set of current elements must therefore form a *closed loop*. However, it is still often useful to decompose a closed loop into discrete or infinitesimally small current elements and sum up their contributions but remember that this is a mathematical construct. Individual current elements cannot exists in a stationary situation. (However, it is possible for the current elements not to form a loop, but to go from infinity and to infinity, such as for an infinitely long wire.)

Introducing the magnetic field. Just like we saw for the electric field, we realize that it is conceptually useful to simplify this experimental expression by introducing a magnetic field, $\mathbf{B}(\mathbf{r})$ so that:

$$\mathbf{F} = Q\mathbf{v} \times \underbrace{\frac{\mu_0}{4\pi} \sum_i \frac{I \Delta \mathbf{l}_i \times \mathbf{R}_i}{R_i^3}}_{=\mathbf{B}} = Q\mathbf{v} \times \mathbf{B} , \tag{11.6}$$

where

$$\mathbf{B} = \frac{\mu_0}{4\pi} \sum_i \frac{I \Delta \mathbf{l}_i \times \mathbf{R}_i}{R_i^3} . \tag{11.7}$$

The magnetic field $\mathbf{B}$ is a result of the current elements. Indeed, we consider the magnetic field to be generated by the current. This formulation of the magnetic field is called Biot-Savart's law and is a fundamental law on the same level as Coulomb's law.

Biot-Savart's law and the force on a moving charge

The magnetic field $\mathbf{B}(\mathbf{r})$ is a vector field generated by loops of discrete ($I \Delta \mathbf{l}_i$) or infinitesimal ($I d\mathbf{l}$) current elements and its value is given by Biot-Savart's law:

Biot-Savart's law on discrete and line-integral form

The magnetic field $\mathbf{B}(\mathbf{r})$ in a point $\mathbf{r}$ due to a current I in a closed loop C is

$$\mathbf{B} = \sum_i \frac{\mu_0}{4\pi} \frac{I \Delta \mathbf{l} \times \mathbf{R}_i}{R_i^3} = \sum_i \frac{\mu_0}{4\pi} \frac{I \Delta \mathbf{l} \times \hat{\mathbf{R}}_i}{R_i^2} \quad \text{(discrete form)} . \tag{11.8}$$

for a set of discrete current elements $I \Delta \mathbf{l}_i$ at positions $\mathbf{r}_i$, where $\mathbf{R}_i = \mathbf{r} - \mathbf{r}_i$, or

$$\mathbf{B} = \oint_C \frac{\mu_0}{4\pi} \frac{I d\mathbf{l} \times \mathbf{R}}{R^3} = \oint_C \frac{\mu_0}{4\pi} \frac{I d\mathbf{l} \times \hat{\mathbf{R}}}{R^2} \quad \text{(integral form)} . \tag{11.9}$$

for a continuous curve of current elements $I d\mathbf{l}$ at positions $\mathbf{r}'$ along the curve, where $\mathbf{R} = \mathbf{r} - \mathbf{r}'$ and $\hat{\mathbf{R}} = \mathbf{R}/R$.

For clarity we can write this integral explicitly in terms of a parameterized curve $\mathbf{r}'(s)$ describing the closed loop C:

$$\mathbf{B} = \oint_C \frac{\mu_0}{4\pi} \frac{I\mathrm{d}\mathbf{l} \times \mathbf{R}}{R^3} = \oint_C \frac{\mu_0}{4\pi} \frac{I\frac{\mathrm{d}\mathbf{r}'(s)}{\mathrm{d}s} \times (\mathbf{r} - \mathbf{r}'(s))}{|\mathbf{r} - \mathbf{r}'(s)|^3} \, \mathrm{d}s \, . \tag{11.10}$$

A common mistake is not to realize that $\mathbf{R}$ varies along the curve. By now, you should be used to perform similar integrals over the charge density to find the electric field.

Physical constants in Biot-Savart's law. We see that Biot-Savart's law has similarities to Coulomb's law for the electric field. The constant μ_0 is called the *permeability of vacuum* and is given as

$$\mu_0 = 4\pi \cdot 10^{-7} \mathrm{Ns}^2/\mathrm{C}^2 = \mathrm{H/m} \tag{11.11}$$

where H, Henry, is the unit for inductance. The magnetic field is measured in units of Tesla, $\mathrm{T} = \mathrm{N/Am}$. A magnetic field of 1 T is a huge field. For comparison, the magnitude of the Earth's magnetic field on the surface of the Earth is about 50μT.

The current element $I\mathrm{d}\mathbf{l}$. We will think of the element $I\mathrm{d}\mathbf{l}$ in the integral and the element $I\Delta\mathbf{l}_i$ in the sum as a *current element*. Mathematically, we see that the magnetic field is the sum of the contributions to the magnetic fields from each of the current elements:

$$\mathbf{B} = \oint_C \mathrm{d}\mathbf{B} = \sum_i \Delta\mathbf{B}_i \, , \tag{11.12}$$

where the magnetic field field from a current element is:

Biot-Savart's law for a current element

The contribution $\mathrm{d}\mathbf{B}$ to the magnetic field at $\mathbf{r}$ from a current element $I\mathrm{d}\mathbf{l}$ at $\mathbf{r}'$ is

$$\mathrm{d}\mathbf{B} = \frac{\mu_0}{4\pi} \frac{I\,\mathrm{d}\mathbf{l} \times \hat{\mathbf{R}}}{R^2} = \frac{\mu_0}{4\pi} \frac{I\,\mathrm{d}\mathbf{l} \times \mathbf{R}}{R^3} \, , \tag{11.13}$$

where $\mathbf{R}$ is the vector from the current element $I\,\mathrm{d}\mathbf{l}$ at $\mathbf{r}'$ to $\mathbf{r}$: $\mathbf{R} = \mathbf{r} - \mathbf{r}'$.

We will often solve problems by first finding the contribution for a current element and then integrating/summing the contributions from all current elements. However, this is a mathematical way to think about the problem, because a stationary current must always form closed loops.

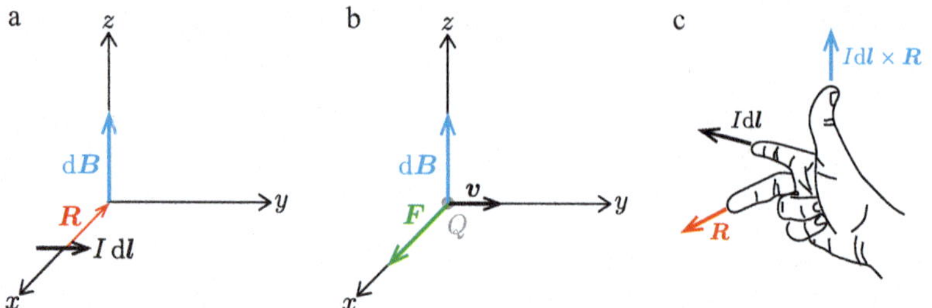

Fig. 11.3 **a** Illustration of the contribution to the magnetic field from a current element. **b** Illustration of the force on a moving charge due to the magnetic field along the z-axis. **c** Illustration of the right-hand rule

Test your understanding

A current element $I\Delta\mathbf{l}$ is placed in the origin and directed as shown in the figure. In what direction does the magnetic field $\mathbf{dB}$ from the current element point in each of the points 1–8?[1]

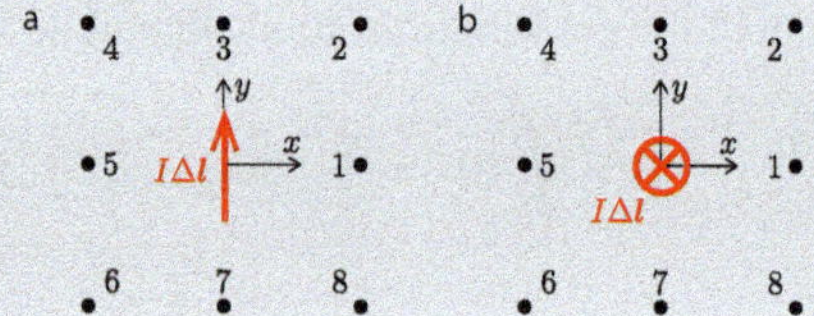

Example: Magnetic field from a current element

Figure 11.3 *illustrates a charge Q moving with a velocity $\mathbf{v} = v\,\hat{\mathbf{y}}$ close to a current element $I\,\mathbf{dl} = I\,\mathrm{d}l\,\hat{\mathbf{y}}$. What is the contribution to the magnetic field at the position of charge Q from the current element and what is the contribution to the force on the charge Q from the current element?*

Approach. We plan to use Biot-Savart's law to find the contribution from the current element by first finding the **R**-vector and then applying Biot-Savart's law directly. Finally, we will find the force on the charge from $Q\mathbf{v} \times \mathbf{B}$.

Solution. We can find the **R**-vector directly from the figure or by vector arithmetic. The **R**-vector points from the current element to the observation point, that is the posi-

[1] (a) 1, 2, 8: $-\hat{\mathbf{z}}$, 3, 7: $\mathbf{0}$, 4, 5, 6: $\hat{\mathbf{z}}$; (b) 1: $-\hat{\mathbf{y}}$, 2: $\hat{\mathbf{x}} - \hat{\mathbf{y}}$, 3: $\hat{\mathbf{x}}$, 4: $\hat{\mathbf{x}} + \hat{\mathbf{y}}$, 5: $\hat{\mathbf{y}}$, 6: $-\hat{\mathbf{x}} + \hat{\mathbf{y}}$, 7: $-\hat{\mathbf{x}}$, 8: $-\hat{\mathbf{x}} - \hat{\mathbf{y}}$.

tion of the charge Q. From the figures we see directly that $\mathbf{R} = -R\hat{\mathbf{x}}$. Alternatively, we can use that $\mathbf{r}' = R\hat{\mathbf{x}}$ and $\mathbf{r} = \mathbf{0}$ so that $\mathbf{R} = \mathbf{r} - \mathbf{r}' = -R\hat{\mathbf{x}}$.

Applying Biot-Savart's law. Second, we use Biot-Savart's law to find the magnetic field

$$\mathrm{d}\mathbf{B} = \frac{\mu_0}{4\pi} \frac{I\,\mathrm{d}\mathbf{l} \times \mathbf{R}}{R^3} \,. \tag{11.14}$$

We therefore need to find the cross product. We recall that to find the cross product we need to apply the *right-hand-rule* as illustrated in Fig. 11.3c. We place the index finger along the first vector, that is along $I\mathrm{d}\mathbf{l}$. Then we orient the hand so that the second vector, $\mathbf{R}$, points along the middle finger. The thumb will then point in the direction of the cross-product, $I\,\mathrm{d}\mathbf{l} \times \mathbf{R}$. The cross product is *normal* (orthogonal) to the two vectors in the cross-product and the right-hand-rule gives the direction of the normal. We see from the figure that the vector points in the positive z-direction, that is, that $\mathrm{d}\mathbf{B} = \mathrm{d}B_z\hat{\mathbf{z}}$:

$$\mathrm{d}\mathbf{B} = \frac{\mu_0}{4\pi} \frac{I\,\mathrm{d}l\,\hat{\mathbf{z}}}{R^2} \,. \tag{11.15}$$

Finding the force. Now, we know the direction and magnitude of the magnetic field at the charge Q. We find the force on the charged particle from $\mathrm{d}\mathbf{F} = Q\mathbf{v} \times \mathrm{d}\mathbf{B}$. Again, we apply the cross product using the right-hand rule. We place the index finger in the y-direction and orient our right hand so that the other three fingers points in the direction of the second vector, that is $\mathrm{d}\mathbf{B}$, which points in the positive z-direction. Our right thumb will then point toward the right, showing that the force is toward the current element:

$$\mathrm{d}\mathbf{F} = Q\mathbf{v} \times \mathrm{d}\mathbf{B} = \frac{\mu_0}{4\pi} \frac{Qv\,I\,\mathrm{d}l\,\hat{\mathbf{x}}}{R^2} \,. \tag{11.16}$$

Superposition principle for magnetic fields

The superposition principle for forces allows us to find the force on a charge $Q\mathbf{v}$ from a set of current loops simply by adding the contributions from each loop:

$$\mathbf{F} = \sum_i \mathbf{F}_i = \sum_i Q\mathbf{v} \times \mathbf{B}_i(\mathbf{r}) = Q\mathbf{v} \times \sum_i \mathbf{B}_i(\mathbf{r}) \,. \tag{11.17}$$

This shows that we can use the superposition principle for magnetic fields just as we did for electric fields: We can add the fields set up by different current elements and current loops to find the total magnetic field.

Superposition principle for magnetic fields

The total magnetic field in a point **r** is the sum of the magnetic fields $\mathbf{B}_i$:

$$\mathbf{B} = \sum_i \mathbf{B}_i \ . \quad (11.18)$$

Transfer of intuition from electric field to magnetic field

We can transfer much of the intuition and methods we have built up for electric fields to magnetic fields:

- The magnetic field varies in space because: $\mathbf{B} = \mathbf{B}(\mathbf{r})$.
- We can use the superposition principle to find the magnetic field from many current elements and current loops.
- We can calculate and study the magnetic field itself, just as we did for the electric field.
- We must always keep track of the **R**-vector in the expression for the magnetic field, just as we did with the **R**-vector for electric fields.

Test your understanding

The figure shows a wire with a current I. (a) What direction is the contribution $d\mathbf{B}(P)$ to the magnetic field in the point P from the line segment $d\mathbf{l}$? (b) What direction do you expect $\mathbf{B}$ to have in the point P?[2]

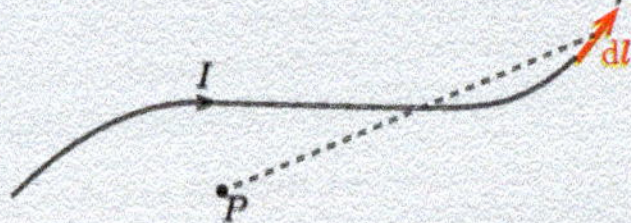

Biot-Savart's law for line, surface and volume currents

We have so far considered currents that flow along a closed loop. However, current distributions may be inhomogeneous in space yet stationary. For example, the current may be described by the volume current density $\mathbf{J}$ or a surface current density $\mathbf{J}_S$.

[2] (a) Up from the plane; (b) Down into the plane.

Biot-Savart's law for a volume current density. In the case of a volume current density $\mathbf{J}$, we can divide space into small current elements in the form of volumes $\mathrm{d}v$ with a surface $\mathrm{d}\mathbf{S}$ and a length $\mathrm{d}\mathbf{l}$. The current through this surface is $I = \mathbf{J} \cdot \mathrm{d}\mathbf{S}$ and the current element is therefore

$$I\,\mathrm{d}\mathbf{l} = \mathbf{J} \cdot \mathrm{d}\mathbf{S}\,\mathrm{d}\mathbf{l}\,. \tag{11.19}$$

If we choose the surface $\mathrm{d}\mathbf{S}$ so that it is normal to $\mathbf{J}$ and $\mathrm{d}\mathbf{l}$, so that it points in the direction of $\mathbf{J}$, we can rewrite this as

$$I\,\mathrm{d}\mathbf{l} = J\,\mathrm{d}S\,\mathrm{d}\mathbf{l} = \mathbf{J}\,\mathrm{d}S\,\mathrm{d}l = \mathbf{J}\,\mathrm{d}v\,. \tag{11.20}$$

The current element is therefore $I\,\mathrm{d}\mathbf{l} = \mathbf{J}\mathrm{d}v$, and Biot-Savart's law for the current element is

$$\mathrm{d}\mathbf{B} = \frac{\mu_0}{4\pi}\frac{\mathbf{J}\,\mathrm{d}v \times \hat{\mathbf{R}}}{R^2}\,. \tag{11.21}$$

Biot-Savart's law for a current density The magnetic field from a current density $\mathbf{J}$ in a volume v is

$$\mathbf{B} = \int_v \mathrm{d}\mathbf{B} = \int_v \frac{\mu_0}{4\pi}\frac{\mathbf{J}\,\mathrm{d}v' \times \hat{\mathbf{R}}}{R^2}\,. \tag{11.22}$$

We can rewrite this with explicit coordinates as

$$\mathbf{B} = \int_v \frac{\mu_0}{4\pi}\frac{\mathbf{J}(\mathbf{r}') \times (\mathbf{r} - \mathbf{r}')}{|\mathbf{r} - \mathbf{r}'|^3}\,\mathrm{d}v'\,, \tag{11.23}$$

where the integral is over the coordinates $\mathbf{r}' = (x', y', z')$ which we write as $\mathrm{d}v' = \mathrm{d}x'\,\mathrm{d}y'\,\mathrm{d}z'$.

Biot-Savart's law for a surface current density. Similarly, for a surface current density $\mathbf{J}_S$ as illustrated in in Fig. 11.4 a current element $I\,\mathrm{d}\mathbf{l}$ becomes $J_S\,\mathrm{d}S\,\mathrm{d}\mathbf{l} = \mathbf{J}_S\,\mathrm{d}x\,\mathrm{d}l = \mathbf{J}_S\,\mathrm{d}S$. The total field is then a surface integral:

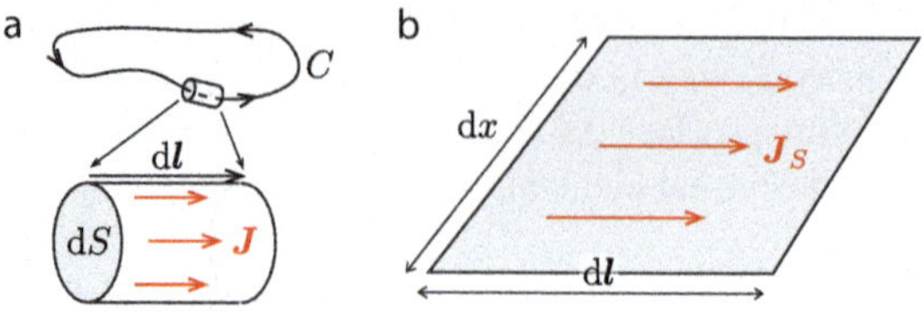

Fig. 11.4 **a** Illustration of a small line element **dl** from a cable along a curve C. **b** Illustration of a small surface element dS

$$\mathbf{B} = \frac{\mu_0}{4\pi} \int_S \frac{\mathbf{J}_S \, \mathrm{d}S \times \hat{\mathbf{R}}}{R^2} . \tag{11.24}$$

This form of Biot-Savart's law is not used that often in practice, but we may use it for theoretical derivations later.

Example: Magnetic field from a circular current

Find the magnetic field from a circular cable with radius a in the xy-plane with the center at the origin. First, find an exact solution along the z-axis and then a numerical solution everywhere in space.

Magnetic field along the z-axis. First, we find the magnetic field along the z-axis. We divide the circle into small elements, find the contribution from a single element, and then integrate over all the elements to find the total field.

We divide the circle into small elements **dl** as illustrated in Fig. 11.5. The contribution from an element **dl** to the magnetic field is:

$$d\mathbf{B} = \frac{\mu_0}{4\pi} \frac{I \, \mathbf{dl} \times \hat{\mathbf{R}}}{R^2} , \tag{11.25}$$

where the **R**-vector is from the element **dl** to the observation point **r**, as illustrated in the figure.

Using symmetry to simplify the integral. We use the symmetry of the system to simplify the integral. From the figure we notice that there are two contributions to the **B**-field. One contribution $\mathrm{d}B_z$ along the z-axis and one contribution $\mathrm{d}B_r$ normal to the z-axis. However, for each element **dl**, there will also be another element **dl**$'$ on the opposite side of the circle. Both these elements will contribute with the same $\mathrm{d}B_z$, but their contributions $\mathrm{d}B_r$ will be of the same size and directed in the opposite directions. Thus, the $\mathrm{d}B_r$ components will cancel each other. Therefore, we only need to sum up the contributions $\mathrm{d}B_z$, and they are the same for all elements **dl**:

$$\mathrm{d}B_z = \frac{\mu_0}{4\pi} \frac{I \, \mathrm{d}l \cos\beta}{R^2} , \tag{11.26}$$

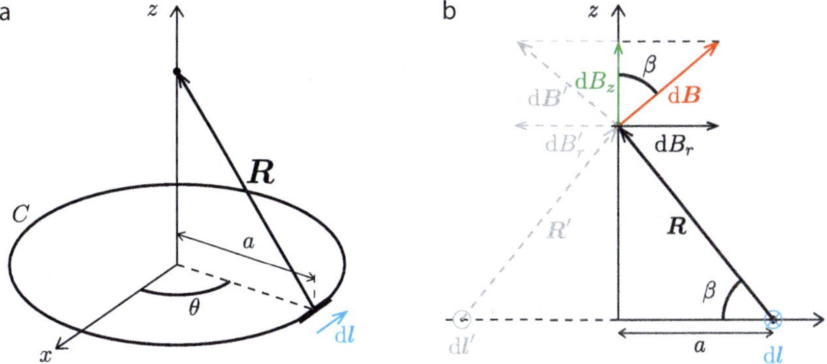

Fig. 11.5 **a** Illustration of a small line element **dl** from a cable along a curve C. **b** Illustration of a cross-section through the circle to show the geometry

where β is the angle between the z axis and $\mathbf{dB}$. We see from the figure that $\cos\beta = a/R$. The integral is therefore simplified to:

$$B_z = \oint_C \frac{\mu_0}{4\pi} \frac{I\,\mathrm{d}l \cos\beta}{R^2} = \frac{\mu_0 I \cos\beta}{4\pi R^2} \oint_C \mathrm{d}l = \frac{\mu_0 I (a/R)}{4\pi R^2} 2\pi a\ , \tag{11.27}$$

where we have used that the length of the curve C, that is the circle, is $2\pi a$. We now use that $R^2 = z^2 + a^2$, getting

$$B_z = \frac{\mu_0 I (a/R)}{4\pi R^2} 2\pi a = \frac{\mu_0 I a^2}{2R^3} = \frac{\mu_0 I a^2}{2\left(z^2 + a^2\right)^{3/2}}\ . \tag{11.28}$$

Using vector algebra to determine the integral. Instead of using a geometric argument, we can use vector algebra to find the magnetic field. To do this, we introduce a vector representation of the curve integral. We start by representing the two vector quantities, $\mathbf{R}$ and $\mathbf{dl}$ in a Cartesian coordinate system. We look at an element $\mathbf{dl}$ which is at an angle θ with the x-axis. This element is at a position $\mathbf{r}' = (a\cos\theta, a\sin\theta, 0)$, and the element points in a direction which is normal to this (which we can find by $\hat{\mathbf{z}} \times \mathbf{r}'$): $\mathbf{dl} = \mathrm{d}l(-\sin\theta, \cos\theta, 0)$. We then find $\mathbf{R} = \mathbf{r} - \mathbf{r}'$ for a point $\mathbf{r} = (0, 0, z)$, that is, $\mathbf{R} = (-a\cos\theta, -a\sin\theta, z)$. Finally, we find

$$\mathbf{dB} = \frac{\mu_0}{4\pi} \frac{I\,\mathbf{dl} \times \mathbf{R}}{R^3}\ , \tag{11.29}$$

and using that $\mathrm{d}l = a\,\mathrm{d}\theta$, we get

$$
\begin{aligned}
\mathrm{d}\mathbf{l} \times \mathbf{R} &= \mathrm{d}l(-\sin\theta, \cos\theta, 0) \times (-a\cos\theta, -a\sin\theta, z) \\
&= a\,\mathrm{d}\theta(z\cos\theta, z\sin\theta, a) \\
&= (a\,\mathrm{d}\theta\, z\cos\theta, a\,\mathrm{d}\theta\, z\sin\theta, a^2\,\mathrm{d}\theta)\ .
\end{aligned}
\tag{11.30}
$$

The integral over the circle is an integral over θ from 0 to 2π. This means that the $\cos\theta$ and $\sin\theta$ terms will be zero when we integrate, and we are left with the z-component only, giving the same result as above:

$$
B_z = \int_0^{2\pi} \frac{\mu_0}{4\pi} \frac{Ia^2\,\mathrm{d}\theta}{R^3} = \frac{\mu_0 I a^2}{2\left(z^2 + a^2\right)^{3/2}}\ .
\tag{11.31}
$$

Finding the magnetic field numerically. To find the magnetic field at any point in space, we need to find the magnetic field numerically. We will do this by first parameterizing the curve carrying the current, in this case the circle, discretizing the curve and finally summing up the contributions to the magnetic field from each discrete element on the circle.

We already parameterized the curve above, describing the curve as $\mathbf{r}' = (a\cos\theta, a\sin\theta, 0)$, where θ is the angle with the x-axis. We divide this into N elements of angular extent $\Delta\theta = 2\pi/N$. Element i has a center at $\theta_i = i\Delta\theta$ and extends an angle $\Delta\theta/2$ in each direction. We assume that the element has a length $\mathrm{d}l$ which is given as the arc length, $\mathrm{d}l = a\Delta\theta$. The direction of $\mathrm{d}\mathbf{l}$ of the element is given as above, $\mathrm{d}\mathbf{l} = \mathrm{d}l(-\sin\theta_i, \cos\theta_i, 0)$.

With this discretization, we find the contribution to the magnetic field $\mathrm{d}\mathbf{B}(\mathbf{r})$ in a point $\mathbf{r}$ from element $\mathrm{d}\mathbf{l}$ at position $\mathbf{r}'$ from Biot-Savart's law:

$$
\mathrm{d}\mathbf{B} = \frac{\mu_0}{4\pi} \frac{I\mathrm{d}\mathbf{l} \times \mathbf{R}}{R^3}\ ,
\tag{11.32}
$$

where $\mathbf{R}$ is the vector from the element at $\mathbf{r}'$ to the observation point, $\mathbf{r}$, that is, $\mathbf{R} = \mathbf{r} - \mathbf{r}'$. We can then simply use the `cross`-function in Python to find the cross product.

First, we write a function `bfield` that finds the magnetic field in a point from the circle:

```
import numpy as np
import scipy.constants as sc
import matplotlib.pyplot as plt
def bfield(r,a,N):
    # Find the magnetic field in the point r
    # N = number of points of resolution
    # a = radius of circle
    B = np.array([0.0, 0.0, 0.0])
    dtheta = 2*np.pi/N
    I, dl = 1.0, a*dtheta
    K = I*sc.mu_0/(4*np.pi)
    for i in range(N):
```

```
        theta = i*dtheta
        rdv = np.array([a*np.cos(theta),a*np.sin(theta),0])
        R = r - rdv
        dlv = dl*np.array([-np.sin(theta),np.cos(theta),0])
        dB = K*np.cross(dlv,R)/np.linalg.norm(R)**3
        B = B + dB
    return B
```

Here, we define `B` as a vector and then sum up the contributions `dB` from the individual elements. We use a notation in the program which is very similar to the mathematical notation. This simplifies the translation from mathematics to computer program and makes it easier for us to check the mathematics in the program.

Using the `bfield`-function, we calculate the magnetic field in the xz-plane. This provides a good visualization of the field because the field is rotationally symmetric around the z-axis. We find the field over a set of x and z values from $-2a$ to $+2a$:

```
N = 100 # Resolution of circle
a = 1.0 # Radius of circle
Lx, Lz, Nx, Nz = 2*a, 2*a, 20, 20
x = np.linspace(-Lx,Lx,Nx)
z = np.linspace(-Lz,Lz,Nz)
rx,rz = np.meshgrid(x,z,indexing="ij")
Bx = np.zeros_like(rx)
Bz = np.zeros_like(rz)
for ix in range(Nx):
    for iz in range(Nz):
        r = np.array([rx[ix,iz],0,rz[ix,iz]])
        Bx[ix,iz],By,Bz[ix,iz] = bfield(r,a,N)
```

Finally, we visualize the field using `quiver` where the magnitudes of the vectors are shown with the length of the vectors and with a logarithmic color scale. The resulting plots are shown in Fig. 11.6.

```
ax1 = plt.subplot(1,2,1)
plt.quiver(rx,rz,Bx,Bz)
plt.xlabel("$x/a$"), plt.ylabel("$z/a$")
ax1.set_aspect("equal", "box")
ax2 = plt.subplot(1,2,2)
nBx = Bx/np.sqrt(Bx**2+Bz**2)
nBz = Bz/np.sqrt(Bx**2+Bz**2)
logB = np.log10(np.sqrt(Bx**2+Bz**2))
plt.quiver(rx,rz,nBx,nBz,logB,cmap="jet")
plt.xlabel("$x/a$"), plt.ylabel("$z/a$")
ax2.set_aspect("equal", "box")
```

Comparing with an electric dipole. This looks similar to the field from an electric dipole. We illustrate this similarity by also plotting the magnetic field from a circular current and the electric field in the xz-plane from a dipole $\mathbf{p} = qd\,\hat{\mathbf{z}}$.

```
Ex, Ez = np.zeros_like(Bx), np.zeros_like(Bz)
d, q = 0.4*a, 1.0 # Dipole distance and charge
r1, q1 = np.array([0,0,d/2]),1.0
```

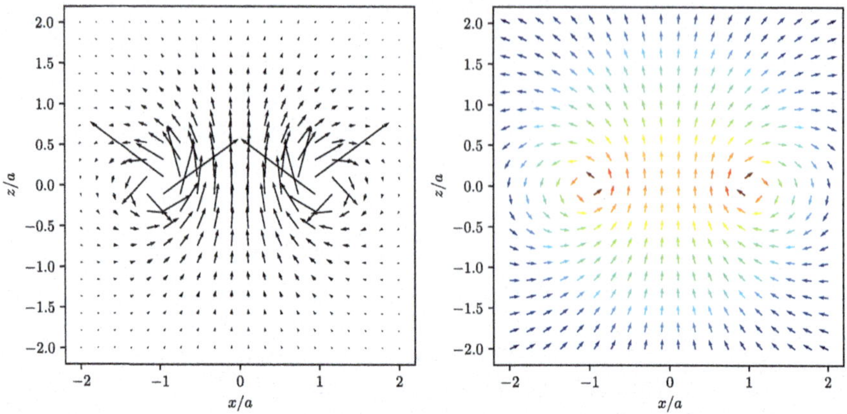

Fig. 11.6 Plot of the magnetic field in the xz-plane from a circular current in the xy-plane

```
r2, q2 = np.array([0,0,-d/2]),-1.0
for ix in range(Nx):
    for iz in range(Nz):
        r = np.array([rx[ix,iz],0,rz[ix,iz]])
        dr1, dr2 = r - r1, r - r2
        Ex[ix,iz],Ey,Ez[ix,iz] = q1*dr1/np.linalg.norm(dr1)**3\
                      + q2*dr2/np.linalg.norm(dr2)**3
plt.subplot(1,2,1)
plt.streamplot(rx.T,rz.T,Bx.T,Bz.T,broken_streamlines=False)
plt.subplot(1,2,2)
plt.streamplot(rx.T,rz.T,Ex.T,Ez.T,broken_streamlines=False)
```

The resulting plots are shown in Fig. 11.7. We see that there are similarities between the electric field and the magnetic field. The magnet indeed looks similar to a dipole. However, there are some very important differences between these two fields. For the magnetic field, all the field lines are closed. This means that for any surface, the flux of the magnetic field into and out of the surface is the same, so that the net flux is zero. Consequently, the divergence of the magnetic field is zero everywhere. On the other hand, for the electric field, all the field lines start from the positive charge and end up on the negative charge. The divergence of the electric field is therefore not zero everywhere. The lack of divergence of the magnetic field will be discussed in more detail in the next chapter.

Example: Magnetic field from arbitrary circuit

We have now found a way to find the magnetic field from a circular circuit. However, what if we instead specify a differently shaped circuit, indeed an arbitrarily shaped circuit? How can we find the magnetic field from such a circuit? We could describe

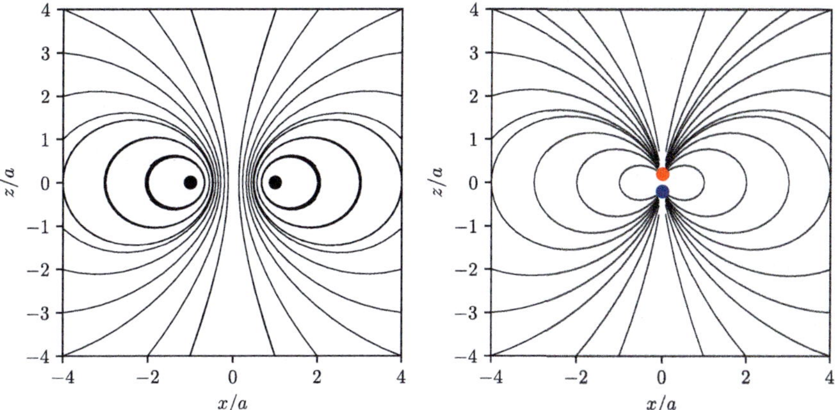

Fig. 11.7 **a** Plot of the magnetic field in the xz-plane from a circular current in the xy-plane with radius a. **b** Plot of the electric field in the xz-plane from an electric dipole from $-(d/2)\hat{\mathbf{z}}$ to $+(d/2)\hat{\mathbf{z}}$ where $d = 0.4a$. (The $y \rightarrow -y$ asymmetries are due to discretization effects in visualization)

the circuit as a set of N discrete points, $\mathbf{r}_i$, $i = 0, 1, \ldots, N-1$, and line segments $\Delta\mathbf{l}_i = \mathbf{r}_{i+1} - \mathbf{r}_i$. However, we would then lack the last line segment from $i = N-1$ to $i = 0$. We can add this segment by adding an additional point to the list, $\mathbf{r}_0$, so that the first and the last element is $\mathbf{r}_0$. We describe the position of a line segment by its midpoint: $\mathbf{r}'_i = (\mathbf{r}_i + \mathbf{r}_{i+1})/2$. The magnetic field is found from Biot-Savart's law:

$$\mathbf{B} \simeq \sum_i \frac{\mu_0}{4\pi} \frac{I\, \Delta\mathbf{l}_i \times \mathbf{R}_i}{R_i^3} \,. \tag{11.33}$$

where $\mathbf{R}_i = \mathbf{r} - \mathbf{r}'_i$

Implementation of curve representation. As a demonstration, we here construct a set of M planar circular circuits each with radius a, uniformly positioned from a height $z = -h/2$ to a height $z = +h/2$. The z-position of circle j is then $z_j = -h/2 + j/(M-1)\,h$, where $j = 0, \ldots, M-1$ indexes the M circles. The position of a point within each circle is $\mathbf{r}_i = a(\cos\theta_i, \sin\theta_i, z_j)$ with $\theta_i = 2\pi(i/(N-1))$ for $i = 0, \ldots, N-1$. (Notice that the last point is $\theta_{N-1} = 2\pi$, which corresponds to $\theta_0 = 0$, as we wanted). First, we define the curve as a list of positions $\mathbf{r}_i$ in `ri_list` to represent this system:

```
N, M = 100, 5  # Nr of points, Nr of circles
a, h = 1.0, 1.0 # radius, height
ri_list = []
for j in range(M):
    zj = -h/2 + (j/(M-1))*h
```

```
    for i in range(N):
        thetai = i/(N-1)*2*np.pi
        ri = np.array([a*np.cos(thetai),a*np.sin(thetai),zj])
        ri_list.append(ri)
```

Implementation of function to find the field. We then write a function to find the magnetic field in a point **r** due to a current I in the curve described by `rilist` in the function `bfielddllist`, where the results are scaled with I. Notice that we use `range(N-1)`, otherwise we would overstep the number of elements in `rilist`:

```
import scipy.constants as sc
def bfielddllist(r,rilist):
    B = np.array([0.0, 0.0, 0.0])
    N = len(rilist)
    for i in range(N-1):
        dli = rilist[i+1]-rilist[i]
        ri = 0.5*(rilist[i+1] + rilist[i])
        R = r - ri
        Rnorm = np.linalg.norm(R)
        dB = np.cross(dli,R)/Rnorm**3
        B = B + dB
    return B*sc.mu_0/(4*np.pi)
```

Find the field in a region in space. We find the magnetic field $\mathbf{B}(\mathbf{r})$ on a grid in the xz-plane, following methods similar to that we introduced for the electric field.

```
Lx, Lz, Nx, Nz = 2*a,2*a,20,20
x = np.linspace(-Lx,Lx,Nx)
z = np.linspace(-Lz,Lz,Nz)
rx,rz = np.meshgrid(x,z,indexing="ij")
Bx, Bz = np.zeros_like(rx), np.zeros_like(rz)
for ix in range(Nx):
    for iz in range(Nz):
        r = np.array([rx[ix,iz],0.0,rz[ix,iz]])
        Bx[ix,iz],By,Bz[ix,iz] = bfielddllist(r,ri_list)
```

Visualization of the magnetic field. Finally, we visualize the magnetic field using the methods we have previously developed. The resulting plot is shown in Fig. 11.8.

```
Bnorm = np.sqrt(Bx*Bx + Bz*Bz)
uBx = Bx/Bnorm
uBz = Bz/Bnorm
Bcolor = np.log10(Bnorm)
plt.figure(figsize=(8,8))
plt.quiver(rx,rz,uBx,uBz,Bcolor,cmap="jet")
plt.xlabel("$x/a$"), plt.ylabel("$z/a$")
plt.axis("equal")
```

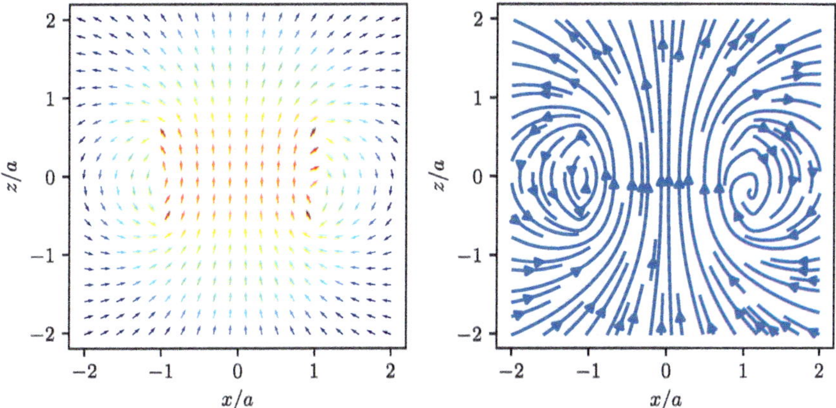

Fig. 11.8 Plot of the magnetic field in the xz-plane from a set of circular currents in the xy-plane

11.2 Magnetic Forces

We have now developed effective analytical and numerical methods to find the magnetic field from a given distribution of currents:

$$\mathbf{B} = \sum_i \frac{\mu_0}{4\pi} \frac{I \Delta \mathbf{l}_i \times \hat{\mathbf{R}}_i}{R_i^2} \ . \tag{11.34}$$

The force on a particle with charge Q and velocity $\mathbf{v}$ moving in a magnetic field $\mathbf{B}$ is:

$$\mathbf{F} = Q\mathbf{v} \times \mathbf{B} \ . \tag{11.35}$$

This expression is called the **Lorentz force**.

Lorentz force

If the charge Q also is moving in an electric field $\mathbf{E}$, the total force is the sum of the forces from the electric and magnetic fields, which we call the Lorentz force:

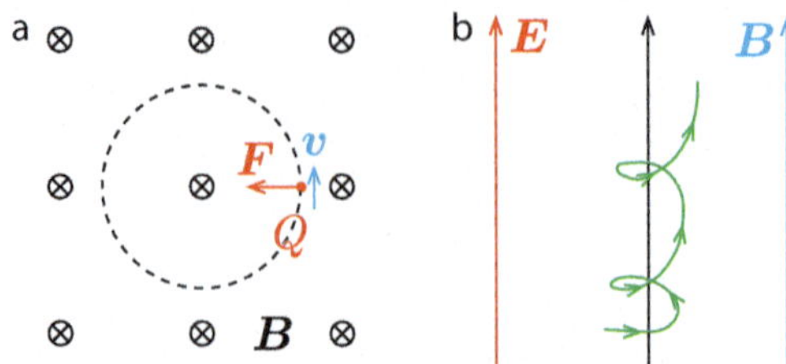

Fig. 11.9 **a** Illustration of a charged particle moving in a magnetic field. **b** Illustration of a charged particle in an electric and a magnetic field

Lorentz force law

The Lorentz force describes the force on a charge Q moving in an electric and a magnetic field:

$$\mathbf{F} = Q\mathbf{E} + Q\mathbf{v} \times \mathbf{B} \, . \tag{11.36}$$

Notice that it is the electric and magnetic field at the position of the charge that enters the expression.

Example: Charged particle in combined fields

Describe the motion of a charged particle moving in a combination of uniform electric and magnetic fields

Approach. We will address two cases illustrated in Fig. 11.9 for a charge moving in a uniform magnetic field and a charge moving in a combined magnetic and electric field. We will describe the motion by first finding the force acting on the particle and then describe the motion through Newton's second law.

A particle in a magnetic field. Fig. 11.9a illustrates a charge Q moving with a velocity $\mathbf{v}$ that is orthogonal to the magnetic field $\mathbf{B}$. In this case the force, $\mathbf{F} = Q\mathbf{v} \times \mathbf{B}$, will be normal to both $\mathbf{B}$ and $\mathbf{v}$. Thus, the charge will move in a circular orbit.

Will the charge slow down due to the interaction with the magnetic field? No, because the force is always normal to the velocity. The *magnetic force therefore does no work on the moving charge*. It will continue in a circular orbit indefinitely. What if the particle starts from rest? Unless there are other forces, the particle will remain at rest.

A particle in an electric and a magnetic field. What if there is an electric field in the same direction as the magnetic field, as illustrated in Fig. 11.9b? If the particle starts at rest, the electric field will start to push it in the direction of the field (if Q is positive). In this case, the $\mathbf{E}$-field will do work, accelerating the particle in the

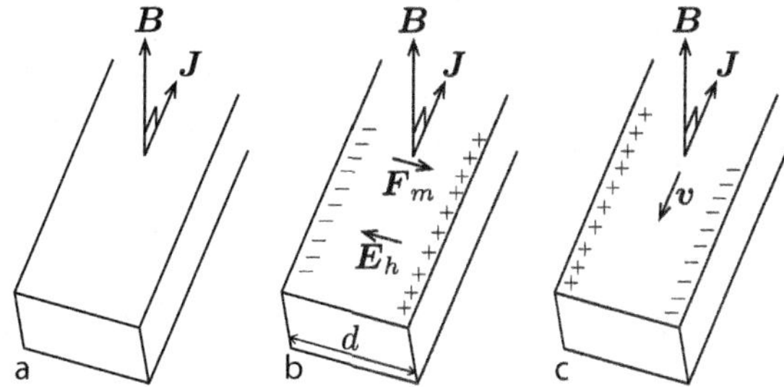

Fig. 11.10 **a** Illustration of the magnetic field and the current density in a semi-conductor. **b** Illustration of behavior of a positive charge. **c** Illustration of the behavior of a negative charge

direction of the field. The magnetic field will still only act in the direction normal to the velocity, which will make the particle go in a spiral. We notice that the velocity in the direction of the electric field, and in this case also in the direction of the magnetic field, will not contribute to the magnetic force (because the component is parallel to **B** and hence will be zero in the cross product). The particle will therefore accelerate along the fields, but the circular motion will remain the same as initially.

Test your understanding

A current I runs through a long, straight wire along the z-axis. A charge $Q > 0$ in $(x, 0, 0)$ is moving with a velocity $\mathbf{v} = v_0\hat{\mathbf{z}}$. (a) What is the direction of the magnetic field at the position of the charge? (b) What is the direction of the magnetic force on the charge Q. (c) Describe the motion of the particle.[3]

Example: Hall effect

The Hall effect is an effect on charge carriers moving in a semi-conductor. In this case, we do not know if the charge carriers are positive or negative. How can we use the behavior of charges in a magnetic field to determine the sign of the charge carriers?

Approach. The trick is to apply a uniform magnetic field normal to the semiconductor as illustrated in Fig. 11.10a. We apply an electric potential along the semiconductor, resulting in a current density **J** along the semiconductor. For practical purposes we can assume that the semiconductor is a conductor where we do not know the sign of the charge carriers. Let us therefore analyze the two possibilities: Positive and negative charge carriers.

Positive charge carriers. If the charge carriers are positive, positive charges are moving in the same direction as the current density **J**. The magnetic force on a

[3] (a) $\hat{\mathbf{y}}$; (b) $-\hat{\mathbf{x}}$; (c) Spiraling around the z-axis.

positive charge is in the direction $Q\mathbf{v} \times \mathbf{B}$, which is to the right in Fig. 11.10b. This means that positive charges will accumulate on the right hand side, and the left hand side will be depleted of positive charges and hence negatively charged. This will result in an electric field, $\mathbf{E}_h$ directed towards the left, which is called the *Hall field*. The charges will continue to build up until the force on a charge from the magnetic force is the same as the force from the electric field, E_h. In equilibrium the two forces will be the same: $QE_h = QvB$, which gives us that $E_h = vB$. We can measure E_h, which is the potential difference between the two sides of the semi-conductor, and therefore estimate v.

Negative charge carriers. What happens if the charge carriers are negative? First, we realize that a negative charge will move in the direction opposite of $\mathbf{J}$. The magnetic force on the charge will therefore be in the same direction, to the right, as for positive charge carriers. But now negative charges will build up on the right-hand side, generating an electric field difference $-E_h$. This means that we can determine if the charge carriers are positive or negative by measuring the Hall field or the Hall potential.

Magnetic forces on a current element

What is the force on a small volume element $\mathrm{d}v$ with a current density $\mathbf{J}$? We assume that each charge Q in the element is moving with a velocity $\mathbf{v}$ and that there are $N\,\mathrm{d}v$ charges in $\mathrm{d}v$. The net force on the element is the sum of the forces on each charge, $\mathrm{d}\mathbf{F} = N\,\mathrm{d}v\,Q\mathbf{v} \times \mathbf{B}$, where we recognize that $NQ\mathbf{v} = \mathbf{J}$, therefore:

Magnetic force on a current density element

The magnetic force from a current element $\mathbf{J}\,\mathrm{d}v$ from a magnetic field $\mathbf{B}$ is

$$\mathrm{d}\mathbf{F} = \mathbf{J}\,\mathrm{d}v \times \mathbf{B}\,. \tag{11.37}$$

Force on a current element. For a current element $I\,\mathrm{d}\mathbf{l} = \mathbf{J}\,\mathrm{d}v$ we get

$$\mathrm{d}\mathbf{F} = I\,\mathrm{d}\mathbf{l} \times \mathbf{B}\,. \tag{11.38}$$

and similarly for a surface current density $\mathbf{J}_S$, we get

$$\mathrm{d}\mathbf{F} = \mathbf{J}_S\,\mathrm{d}S \times \mathbf{B}\,. \tag{11.39}$$

For a piece of a wire of length L normal to the magnetic field, the force is therefore $F = ILB$.

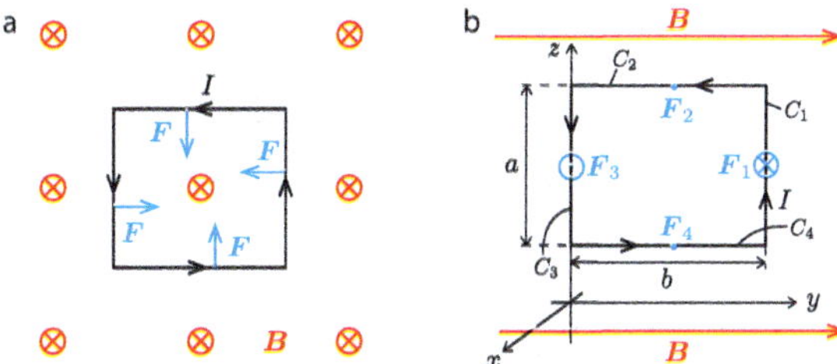

Fig. 11.11 Illustration of a closed circuit in a uniform magnetic field for two orientations of the magnetic field

Force on a closed circuit. We find the force on a closed circuit by integrating the contributions from each circuit element **dl**:

$$\mathbf{F} = \oint_C I \, \mathrm{d}\mathbf{l} \times \mathbf{B} \,. \tag{11.40}$$

For a **uniform magnetic field**, we can place **B** outside the integral, getting:

$$\mathbf{F} = \left(\oint_C I \, \mathrm{d}\mathbf{l} \right) \times \mathbf{B} = 0 \times \mathbf{B} = 0 \,. \tag{11.41}$$

There is therefore no net force on a closed circuit in a uniform magnetic field.

Test your understanding

Two long, straight wires 1 and 2 are parallel with the x-axis a distance d apart and go through $(0, 0, 0)$ and $(0, d, 0)$ respectively. Currents I_1 and I_2 are running through the wires in the positive x-direction. (a) What is the direction of the magnetic field from wire 1 in the position of wire 2? (b) What is the direction of the force on wire 2?[4]

Torque on a closed circuit in a uniform field

Figure 11.11 shows two different situations for a closed circuit with a current I in a magnetic field.

Force on a circuit normal to the field. In Fig. 11.11a we see that the forces on all the elements are pointing into the center of the circuit, attempting to compress the circuit. The net force is zero.

[4] (a) $\hat{\mathbf{z}}$; (b) Toward the other wire.

Force on a circuit parallel to the field. In Fig. 11.11b the forces along the two parts of the circuit (C_2 and C_4) that are parallel to the magnetic field are zero, whereas the forces on the two other sides (C_1 and C_3) act out of the plane of the circuit. In this case the net force is zero, but the net torque around the z-axis is not zero.

Torque on a circuit parallel to the field. The torque around an axis z of a force $\mathbf{F}$ is $\tau = \mathbf{r} \times \mathbf{F}$, where $\mathbf{r}$ is a vector from the axis to the point where the force acts and $\mathbf{F}$ is the force. The torque of $\mathbf{F}_2$ and $\mathbf{F}_4$ around the z-axis is zero, because $\mathbf{F}$ and $\mathbf{r}$ are parallel. The torque of $\mathbf{F}_3$ is zero since $\mathbf{r}$ is zero in this case. For $\mathbf{F}_1$, we see that $\mathbf{r} = b\hat{\mathbf{y}}$ and the torque is

$$\tau = b\hat{\mathbf{y}} \times \mathbf{F}_1 , \tag{11.42}$$

where b is the length of C_4 and

$$\mathbf{F}_1 = \int_{C_1} I \, \mathrm{d}\mathbf{l} \times \mathbf{B} = I \int_{C_1} \mathrm{d}\mathbf{l} \times \mathbf{B} = -IaB\hat{\mathbf{x}} , \tag{11.43}$$

where a is the length of C_1. This gives

$$\tau = b\hat{\mathbf{y}} \times \left(-IaB\hat{\mathbf{x}}\right) = bIaB\hat{\mathbf{z}} = ISB\hat{\mathbf{z}} , \tag{11.44}$$

where $S = ab$ is the area of the circuit, and $\mathbf{S} = S\hat{\mathbf{x}}$ is the oriented area of the circuit, with a direction given by the right-hand rule. The torque, τ, can therefore be written as

$$\tau = \underbrace{I\mathbf{S}}_{=\mathbf{m}} \times \mathbf{B} . \tag{11.45}$$

We call the term $\mathbf{m} = I\mathbf{S}$ the *magnetic moment* of the circuit.

Magnetic moment

We call the vector $\mathbf{m} = I\mathbf{S}$ the **magnetic moment** of the circuit. The torque of a circuit with magnetic moment $\mathbf{m}$ in a homogeneous field $\mathbf{B}$ is:

$$\tau = \mathbf{m} \times \mathbf{B} . \tag{11.46}$$

Notice that we simply say the torque of the magnetic moment without specifying the axis, because the torque is the same around any point. We can show this for a loop of points $\mathbf{r}_i$ with forces $\mathbf{F}_i$, where $\sum_i \mathbf{F}_i = 0$. The torque with respect to the origin is $\tau = \sum_i \mathbf{r}_i \times \mathbf{F}_i$. For the torque around another point $\mathbf{r}_0$ we introduce $\mathbf{r}_i' = \mathbf{r}_i - \mathbf{r}_0$ and the torque around $\mathbf{r}_0$ is

$$\tau_0 = \sum_i \mathbf{r}'_i \times \mathbf{F}_i = \sum_i (\mathbf{r}_i - \mathbf{r}_0) \times \mathbf{F}_i = \sum_i \mathbf{r}_i \times \mathbf{F}_i - \sum_i \mathbf{r}_0 \times \mathbf{F}_i$$
$$= \tau - \mathbf{r}_0 \times \underbrace{\sum_i \mathbf{F}_i}_{=0} = \tau \ . \tag{11.47}$$

This result is a general result for the torque of a current loop with any shape and size in a uniform magnetic field.

Torque aligns magnetic moment with field. We notice from Fig. 11.11b that the torque tends to turn the current loop so that the magnetic moment, **m**, aligns with the magnetic field, **B**. Notice also that the magnetic moment points in the same direction as the magnetic field set up by the current in the loop, so that the torque tends to align the field set up by the loop with the external magnetic field.

11.3 Vector Potential for the Magnetic Field

For the electric field, we introduced the scalar potential V, which proved a useful conceptual and computational tool through $\mathbf{E} = -\nabla V$. Can we introduce a similar potential for the magnetic field? For a set of point charges, we found the electric potential from

$$V(\mathbf{r}) = \sum_i \frac{Q_i}{4\pi \epsilon_0 R_i} \ . \tag{11.48}$$

By analogy, for a set of circuit elements $\Delta \mathbf{I}_i = I \Delta \mathbf{l}_i$, we introduce the magnetic vector potential $\mathbf{A}$ as

$$\mathbf{A} = \sum_i \frac{\mu_0 I \Delta \mathbf{l}_i}{4\pi R_i} \ . \tag{11.49}$$

The magnetic vector potential does not have a similarly simple interpretation as the electric scalar potential. It is not directly related the potential energy of the system. However, we will instead show that we can find the magnetic field from the magnetic vector potential from $\mathbf{B} = \nabla \times \mathbf{A}$. This makes $\mathbf{A}$ a useful tool to find the magnetic field from current distributions.

Proving that $\mathbf{B} = \nabla \times \mathbf{A}$. We will prove that $\mathbf{B} = \nabla \times \mathbf{A}$ by proving that the magnetic field from a small current element $I \Delta \mathbf{l}_i$ is given as $\Delta \mathbf{B}_i = \nabla \times \mathbf{A}_i$ where $\mathbf{A}_i$ is the vector potential from the current element $I \Delta \mathbf{l}_i$. If this is true, we can find the total vector potential by summing the contributions from each current element. We start by placing the coordinate system so that element $I \Delta \mathbf{l}_i$ is in the origin and we orient the z-axis along $\Delta \mathbf{l}_i$ so that $\Delta \mathbf{l}_i = \Delta l_i \hat{\mathbf{z}}$. We find the contribution to the vector potential in a point $\mathbf{r}$ from this current element from

$$\Delta \mathbf{A}_i = \frac{\mu_0 I \Delta l_i \hat{\mathbf{z}}}{4\pi R_i} , \tag{11.50}$$

where $\mathbf{R}_i = \mathbf{r} - \mathbf{r}_i$ and $\mathbf{r}_i$ is zero since the element is placed in the origin, so that $\mathbf{R}_i = \mathbf{r}$ and $R_i = r$. We find the curl of this expression

$$\nabla \times \Delta \mathbf{A}_i = \nabla \times \frac{\mu_0 I \Delta l_i \hat{\mathbf{z}}}{4\pi r} . \tag{11.51}$$

The only part of this expression that depends on x, y or z is $r = (x^2 + y^2 + z^2)^{1/2}$. We notice that

$$\frac{\partial}{\partial x}\frac{1}{r} = \frac{\partial}{\partial x}(x^2 + y^2 + z^2)^{-1/2} = -x(x^2 + y^2 + z^2)^{-3/2} = -\frac{x}{r^3} . \tag{11.52}$$

and similarly for the derivative with respect to y. We therefore find that

$$\nabla \times \frac{\mu_0 I \Delta l_i \hat{\mathbf{z}}}{4\pi r} = \frac{\mu_0 I \Delta l_i}{4\pi} \nabla \times \frac{\hat{\mathbf{z}}}{r} = \frac{\mu_0 I \Delta l_i}{4\pi} \frac{(-y, x, 0)}{r^3} . \tag{11.53}$$

We recognize that $(-y, x, 0) = \hat{\mathbf{z}} \times (x, y, z) = \hat{\mathbf{z}} \times \mathbf{r}$. (Ok, this is not easy to recognize, but you can check that this is correct). We have therefore found that

$$\nabla \times \Delta \mathbf{A}_i = \frac{\mu_0 I \Delta l_i \hat{\mathbf{z}} \times \mathbf{r}}{4\pi r^3} = \frac{\mu_0 I \Delta \mathbf{l}_i \times \mathbf{R}_i}{4\pi R_i^3} . \tag{11.54}$$

We recognize this as the contribution to the magnetic field in the point $\mathbf{r}$ from the current element $I \Delta \mathbf{l}_i$ at $\mathbf{r}_i$. This result is independent of the choice of coordinate system. We have therefore proven that

$$\Delta \mathbf{B}_i = \nabla \times \Delta \mathbf{A}_i . \tag{11.55}$$

From the superposition principle and the linear property of the cross product we find that

$$\mathbf{B} = \sum_i \Delta \mathbf{B}_i = \sum_i \nabla \times \Delta \mathbf{A}_i = \nabla \times \sum_i \Delta \mathbf{A}_i = \nabla \times \mathbf{A} . \tag{11.56}$$

We can extend this result to a current distribution in a plane or in a volume.

Magnetic vector potential

We can find the magnetic vector potential $\mathbf{A}(\mathbf{r})$ in a position $\mathbf{r}$ from a current distribution given by the current elements $\Delta \mathbf{l}_i$ at positions $\mathbf{r}_i$ from

$$\mathbf{A} = \sum_i \frac{\mu_0 I \Delta \mathbf{l}_i}{4\pi R_i} . \tag{11.57}$$

The magnetic field is then given as

$$\mathbf{B} = \nabla \times \mathbf{A} . \tag{11.58}$$

We find the magnetic vector potential from a circuit, a surface or a volume current from:

$$\mathbf{A} = \frac{\mu_0}{4\pi} \int_C \frac{I \, \mathrm{d}\mathbf{l}}{R} , \quad \mathbf{A} = \frac{\mu_0}{4\pi} \int_S \frac{\mathbf{J}_s \, \mathrm{d}S}{R} , \quad \mathbf{A} = \frac{\mu_0}{4\pi} \int_v \frac{\mathbf{J} \, \mathrm{d}v}{R} . \tag{11.59}$$

For any given current distribution, this defines a unique magnetic vector potential. However, we realize that we can add any vector constant to the vector potential without changing the magnetic field.

Notice that because $\mathbf{B}$ is given as $\nabla \times \mathbf{A}$, we also known that $\nabla \cdot \mathbf{B} = 0$ because the divergence of a curl is always zero: $\nabla \cdot \mathbf{B} = \nabla \cdot \nabla \times \mathbf{A} = 0$. We can therefore also consider this a proof that the divergence of the magnetic field is zero. We will address this feature of the magnetic field in the next chapter.

Summary

The contribution to the **magnetic field** in a point $\mathbf{r}$ from a current element $I \, \mathbf{dl}$ at $\mathbf{r}'$ is given by Biot-Savart's law for a current element:

$$\mathrm{d}\mathbf{B} = \frac{\mu_0}{4\pi} \frac{I \, \mathrm{d}\mathbf{l} \times \hat{\mathbf{R}}}{R^2} ,$$

where $\mathbf{R} = \mathbf{r} - \mathbf{r}'$.

The **magnetic field B** from a current loop C is given by **Biot-Savart's** law for current loops:

$$\mathbf{B} = \frac{\mu_0}{4\pi} \oint_C \frac{I \, \mathrm{d}\mathbf{l} \times \mathbf{R}}{R^3} .$$

Magnetic fields obey the **superposition principle**:

$$\mathbf{B} = \sum_i \mathbf{B}_i$$

Biot-Savart's law on differential form states that the contribution $\mathbf{dB}$ to the magnetic field at $\mathbf{r}$ from a current density $\mathbf{J}$ at $\mathbf{r}'$ is

$$\mathrm{d}\mathbf{B} = \frac{\mu_0}{4\pi}\frac{\mathbf{J}\,\mathrm{d}v \times \hat{\mathbf{R}}}{R^2} = \frac{\mu_0}{4\pi}\frac{\mathbf{J}\,\mathrm{d}v \times \mathbf{R}}{R^3} ,$$

where $\mathbf{R}$ is the vector from the volume element $\mathrm{d}v$ at $\mathbf{r}'$ to $\mathbf{r}$: $\mathbf{R} = \mathbf{r} - \mathbf{r}'$.

The **Lorentz force** describes the force on a charge Q moving in an electric and a magnetic field:

$$\mathbf{F} = Q\mathbf{E} + Q\mathbf{v} \times \mathbf{B} .$$

The **magnetic force** on a current element $I\,\mathbf{dl}$ from the interaction with the magnetic field $\mathbf{B}$ is

$$\mathrm{d}\,\mathbf{F} = I\,\mathrm{d}\mathbf{l} \times \mathbf{B} .$$

We call $\mathbf{m} = I\mathbf{S}$ the **magnetic moment** of a circuit. The torque of a magnetic moment $\mathbf{m}$ in the field $\mathbf{B}$ is:

$$\tau = \mathbf{m} \times \mathbf{B} .$$

Exercises

Discussion exercises

11.1 Minimal field. You want to construct a circuit with a battery and a resistor with a constant current. How can you construct the circuit so that the magnetic field set up by the circuit becomes as small as possible so as not to interfer with other circuits close by.

11.2 Field in capacitor. You want to construct a circuit with a battery, a resistor and a capactor in series. As you turn the circuit on, a gradually reducing current is flowing in the circuit. Your friend Q states that since there is no current flowing across the capacitor, there cannot be a magnetic field inside the capacitor. Is this correct?

11.3 Field-free box. On the surface of the Earth, the magnetic field from the Earth is approximately uniform over short distances. How would you build a box that has essentially zero magnetic field inside it?

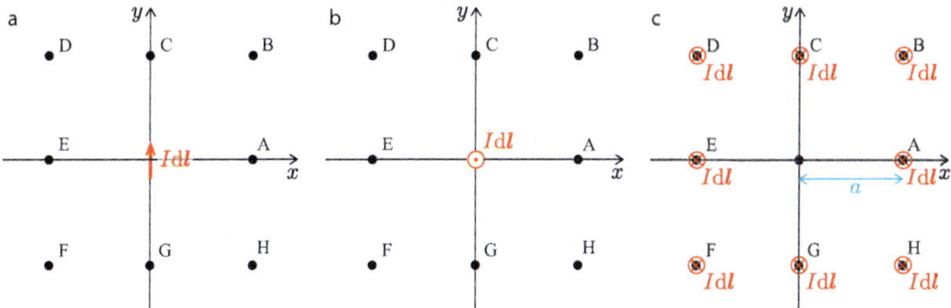

Fig. 11.12 **a, b** The current element is indicated. What is the magnetic field in the points shown in the figure

11.4 Model for permanent magnet. Let us consider a model for a permanent magnet is that the magnetic field is due to charges (for example electrons) moving around the nucleus of the atoms. How could you use such a model to (i) say something about the magnetic field around a permanent magnet, (ii) say something about what happens if we divide a permanent magnet into two pieces.

11.5 Magnetic North. How can you explain the magnetic field of the Earth using the model for electromagnets we have developed in this chapter? (You may need to use that the Earth has a liquid, metallic core).

Tutorials

11.6 Magnetic field from a current element. Figure 11.12 show a current element.
(a) The current element $I\mathrm{d}\mathbf{l} = I\mathrm{d}l\hat{\mathbf{y}}$ is placed in the origin. Draw the **R**-vector and find the direction of the magnetic field for each of the points (A-H) in the figure.
(b) The current element $I\mathrm{d}\mathbf{l} = I\mathrm{d}l\hat{\mathbf{z}}$ is placed in the origin as illustrated in Fig. 11.12b. Draw the **R**-vector and find the direction of the magnetic field for each of the points (A-H) in the figure.
(c) We change the problem slightly and place a current element $I\mathrm{d}\mathbf{l} = -I\mathrm{d}l\hat{\mathbf{z}}$ in the various positions (A-H) shown in Fig. 11.12c. You will here only address one element at a time. First, the element is placed in A, then in B and so on. Draw the **R**-vector and find the direction of the magnetic field in the origin for each of the placements (A–H) in the figure.

11.7 Two current elements. Figure 11.13 shows systems where there are two current elements.
(a) For system A, what is the direction of the magnetic field in the point midway between the elements?
(b) Current elements cannot exist alone for a stationary system. How do you think that it is natural to draw a full circuit that contains the two elements in system A?
(c) For system B, what is the direction of the magnetic field in the point midway between the elements?

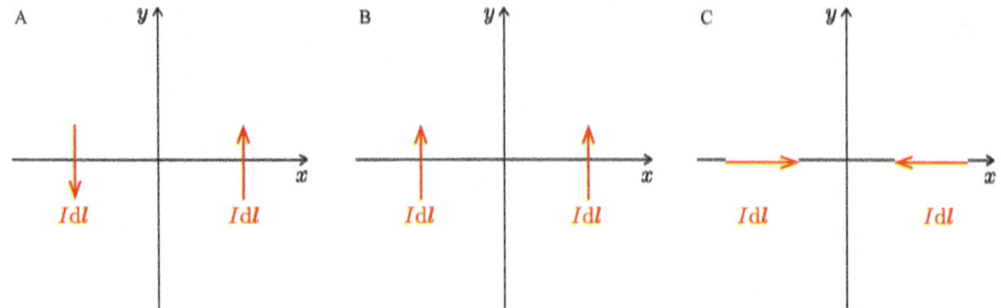

Fig. 11.13 Two current elements

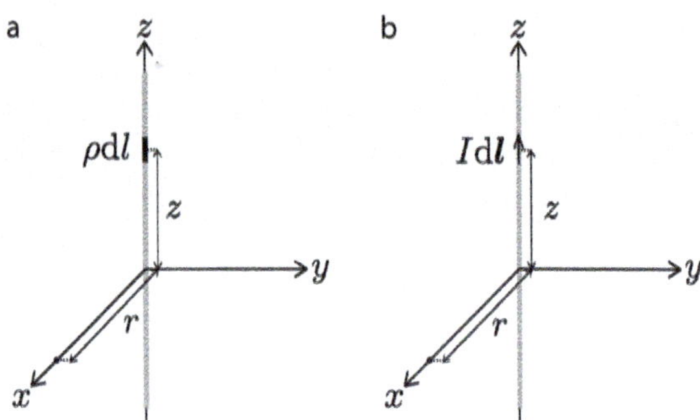

Fig. 11.14 A line charge and a line current

(d) Can you draw one or more circuits that include the two current elements in system B so that your answer remains the same? There may be more than one possible solution, can you find more than one answer?
(e) For system B, what is the direction of the magnetic field in the point midway between the elements?
(f) Can you draw one or more circuits that include the two current elements in system C so that your answer remains the same?

11.8 Comparing a line charge and a line current. Figure 11.14 shows two systems. Part a shows a line charge, and part b shows a line current. We will compare the two systems to see how we can transfer the skills we have developed in solving line charge problems to line currents.
(a) We start by repeating how we find the field for the line charge in the system on the left. Let us find the contribution to the electric field in the point $(r, 0, 0)$ from a charge element $\rho\,dl$ in the point $(0, 0, z)$. Draw the **R**-vector into the figure and find an expression for the **R**-vector.
(b) What is the argument for the electric field only to have a radial component, that is, there there only will be a component in the x-direction on a point on the x-axis?
(c) Write down the contribution dE_x to the electric field in the point $(r, 0, 0)$.
d) Write down the integral you need to solve to find the total electric field. What is the result of the integral?

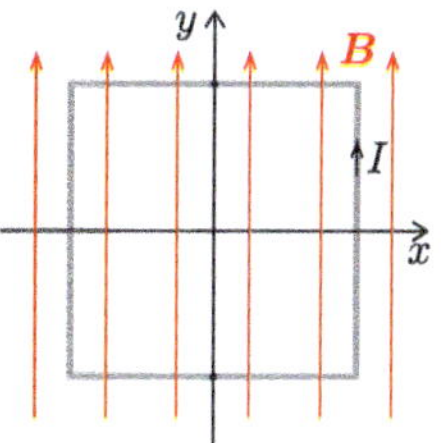

Fig. 11.15 A quadratic loop in a magnetic field

(e) Then we are ready to address a similar problem, a line current, as shown in Fig. 11.14b. We want to find the magnetic field in the point $(r, 0, 0)$ (along the x-axis) from a current element $I\,\mathbf{dl} = I\,\mathrm{d}l\hat{\mathbf{z}}$ in the point $(0, 0, z)$. Draw the **R**-vector in the figure and find an expression for the **R**-vector.
(f) What is the oriention of the contributions to the **B**-field from all the current elements along the z-axis?
(g) Find the contribution $\mathrm{d}\mathbf{B}$ to the magnetic field in the point $(r, 0, 0)$ from the current element $I\,\mathbf{dl} = I\,\mathrm{d}l\hat{\mathbf{z}}$ in the point $(0, 0, z)$.
(h) Write down the integral you need to solve to find the total magnetic field.
(i) Can you write down the solution to the integral directly by comparing with the result you got for the electric field?
(j) Briefly sum up the differences between the two approaches.

11.9 Current loop in a magnetic field. A small quadratic current loop with side a is placed in the xy-plane in a magnetic field $\mathbf{B} = B_0\hat{\mathbf{y}}$ as shown in Fig. 11.15.
(a) What is the net force on the loop?
(b) What is the torque around the x-axis?
(c) Explain how you can use such as current loop to measure the direction or the magnitude of a magnetic field.

11.10 Charges in a tokamak. In this exercise we will study charges in a tokomak, which is a magnetic container designed to contain a plasma, which is a gas of charged particles. The tokamak is one of several possible geometries that are explored for containing the plasma in a fusion reactor. You can read more about the tokamak geometry on e.g. Wikipedia. Currently, different geometries are being developed, such as advanced tokamak geometries or Stellarator-geometries which are based on designing inhomogeneous magnetic fields.
(a) We start by addressing a homogeneous field $\mathbf{B} = B_0\hat{\mathbf{z}}$. What happens to a particle of mass m and charge Q which is moving in the same direction as the magnetic field?
(b) Find the path of a particle with mass m and charge Q with initial velocity $\mathbf{v} = v_0\hat{\mathbf{x}}$.

(c) We will now address a simplified magnetic field inside the Tokamak: a homogeneous circular magnetic field $\mathbf{B} = B_0\hat{\boldsymbol{\phi}}$ in cylindrical coordinates which are directed with the z-axis along what we call the tokamak axis. If the particle starts with a velocity along the magnetic field, will it continue in a circular path around the z-axis? The following program solves the equations of motion for a charged particle, but lacks a few elements:

```
import numpy as np
import matplotlib.pyplot as plt
def B_vec(B0,r):
    phihat = np.cross(np.array([0,0,1]),r)
    return B0*phihat/np.linalg.norm(phihat)
r0 = np.array([2,0,0])
v0 = np.array([0,1,0])
T, dt = 100.0, 0.01
m, Q = 1.0, 1.0
B0 = 1.0
nstep = int(T/dt)
v = np.zeros((nstep,3))
r = np.zeros((nstep,3))
v[0] = v0
r[0] = r0
for i in range(nstep-1):
    F = # Fill in the expression for the force here
    a = F/m
    v[i+1] = v[i] + a*dt
    r[i+1] = r[i] + v[i+1]*dt
from mpl_toolkits import mplot3d
plt.figure(figsize=(6,6))
ax = plt.axes(projection="3d")
ax.plot3D(r[:,0],r[:,1],r[:,2],"gray")
```

(d) Complete the code by inserting the correct code for the force. Run the program. What do you observe? Vary $\mathbf{v}_0$ and observe what happens.
(e) How can we modify the magnetic field to contain the particle inside a donut-shaped tokamak? Test various simple modifications to the magnetic field and observe what happens. (Hint: Find illustrations of tomamaks for inspiration.)
(f) An example of a field called a magnetic trap is $\mathbf{B} = (B_0/b)(x, y, -2z)$ where you may use $B_0/b = 0.1$ with the values provided above. Describe the motion of a charge in this field.

11.11 An explicitely parameterized curve integral. In this exercise we will explicitely write down and solve a parameterized curve integral along a current loop. We address a circular current loop in the xy-plane with its center in the origin, radius a and a current I.
(a) Find a parameterization that describes the current loop, $\mathbf{r}'(\phi)$, where ϕ is the angle with the x-axis.
(b) We want to find the magnetic field along the z-axis. What is the $\mathbf{R}$-vector for a current element in the point $\mathbf{r}'(\phi)$?
(c) What is $\mathbf{dl}$ expressed in terms of ϕ and $d\phi$?

(d) What is the contribution to the magnetic field in $(0, 0, z)$ from a current element along the curve in the point $\mathbf{r}'(\phi)$?
(e) What is the curve integral over the current loop? Write down the integral as an integral over ϕ.
(f) How would these calculations change if we instead parameterized the curve using the arc length s so that $\mathbf{r}'(s) = a(\cos(s/a), \sin(s/a), 0)$?

Exercises

11.12 Finding B from a line segment. In this exercise we will focus on building skills in setting up the integrals needed to find the magnetic field from a current distribution using Biot-Savart's law. We will see how we can find the magnetic field from a line current. There is a current I in a wire along the x-axis. The current runs in the positive x-direction. We want to calculate the resulting field at a position $\mathbf{r} = (x, y, z) = (0, 0, z)$.
(a) Make a drawing of the system. Include a line element $d\mathbf{l}$ in your drawing, and the vector $\mathbf{R}$. Also express $\mathbf{R}$ on coordinate form $\mathbf{R} = (R_x, R_y, R_z)$.
(b) Find an expression for the contribution $\mathrm{d}\mathbf{B}$ to the magnetic field at $\mathbf{r}$ from the line element $\mathrm{d}\mathbf{l}$. First do it geometrically (without writing vector on Cartesian form), then do it using the Cartesian coordinate form for $\mathbf{R}$. (Notice that vector algebra actually works).
(c) Set up the integral to find the magnetic field $\mathbf{B}$ in the case when (i) the line is infinitely long, (ii) the line goes from minus infinity to $x = 0$, (iii) the line goes from $x = -L/2$ to $x = L/2$. (For case (ii) and (iii) we would need to connect the wire to other wires for the system to be physically reasonable with a constant current).

You may use that

$$\int \frac{1}{\left(x^2 + z^2\right)^{3/2}} \, \mathrm{d}x = \frac{x}{z^2\sqrt{x^2 + z^2}} \, . \tag{11.60}$$

(d) Find the $\mathbf{B}$-field in case (i), (ii) and (iii).
(e) Check that when $z \gg L$ your result in (iii) gives back the result for the magnetic field from a current element, that is, Biot-Savart's law.

11.13 Force on a current element. An infinitely long, straight wire that carries a current I_1 is partially surrounded by the loop shown in Fig. 11.16. The loop has a length L and a radius R, and carries a current I_2. The axis of the loop coincides with the infinite wire. Calculate the force that acts on the loop by dividing the loop into four separate segments and finding the force on each segment.

11.14 Magnetic field from a thin sheet. The surface current density in an infinite thin sheet in the xy-plane at $z = 0$ is $\mathbf{J}_S = J_S\hat{\mathbf{x}}$. We want to find the magnetic field in a position $\mathbf{r}$.
(a) Why is the magnetic field in the position $\mathbf{r} = (x, y, z)$ the same as in $(0, 0, z)$?

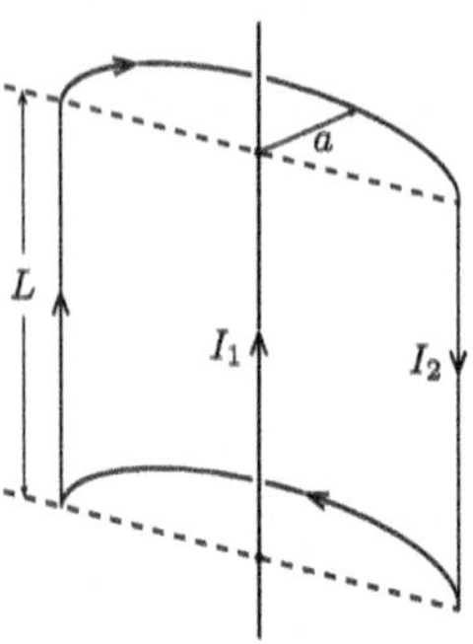

Fig. 11.16 An infinitely long, straight wire with a current I

(b) We introduce a surface element $\mathrm{d}S = \mathrm{d}x\mathrm{d}y$ at a position $\mathbf{r}' = (x', y', 0)$. Draw the $\mathbf{R}$-vector used to find the contribution from the element $\mathrm{d}S$ to the magnetic field at $(0, 0, z)$.
(c) Find the contribution $\mathbf{dB}$ from the current in the surface element $\mathrm{d}S$ to the magnetic field at $(0, 0, z)$.
(d) Explain why the magnetic field only has a y-component.
(e) Show that the magnetic field is $B_y = -\mu_0 J_s/2$ for $z > 0$. (Hint: It may be useful to replace the integral over x' and y' with an integral over polar (cylindrical) coordinates r and ϕ instead.)

Homework

11.15 Magnetic field above a circular circuit. We will study a circular circuit in the xy-plane, with radius a and with its center in the origin. A current I runs in the positive rotational direction (counter clockwise).
(a) Find an expression for the vector $\mathbf{R}$ from a point $(a, \phi, 0)$ on the circuit to a point $(0, 0, z)$ on the z-axis, where both points are given in cylindrical coordinates. Also find an expression for $R^2 = |\mathbf{R}|^2$ og $\hat{\mathbf{R}} = \frac{\mathbf{R}}{|\mathbf{R}|}$
(b) Find an expression from the contribution $\mathbf{dB}$ to the magnetic field in the point $(0, 0, z)$ from an infinitesimal current element $I\mathrm{d}\,\mathbf{l}$ in the circular circuit.
(c) Find the magnetic field in $(0, 0, z)$.

11.16 Infinite bent wire. An infinite thin wire is bent about a point P in a half-circle of radius a with P as its center as shown in Fig. 11.17. A constant current I runs through the wire.
(a) Find the magnetic field at point P by adding the contributions from the three circuit elements C_1, C_2 and C_3.
(b) Now that you have found the analytical expression for the magnetic field at P, write a code that computes the magnetic field at P numerically. You should use the answer you found analytically in the last problem to check that your code produces the correct result.

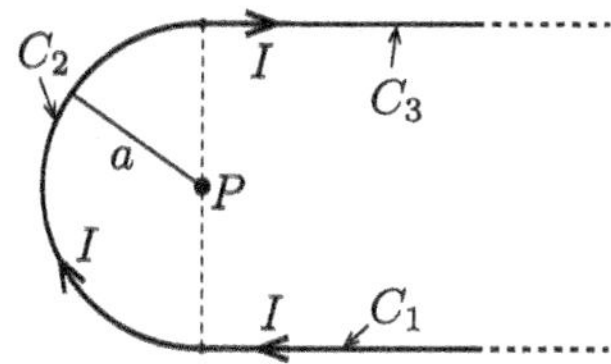

Fig. 11.17 Illustration of a bent wire circuit

(c) Extend your code so that it can compute the magnetic field at any point $\mathbf{r} = x\hat{\mathbf{x}} + y\hat{\mathbf{y}} + z\hat{\mathbf{z}}$.
(d) Extend the code you wrote so that it plots the streamlines of the magnetic field in the yz-plane at $x = 0$.

11.17 Field from a rotating circle. A circular wire with radius a and line charge density ρ is placed in the xy-plane with its center in the origin. The wire rotates around the z-axis with an angular frequency ω.
(a) Find the current I through the xz-plane.
(b) Find the magnetic field along the z-axis from the rotating wire.
(c) A circular disk of radius a and surface charge density ρ_S is placed in the xy-plane with its center in the origin. The disk rotates around the z-axis with an angular frequency ω. Show that the magnetic field along the z-axis for $z > 0$ is:

$$B_z = \frac{\mu_0 \rho_S \omega}{2} \left(\frac{r^2 + 2z^2}{\sqrt{r^2 + z^2}} - 2z \right) .$$

11.18 Visualize the field from a dipole. In this exercise we are going to visualize the field from a magnetic dipole with a streamplot. The field from a dipole is given by the equation

$$\mathbf{B}(\mathbf{r}) = \frac{\mu_0}{4\pi} \left(\frac{3\mathbf{r}(\mathbf{m} \cdot \mathbf{r})}{r^5} - \frac{\mathbf{m}}{r^3} \right) \tag{11.61}$$

Make a function that takes magnetic moment $\mathbf{m}$, its location and the position $\mathbf{r}$ where you want to evaluate the field. The output should be the resulting magnetic field. You only need to do it in two dimensions. Make a streamplot to visualize the field.

11.19 Two semi-infinite line currents. In this exercise we will study a wire that consists of two semi-infinite lines: One wire along the x-axis from $+\infty$ to the origin, and one wire along the y-axis from the origin to $+\infty$. We want to find the magnetic field from this wire when there is a stationary current I through it from $+\infty$ along the x-axis to $+\infty$ on the y-axis.
(a) Make a sketch of the system including the current I.
(b) First, we find the magnetic field from a current along the x-axis. What is the contribution $d\mathbf{B}$ to the magnetic field in the point $\mathbf{r} = (x, y, 0)$ from a current element in the point $\mathbf{r}' = (x', 0, 0)$?

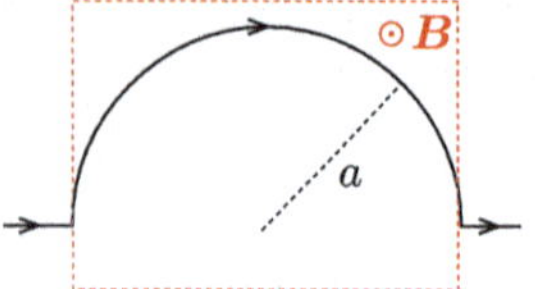

Fig. 11.18 A wire with a current I is placed partly in a magnetic field pointing out of the sheet

(c) What is the magnetic field in the point $\mathbf{r} = (x, y, 0)$ from the current I from $x' = +\infty$ to $x' = 0$? (Hint: You may need the integral $\int du/(u^2 + a^2)^{3/2} = u/(a^2(u^2 + a^2)^{1/2})$.)
(d) What is the magnetic field in the point $\mathbf{r} = (x, y, 0)$ from a wire from $y' = 0$ to $y' = \infty$ when there is a current I flowing from the origin?
(e) What is the total magnetic field in the point $\mathbf{r} = (x, y, 0)$?
(f) Visualize the magnetic field in the xy-plane.

11.20 Force from magnetic field on curved wire. Figure 11.18 shows a wire with a current I. The part of the wire that is shaped like a half cricle is in a region with a uniform magnetic field pointing up through the sheet. Find the force on the wire from the magnetic field.

11.21 Force on a loop near a wire. A long wire with a current I lies along the z-axis. A square circuit with dimensions $a \times a$ lies in the xz-plane at a distance b from the z-axis and carries a current I in the positive direction around the y-axis.
(a) Find the magnetic field from the long wire in the xz-plane.
(b) Find the net force on the square circuit.

Modeling projects

11.22 A magnetic dipole. In this exercise we will study the magnetic field around a magnetic dipole in the shape of a circular circuit with current I and radius a in the xy-plane and centered in the origin.
(a) Show that the magnetic field $\mathbf{B}(x, 0, 0)$ from an infinitely long wire along the y-axis with current I is given as $\mathbf{B}(x, 0, 0) = \mu_0 I(-\hat{\mathbf{z}})/(2\pi x)$.
(b) Write a program to find the magnetic field from a circular circuit with current I and radius a as described above. Visualize the magnetic field in the xz-plane both in a region around the origin and in a region around $(a, 0, 0)$.
(c) Compare the magnetic field from the circular circuit with the result from the infinite wire along the y-axis. Choose a reasonable region for the comparison so that different elements from the physical behavior of the system is evident. Comment on the results.

11.23 A finite solenoid. In this exercise we will study the magnetic field in and around a finite solenoid with current I and M windings. We start by studying a single circuit and then add several circuits together to form a solenoid.

(a) A circular circuit with radius a is in the xy-plane with its center in the origin. A current I runs through the circle in counter-clockwise direction. Show that the magnetic field in the origin is: $\mathbf{B}(0, 0, 0) = \mu_0 I/(2a)\hat{\mathbf{z}}$.
(b) Write a function `Bcircle(r,r0,a,I,N)` which finds the magnetic field $\mathbf{B}(\mathbf{r})$ in the point $\mathbf{r}$ from a circular circuit with current I, radius a, in a plane parallel with the xy-plane with its center in $\mathbf{r}_0$. Here, N is the number of discrete elements used to describe the circle. ($N = 100$ is sufficient).
(c) Check your function by comparing the result with the exact solution you found above by using $a = 1$ m and $I = 1$ A.
(d) Visualize the magnetic field in the xz-plane. Choose a suitable region.
(e) We will now address a finite solenoid consisting of M circular circuits each in planes parallel to the xy-plane. They all have radius a and centers along the z-axis. The circuits are uniformly spaced from $z = -a$ to $z = a$. The same current I runs in all the M circuits. Write a function `Bsolenoid(r,a,I,N,M)` which find the magnetic field $\mathbf{B}(\mathbf{r})$ from the solenoid.
(f) Use the function `Bsolenoide` to visualize the magnetic field in the xz-plane for $M = 5$.
(g) Plot the magnetic field in the z-direction along three lines that are parallel with the z-axis, that is, $B_z(0, 0, z)$, $B_z(a/2, 0, 0)$, and $B_z(2a, 0, 0)$ in the region from $z = -a$ to $z = a$. Comment on the results. Does this correspond to what you expected from a solenoid?

Chapter 12
Ampere's Law

12.1 Magnetic Flux

For electric fields, we found Gauss' law, which related the flux of the electric field to the net free charge: $\int_S \mathbf{E} \cdot d\mathbf{S} = Q_{\text{in}}/\epsilon_0$ on integral form and $\nabla \cdot \mathbf{E} = \rho/\epsilon_0$ on differential form. We derived this law from Coulomb's law. Electric fields are generated by charges, whereas magnetic fields are generated by current elements. Do we still have a similar law for magnetic fields?

We define the magnetic flux through a surface S as illustrated in Fig. 12.1 as:

$$\Phi_S = \int_S \mathbf{B} \cdot d\mathbf{S}. \tag{12.1}$$

A fundamental property of magnetic fields is that for any *closed* surface, this flux is zero! (We demonstrate this below). We can rewrite this on differential form using the divergence theorem: The flux of $\mathbf{B}$ through a closed surface S is equal to the integral of the divergence of $\mathbf{B}$ over the volume v enclosed by S:

$$\oint_S \mathbf{B} \cdot d\mathbf{S} = \int_v \nabla \cdot \mathbf{B} \, dv = 0. \tag{12.2}$$

This is true for any closed surface S, consequently, the argument of the integral must be equal to zero, $\nabla \cdot \mathbf{B} = 0$.

Law of conservation of magnetic flux

The flux of the magnetic field $\mathbf{B}$ through any *closed* surface S is always zero:

$$\oint_S \mathbf{B} \cdot d\mathbf{S} = 0. \tag{12.3}$$

© The Author(s), under exclusive license to Springer Nature Switzerland AG 2026

A. Malthe-Sørenssen, *Elementary Electromagnetism Using Python*, Undergraduate Texts in Physics, https://doi.org/10.1007/978-3-032-19876-1_12

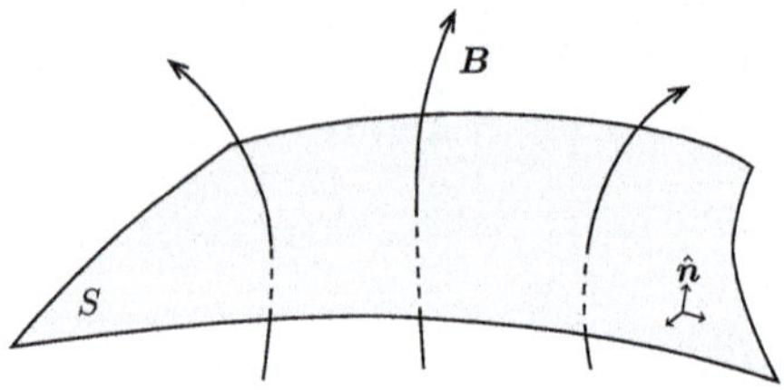

Fig. 12.1 The flux of the magnetic field through a surface S

This also means that the divergence of the magnetic field is zero:

$$\nabla \cdot \mathbf{B} = 0. \tag{12.4}$$

We do not know of any situations where these equations are not satisfied. This also implies that magnetic monopoles do not seem to exist.

Test your understanding
(a) The figure shows four different magnetic fields. Is the flux through the surface S (in red) positive, negative or zero? (b) Which of the fields in the figure can be magnetic fields? (c) Can the field $\mathbf{B} = x\hat{\mathbf{x}} - y\hat{\mathbf{y}}$ be a magnetic field? (d) Can the field $\mathbf{B} = z\hat{\mathbf{z}}$ be a magnetic field?[1]

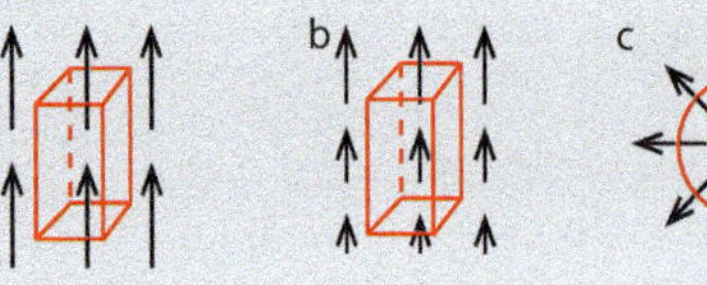

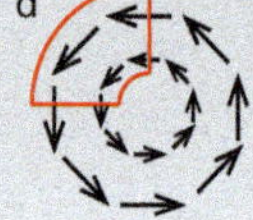

Proof that the Divergence of the Magnetic Field is Zero

Our plan is to demonstrate that the divergence of the magnetic field contribution from a single current element is zero and then argue that the sum of many such current elements will also have zero divergence. The contribution to the magnetic field $\mathbf{B}$ in a point $\mathbf{r}$ from a current element $\mathbf{J}\,\mathrm{d}v$ in $\mathbf{r}'$ is given by Biot-Savart-s law:

$$\mathrm{d}\mathbf{B} = \frac{\mu_0}{4\pi}\frac{\mathbf{J}\,\mathrm{d}v \times \hat{\mathbf{R}}}{R^2}, \tag{12.5}$$

where $\mathbf{R} = \mathbf{r} - \mathbf{r}'$. For simplicity we place the current element in the origin so that $\mathbf{r}' = 0$ and $\mathbf{R} = \mathbf{r} - \mathbf{r}' = \mathbf{r}$. The contribution $\mathrm{d}\mathbf{B}(\mathbf{r})$ from this current element to the magnetic field in $\mathbf{r}$ is then

[1] (a) a: 0, b: >0, c: >0, d: 0; (b) a, d; (c) Yes, $\nabla \cdot \mathbf{B} = 0$; (d) No, $\nabla \cdot \mathbf{B} \neq 0$.

$$\mathrm{d}\mathbf{B} = \frac{\mu_0}{4\pi}\frac{\mathbf{J}\,\mathrm{d}v \times \hat{\mathbf{r}}}{r^2} = \frac{\mu_0\,\mathrm{d}v}{4\pi}\frac{\mathbf{J}\times\hat{\mathbf{r}}}{r^2}. \tag{12.6}$$

We find the divergence of $\mathrm{d}\mathbf{B}$ by using the product rule: $\nabla\cdot(\mathbf{a}\times\mathbf{b}) = \mathbf{b}\cdot\nabla\times\mathbf{a} - \mathbf{a}\cdot\nabla\times\mathbf{b}$:

$$\nabla\cdot\mathrm{d}\mathbf{B} = \frac{\mu_0\,\mathrm{d}v}{4\pi}\nabla\cdot\left(\mathbf{J}\times\frac{\hat{\mathbf{r}}}{r^2}\right) = \frac{\mu_0\,\mathrm{d}v}{4\pi}\left(\frac{\hat{\mathbf{r}}}{r^2}\cdot(\nabla\times\mathbf{J}) - \mathbf{J}\cdot\left(\nabla\times\frac{\hat{\mathbf{r}}}{r^2}\right)\right). \tag{12.7}$$

Notice that $\mathbf{J}$ here is the value for $\mathbf{J}$ in the origin. This value does not vary if we vary the reference point $\mathbf{r}$. $\mathbf{J}$ is therefore a constant and $\nabla\times\mathbf{J}$ is zero. What about the second term? We recognize that $\hat{\mathbf{r}}/r^2$ has the same form as the electric field from a single point charge in the origin, and we know that the curl of this field is zero. Therefore, $\nabla\times\hat{\mathbf{r}}/r^2$ must be zero. We therefore conclude that the divergence of $\mathrm{d}\mathbf{B}$ is zero for a single current element. Using the superposition principle, we conclude that the divergence of any circuit which is composed of individual current elements also must be zero. This shows that the divergence of any magnetic field is zero and that the magnetic flux through any closed surface is zero.

12.2 Ampere's Law

We found that electrostatics could be defined from the divergence and curl of the electric field. Similarly, the laws of magnetostatics can be defined through the divergence and curl of the magnetic field. For the electric field, we have learned to apply Gauss' law to find the electric field in symmetric cases: $\nabla\cdot\mathbf{E} = \rho/\epsilon_0$. The corresponding law for magnetic fields is Ampere's law, which describes the curl of the magnetic field.

Ampere's law on differential form states that the curl of the magnetic field is $\nabla\times\mathbf{B} = \mu_0\mathbf{J}$. We can transform the law to integral form using Stokes' theorem:

$$\int_S(\nabla\times\mathbf{B})\cdot\mathrm{d}\mathbf{S} = \oint_C\mathbf{B}\cdot\mathrm{d}\mathbf{l} = \mu_0\int_S\mathbf{J}\cdot\mathrm{d}\mathbf{S}. \tag{12.8}$$

where we recognize that the curve C is the perimeter of S. We recall that $\int_S\mathbf{J}\cdot\mathrm{d}\mathbf{S}$ is the net current I passing through the surface S. We can therefore rewrite Ampere's law as:

$$\oint_C\mathbf{B}\cdot\mathrm{d}\mathbf{l} = \mu_0 I_{\mathrm{net}}. \tag{12.9}$$

Ampere's law

Ampere's law on differential form states that

$$\nabla \times \mathbf{B} = \mu_0 \mathbf{J} . \tag{12.10}$$

Ampere's law on integral form states that:

$$\oint_C \mathbf{B} \cdot d\mathbf{l} = \mu_0 \int_S \mathbf{J} \cdot d\mathbf{S} . \tag{12.11}$$

where S is a surface with C as its perimeter.

Ampere's law for a current loop states that:

$$\oint_C \mathbf{B} \cdot d\mathbf{l} = \mu_0 I_{\text{net}} . \tag{12.12}$$

where the current I_{net} is the net current through a (any) surface that has the curve C as its perimeter.

Let us address the various terms in Ampere's law in detail.

- The integral $\oint_C \mathbf{B} \cdot d\mathbf{l}$ is a line integral of the magnetic field along a closed curve C. You have seen such integrals before. However, just as with Gauss' law, we will usually not perform this integral but instead take the integral along a path where $\mathbf{B}$ is constant or where $\mathbf{B} \cdot d\mathbf{l}$ is constant. Hence, we can calculate the value of B along the path.
- We recognize the integral $\int_S \mathbf{J} \cdot d\mathbf{S}$ as the net current flowing through the surface S. The surface S in Ampere's law is not an arbitrary surface. It must be a surface which has the curve C as a boundary. In addition, the orientation of the surface S must correspond to the orientation of the path C according to the right-hand rule as illustrated in Fig. 12.2. Notice that there are many surfaces S that have C as a boundary. We will therefore choose a surface that simplifies the calculation.
- Instead of the integral $\int_S \mathbf{J} \cdot d\mathbf{S} = I$, we can use the current I. This current is the *net current* through the surface S. This means that we must add together all the currents going through the surface, taking into account the directions of the currents.

Test your understanding

A magnetic field is $\mathbf{B} = -B_0(y/a)\hat{\mathbf{x}}$ in a region in space. Is there a current density in this region or could this field be set up by currents in other regions in space?[2]

[2] Ampere's law gives $\nabla \times \mathbf{B} = B_0/a\hat{\mathbf{z}} = \mu_0 \mathbf{J}$ so there is a current density along $\hat{\mathbf{z}}$.

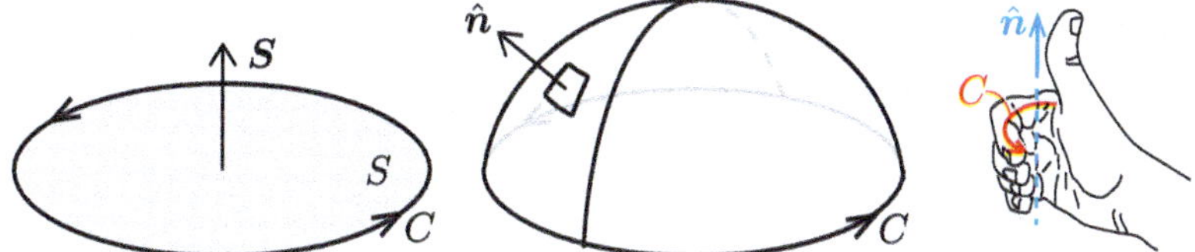

Fig. 12.2 Illustration of the relation between the curve C and the surface S. The direction of the surface corresponds to the direction of the curve according to the right-hand rule. If you place your right hand so that your fingers curve along the curve in positive direction, your thumb is pointing in the positive direction of the surface S. We call this the orientation of the surface, which is described by the direction of the normal vector of the surface

Fig. 12.3 Illustration of currents going along various wires and curves

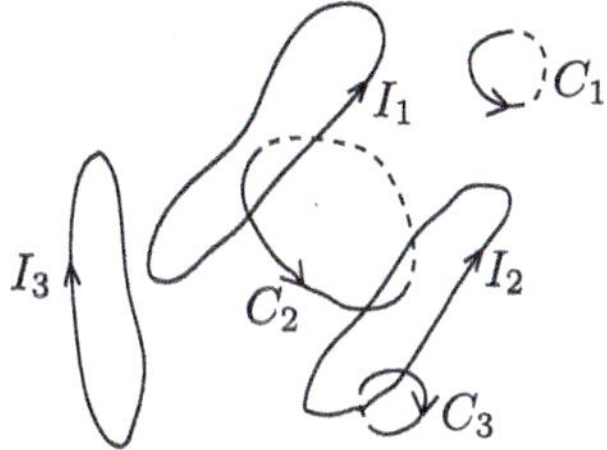

Example: Current Through Curves

Figure 12.3 *shows three currents I_1, I_2 and I_3 along the illustrated circuits. What is the net current, and therefore also the Ampere integral $\oint \mathbf{B} \cdot d\,\mathbf{l}$, through each of the curves C_1, C_2, and C_3?*

Curve C_1. We see that for curve C_1 there are no currents passing through a planar surface through C_1. Therefore $I = 0$ and:

$$\oint_{C_1} \mathbf{B} \cdot d\,\mathbf{l} = \mu_0 I = 0. \tag{12.13}$$

Curve C_2. For curve C_2 there are two currents. First, we find the orientation of the planar surface S_2 with C_2 as the boundary. We see that current I_1 is going through a surface in the direction of the surface normal. Thus I_1 contributes positively to the total current through C_2. However, current I_2 has a direction that is opposite that of the surface normal to S_2. Hence, I_2 contributes negatively to the total current. We find that:

$$\oint_{C_2} \mathbf{B} \cdot d\,\mathbf{l} = \mu_0 (I_1 - I_2) \tag{12.14}$$

Curve C_3. For curve C_3 there are only one current that passes through the curve, I_2. We find the orientation of the surface S_3 and see that I_2 is in the same direction as the surface normal, hence the current I_2 contributes positively to the total current through C_3. This gives that:

$$\oint_{C_3} \mathbf{B} \cdot \mathrm{d}\mathbf{l} = \mu_0 I_2 \tag{12.15}$$

Test your understanding
(a) The figure shows two circuits C_1 and C_2 and two wires both with a current I, one going into the plane (with a cross) and one out of the plane (with a dot). What is $\oint \mathbf{B} \cdot \mathrm{d}\mathbf{l}$ for C_1 and for C_2? (b) The figure shows a circuit C and two wires: one wire with a current I_2 coming out of the plane and one wire with a current I_1 with an angle θ with the plane. Assume that C is the in the plane of the sheet. What is $\oint \mathbf{B} \cdot \mathrm{d}\mathbf{l}$?[3]

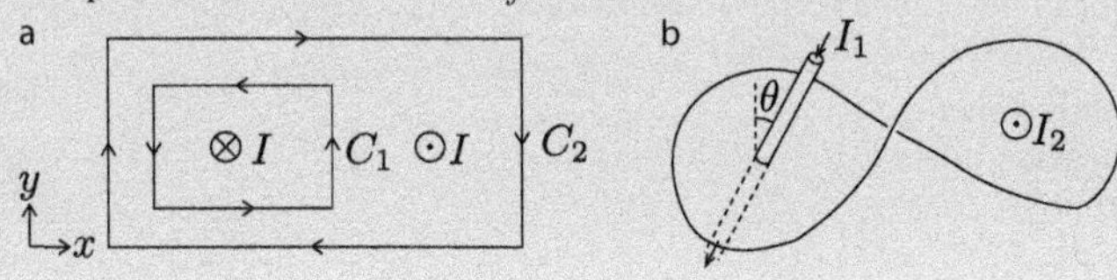

Example: Infinitely Long Cylindrical Conductor

Find the magnetic field **B** *around a cylindrical conductor using Ampere's law. The cylinder has a radius a and is effectively infinitely long. The cylinder carries a current I along the cylinder axis.*

Approach. The idea is to find an integration loop C, called an Amperian loop, so that $\mathbf{B} \cdot \mathrm{d}\mathbf{l}/|\mathrm{d}l| = B$ is a constant along this loop so that $\oint_C \mathbf{B} \cdot \mathrm{d}\mathbf{l} = B \oint_C \mathrm{d}l = BL$, where L is the length of the loop. We can then find B from Ampere's law using $\oint_C \mathbf{B} \cdot \mathrm{d}\mathbf{l} = \mu_0 I$ by finding the current I through a surface with C as its boundary. However, to realize this plan we first need to find the symmetry of the magnetic field. We start by drawing the system in Fig. 12.4. What does it mean that the cylinder is very long? It means that we can effectively assume that it is infinitely long and that there are no edge effects.

Symmetry of the current density. The magnetic field **B** is set up by the currents. We therefore start by describing the symmetry of the current density in the cylinder. The current density must be the same everywhere along the cylinder, because the cylinder has *translational symmetry*. Also, the current flows along the cylinder axis. We therefore rule out any radial or azimuthal components of the current density. Consequently, $\mathbf{J} = J(r)\,\hat{\mathbf{z}}$, where r is the distance from the cylinder axis.

Symmetry of the magnetic field from the symmetry of the cylinder. The *translational* symmetry of the infinite cylinder implies that the magnetic field cannot depend on z. The rotational symmetry around the cylinder implies that the magnetic field

[3] (a) C_1: $-\mu_0 I$, C_2: $-\mu_0 I$; (b) $\mu_0(I_1 + I_2)$ (θ is not important).

cannot depend on the azimuthal angle ϕ. Thus, the field can only depend on the distance r from the axis, $\mathbf{B} = \mathbf{B}(r)$. However, the translational and rotational symmetry still opens for the $\mathbf{B}$-field to have components in the r-, ϕ- and z-direction: $\mathbf{B} = B_r(r)\hat{\mathbf{r}} + B_\phi(r)\hat{\boldsymbol{\phi}} + B_z(r)\hat{\mathbf{z}}$

Symmetry of the magnetic field from Biot-Savart's law. What arguments can we use to constrain the direction of the magnetic field? We know that the magnetic field is set up by the currents. From Biot-Savart's law we know that the contribution from a volume element $\mathrm{d}v$ is

$$\mathrm{d}\mathbf{B} = \mu_0 \frac{\mathbf{J}\,\mathrm{d}v \times \mathbf{R}}{4\pi R^3}, \tag{12.16}$$

where the vector $\mathbf{R}$ points from the element $\mathrm{d}v$ to the observation point. We argued above that $\mathbf{J} = J\hat{\mathbf{z}}$. This means that $\mathrm{d}\mathbf{B}$ will not have any component along z due to the cross product (the cross product is normal to the z-axis). Therefore $B_z(r) = 0$.

Symmetry of the magnetic field from zero-divergence condition. Can the magnetic field have a component in the r-direction, $B_r(r)$? No, because this would imply that there is a net flux of the magnetic field out of a cylindrical surface. If we place a cylindrical (Gauss) surface around the z-axis, there would be no flux of the magnetic field out of the top or bottom surfaces (because $B_z = 0$). The flux out of the surface would then be $\Phi_B = B_r(r)2\pi r L$. This must be zero because $\Phi_B = \oint_S \mathbf{B} \cdot d\,\mathbf{S} = 0$ is always zero. Therefore, $B_r(r) = 0$.

The simplified magnetic field. The argument presented here was long and systematic. When you get used to this, you should be able to make these types of considerations quickly. We have now found that the magnetic field must have the form

$$\mathbf{B} = B_\phi(r)\hat{\boldsymbol{\phi}}. \tag{12.17}$$

Application of Ampere's law to find the magnetic field. We are now ready to use Ampere's law to find the magnetic field. Because the field only depends on r we should choose a curve with a constant r so that the magnitude of the field is constant

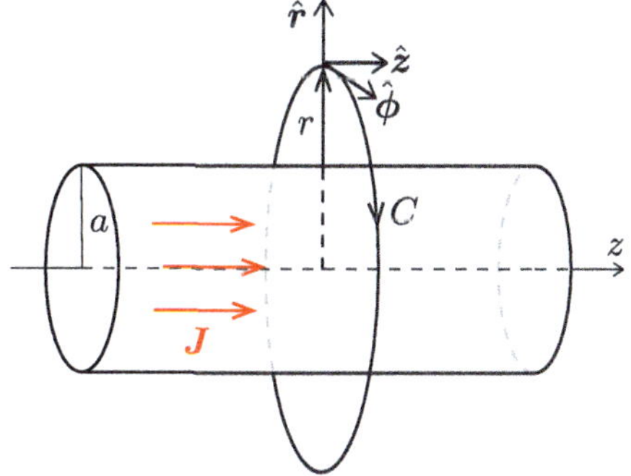

Fig. 12.4 Illustration of the current in and the magnetic field around an infinite conducting cylinder

in the Ampere integral. We should also choose a curve so that the magnetic field and the tangential vector $\mathbf{dl}$ along the curve always has the same relative angle. We do this by choosing a circular curve in the xy-plane with its center at the z-axis. We choose the curve to be oriented in the positive rotational direction around the z-axis (according to the right-hand-rule), so that $\mathbf{dl} = dl\hat{\boldsymbol{\phi}}$ and

$$\oint_C \mathbf{B} \cdot \mathbf{dl} = \oint_C B_\phi(r)\, dl = B 2\pi r = \mu_0 I(r). \tag{12.18}$$

where I is the current going through a cross section of the cylinder (inside the radius r). If $r > a$ then all the current I goes through the curve and $I(r) = I$ is the total current in the cylinder. The magnetic field is therefore:

$$\mathbf{B} = B_\phi \hat{\boldsymbol{\phi}} = \frac{\mu_0 I}{2\pi r} \hat{\boldsymbol{\phi}}. \tag{12.19}$$

What if $r < a$? There is nothing in our arguments that implies that r must be larger than a. However, if $r < a$, we must ensure that $I(r)$ only includes the part of the current that is inside the loop of radius r. If we assume that the current density $\mathbf{J}$ is uniform throughout the cylinder we can calculate the current $I(r)$:

$$I(r) = \int_S \mathbf{J} \cdot \mathbf{dS} = J\pi r^2 \,. \tag{12.20}$$

But what is J? We can find this by setting $r = a$, for which $I(a) = I = J\pi a^2$, hence $J = I/(\pi a^2)$, and

$$I(r) = \frac{I}{\pi a^2}\pi r^2 = I\frac{r^2}{a^2}. \tag{12.21}$$

The magnetic field for $r < a$ is then:

$$\mathbf{B} = \frac{\mu_0 I(r)}{2\pi r}\hat{\boldsymbol{\phi}} = \frac{\mu_0 I r}{2\pi a^2}\hat{\boldsymbol{\phi}}. \tag{12.22}$$

and the complete field is

$$\mathbf{B} = \begin{cases} \frac{\mu_0 I}{2\pi r}\hat{\boldsymbol{\phi}} & , r \geq a \\ \frac{\mu_0 I r}{2\pi a^2}\hat{\boldsymbol{\phi}} & , r < a \end{cases} \tag{12.23}$$

Typical numbers for the magnetic field. If a current of $I = 10\text{A}$ is carried by the wire (the cylindrical conductor), what is the magnetic field at a distance of $r = 10\text{cm}$ from the wire? We find that

$$B = \frac{4\pi\, 10^{-7}\, 10}{2\pi\, 0.1}\text{T} = 2 \cdot 10^{-5}\text{T} = 20\mu\text{T}. \tag{12.24}$$

This is a typical magnitude of a magnetic field from the wires in your house.

Method: Applying Ampere's law to find the magnetic field
Based on this example, we propose a general strategy for how to find the magnetic field for a current distribution (from a current loop) using Ampere's law:

- Find a (integration) Amperian loop C that bounds a surface so that the component of $\mathbf{B}$ along $\mathbf{dl}$ is a constant on this loop. (It may be zero on some parts of the loop—zero is a constant). We call this loop the *Amperian loop* for the system.
- This usually requires that you find a simplified description of the magnetic field in a coordinate system that has the symmetry of the system. For example, for a cylindrical system we often find that $\mathbf{B} = B_\phi(r)\hat{\boldsymbol{\phi}}$.
- You also often have to use that the divergence of $\mathbf{B}$ is zero to argue for the symmetry of the $\mathbf{B}$-field.
- Find the curve integral, which often can be simplified to only include the length of the curve: $\oint_C \mathbf{B} \cdot \mathrm{d}\mathbf{l} = B_l \oint_C \mathrm{d}l$
- Use Ampere's law $\mu_0 I = B \oint_C \mathrm{d}l$, to find the magnetic field as a function of current and position.
- Notice that the curve sometimes will enclose the current several times, for example when the wire is twisted several times around the magnetic field as in a solenoid. You must then include the current through all the loops or multiply the current by the number of loops.

Example: Toroid

There are several ways to increase the current going through a surface: We can increase the current carried in a single wire or we can increase the number of wires. There are two main geometries where the number of wires is increased systematically to generate a well-controlled magnetic field in a region of space: the *toroid* and the *solenoid*. The toroid consists of a donut shaped material of rectangular cross section as illustrated in Fig. 12.5a. The wire is uniformly wound N times around the toroid and carries a current I. The toroid has an inner radius a, an outer radius b and a height h. The wire is usually much more densely wound than illustrated in the figure.

Find the magnetic field inside and outside the toroid.

Approach. We plan to use Ampere's law to find the magnetic field and start by finding the symmetry of the magnetic field and from there a find reasonable Amperian loop along which to integrate.

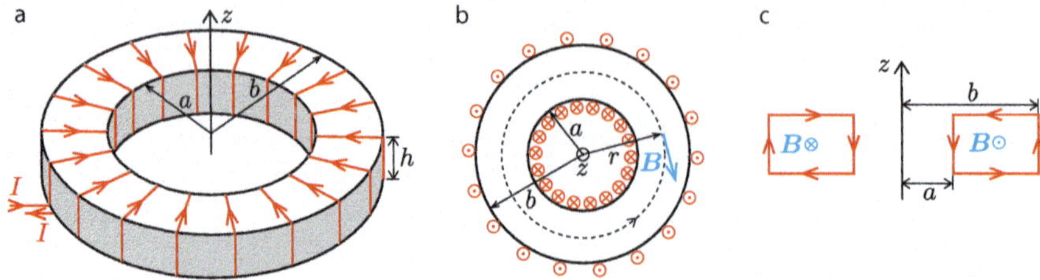

Fig. 12.5 Illustration of the toroid geometry. **a** Perspective view. **b** Top view. Currents going into the plane are shown with a cross inside a circle and currents going out of the plane are dots inside circles

Symmetry. What is the expected symmetry of the **B**-field in this case? We expect it to have some sort of cylindrical symmetry because the system does not change when rotated around the z-axis. We use cylindrical coordinates to describe the system $\mathbf{B} = \mathbf{B}(r, \phi, z)$. First, we do not expect the field to depend on ϕ, because of the rotational symmetry around the z-axis. We approximate the wound wire with N windings as a set of N rectangular circuits. Each circuit is in the rz-plane as illustrated in Fig. 12.5c. For each such circuit, the magnetic field in the plane of the circuit will be normal to the plane of the circuit, because from Biot-Savart's law we know that the field will be normal to the current elements, which are in the rz-plane, and the **R**-vector, which also is in the rz-plane. Consequently, the magnetic field will point in the $-\phi$ direction. We also assume that the field does not depend on z. The symmetry of **B** is therefore:

$$\mathbf{B} = B(r)\hat{\boldsymbol{\phi}} \,. \tag{12.25}$$

Applying Ampere's law. We can now apply Ampere's law along a path where $\mathbf{B} \cdot \mathrm{d}\mathbf{l}$ is constant, that is, for a circle with radius r as illustrated in Fig. 12.5c. For this path we have:

$$\oint_C \mathbf{B} \cdot \mathrm{d}\mathbf{l} = B 2\pi r = -N I \mu_0 , \tag{12.26}$$

where we have used that the total current through the loop C is the sum of the currents in all the N windings of the wire around the toroid. We see from Fig. 12.5b that the surface normal for C points in the positive z-direction, whereas all the currents inside C points in the negative z-direction. The total current is therefore $-NI$. This is only true inside the toroid. Outside the toroid there is no net current through the circular path C and the field is zero. This means that the field is:

$$\mathbf{B} = \begin{cases} -\frac{\mu_0 N I}{2\pi r}\hat{\boldsymbol{\phi}} & \text{inside the toroid} \\ 0 & \text{outside} \end{cases} \tag{12.27}$$

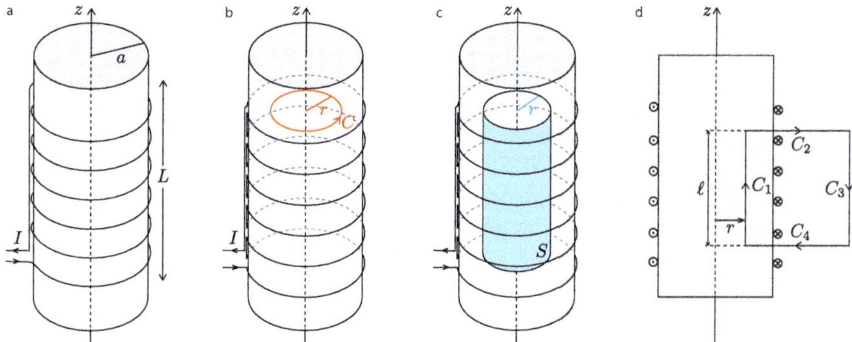

Fig. 12.6 Illustration of the solenoid geometry

Example: Solenoid

A solenoid is what we get if we wind a wire around a long cylinder as illustrated in Fig. 12.6a. We assume that the cylinder is effectively infinitely long and that there are N windings per length L of the cylinder.

Find the magnetic field inside and outside the solenoid.

Solenoid and Toroid. We could assume that the solenoid is a toroid in the limit when the radius of the toroid becomes very huge. The field inside the toroid is $\mathbf{B} = \mu_0 I N/(2\pi r)\hat{\boldsymbol{\phi}}$. We see that the number of windings per length for the toroid is $N/(2\pi r)$ and that the direction of the field is tangential to the toroid, which corresponds to being along the z-axis for the solenoid. The magnetic field of the solenoid is therefore expected to be $\mathbf{B} = \mu_0 I N/L\hat{\mathbf{z}}$, where z is the axis long the solenoid.

Approach. Let us instead do a direct calculation for the solenoid starting from the symmetry of the magnetic field in the solenoid and then apply Ampere's law for a well-chosen curve.

Detailed geometry of the solenoid. We look at a length L along the cylinder and approximate the windings as N circular paths that are connected by short linear wires directed along the z-axis as illustrated in Fig. 12.6b. The current returns through a wire placed along the short linear wires. The magnetic field from these two wires that are placed closely together therefore cancels, and we only look at the magnetic field from the N circles.

Symmetry of the solenoid. We use cylindrical coordinates to describe the system because the system does not change if we rotate it around the z-axis. Also, for a long system, the system does not change if we translate it along the z-axis. This means that the field cannot depend on z or ϕ, but only on r.

Components of the field. The magnetic field may in general have three components: $\mathbf{B} = B_r(r)\hat{\mathbf{r}} + B_\phi(r)\hat{\boldsymbol{\phi}} + B_z(r)\hat{\mathbf{z}}$. To determine the ϕ-component of the field, we can apply Ampere's law for a circular path C_1 around the cylinder in Fig. 12.6b. Because there is no net current through C_1, Ampere's law gives:

$$\oint_{C_1} \mathbf{B} \cdot d\mathbf{l} = \oint_{C_1} B_\phi \, dl = B_\phi 2\pi r = 0 \quad \Rightarrow \quad B_\phi = 0. \tag{12.28}$$

To determine the r-component in the direction radially out from the z-axis, we can apply the divergence theorem for the **B**-field for a cylindrical surface around the cylinder as illustrated in Fig. 12.6c:

$$\int \mathbf{B} \cdot d\mathbf{S} = B_r 2\pi r L = 0 \quad \Rightarrow \quad B_r = 0. \tag{12.29}$$

We are therefore left with $\mathbf{B} = B_z(r)\hat{\mathbf{z}}$. We determine $B_z(r)$ from Ampere's law by selecting the Amperian loop in Fig. 12.6d and find $\oint \mathbf{B} \cdot d\mathbf{l}$. The loop consists of four parts. A part C_1 of length ℓ inside the cylinder along the z-axis. The contribution from this to the integral is $B_z \ell$. Parts C_2 and C_4 do not contribute to the integral because the magnetic field $B_z(r)\hat{\mathbf{z}}$ is normal to the path along the r-direction. We can place part C_3 far away from the axis, where we expect the magnetic field to decay to zero, and there is therefore not contribution from C_3 to the integral. Ampere's law gives

$$\oint_C \mathbf{B} \cdot d\mathbf{l} = \int_{C_1} B_z dl = B_z \ell = \mu_0 I_{\text{net}}. \tag{12.30}$$

In this case, $I_{\text{net}} = (N/L)\ell I$, where N/L is the number of windings per unit length so that $(N/L)\ell$ is the number of windings inside the curve of length ℓ. The field is therefore

$$\mathbf{B} = B_z \hat{\mathbf{z}} = \mu_0 I \frac{N}{L} \hat{\mathbf{z}}. \tag{12.31}$$

Consequently, the magnetic field is uniform inside the solenoid. The solenoid is therefore the magnetic equivalent of the plane capacitor: a simple geometry where the magnetic field is uniform inside.

Test your understanding
A conductor consists of two cylindrical, massive parts with the same axis as shown in the figure. A constant current I is running through the conductor. We measure the magnitude of the magnetic field in the two points 1 and 2. You can assume that 1 and 2 are far from each other so that the field in 1 is not affected by the current in the thin cylinder and that the field in 2 is not affected by the current in the thick cylinder. (a) Is B_1 larger, equal or smaller than B_2? (b) What would the answer be is both points 1 and 2 are inside the radius of the smallest cylinder? (c) What would the answer be if the points 1 and 2 are in a distance r which is smaller than the radius of the largest cylinder, but larger than the smallest cylinder?[4]

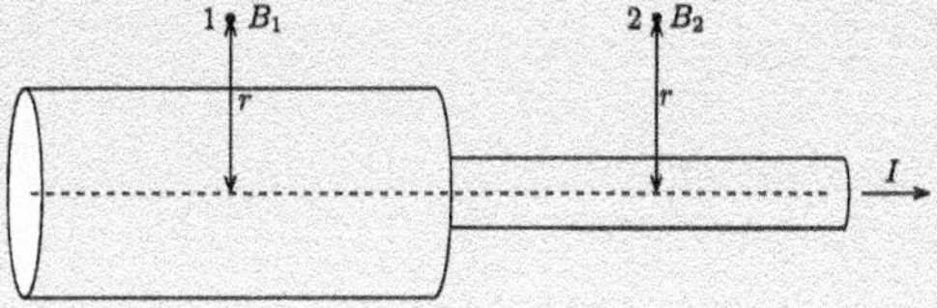

12.3 Comparison of Magnetostatics and Electrostatics

We have now discovered the main laws of electro- and magneto-*statics*, and we have found that they have a nice kind of symmetry. For electrostatics we found that

$$\nabla \cdot \mathbf{E} = \frac{1}{\epsilon_0}\rho \quad \text{(Gauss' law)} \tag{12.32}$$

$$\nabla \times \mathbf{E} = 0 \tag{12.33}$$

These two laws are called **Maxwell's equations** for electrostatics. We can use these to find the electric field for any charge distribution. It is only the second equation that will be modified (with Faraday's law) when we introduce a time-varying magnetic field. These equations can be found from Coulomb's law, and we can show that they reproduce Coulomb's law.

For magnetostatics we have that

$$\nabla \cdot \mathbf{B} = 0 \tag{12.34}$$

$$\nabla \times \mathbf{B} = \mu_0 \mathbf{J} \quad \text{(Ampere's law)} \tag{12.35}$$

These two laws are called **Maxwell's equations** for magnetostatics. Together with Maxwell's equations for electrostatics, they form the four Maxwell's equations for stationary systems.

In addition to Maxwell's equations, Lorentz' law shows how the fields interacts with charges: The force on a charge Q moving with a velocity $\mathbf{v}$ in an electric field

[4] (a) $B_1 = B_2$; (b) $B_1 < B_2$; (c) $B_1 < B_2$.

E and a magnetic field **B** is:

$$\mathbf{F} = Q\left(\mathbf{E} + \mathbf{v} \times \mathbf{B}\right). \tag{12.36}$$

These laws essentially provides the entire description of (static) electromagnetism. We will modify them as we introduce dynamics, that is, time-varying fields.

12.4 Proof of Ampere's Law

To prove Ampere's law we will follow Notaros (2011) and use the vector potential representation of the magnetic field, where

$$\mathbf{A} = \frac{\mu_0}{4\pi} \int \frac{\mathbf{J}}{R}\, \mathrm{d}v', \tag{12.37}$$

and $\mathbf{B} = \nabla \times \mathbf{A}$. We want to prove Ampere's law, that is, to show that $\nabla \times \mathbf{B} = \mu_0 \mathbf{J}$.

Using the vector relation for $\mathbf{a} \times (\mathbf{b} \times \mathbf{c})$. We start by calculating $\nabla \times \mathbf{B}$ using vector relations. We know that the general vector relation

$$\mathbf{a} \times (\mathbf{b} \times \mathbf{c}) = \mathbf{b}\left(\mathbf{a} \cdot \mathbf{c}\right) - \mathbf{c}\left(\mathbf{a} \cdot \mathbf{b}\right). \tag{12.38}$$

We use this to find $\nabla \times (\nabla \times \mathbf{A})$, getting:

$$\nabla \times (\nabla \times \mathbf{A}) = \nabla\left(\nabla \cdot \mathbf{A}\right) - \left(\nabla \cdot \nabla\right)\mathbf{A} = \nabla\left(\nabla \cdot \mathbf{A}\right) - \nabla^2 \mathbf{A}. \tag{12.39}$$

Finding an expression for $\nabla^2 \mathbf{A}$. Let us start with the second term, $\nabla^2 \mathbf{A}$. This is the Laplacian of a vector function **A**. In Cartesian coordinates this corresponds to the Laplacian for each of coordinates:

$$\nabla^2 \mathbf{A} = \nabla^2 A_x \hat{\mathbf{x}} + \nabla^2 A_y \hat{\mathbf{y}} + \nabla^2 A_z \hat{\mathbf{z}}, \tag{12.40}$$

where

$$A_x = \frac{\mu_0}{4\pi} \int \frac{J_x}{R}\, \mathrm{d}v' \tag{12.41}$$

so that

$$\nabla^2 A_x = \frac{\mu_0}{4\pi} \int \nabla^2 \frac{J_x}{R} \mathrm{d}v' \tag{12.42}$$

Now, we will use a special trick, where we use our experience from electrostatics to solve this equation. From electrostatics we remember that the potential V is given by the integral:

$$V = \frac{1}{4\pi\epsilon_0} \int \frac{\rho \, \mathrm{d}v'}{R} \tag{12.43}$$

and that V must satisfy Poisson's equation $\nabla^2 V = -\rho/\epsilon_0$. We insert the integral expression for V into Poisson's equation:

$$\nabla^2 V = \frac{1}{4\pi\epsilon_0} \int \nabla^2 \frac{\rho}{R} \, \mathrm{d}v' = -\frac{\rho}{\epsilon_0} \tag{12.44}$$

and

$$\frac{1}{4\pi} \int \nabla^2 \frac{\rho}{R} \, \mathrm{d}v' = -\rho \tag{12.45}$$

This must be true for any ρ, including for $\rho = J_x$. That is, we replace ρ with J_x and find that

$$\frac{1}{4\pi} \int \nabla^2 \frac{J_x}{R} \, \mathrm{d}v' = -J_x \, . \tag{12.46}$$

The same is true for the other component of $\mathbf{J}$. We therefore get that

$$\nabla^2 \mathbf{A} = -\mu_0 \mathbf{J} \tag{12.47}$$

Finding an expression for $\nabla (\nabla \cdot \mathbf{A})$. Now, we have found one of the two terms in (12.39). What about the first term, $\nabla (\nabla \cdot \mathbf{A})$? Let us look at $\nabla \cdot \mathbf{A}$ in detail. It is:

$$\nabla \cdot \mathbf{A} = \frac{\mu_0}{4\pi} \nabla \cdot \int \frac{\mathbf{J}(\mathbf{r}') \, \mathrm{d}v'}{R} = \frac{\mu_0}{4\pi} \int \nabla \cdot \left(\frac{\mathbf{J}}{R} \right) \mathrm{d}v' . \tag{12.48}$$

We have moved ∇ inside the integral, since the integral is with respect to $\mathbf{r}'$ and ∇ is the derivative with respect to $\mathbf{r}$, which is a constant in the integration. We can now use that $\nabla(1/R) = -\nabla'(1/R)$ and the divergence theorem to rewrite this integral as

$$\frac{\mu_0}{4\pi} \int \nabla' \cdot \left(\frac{\mathbf{J}}{R} \right) \mathrm{d}v' = \frac{\mu_0}{4\pi} \oint_S \frac{\mathbf{J}}{R} \cdot \mathrm{d}\mathbf{S} \, . \tag{12.49}$$

Here, the integral is over a volume v and the surface S enclosing the volume. If we let the volume v increase and hence the surface S becomes larger and larger, all the currents will eventually only be inside the volume. Therefore, no currents will cross the surface S and the integral will approach zero. Consequently, we found that $\nabla \cdot \mathbf{A} = 0$.

Putting it all together. We have therefore found that

$$\nabla \times (\nabla \times \mathbf{A}) = \mu_0 \mathbf{J} \, , \tag{12.50}$$

and therefore that

$$\nabla \times \mathbf{B} = \mu_0 \mathbf{J}, \tag{12.51}$$

which is Ampere's law on differential form. We have also derived the *vector Poisson's equation* for the magnetic vector potential:

$$\nabla^2 \mathbf{A} = -\mu_0 \mathbf{J}. \tag{12.52}$$

We can use the vector Poisson's equation to find the magnetic field given the current distributions in the same way as we used Poisson's equation to find the electric field from the charge distribution. However, the equation is somewhat more complicated because it is a vector equation, but we can in principle use many of the same tools as we have developed for Laplace's and Poisson's equation.

Summary

The **magnetic flux** through a closed surface and the divergence of the magnetic field is zero

$$\oint_A \mathbf{B} \cdot d\mathbf{S} = 0 \quad \Leftrightarrow \quad \nabla \cdot \mathbf{B} = 0.$$

Ampere's law states

$$\nabla \times \mathbf{B} = \mu_0 \mathbf{J} \quad \Leftrightarrow \quad \oint_C \mathbf{B} \cdot d\mathbf{l} = \mu_0 \int_S \mathbf{J} \cdot d\mathbf{S},$$

Applied to an *Amperian loop* C through which a net current I_{net} flows, Ampere's law is:

$$\oint_C \mathbf{B} \cdot d\mathbf{l} = \mu_0 I_{\text{net}}.$$

We find the magnetic field using Ampere's law by:

- Finding an integration loop (Amperian loop) C where $\mathbf{B} \cdot d\mathbf{l}$ is a constant due to symmetry
- Apply Ampere's law to find the magnetic field as a function of current and position.

There exists a **vector potential A** so that $\mathbf{B} = \nabla \times \mathbf{A}$. The vector potential is defined as

$$\mathbf{A}(\mathbf{r}) = \frac{\mu_0}{4\pi} \int_v \frac{\mathbf{J}(\mathbf{r}')\, dv'}{|\mathbf{r} - \mathbf{r}'|},$$

or from the **vector Poisson's equation**: $\nabla^2 \mathbf{A} = -\mu_0 \mathbf{J}$. The divergence of the vector potential is zero: $\nabla \cdot \mathbf{A} = 0$.

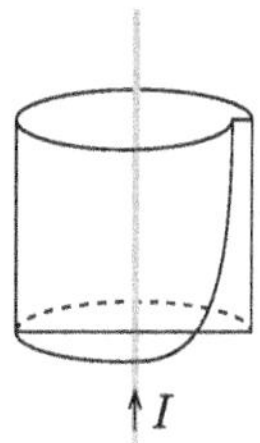

Fig. 12.7 A creative surface around a current-carrying wire

Exercises

Discussion Exercises

12.1 Monopoles. Let us assume that your friend Q claims to have found a magnetic monopole? How would you go about to document or disprove this? What properties would a magnetic monopole have?

12.2 Magnetic storm. Electrically charged particles flowing out from the Sun may in periods with huge solar activity create disturbances in the Earth's magnetic field. Can you explain this phenomenon?

12.3 Field lines. The magnetic field on the inside of a toroid is non-zero, but outside it is zero. Argue why this could be the case using both symmetry and an Amperian loop. What does this mean for the magnetic field of an infinite solenoid?

Tutorials

12.4 Possible and impossible fields. Which of these fields may be magnetic fields?
(a) $\mathbf{B} = x\hat{\mathbf{x}} + y\hat{\mathbf{y}} - 2z\hat{\mathbf{z}}$
(b) $\mathbf{B} = B_r(r)\hat{\mathbf{r}}$ (in spherical coordinates)
(c) $\mathbf{B} = B_\phi\hat{\boldsymbol{\phi}}$ (in cylindrical coordinates)
(d) $\mathbf{B} = -y\hat{\mathbf{x}} + x\hat{\mathbf{y}}$

12.5 It cannot possibly be a field. Could you sketch a field, which is not on the form $\mathbf{B} = B_r\hat{\mathbf{r}}$, and which cannot be a magnetic field?

12.6 Is something wrong with the law. Your friend Q does not believe that there cannot be magnetic monopoles. *Look here*, he says, *if there is a current through a long wire, there is a magnetic field around the wire. If I place the surface in* Fig. 12.7 *around it, then there will be a flux in through the vertical surface, but no flux out. Therefore there is a net flux in through the surface and your theory is disproven!* Is Q's argument sound?

12.7 Net current. Figure 12.8 shows two different Amperian loops A and B. Loop A passes through the points $(0, 0, 0)$, $(0, a, 0)$, $(0, a, 2a)$ and $(0, 0, 2a)$. Loop B passes through the points $(0, 0, 0)$, $(0, a, 0)$, $(a, a, 2a)$, and $(a, 0, 2a)$. There is a uniform current density $\mathbf{J} = J_0\hat{\mathbf{x}}$ everywhere.

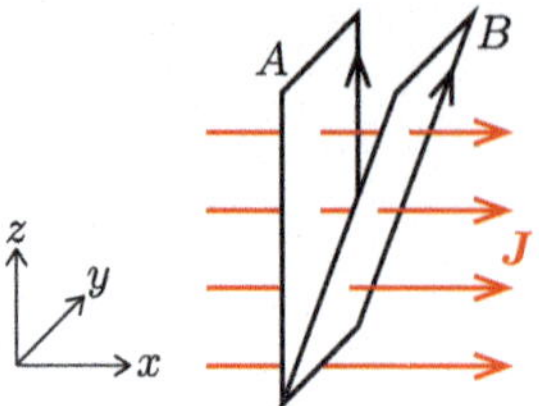

Fig. 12.8 Two current loops

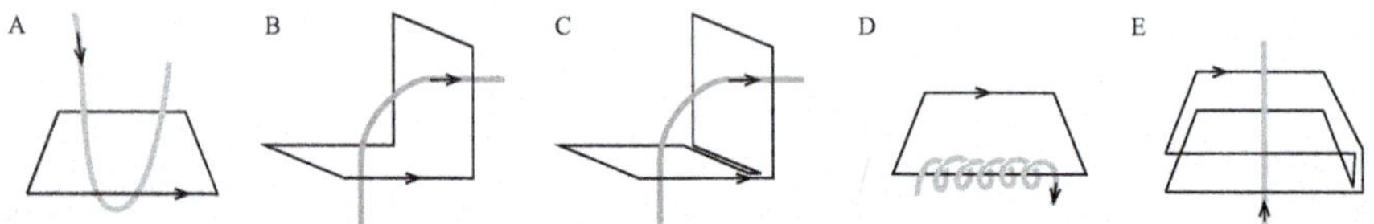

Fig. 12.9 A selection of Amperian loops

(a) What is the net current through loop A? Be extra careful in checking the sign of the current. What determines this sign?
(b) What is the net current through loop B?
(c) Based on this, can you come up with a smart rule for how to find the net current through a loop?

12.8 Strange Amperian loops. Figure 12.9 shows different Amperian loops and a wire that carries a current I. We want to find the net current through each of the Amperian loops.

12.9 Ampere's law and symmetries. In this exercise we train how to use Ampere's law to find the magnetic field. We study an infinitely large, thin sheet in the xy-plane with a uniform surface current density $\mathbf{J} = J_0\hat{\mathbf{x}}$.
(a) What coordinate system should we use to describe this system?
(b) What coordinates may $\mathbf{B}$ depend on?
(c) Can the magnetic field have a component in the x-direction? Provide an argument for your answer.
(d) What happens with the magnetic field if we replace $-z$ with z? Carefully consider what happens with a possible y- and z-component of the $\mathbf{B}$-field.
(e) Can the magnetic field have a component in the z-direction. Provide an argument for your answer.
(f) How would you choose an Amperian loop so that the magnetic field along the Amperian loop is constant (in the direction along the loop)?
(g) Write down Ampere's law for this loop. Check the sign of the current, the signs of the magnetic field and the direction of the line elements along the loop.
(h) Find the magnetic field.

12.10 Surface current on a cylinder. An infinitely long cylindrical conductor with radius a has a surface current density $\mathbf{J}_S$ in the azimuthal direction $\mathbf{J}_S = J_0\hat{\boldsymbol{\phi}}$ as shown in Fig. 12.10.

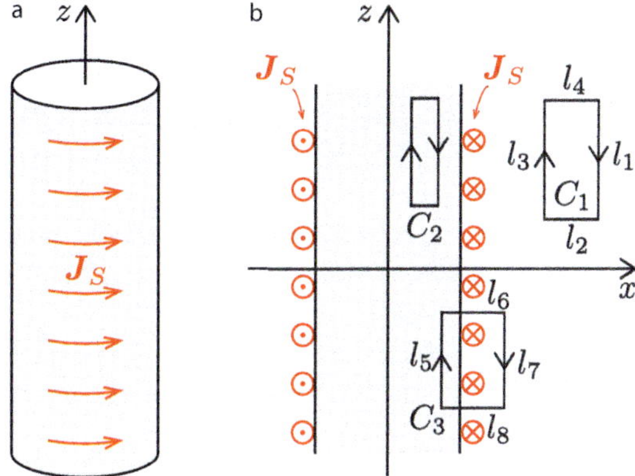

Fig. 12.10 Cylinder with surface current density

(a) From the symmetry of the system, which variables r, ϕ, z may $\mathbf{B}$ depend on, $\mathbf{B} = \mathbf{B}(r, \phi, z)$?

(b) Can the $\mathbf{B}$-field have a component in the radial direction?

(c) Can the $\mathbf{B}$-field have a component B_ϕ in the azimuthal direction?

(d) We use Ampere's law on the circuit C_1 in the xz-plane. What does this tell us about B_z?

(e) If we assume that $|\mathbf{B}| \to 0$ when $r \to \infty$. What is the magnetic field outside the cylinder, that is, for $r > a$?

(f) What does Ampere's law applied to the Amperian loop C_2 tell us about $\mathbf{B}$?

(g) Apply Ampere's law on circuit C_3 to find $\mathbf{B}$ inside the cylinder.

Homework

12.11 Coaxial cable.

(a) Find the $\mathbf{B}$-field from an infinite, straight, cylindrical conductor with radius a which carries a current I which is uniformly distributed across the conductors cross section.

(b) Find the magnetic field from an infinitely long cylindrical shell with inner radius b and outer radius c when it carries a current I which is uniformly distributed across its cross section.

(c) Find the magnetic field from a coaxial cable where the inner conductor has a radius a and the outer conductor has inner radius b and outer radius c. There is a current $+I$ in the inner cable and a current $-I$ in the outer cable.

12.12 Cylindrical cable with cavity. An infinitely long cylindrical conductor has a radius a and carries a current I. The cable has an infinitely long cylindrical cavity, a hole, with radius b with its center a offset distance d from the center of the conductor $(d + b < a)$.

(a) Make a sketch of the system.

(b) Show that the magnetic field in the cylinder without a cavity is $\mathbf{B}(r) = \frac{\mu_0 I r}{2\pi a^2}\hat{\boldsymbol{\phi}}$.

(c) Find the magnetic field inside the cavity and show that it is uniform within the cavity. (Hint: Use superposition and rewrite $\hat{\boldsymbol{\phi}}$ as a cross product of e.g. $\hat{\mathbf{z}}$ and $\mathbf{r}$.)

12.13 A thick plate. In this exercise we will address the magnetic field from a plate with finite thickness. The plate stretches to infinity in the x- and y-direction and extends from $-h$ to $+h$ in the z-direction. The plate has a current density $\mathbf{J} = J_0|z|\hat{\mathbf{x}}$ for $z \in (-h, h)$ and 0 otherwise (outside the plate).
(a) Why does the **B**-field not have a component in the x-direction?
(b) What symmetries do you find for **B**?
(c) Why does the **B**-field not have a component in the z-direction?
(d) Find the magnetic field everywhere in space.

12.14 The field from a rotating charged cylinder. A long, cylindrical shell with axis along the z-axis has radius a and surface charge density ρ_s. The cylinder rotates around its axis with an angular velocity ω.
(a) What are the relevant symmetries in this system? What do we know about the magnetic field based on these symmetries?
(b) What components of the magnetic field are non-zero?
(c) What would be a good choice for an Amperian loop to find the magnetic field from the cylinder? (Hint: A moving charge can be considered as a current.)
(d) Find the magnetic field from the rotating cylinder.

The magnetic field in a region in space centered around the origin has cylindrical symmetry and is given as $\mathbf{B} = B_0\hat{\boldsymbol{\phi}}$, where B_0 is a constant.
(e) What is the current density in this region of space?
(f) Assume that the current density is given by the expression from the previous exercise out to a radius a, that is for $r < a$, and then is 0 beyond that, that is, for $r > a$. What is the magnetic field for $r > a$?

Reference

Branislav M. Notaros, *Electromagnetics*, Pearson, 2011.

Chapter 13
Magnetization

We have now introduced Ampere's law for magnetic fields, which plays a role similar to that of Gauss' law for electric fields. However, so far we have only addressed magnetic fields in vacuum. For electric fields, we found that polarization of dielectric materials introduces bound charges in addition to the free charges. We could either treat the bound charges explicitly in Gauss' law, or we could introduce a new concept, the displacement field $\mathbf{D}$ and reformulate Gauss' law in terms of the displacement field to simplify the treatment of polarized materials. In this chapter, we will introduce a similar type of approach to address magnetization.

While polarization occurred by aligning dipoles or inducing electron cloud distortions due to the electric field, magnetization occurs due to alignment of microscopic, permanent currents, which in a classical picture can be interpreted as the motion of electrons spinning around their nuclei. Each of these microcurrents act as electromagnets, generating magnetic fields. These fields will usually cancel each other because they are aligned in random directions, but when a magnetic field is applied, they will align, and the material becomes magnetized. Unlike for polarization, the magnetic field set up by magnetization may be parallel to the magnetic field (for paramagnetic materials) or opposite to the magnetic field (for diamagnetic materials). For some magnetic materials, the magnetization remains also after the applied magnetic field is turned off (for ferromagnetic materials).

13.1 Magnetization and Bound Currents

Classical model

How can we understand permanent magnets? Let us start from a simple classical model where electrons are orbiting the nucleus as illustrated in Fig. 13.1. The electrons have spin and there may be other contributions to the angular momentum of

© The Author(s), under exclusive license to Springer Nature Switzerland AG 2026

A. Malthe-Sørenssen, *Elementary Electromagnetism Using Python*, Undergraduate Texts in Physics, https://doi.org/10.1007/978-3-032-19876-1_13

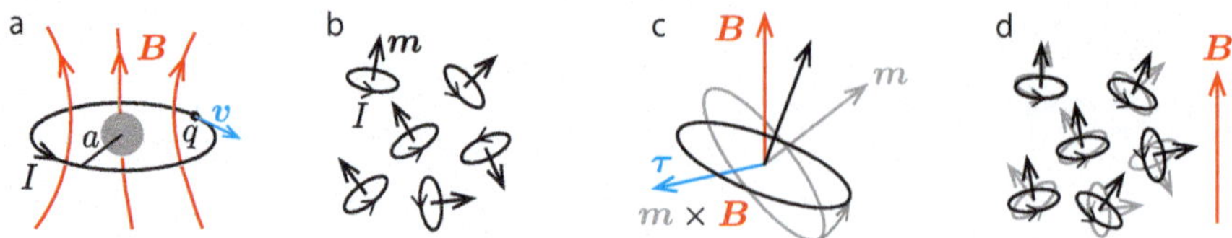

Fig. 13.1 **a** Illustration of an electron orbital in a classical interpretation and the resulting magnetic field. **b** A magnetic material consists of many, randomly oriented current loops. **c** When a magnetic moment is placed in a magnetic field, a torque will act on the current loop to align it with the magnetic field. **d** When a magnetic field is applied to a magnetic material with many current loops, they tend to align with the applied magnetic field, modifying the total magnetic field

the atom but let us first address the effect of a moving electron. It moves in a circular orbit with radius a, charge $q = -e$ and a period $T = 2\pi a/v$, where v is the velocity. This corresponds to a current of $I = e/T = ev/(2\pi a)$ in the opposite direction to the motion of the electron. We have illustrated this in Fig. 13.1a by showing that the direction of the current is opposite the direction of the electron. The magnetic moment of this orbital is

$$\mathbf{m} = \frac{1}{2} e v a \hat{\mathbf{z}} , \tag{13.1}$$

where $\hat{\mathbf{z}}$ points in the direction of the axis of the current.

The orbiting electron corresponds to a small microcircuit that will set up a magnetic field $\mathbf{B}_m$ as illustrated in Fig. 13.1a. The magnetic field will point in the direction of the magnetic moment. A simple model of a magnetic material is that it consists of many such microscopic magnetic moments, where each moment corresponds to a small localized or bound current. The net magnetic field from all these microscopic magnetic moments will typically be zero if they are not aligned, but point in random directions as illustrated in Fig. 13.1b.

If we apply a magnetic field to such a system, there will be a torque $\boldsymbol{\tau} = \mathbf{m} \times \mathbf{B}$ acting on each such magnetic moment $\mathbf{m}$, as illustrated in Fig. 13.1c. The torque will act to rotate the dipole so that the magnetic moment aligns with the applied magnetic field. Consequently, the magnetic field $\mathbf{B}_m$ set up by the magnetic moments will act in the same direction as the applied magnetic field $\mathbf{B}$ as illustrated in Fig. 13.1d.[1] This effect is similar to what we found when we addressed polarization for electric fields, and we will now introduce a magnetization $\mathbf{M}$ in analogy with the polarization $\mathbf{P}$. Just as for polarization, we note that the laws we have seen so far in magnetostatics, Biot-Savart's law and Ampere's law, are valid in a magnetic medium, but we need to include all currents, including the effects of the microscopic current loops. Just

[1] Notice that there are also mechanisms that act in the opposite direction, so that the magnetic moment may be turned in a direction opposite the applied magnetic field, an effect we call diamagnetism.

like for polarization, this is impractical, and we modify the theory by introducing a magnetization $\mathbf{M}$ to take care of the bound currents. Let us see how we can develop that theory.

Magnetization

A magnetic material consists of many microscopic current loops representing the bound currents with magnetic moments $\mathbf{m}$ as illustrated in Fig. 13.1. We interpret these magnetic moments as small circuits with current I so that their magnetic moment is

$$\mathbf{m} = I\mathbf{S}_m \ , \tag{13.2}$$

where $\mathbf{S}_m$ describes both the area and the direction of the magnetic moment (and therefore also the direction of the magnetic field from the current loop). In a small volume element $\mathrm{d}v$ of the magnetic material, there will by many magnetic moments. We therefore describe the system by the *magnetization*, which is the sum of the magnetic moments in the volume element divided by the volume of the element (the volume density of magnetic moments):

Magnetization
The magnetization $\mathbf{M}$ is defined as the net magnetic moment per unit volume:

$$\mathbf{M} = \frac{1}{\mathrm{d}v} \sum_{i \text{ in } \mathrm{d}v} \mathbf{m}_i \ , \tag{13.3}$$

which means that the total magnetic moment in $\mathrm{d}v$ is $\mathbf{M}\,\mathrm{d}v$. The magnetization $\mathbf{M}$ plays a role similar to the polarization $\mathbf{P}$. We interpret $\mathbf{M}$ as the *net* magnetic moment in a volume, which we expect to increase if the microscopic magnetic moments $\mathbf{m}_i$ are aligned.

First, we will simply assume that the material may be magnetized with a given magnetization $\mathbf{M}$, find the associated current, and then use Ampere's law to find the corresponding magnetic fields. Later, we will modify Ampere's law to include the effects of these bound currents.

Test your understanding

(a) The figure shows a circuit with a current I in a constant, homogeneous magnetic field $\mathbf{B} = B_0\hat{\mathbf{z}}$. What is the direction of the torque on the circuit? (b) The figure shows a square circuit of dimensions $d \times d$. A region in space is filled with such circuits placed a distance d from each other in the x, y and z-direction. What is the magnetization of the system?[2]

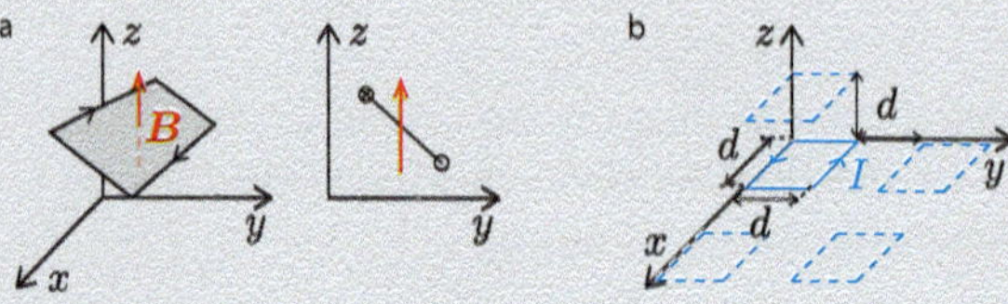

Bound currents in a magnetized material

What is the net effect of all the microloops of currents on the macroscopic current? We define the current as the net flux of charges through a surface S enclosed by a curve C. How does the microloops contribute? The local magnetization $\mathbf{M}(\mathbf{r})$ is the effect of many microloops in a small region around $\mathbf{r}$. We assume that we can approximate $\mathbf{M}$ as the effect of identical microloops with magnetic moment $\mathbf{m}$ so that $\mathbf{M} = N\mathbf{m}$ where N is the number of microloops per unit volume. Figure 13.2a illustrates a surface S enclosed by a curve C as well as several microloops with varying orientation depending on how $\mathbf{M}$ varies in space.

First, we notice that it is only the microloops that enclose (run around) the curve C that will contribute to the net current through C. Loops that do not enclose C will either not intersect the surface S spanned by C at all, or they will intersect the surface in two points, both inward and outward, so that the net contribution to the current through the surface will be zero. To sum up all the contributions from the microloops to the current through C, we therefore only need to sum up the contributions from loops that enclose C.

We divide the curve C into small elements $d\mathbf{l}$. Which current loops enclose $d\mathbf{l}$? This depends on the orientation of the loops relative to the orientation of $d\mathbf{l}$. The orientation of the loops is given by the direction of $\mathbf{M}$. We interpret each small magnetic moment $\mathbf{m}$ as due to a current I around a loop with area S_m and normal vector $\hat{\mathbf{n}}$ so that $\mathbf{S}_m = S_m\hat{\mathbf{n}}$ and $\mathbf{m} = I\mathbf{S}_m$. All the current loops with centers that are within the tilted cylinder with area S_m along $d\mathbf{l}$ will intersect the curve C only once, as illustrated in Fig. 13.2b. The volume of this cylinder is the base area S_m multiplied with the height, $\hat{\mathbf{n}} \cdot d\mathbf{l}$, as illustrated in Fig. 13.2c. That is, $dv = S_m\hat{\mathbf{n}} \cdot d\mathbf{l} = \mathbf{S}_m \cdot d\mathbf{l}$. The number of magnetic moments in this volume is $N dv$, and each magnetic moment will contribute with a current I.

[2] (a) $\hat{\mathbf{x}}$; (b) $\mathbf{M} = \sum_i \mathbf{m}_i / V = d^2 I\hat{\mathbf{z}}/(d^3) = I/d\hat{\mathbf{z}}$.

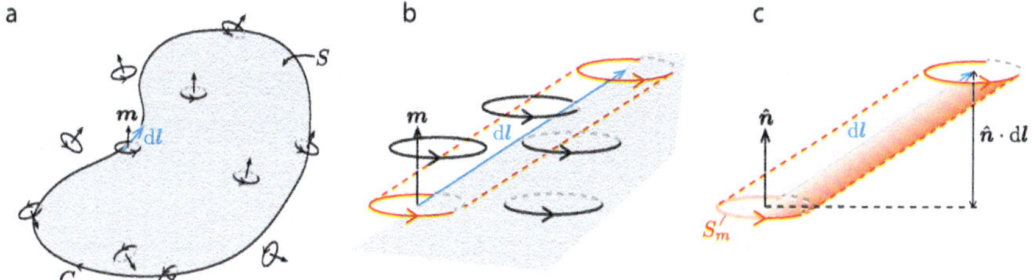

Fig. 13.2 **a** Illustration of a loop C and a corresponding surface S. **b** Current loops around a small element **dl** along the loop. **c** Illustration of base area and height of the cylindrical volume

We call the current from the microloops the *bound current*, $I_{b,C}$, through C. The contribution $\mathrm{d}I_{b,C}$ to the bound current from microloops around **dl** is therefore

$$\mathrm{d}I_{b,C} = I\,N\,\mathrm{d}v = I\,N\mathbf{S}_m \cdot \mathbf{dl} = N\,I\mathbf{S}_m \cdot \mathbf{dl}\,. \tag{13.4}$$

We insert $\mathbf{m} = I\mathbf{S}_m$:

$$\mathrm{d}I_{b,C} = N\,\mathbf{m} \cdot \mathbf{dl}\,, \tag{13.5}$$

and recognize that $N\mathbf{m} = \mathbf{M}$. We therefore get:

$$dI_{b,C} = \mathbf{M} \cdot \mathbf{dl}\,. \tag{13.6}$$

The total contribution from the bound currents along the whole curve C is found by summing up the contributions from each **dl**, that is, by integrating along the curve C:

$$I_{b,C} = \oint_C \mathbf{M} \cdot \mathbf{dl}\,. \tag{13.7}$$

The total current through C is the sum of bound currents and the free currents going through the curve C: $I = I_{b,C} + I_C$, where I_C is the free (non-bound) current through C.

Bound current
The bound current through a surface S enclosed by a curve C is

$$I_{b,C} = \oint_C \mathbf{M} \cdot \mathbf{dl}\,, \tag{13.8}$$

where $\mathbf{M}(\mathbf{r})$ is the magnetization in the material.

Test your understanding

Assume that the magnetization in a region is $\mathbf{M} = M_0\hat{\mathbf{z}}$. What is the bound current through a circle with radius a in the xy-plane with its center in the origin?[3]

Bound current density

We have found an expression for the bound current. Can we find a corresponding expression for the bound current density? We know that the current I is related to the current density $\mathbf{J}$ of the free current. We define the bound current density $\mathbf{J}_b$ through a similar relation for the bound current:

$$I = \int_S \mathbf{J} \cdot d\mathbf{S} \quad , \quad I_{b,C} = \int_S \mathbf{J}_b \cdot d\mathbf{S} \,. \tag{13.9}$$

where we insert the integral for the bound current and rewrite it using Stokes' theorem:

$$I_{b,C} = \oint_C \mathbf{M} \cdot d\mathbf{l} = \int_S \mathbf{J}_b \cdot d\mathbf{S} = \int_S \nabla \times \mathbf{M} \cdot d\mathbf{S} \,. \tag{13.10}$$

Because this is true for any surface S, it must also be true for the arguments of the integral, that is, we have found an expression for the bound current density: $\mathbf{J}_b = \nabla \times \mathbf{M}$.

Bound surface current density

Figure 13.3a illustrates a cylindrical permanent magnet with a uniform magnetization $\mathbf{M} = M_0\hat{\mathbf{z}}$. The bound current density inside the magnet is then $\mathbf{J}_b = \nabla \times \mathbf{M} = 0$. Where did the bound currents go? Fig. 13.3b illustrates the many parallel microscopic current loops in the material. We see that anywhere inside the material, a local current I along a loop is counteracted by an equal and opposite current I from an adjacent loop. Inside the material, the currents from these loops cancel, so that there is no net bound current density. However, at the external boundaries of the magnetic material, there are no counteracting currents. As a result, there is a surface current on the outer surface of the permanent magnet, which points in the azimuthal direction.

How can we find an expression for the bound surface current density? Fig. 13.3c shows a small part of the surface of a magnetic material with surface normal $\hat{\mathbf{n}}$ and magnetization $\mathbf{M}$. We form a current loop C in the plane formed by $\mathbf{M}$ and $\hat{\mathbf{n}}$. The

---

[3] $I_{C,b} = \int_C \mathbf{M} \cdot d\mathbf{l} = 0$.

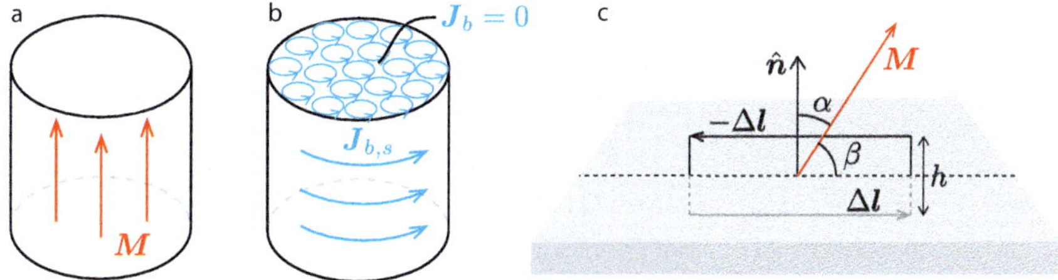

Fig. 13.3 **a, b** Illustration of a permanent magnet with a uniform magnetization $\mathbf{M} = M_0\hat{\mathbf{z}}$ inside. **c** Illustration of a curve through the surface of a magnetized material

loop has a length Δl and a thickness Δh, where $\Delta h \to 0$. Inside the material, there is a magnetization $\mathbf{M}_{\text{in}} = \mathbf{M}$, while outside the material, the magnetization is zero, $\mathbf{M}_{\text{out}} = \mathbf{0}$. The bound current flowing through the loop C is:

$$I_b = \int_C \mathbf{M} \cdot d\mathbf{l} = \mathbf{h} \cdot \mathbf{M} + \Delta\mathbf{l} \cdot \mathbf{M}_{\text{in}} - \mathbf{h} \cdot \mathbf{M} - \Delta\mathbf{l} \cdot \mathbf{M}_{\text{out}} , \tag{13.11}$$

where we let $|\mathbf{h}| \to 0$. We therefore get

$$I_b = \Delta\mathbf{l} \cdot \mathbf{M} = M\Delta l \cos\beta = M\Delta l \sin\alpha , \tag{13.12}$$

where β is the angle between $\mathbf{M}$ and $\Delta\mathbf{l}$ and α is the angle between $\hat{\mathbf{n}}$ and $\mathbf{M}$. The direction of this current is out of the plane, and we recognize it as a bound surface current density $J_{b,s}$ multiplied with the length Δl:

$$I_b = J_{b,s}\Delta l . \tag{13.13}$$

We therefore see that

$$J_{b,s}\Delta l = M\Delta l \sin\alpha \quad \Rightarrow \quad J_{b,s} = M \sin\alpha . \tag{13.14}$$

The surface current density points out of the plane of the curve C, that is, in the direction of $\mathbf{M} \times \hat{\mathbf{n}}$. We also notice that $\mathbf{M} \times \hat{\mathbf{n}} = M \sin\alpha$, and therefore we write:

$$\mathbf{J}_{b,s} = \mathbf{M} \times \hat{\mathbf{n}} . \tag{13.15}$$

This is the bound surface current density.

Bound current densities

The *bound volume current density* $\mathbf{J}_b$ inside a magnetic material with magnetization $\mathbf{M}$ is:

$$\mathbf{J}_b = \nabla \times \mathbf{M} \,. \tag{13.16}$$

and the *bound surface current density* $\mathbf{J}_{b,s}$ on the surface of a magnetic material (the interface between the magnetic material and vacuum/air) is

$$\mathbf{J}_{b,s} = \mathbf{M} \times \hat{\mathbf{n}} \,, \tag{13.17}$$

where $\hat{\mathbf{n}}$ is the local surface normal.

Test your understanding

A cylinder along the z-axis is made from a material that has been permanently magnetized with a magnetization $\mathbf{M}$. What are the bound surface current densities if (a) $\mathbf{M} = M_0\hat{\mathbf{z}}$; (b) $\mathbf{M} = M_0\hat{\boldsymbol{\phi}}$; (c) $\mathbf{M} = M_0\hat{\mathbf{x}}$?[4]

Example: Permanent magnet

Find the current densities and the magnetic field of an infinitely long cylindrical permanent magnet along the z-axis with a uniform magnetization $\mathbf{M} = M_0\hat{\mathbf{z}}$.

Approach. First, we find the current densities inside and on the boundaries of the magnet. We use these currents in Ampere's law to find the magnetic field.

Current densities. We know that inside the magnet, the bound volume current density is $\mathbf{J}_b = \nabla \times \mathbf{M} = 0$. At the top and bottom surfaces of the cylinder, the surface normal vector is $\pm\hat{\mathbf{z}}$, and therefore the bound surface current density is zero since $\mathbf{M}$ and $\hat{\mathbf{n}}$ are parallel and their cross products are zero. At the outer surface of the cylinder, the surface normal is $\hat{\mathbf{n}} = \hat{\mathbf{r}}$. The bound surface current density is therefore

$$\mathbf{J}_{b,s} = \mathbf{M} \times \hat{\mathbf{n}} = M_z\hat{\mathbf{z}} \times \hat{\mathbf{r}} = M_z\hat{\boldsymbol{\phi}} \,. \tag{13.18}$$

Thus, there is a bound surface current going around the cylinder.

Magnetic field. Ampere's law can be used for current distributions, also for bound currents. We start by addressing the expected symmetry of the magnetic field: We use cylindrical coordinates where $\mathbf{B}$ can only depend on r, because of rotational and translational symmetry along the z-axis. There cannot be any component in the

[4] (a) Top/bottom: $\mathbf{J}_{b,s} = 0$, cylinder surface: $\mathbf{J}_{b,s} = M_0\hat{\boldsymbol{\phi}}$; (b) Top/bottom: $\mathbf{J}_{b,s} = \pm M_0\hat{\mathbf{r}}$, cylinder surface: $\mathbf{J}_{b,s} = -M_0\hat{\mathbf{z}}$; (c) Top/bottom: $\mathbf{J}_{b,s} = \pm(-M_0)\hat{\mathbf{y}}$, cylinder surface: $\mathbf{J}_{b,s} = M_0 \sin\phi\hat{\mathbf{z}}$.

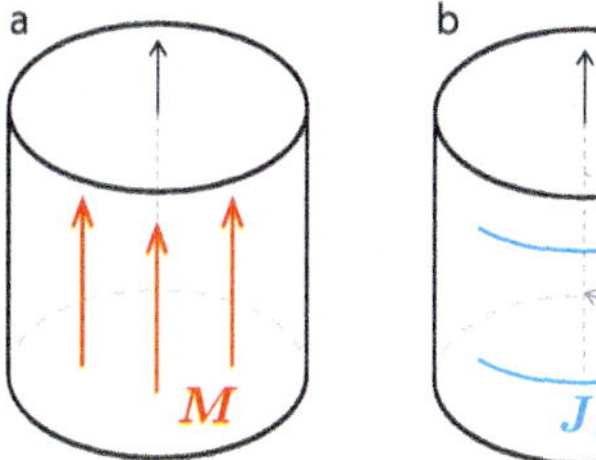

Fig. 13.4 Illustration of the magnetization, the bound surface current density and the integration loop C for a permanent magnet with a uniform magnetization $\mathbf{M} = M_0\hat{\mathbf{z}}$

$\hat{\mathbf{r}}$ direction, as this would lead to a non-zero flux out of a cylindrical surface, and there cannot be any component in the ϕ-direction because there is no current, bound or free, through a circle around the cylinder axis. The magnetic field is therefore $\mathbf{B} = B_z(r)\hat{\mathbf{z}}$.

In Fig. 13.4 we construct an Amperian loop C that runs a length L parallel to the z-axis at a distance r to the axis, then out radially to infinity, down again a length L parallel to the z-axis, and then into the axis along $-\hat{\mathbf{r}}$. We notice that $\mathbf{B} \cdot \mathrm{d}\mathbf{l}$ is zero along $\hat{\mathbf{r}}$ because $\mathbf{B}$ does not have a radial component. In addition, we assume that $\mathbf{B}$ approaches zero far away from the cylinder. The only contribution to the Ampere integral is from the length L at a distance r:

$$\oint_C \mathbf{B} \cdot \mathrm{d}\mathbf{l} = \int B_z\hat{\mathbf{z}} \cdot \hat{\mathbf{z}}\,\mathrm{d}z = B_z L = \mu_0 I \ . \tag{13.19}$$

When C encloses the cylinder surface, that is when r is smaller than the cylinder radius, the current through C is $J_{s,b}L$, where $J_{s,b} = M_z$, otherwise the current is zero:

$$\oint_C \mathbf{B} \cdot \mathrm{d}\mathbf{l} = B_z L = \mu_0 I = \mu_0 J_{s,b}\Delta L = \mu_0 M_z \Delta L \ . \tag{13.20}$$

That is, we have found that $\mathbf{B} = \mu_0 M_0\hat{\mathbf{z}}$ inside a permanent magnet and zero outside the magnet when we are far away from the edges. However, for a finite permanent magnet, the magnetic field will extend beyond the finite size of the magnet.

13.2 Ampere's Law for Magnetic Materials

Ampere's law states that

$$\oint_C \mathbf{B} \cdot \mathrm{d}\mathbf{l} = \mu_0 I \ , \tag{13.21}$$

for any loop C. The current I includes both the free current through C, I_C, and the bound currents, $I_{b,C}$, $I = I_C + I_{C,b}$ where

$$I_{b,C} = \oint_C \mathbf{M} \cdot d\mathbf{l} \,. \tag{13.22}$$

We insert this back into Ampere's law

$$\oint_C \mathbf{B} \cdot d\mathbf{l} = \mu_0 I = \mu_0 \left(I_C + I_{b,C} \right) = \mu_0 I_C + \mu_0 \oint_C \mathbf{M} \cdot d\mathbf{l} \,, \tag{13.23}$$

which can be rewritten as

$$\oint_C (\mathbf{B} - \mu_0 \mathbf{M})\, d\mathbf{l} = \mu_0 I_C \,. \tag{13.24}$$

It is common to divide by μ_0 on both sides, giving:

$$\oint_C \underbrace{\left(\frac{1}{\mu_0} \mathbf{B} - \mathbf{M} \right)}_{\mathbf{H}} \cdot d\mathbf{l} = I_C \,. \tag{13.25}$$

Here, we have defined the **H**-field:

$$\mathbf{H} = \frac{1}{\mu_0} \mathbf{B} - \mathbf{M} \,. \tag{13.26}$$

We have now reformulated Ampere's law for a magnetic material:

Ampere's law in a magnetic material
Ampere's law in a magnetic material is:

$$\oint_C \mathbf{H} \cdot d\mathbf{l} = I_{C,\text{free}} \,, \tag{13.27}$$

for any closed loop C. The current $I_{C,\text{free}}$ is the free current passing through the closed loop C. The **H**-field is defined as

$$\mathbf{H} = \frac{1}{\mu_0} \mathbf{B} - \mathbf{M} \,, \tag{13.28}$$

where **B** is the magnetic field and **M** is the magnetization.

Ampere's law on differential form

We can rewrite Ampere's law on differential form using Stokes' theorem. We start from Ampere's law:

$$\oint_C \mathbf{H} \cdot d\mathbf{l} = I_{C,\text{free}} = \int_S \mathbf{J} \cdot d\mathbf{S} \,. \tag{13.29}$$

We apply Stokes' theorem to get:

$$\oint_C \mathbf{H} \cdot d\mathbf{l} = \int_S \nabla \times \mathbf{H} \cdot d\mathbf{S} = \int_S \mathbf{J} \cdot d\mathbf{S} \,. \tag{13.30}$$

Because this is true for any surface S, the arguments of the integral must be identical: $\nabla \times \mathbf{H} = \mathbf{J}$, where $\mathbf{J}$ now only include the free currents.

Ampere's law on differential form
Ampere's law on differential form states that

$$\nabla \times \mathbf{H} = \mathbf{J} \,, \tag{13.31}$$

where $\mathbf{J}$ is the current density of free (not bound) charges and

$$\mathbf{H} = \frac{1}{\mu_0}\mathbf{B} - \mathbf{M} \,, \tag{13.32}$$

where $\mathbf{M}$ is the magnetization.

13.3 Magnetic Materials

Ampere's law for a magnetic material is analogous to the extension of Gauss' law to dielectric materials. We can use Ampere's law to find the $\mathbf{H}$ field for a system, but to find the magnetic field $\mathbf{B}$ we need to have a theory to relate the magnetization to the magnetic field or a model for the magnetization. Here, we will introduce models for linear magnetic materials and non-linear magnetic materials.

Linear materials

For many magnetic materials, the magnetization is proportional to **H**:

$$\mathbf{M} = \chi_m \mathbf{H} \,, \tag{13.33}$$

where χ_m is called the *magnetic susceptibility* of the material. We call such material *linear (magnetic) materials*. For these materials, the magnetic field **B** is proportional to **H**, because

$$\mathbf{H} = \frac{1}{\mu_0}\mathbf{B} - \mathbf{M} \quad \Rightarrow \quad \mathbf{B} = \mu_0 \left(\mathbf{H} + \mathbf{M} \right) \,, \tag{13.34}$$

which when $\mathbf{M} = \chi_m \mathbf{H}$ gives

$$\mathbf{B} = \mu_0 \left(\mathbf{H} + \mathbf{M} \right) = \mu_0 \left(1 + \chi_m \right) \mathbf{H} = \mu_r \mu_0 \mathbf{H} = \mu \mathbf{H} \,, \tag{13.35}$$

where we have introduced a notation similar to what we used for polarization. The relative permeability is $\mu_r = 1 + \chi_m$ and the absolute permeability is $\mu = \mu_r \mu_0$. In vacuum, we have that $\mu = \mu_0$ and $\mathbf{B} = \mu_0 \mathbf{H}$.

Linear magnetic materials
For *linear magnetic materials* the magnetic field is proportional to **H**:

$$\mathbf{B} = \mu_r \mu_0 \mathbf{H} = \mu \mathbf{H} \,, \tag{13.36}$$

where μ_r is called the *relative permeability* and μ is called the *absolute permeability*.

Material properties

The relative permeability is a material property and we list values for some materials in Table 13.1. We classify magnetic materials as *diamagnetic* when $\mu_r < 1$, *paramagnetic* when $\mu_r > 1$ while $\mu_r \simeq 1$, and *ferromagnetic* when $\mu_r \gg 1$.

Hysteresis and Non-linear Materials

Notice that many ferromagnetic materials have a significant magnetic response, but that the behavior often are *non-linear* and depend on the history of the material. A

Table 13.1 Values for the relative permeability, μ_r, for various materials

Material	μ_r	Material	μ_r
Ferrite (manganese zinc)	350–20000	Air	1.0000004
Nickel	100–600	Aluminum	1.00002
Copper	0.999994	Iron (99.8%)	5000
Water	0.999992	Pure iron	200000
Vacuum	1	Metglas 2714A	10^6

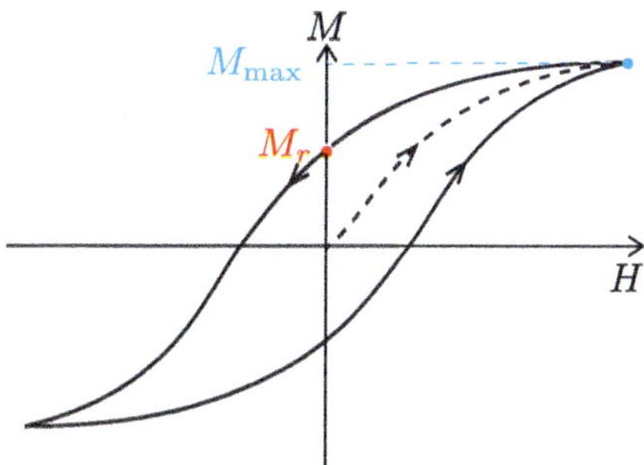

Fig. 13.5 Illustration of hysteresis loop for magnetization

typical behavior of the magnetization of a ferromagnetic material as a function of **H** is illustrated in Fig. 13.5. The magnetization does not only depend on the **H**-field, but also on the history of the magnetization:

- Let us assume that both the magnetization M and the applied field H starts as zero.
- As we increase the applied field, H, the magnetization will also increase. First it increases linearly, but eventually it saturates and does not increase much more, because all domains (all microcurrents) are aligned, and the magnetization cannot increase further even if the H-field is increased. At this point the system has reached maximum magnetization M_{max}.
- If we reduce H, the magnetic field will lag. When we have reduced the magnetic field H to zero, the magnetization is not zero, which means that the system is now a *permanent magnet* with a residual magnetization M_r
- When H becomes negative, M will reduce further, eventually becoming zero and negative, until the magnetization again saturates at its maximum value in the opposite direction.
- Notice that at some points along the curve, H and M might have opposite signs, showing that they point in different directions.

Test your understanding

There is a magnetic field $\mathbf{B}_0 = B_0(I)\hat{\mathbf{z}}$ inside a cylindrical solenoid with a current I when there is vacuum inside the solenoid. Then we fill it with a linear magnetic material with permeability μ. (a) What is $\mathbf{H}_0$ inside the solenoid before it is filled with the magnetic material? (b) What is $\mathbf{H}$ inside the solenoid after it if filled with the magnetic material? (c) What is the magnetization $\mathbf{M}$ inside the solenoid? (d) What is the magnetic field $\mathbf{B}$ inside the solenoid?[5]

Example: Coaxial conductor in a magnetic material

What is the magnetic field inside a coaxial cable with a magnetic material?
Approach. We find the symmetry of the system and apply Ampere's law to find the **H**-field and from this the **B**-field. We assume that the coaxial system consists of an inner cylinder of radius a and an outer cylindrical shell with inner radius b. We also assume that the current is on the outer surface of the inner cylinder and on the inner surface of the outer cylinder.
Symmetry. First, we realize that that symmetry of the **H**-field is the same as the symmetry of the **B**-field. What is the symmetry of the magnetic field? We realize that this system must have the same symmetry as for a current along a cylindrical wire, where we found that the magnetic field has cylindrical symmetry with an axis along the axis of the cylinder: $\mathbf{B} = B_\phi(r)\,\hat{\boldsymbol{\phi}}$. The **H**-field must have the same symmetry:

$$\mathbf{H} = H_\phi(r)\hat{\boldsymbol{\phi}} \,, \tag{13.37}$$

Applying Ampere's law. We apply Ampere's law to a circular loop of radius r around the cylinder. For this loop, the **H**-field points along the **dl**-vector:

$$\oint_C \mathbf{H}\cdot\mathrm{d}\mathbf{l} = \oint_C H_\phi(r)\,\mathrm{d}l = H\oint_C \mathrm{d}l = H\,2\pi r = I \,, \tag{13.38}$$

The current I is only non-zero in the range $a < r < b$. For $r < a$ there is no current inside the loop, since all the current is on the surface at $r = a$. For $r > b$, the net current inside r is zero because the currents on the inner and the outer surfaces are in opposite direction and therefore have opposite signs. For $a < r < b$, we find that

$$\mathbf{H} = \frac{I}{2\pi r}\hat{\boldsymbol{\phi}} \,, \tag{13.39}$$

The magnetic field is then

[5] (a) $\mathbf{H}_0 = \mathbf{B}_0/\mu_0$, (b) $\mathbf{H} = \mathbf{H}_0$, (c) $\mathbf{M} = \chi_m\mathbf{H} = \chi_m B_0/\mu_0\,\hat{\mathbf{z}}$, (d) $\mathbf{B} = B_0(1+\chi_m)\,\hat{\mathbf{z}}$.

$$\mathbf{B} = \mu \mathbf{H} = \frac{\mu I}{2\pi r} \hat{\boldsymbol{\phi}} . \tag{13.40}$$

This is a general result, which is valid both for a magnetic material and for vacuum, for which $\mu = \mu_0$.

13.4 Boundary Conditions for Magnetic Fields

For the electric field we derived boundary conditions at the interface between materials with different dielectric constants by applying Gauss' law on small volumes or that $\oint_C \mathbf{E} \cdot \, \mathrm{d}\mathbf{l} = 0$ for small, closed curves. Let us apply a similar method to find the boundary conditions for the magnetic field across a boundary.

Normal boundary conditions

Figure 13.6a illustrates a boundary between two materials 1 and 2 with differing magnetic properties. We relate the normal component of the magnetic field in material 1 and material 2 by introducing a cylindrical Gauss surface with a small height Δh and a surface area ΔS. The axis of the cylinder is oriented along the local surface normal $\hat{\mathbf{n}}$ of the interface. We apply the flux integral:

$$\int_S \mathbf{B} \cdot \mathrm{d}\mathbf{S} = 0 , \tag{13.41}$$

where S is the surface of the cylinder. We assume that $\Delta h \to 0$ so that the flux contribution through the cylinder side is zero. The integral is then

$$\int_S \mathbf{B} \cdot \, \mathrm{d}\mathbf{S} \simeq \mathbf{B}_1 \cdot \hat{\mathbf{n}} \Delta S - \mathbf{B}_2 \cdot \hat{\mathbf{n}} \Delta S = 0 , \tag{13.42}$$

which gives that

$$\mathbf{B}_1 \cdot \hat{\mathbf{n}} = \mathbf{B}_2 \cdot \hat{\mathbf{n}} . \tag{13.43}$$

The normal component of the magnetic field is therefore continuous across an internal interface, or equivalently, the normal components of the magnetic field is the same on both sides of an internal interface. (Notice that for electric fields, it was the tangetial component of the electrical field that was the same on both sides. $E_{1,t} = E_{2,t}$.)

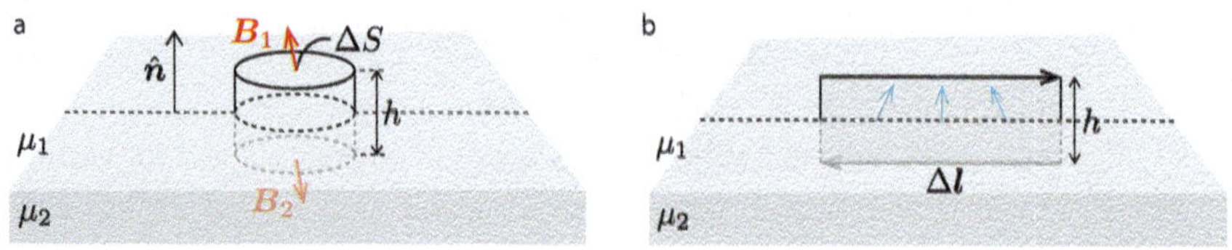

Fig. 13.6 Illustration of a boundary between two magnetic materials used to determine the normal and tangential boundary conditions for the magnetic fields

Tangential boundary conditions

Figure 13.6b illustrates a boundary between two materials 1 and 2 with differing magnetic properties. To compare the tangential components on each side of the interface, we introduce a small current loop C where the thickness h of the loop goes to zero, $h \rightarrow 0$. We apply Ampere's law on this loop:

$$\oint_C \mathbf{H} \cdot d\mathbf{l} = \underbrace{\oint_h \mathbf{H} \cdot d\mathbf{l}}_{=0} + \mathbf{H}_1 \cdot \Delta\mathbf{l} + \underbrace{\oint_h \mathbf{H} \cdot d\mathbf{l}}_{=0} + \mathbf{H}_2 \cdot (-\Delta\mathbf{l}) = J_s \Delta l \ , \tag{13.44}$$

where J_s is the free surface current normal to the tangential direction. In the limit when $h \rightarrow 0$ we find

$$H_{1,t} - H_{2,t} = J_s \ , \tag{13.45}$$

which we can rewrite in vector notation as:

$$\hat{\mathbf{n}} \times \mathbf{H}_1 - \hat{\mathbf{n}} \times \mathbf{H}_2 = \mathbf{J}_s \ . \tag{13.46}$$

Boundary conditions for magnetic fields
At an interface between a magnetic material 1 and a magnetic material 2 we have the following boundary conditions:

$$\mathbf{B}_1 \cdot \hat{\mathbf{n}} = \mathbf{B}_2 \cdot \hat{\mathbf{n}} \ , \tag{13.47}$$

and

$$\hat{\mathbf{n}} \times \mathbf{H}_1 - \hat{\mathbf{n}} \times \mathbf{H}_2 = \mathbf{J}_s \ . \tag{13.48}$$

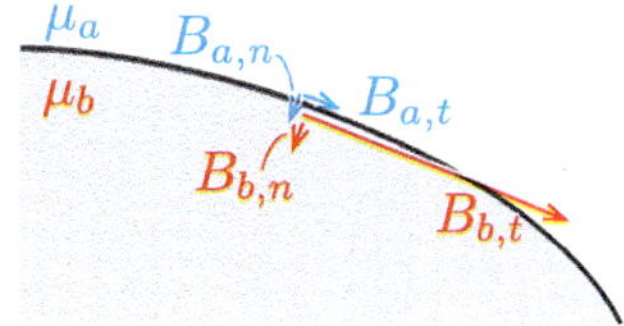

Fig. 13.7 Illustration of a boundary between a ferromagnetic material (**b**) and air (**a**)

Example: Air-ferromagnet transition

Figure 13.7 *illustrates a curved interface between air (material a) and a ferromagnet with $\mu_b/\mu_0 \gg 1$ (material b). What is the implications for the tangential and normal magnetic fields? There are no free surface currents.*

The tangential fields are related by

$$H_{a,t} = H_{b,t} \quad \Rightarrow \quad \frac{B_{a,t}}{\mu_a} = \frac{B_{b,t}}{\mu_b} , \tag{13.49}$$

where we have used that $\mathbf{B} = \mu\mathbf{H}$. In this case, $\mu_a/\mu_0 \simeq 1$, so that:

$$B_{b,t} \simeq \frac{\mu_b}{\mu_0} B_{a,t} . \tag{13.50}$$

For the case of a ferromagnetic material, where $\mu_b \gg \mu_a$, the magnetic field outside the ferromagnet will be negligible compared with the field inside the material. The magnetic field therefore tends to follow a ferromagnetic material.

13.5 Magnetic Circuits

This last example demonstrates that the magnetic field tends to follow ferromagnetic materials. A *magnetic circuit* is a circuit made from materials that "conduct" magnetic flux so that the field lines close upon themselves.

Magnetic flux in a toroidal circuit

Figure 13.8a illustrates a magnetic circuit in the form of a toroidal magnet with a small gap. The magnetic field is set up by a wire wrapped N times around a part of the toroid. In this region, we expect the magnetic field to be approximately uniform because the thickness of the toroid is small compared with the radius of the toroid, so the toroid is like a solenoid in this region.

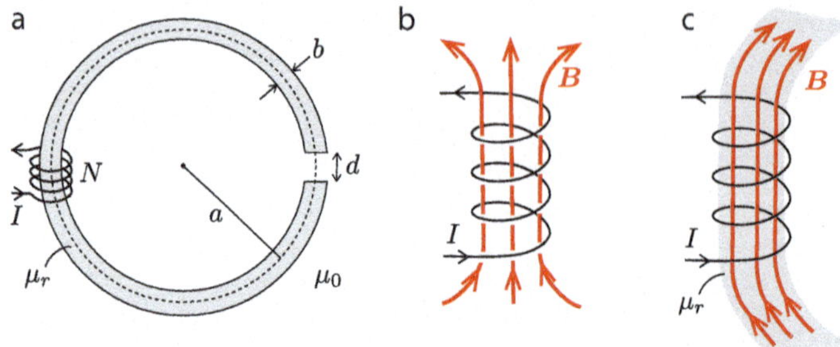

Fig. 13.8 **a** Illustration of magnetic circuit, **b**, **c** Illustration of the magnetic field from N winded wires with and without a magnetic material. The field tends to follow the magnetic material when $\mu_r \gg 1$

If there was no toroid-shaped magnetic material, we would expect the magnetic field to spread out outside the region where the wires are winded as illustrated in Fig. 13.8b. However, due to the magnetic material and the large ratio in magnetic permeability between the air and the magnetic material, we expect the magnetic field to follow the magnetic material instead. This is indeed why we call the system a magnetic circuit: The magnetic field follows the magnetic material also outside the region where the wired are winded.

Because the magnetic field follows the material without spreading out, we expect the magnetic flux, Φ_B, to be conserved across any cross section along the wire. The magnetic flux is

$$\Phi_B = \int_S \mathbf{B} \cdot d\mathbf{S} = \pi b^2 B \ , \tag{13.51}$$

where we have used that the magnetic field points along the toroid and is therefore normal to the surface S and that the magnetic field is approximately uniform across the cross section.

We also assume that the magnetic field does not spread out significantly in the air gap, therefore the magnetic flux is also conserved in the air gap. Assuming that the magnetic field also is uniform in the air gap, we get that (13.51) also holds in the air gap and that the magnetic field B is the same all along the circuit. However, the H field will vary between the magnetic material and the air gap.

How can we find B? We can use Ampere's law along an integration path C following the circuit as illustrated in Fig. 13.8a:

$$\oint_C \mathbf{H} \cdot d\mathbf{l} = NI \ . \tag{13.52}$$

Notice that the current through S is still NI, even if this current only passes through one part of the circuit C. What is $\mathbf{H}$ along the circuit? We will assume that $\mathbf{H}$ is aligned with $\mathbf{dl}$ everywhere: The field follows the circuit and does not leak out into the surrounding air. Inside the magnetic material we have that $B = \mu H = \mu_0 \mu_r H$, so that $H = B/(\mu_0 \mu_r)$ and inside the air gap we have that $B = \mu_0 H$ so that $H = B/\mu_0$, where B is the same along the whole path, as we argued above. The length of the path inside the material is $2\pi a - d$ and the length inside the air gap is d, so that:

$$\oint_C \mathbf{H} \cdot \mathbf{dl} = NI = \frac{B}{\mu_0} d + \frac{B}{\mu_0 \mu_r} (2\pi a - d) \ . \tag{13.53}$$

We find B from this equation:

$$B = \frac{\mu_0 NI}{d + \frac{1}{\mu_r}(2\pi a - d)} \ . \tag{13.54}$$

And the flux is

$$\Phi_B = BS = \frac{NI}{\frac{d}{\mu_0 S} + \frac{2\pi a - d}{\mu_0 \mu_r S}} \ . \tag{13.55}$$

We can rewrite this in a way that demonstrates the similarity with the electrical circuit by introducing a *reluctance* $R_{m,t}$ for the toroid and $R_{m,g}$ for the air gap:

$$R_{m,t} = \frac{2\pi a - d}{\mu_0 \mu_r S} \ , \quad R_{m,g} = \frac{d}{\mu_0 S} \ . \tag{13.56}$$

The expression for the flux becomes

$$\Phi_B = \frac{NI}{R_{m,t} + R_{m,g}} \ , \tag{13.57}$$

This is very similar to the behavior of an electrical circuit with two resistances $R_t + R_g$ in series, for which the current would be:

$$I = \frac{V}{R_t + R_g} \ . \tag{13.58}$$

Test your understanding

The figure shows a magnetic circuit constructed from a ferromagnetic material with a large μ_r. The circuit consists of one part with cross section with radius a and another part with cross section with radius b. The magnetic field at S_1 is B_1. What is the flux through (a) S_1, (b) S_2, (c) S_3, (d) S_4? (e) What is the magnetic field at S_3?[6]

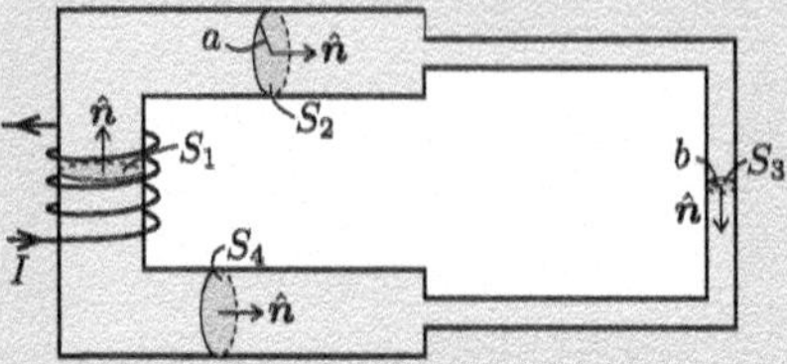

Analogy between magnetic flux and electric current density

The example of the toroidal circuit illustrates similarities between magnetic and electric circuits. However, there are also important differences. In an electric circuit, the current density **J** is restricted to follow the conducting materials, with essentially zero current outside the conductors, because the conductivity of air or vacuum is practically zero. However, in a magnetic circuit, the permeability (μ) is not zero outside the ferromagnetic conductors, so there is some leakage of the field out from the conductors into the surrounding air or vacuum. In addition, a magnetic circuit will typically contain air gaps, where we will assume that the gap is so narrow that the fringe effects near the gap edges will not impact the way the flux is flowing in the circuit.

The similarities and differences between magnetic and electric circuits include:

- **Current and flux**: In the *electric circuit*, there is a current $I = \int_S \mathbf{J} \cdot d\mathbf{S}$ through any cross-section of the circuit. The current is conserved because the conductivity of the region outside the circuit is zero. (The current density **J** does not have a normal component on the boundary). In the *magnetic circuit*, there is a magnetic flux, $\Phi = \int_S \mathbf{B} \cdot d\mathbf{S}$ through any cross section of the circuit. We will assume that the flux is conserved because the magnetic field lines tend to follow the magnetic material.
- **Conservation of current and flux**: Both the current out of a closed surface S and the flux out of a closed surface S is zero: $\oint_S \mathbf{J} \cdot d\mathbf{S} = 0$ and $\oint_S \mathbf{B} \cdot d\mathbf{S} = 0$.
- **Relation to conductivity and permeability**: The current density is related to the electric field through Ohm's law on microscopic level: $\mathbf{J} = \sigma\mathbf{E}$, while the magnetic field is related to the **H**-field through the law for magnetic materials: $\mathbf{B} = \mu\mathbf{H}$.

[6] (a,b,c) $B_1\pi a^2$; (d) $-B_1\pi a^2$; (e) $B_3 = (a/b)^2 B_1$.

- **Drivers for E and H**: The voltage source is driving the electric field in the electric circuit, $\oint_C \mathbf{E} \cdot \mathrm{d}\mathbf{l} = V(t)$, while windings of a current-carrying wire is driving the magnetic field, $\oint_C \mathbf{H} \cdot \mathrm{d}\mathbf{l} = NI$.
- **Resistance and reluctance**: The *resistance* R relates the driving voltage to the current in an electric circuit: $V = I \sum R_i$. There is a similar relationship for the *reluctance* of a magnetic circuit: $NI = \Phi \sum R_{m,i}$

Summary

The **magnetization M** is the net magnetic moment per unit volume:

$$\mathbf{M} = \frac{\sum_i \mathbf{m}_i}{\mathrm{d}v} ,$$

where the sum is over all the microloops i in the volume element $\mathrm{d}v$.
Ampere's law in a magnetic material is

$$\oint_C \mathbf{H} \cdot \mathrm{d}\mathbf{l} = I_{C,\mathrm{free}} ,$$

where the **H**-field is defined as

$$\mathbf{H} = \frac{1}{\mu_0}\mathbf{B} - \mathbf{M}$$

The **bound current density** $\mathbf{J}_b$ is related to the magnetization through

$$\mathbf{J}_b = \nabla \times \mathbf{M}$$

and the **bound surface current density** $\mathbf{J}_{b,s}$ on the interface between a magnetic material and vacuum (air) with surface normal $\hat{\mathbf{n}}$ pointing from the magnetic material to the air is:

$$\mathbf{J}_{b,s} = \mathbf{M} \times \hat{\mathbf{n}} .$$

Ampere's law on differential form is:

$$\nabla \times \mathbf{H} = \mathbf{J} .$$

For **linear magnetic materials** the magnetic field is proportional to **H**:

$$\mathbf{B} = \mu_r \mu_0 \mathbf{H} = \mu \mathbf{H}$$

where μ_r is called the relative permeability and μ is called the absolute permeability.

The **boundary conditions** for the magnetic fields are:

$$\mathbf{B}_1 \cdot \hat{\mathbf{n}} = \mathbf{B}_2 \cdot \hat{\mathbf{n}} \,,$$

and

$$\hat{\mathbf{n}} \times \mathbf{H}_1 - \hat{\mathbf{n}} \times \mathbf{H}_2 = \mathbf{J}_s \,.$$

For **magnetic circuits** the magnetic flux, $\Phi = \int_S \mathbf{B} \cdot d\mathbf{S}$ is conserved along the circuit as long as $\mu \gg \mu_0$ for the magnetic material. We can use concepts from electrical circuits as an approximation to understand and describe magnetic circuits.

Exercises

Discussion exercises

13.1 Hot paramagnet. Why do we expect that the permeability of a paramagnetic material decreases at higher temperatures?

13.2 The Einstein-de-Haas effect. An iron cylinder is placed so that it may rotate freely about its axis. The cylinder starts from rest and a magnetic field is turned on so that the cylinder is magnetized in a direction parallel to its axis. If the direction of the *external* field suddenly is reversed, the direction of the magnetization will also reverse and the cylinder will start rotating around its axis. (This is called the *Einstein-de-Haas effect*). Explain why the cylinder starts rotating.

13.3 Infinite magnet. Q has just received a new, cylindrical permanent magnet. He has learned that the magnet has a magnetic field because there is a net current along the surface of the magnet. But he has also learned that a current I in a material with finite resistivity will lead to a power loss of RI^2. He measures R between two points on the surface of the magnet and finds that this is not zero. Why does the magnet not heat up?

13.4 Drops of flying air. If a magnet hangs above a container with liquid air, drops will be attracted to the poles of the magnet. The drops contain only liquid oxygen. Even if nitrogen is the main constituent in air, it is not attracted to the magnet. Explain what this tells us about the magnetic susceptibilities of oxygen and nitrogen and explain why the poles of a magnet in air at room temperature do not attract molecules of oxygen gas.

13.5 Two halves of a magnet. Q has a permanent magnet shaped as a rectangular, flatt prism. One half is colored red and is marked by the letter N, the other is colored black and is marked by the letter S. Q divided the magnet into two pieces, a red and a black piece, but is surprised when each of the pieces also behaves as a magnet with a north and a south pole. How can you explain this?

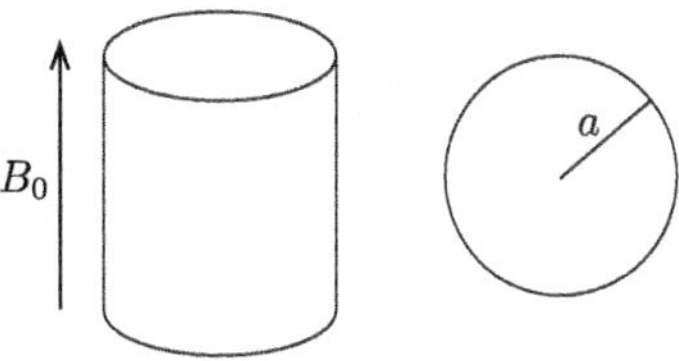

Fig. 13.9 A cylindrical magnet

Tutorials

13.6 Cylindrical magnet. A cylindrically shaped linearly magnetic material with magnetic susceptibility χ_m is placed in an external, homogeneous magnetic field $\mathbf{B}_0$. The axis of the cylinder is directed along the magnetic field as illustrated in Fig. 13.9. You can assume that the magnetic field is homogeneous and parallel to $\mathbf{B}_0$ inside the cylinder.
(**a**) What is **H** outside the cylinder?
(**b**) Write down Ampere's law for a magnetic material.
(**c**) What type of Amperian loop would you choose to find **H** inside the cylinder?
(**d**) What is **H** inside the cylinder?
(**e**) Check that your result is consistent with the boundary conditions at the boundary between the cylinder and the vacuum outside.
(**f**) What is **M** inside the cylinder for a diamagnetic and a paramagnetic material?
(**g**) What is the magnetic field **B** inside the cylinder for a diamagnetic and a paramagnetic material?
(**h**) How would you argue that the magnetic fields are homogeneous inside the cylinder?

13.7 Similarities and differences between Gauss and Ampere. *In this tutorial we will focus on the commonalities in the approaches to find the electric field using Gauss' law for a dielectric and to find the magnetic field using Ampere's law for a magnetizable material.*

We will study an infinitely long cylinder of radius a in two scenarios. In scenario A the cylinder has a constant charge density ρ and consists of a dielectric material with dielectric constant ϵ, while the space outside the cylinder is vacuum. In scenario B the cylinder has a constant current density **J** directed along the axis of the cylinder with a magnetic permeability μ, while the space outside the cylinder is vacuum.
(**a**) Write down Gauss' and Ampere's laws.
First, we will address the fields **D** (scenario A) and **H** (scenario B) outside the cylinder.
(**b**) What symmetries would you use to apply Gauss' law to scenario A? What is the free charge? Find the field **D**.
(**c**) What symmetries would you use to apply Ampere's law to scenario B? What is the free current? Find the field **H**.
Second, we will address the fields **D** (scenario A) and **H** (scenario B) inside the cylinder.

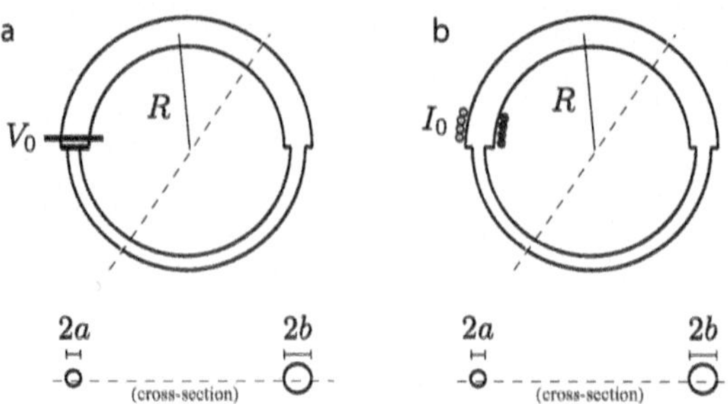

Fig. 13.10 A magnetic and an electric circuit

(d) What symmetries would you use to apply Gauss' law to scenario A? What is the free charge? Find the field **D**.
(e) What symmetries would you use to apply Ampere's law to scenario B? What is the free current? Find the field **H**.

13.8 Magnetic circuits. *The focus on this tutorial is to develop an understanding of the concept of magnetic circuits and how we can apply our intuition from electric circuits to address magnetic circuits.*
In this tutorial we will address the correspondence between current flow in an electrical circuit (scenario A) with flux in a magnetic circuit (scenario B) as illustrated in Fig. 13.10. The circuits consists of cylindrical half-donuts. The electric circuit has a conductivity σ and the magnetic circuit has a magnetic permeability $\mu \gg \mu_0$. You may assume there to be air or vacuum outside the circuit. The electric circuit is driven by a constant emf V_0 from a battery. The magnetic circuit is driven by a constant current I_0 which is wrapped N times around the circuit cylinder along a very short length along the cylinder as illustrated in the figure.
(a) Sketch the electric field **E** along the circuit in scenario A and the **H**-field in scenario B.
(b) What line integral along the circuit would you use to find the electric field in scenario A and the H-field in scenario B?
(c) Use your results from above to show that $E_1 \pi R + E_2 \pi R = V_0$ and that $H_1 \pi R + H_2 \pi R = N I_0$ where index 1 refers to the thick part of the cylinder and index 2 refers to the thin part of the cylinder.
(d) How can you relate **J** to **E** and similarly **B** to **H**?
(e) What are the currents I_1 and I_2 expressed in terms of σ, E_i and the cross-sectional areas S_i? What are the fluxes Φ_1 and Φ_2 expressed in terms of μ, H_i and the cross-sectional areas S_i? (You can assume that **J** and **H** are uniform across a cross-section of the circuit).
(f) What conservation law relates the currents along the circuit in scenario A and the fluxes along the circuits in scenario B?
(g) Find the electric fields E_i expressed in terms of I, σ and S_i. Similarly, for scenario B find the magnetic fields H_i expressed in terms of Φ, μ and S_i.

(h) Finally, put your results back into the integral around the circuit to find Kirchhoff's law of voltage drops (scenario A):

$$\left(\frac{\pi R}{\sigma S_1}\right) I + \left(\frac{\pi R}{\sigma S_2}\right) I = V_0 \tag{13.59}$$

and similarly for scenario B:

$$\left(\frac{\pi R}{\mu S_1}\right) \Phi + \left(\frac{\pi R}{\mu S_2}\right) \Phi = N I_0 \tag{13.60}$$

Homework

13.9 A non-uniformly magnetized cylinder. An infinitely long ferromagnetic cylinder of radius a has a magnetization that depends on the distance r to the center of the cylinder: $\mathbf{M} = M_0(1 - (r/a)^2)\hat{\mathbf{z}}$.
(a) Find the bound current density and the bound surface current density of the cylinder.
(b) Find the magnetic field inside the cylinder by integrating the surface current density.

13.10 A magnetized sphere. A ferromagnetic sphere of radius a has a uniform magnetization $\mathbf{M} = M_0\hat{\mathbf{z}}$.
(a) Find the bound surface current density on the surface of the sphere.
(b) Find the magnetic field at the center of the sphere by integrating the bound surface current density.

13.11 Surface current. Assume that there is a uniform surface current density in the x-direction, $\mathbf{J}_S = J_S\hat{x}$, everywhere in the xy-plane. This means that a current $I = LJ_S$ passes through a line section from $y = 0$ til $y = L$ in the positive x-direction.
(a) Find the magnetic field $\mathbf{B}$ everywhere in space.
Assume that there is a uniform surface current density $\mathbf{J} = J_0\hat{\phi}$ in cylindrical coordinates on an infinitely long cylinder surface with radius a and axis along the z-axis.
(b) Find the $\mathbf{H}$-field and the magnetic field $\mathbf{B}$ everywhere in space when there is vacuum inside the cylinder.
(c) Assume that there instead is a permanent magnet with magnetization $\mathbf{M} = -J_0\hat{\mathbf{z}}$ inside the cylinder. What is now $\mathbf{B}$ and $\mathbf{H}$ inside the cylinder? Provide a physical explanation for the result.
If the length L of the cylinder is much smaller than the radius a, we can describe the cylinder as a circular circuit with radius a and current $I = J_0L$. Assume that the cylinder is in vacuum.
(d) What is contribution $d\mathbf{B}$ to the magnetic field $\mathbf{B}$ in in the point $\mathbf{r} = (x, y, z)$ from a small element $d\mathbf{l}$ of the current loop in the position $a(\cos\phi, \sin\phi, 0)$ where ϕ is the angle from the x-axis?

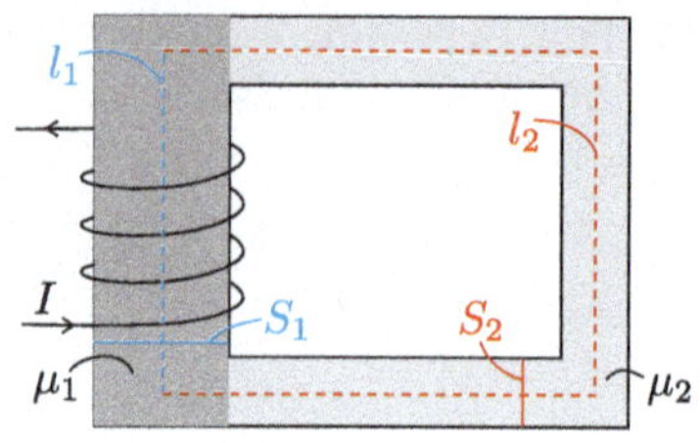

Fig. 13.11 Illustration of magnetic circuit of two materials 1 and 2 with different cross sections S_1 and S_2, different permeability μ_1 and μ_2 and different lengths l_1 and l_2

(e) Write a program to find the magnetic field **B** in a point $\mathbf{r} = (x, y, z)$ and visualize the field in the xz-plane.

13.12 A simple magnetic trap. In this exercise we will study the magnetic field $\mathbf{B} = \frac{B_0}{b}\left(x\hat{\mathbf{x}} + y\hat{\mathbf{y}} - 2z\hat{\mathbf{z}}\right)$, where B_0 is a constant and b is a length.
(a) Sketch the **B**-field in the xy-plane.
(b) What is the divergence, $\nabla \cdot \mathbf{B}$, of the magnetic field?
We place a quadratic circuit with side L in the xy-plane centered in the origin and with surface normal in the positive z-direction.
(c) What is the magnetic flux, Φ, through the circuit from the magnetic field?
(d) The circuit is moved along the z-axis with the velocity. Find the emf, e, induced in the circuit.
(e) Assume the circuit carries a current I. What is the force on the circuit from the magnetic field when $z = 0$?
(f) What is the energy density in this magnetic field?

13.13 Rectangular magnetic circuit. Figure 13.11 illustrates a magnetic circuit consisting of two magnetic materials with permeabilities μ_1 and μ_2 and with two different cross-sections, S_1 and S_2. The length of the part of the circuit with cross section S_1 is l_1 and the length of the part of the circuit with cross section S_2 is l_2. We assume that $\mu_1 \gg \mu_0$ and $\mu_2 \gg \mu_0$ so that the magnetic flux is constrained to the circuit (leakage is negligible).
(a) Show that the magnetic flux in the circuit is $\Phi = NI/(l_1/(\mu_1 S_1) + l_2/(\mu_2 S_2))$.
(b) Show that the result can be rewritten as $\Phi = NI/(R_{m,1} + R_{m,2})$ in terms of the reluctances of the system and find the reluctances.
(c) Use this example to formulate a general law for the reluctances connected in series.

13.14 A quadratic circuit. In this exercise we will study a quadratic circuit, such as you will find in most circuitboards. The circuit is quadratic with sides a and is placed in the xy-plane as in Fig. 13.12. We will first find the magnetic field from the lines and then add them to find the field from all four lines making up the square.
(a) What is the magnetic field from an infinitely long lines along the x-axis with a current I in the positive x-direction? Find the field, **B**, in a point $(x, y, 0)$
(b) We will now address a line along the x-axis from $x = -a/2$ to $x = a/2$. A current I is running in the positive x-direction. What is the contribution $\mathbf{dB}$ to the magnetic field in the point $(0, y, 0)$ from a current element $I\,\mathbf{dl}$ in the point $(x', 0, 0)$?

Fig. 13.12 Illustration of a square circuit

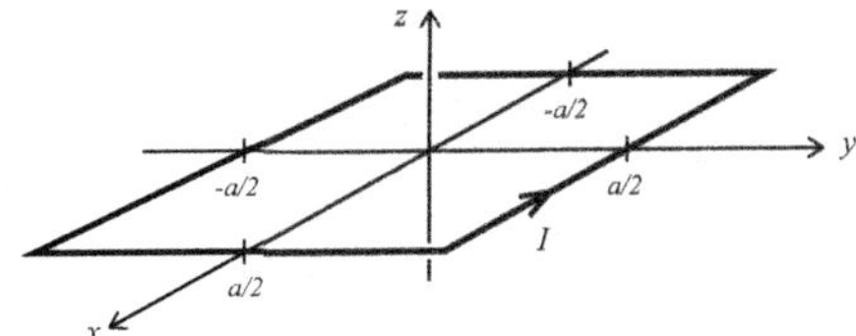

(c) Solve the integral to find the magnetic field in the point $(0, y, 0)$ for a line from $-a/2$ to $a/2$. (Hint: $\int \frac{du}{(a^2+u^2)^{3/2}} = \frac{u}{a^2\sqrt{a^2+u^2}}$)
(d) We will now compare a numerical solution of the magnetic field around a line along the x-axis from $-a/2$ to $a/2$ with the result we found above. Write a program to find the magnetic field from the current I along a line starting in $\mathbf{a}_0$ to the point $\mathbf{a}_0 + \Delta\mathbf{a}$.
(e) Use the program to find the magnetic field from the current I along the line from $-a/2$ to $a/2$ along the x-axis. Find the field along a line parallel to the x-axis through $(0, a/10, 0)$ and compare with your results for an infinite line and from the finite line. Compare by plotting the numerical results and the theoretical results in the same plot.
(f) Find the magnetic field B_z inside a quadratic circuit with side a. Visualize the field.
(g) Visualize the field in the xz-plane (that is, for $y = 0$).

Chapter 14
Faraday's Law

14.1 Emf: Electromotoric Force

What is driving the current flowing in a circuit? Figure 14.1 illustrates a circuit C with a parallel-plate capacitor, a wire with a finite conductivity and a resistor R. If the capacitor is charged up so that there are positive charges on the top side of the capacitor, there will be an electric field along the wire. As we discussed before, after a brief initial reorganization of charges on the surface of the wire, the electric field in the wire will point along the wire. In the resistor, the electric field is pointing from high to low potential as illustrated. The current is driven by the potential difference of the capacitor but will eventually decay as charge is transported through the wire from the positive to the negative side of the capacitor.

Instead of a capacitor, we could insert a mechanical battery—a device that moves a positive charge from the lower side to the upper side of the capacitor. This would require an external force. Inside the capacitor, there is an electric field $\mathbf{E}$. To move a charge from the lower side to the upper side of the capacitor, we need a force per unit charge, $\mathbf{f}$, which is at least $\mathbf{f} = -\mathbf{E}$. This force would be a property of the internal mechanics of the battery, or more generally, of any another mechanism that has the same effect as a battery of providing an external force per unit charge.

Driving forces. In this case, there are two forces *driving* charges around the circuit, $\mathbf{f}_d = \mathbf{f} + \mathbf{E}$: One force from the electric field $q\mathbf{E}$ and an external force $q\mathbf{f}$. The external force may be due to various mechanisms and there may be several such forces acting around the circuit, for example, a chemical force in a battery, a temperature gradient in a thermocouple, or light in a photoelectric cell. We call such a force a *non-Coulombic force*, because it is not due to the electric field, but due to other, non-Coulombic interactions. Notice that there are also forces *resisting* the movement of charges. In a wire with finite conductivity and in the resistor, the resisting force is due to the interaction with the material. In equilibrium, the driving and the resisting forces are equal, so that the charges do not accelerate through the wire, as we discussed in Chap. 9.

© The Author(s), under exclusive license to Springer Nature Switzerland AG 2026

A. Malthe-Sørenssen, *Elementary Electromagnetism Using Python*, Undergraduate Texts in Physics, https://doi.org/10.1007/978-3-032-19876-1_14

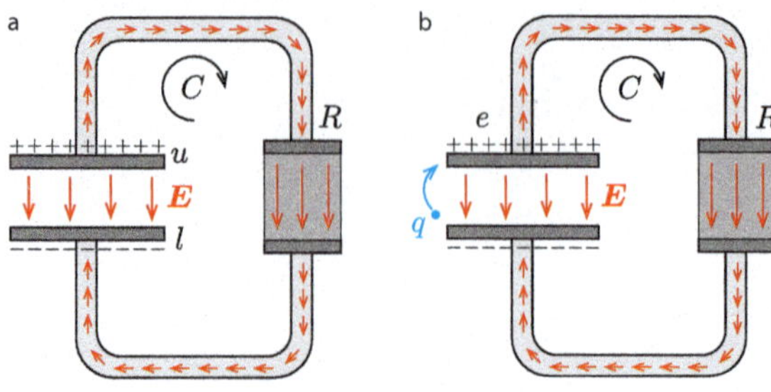

Fig. 14.1 Illustration of a circuit driven by a charged capacitor (left) and driven by a battery (right)

Electromotive force—emf. We introduce the term *electromotive force* or *emf* of the circuit as the line integral of the driving forces around the circuit:

$$e = \oint_C (\mathbf{f} + \mathbf{E}) \cdot d\mathbf{l}. \tag{14.1}$$

It is unfortunate that it is called electromotive *force* even if it is not a force, but this term is so ingrained in electromagnetics that we will use it, even if it is a bit of a misnomer.

Emf of ideal battery. Let us see how we can interpret the emf by finding the emf of the external non-Coulombic force $\mathbf{f}$ in the *ideal battery* illustrated in Fig. 14.1. The emf is defined as

$$e = \oint_C (\mathbf{f} + \mathbf{E}) \cdot d\mathbf{l} = \oint_C \mathbf{f} \cdot d\mathbf{l} + \oint_C \mathbf{E} \cdot d\mathbf{l} = \oint_C \mathbf{f} \cdot d\mathbf{l}, \tag{14.2}$$

where we have used that $\oint_C \mathbf{E} \cdot d\mathbf{l} = 0$. Because $\mathbf{f}$ is zero outside the battery, it is only the line integral of $\mathbf{f}$ inside the battery that contributes to the integral, that is, the line integral between the upper and the lower terminals of the battery (the upper, u, and lower, l, plates in the figure):

$$e = \oint_C \mathbf{f} \cdot d\mathbf{l} = \int_l^u \mathbf{f} \cdot d\mathbf{l}. \tag{14.3}$$

Inside the battery, we assume that the force is just large enough to move the charge from the lower to the upper side of the battery. This corresponds to the charge moving at a constant velocity and that the net force on the charge is zero: $\mathbf{f} + \mathbf{E} = 0$ and $\mathbf{f} = -\mathbf{E}$. We insert this and get

$$e = \int_l^u \mathbf{f} \cdot d\mathbf{l} = -\int_l^u \mathbf{E} \cdot d\mathbf{l} = V. \tag{14.4}$$

We recognize this integral as the potential difference V across the battery, where the upper side is at a higher potential than the lower side. We therefore see that we can interpret the emf as the potential increment across the battery, that is, the voltage the battery supplies to the circuit. It is useful to think of the emf as the driver of the current in the circuit.

Electromotoric force

The electromotoric force, emf, is defined as

$$e = \oint_C (\mathbf{f} + \mathbf{E}) \cdot d\mathbf{l}, \tag{14.5}$$

where $\mathbf{f}$ is an external force per unit charge acting on charges along some part of the circuit. There may be several external forces acting along the circuit, and we interpret $\mathbf{f}$ as the superposition of these forces.

Emf Sources

What determines the voltage difference V and therefore the emf of a battery? It is determined by an internal property of the battery, which is the force $\mathbf{f}$ per unit charge that the battery provides.

Mechanical battery. In Fig. 14.2 we illustrate a *mechanical battery* illustrated as a conveyor belt that pushes a charge with a force $\mathbf{f}_{NC}$, where we have used the subindex $_{NC}$ to indicate that the force is non-Coulombic, that is, an external force. Let us see what happens if we start this system from a neutral state and then suddenly turn on a force that moves electrons to the right. The force on the first electron will only be from the external force. After a while, electrons will have moved from the left to the right, so that the right side will be negatively charged and the left side positively charged. Thus, there will be an additional force acting on the charges from the electric field. Eventually, the system will reach an equilibrium, so that the external force is equal to the electric field, that is, $\mathbf{E} = -\mathbf{f}_{NC}$. In this case, the potential difference will be $V = F_{NC}s/e = f_{NC}s$, where s is the distance between

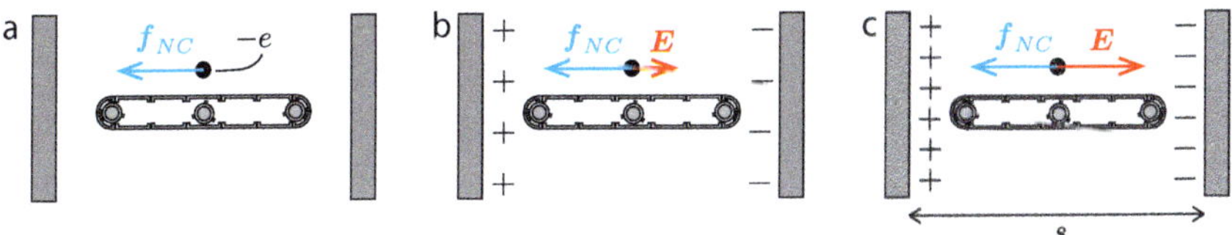

Fig. 14.2 Illustration of a mechanical battery. **a** When the battery is turned on. **b** After a short time. **c** When an equilibrium has been reached

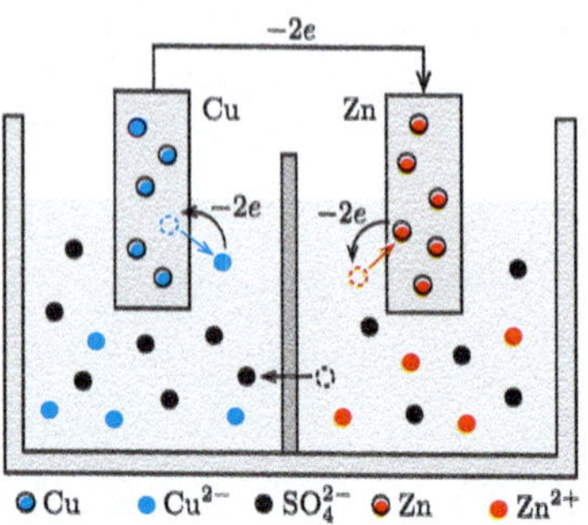

Fig. 14.3 Illustration of a chemical Daniell-cell battery

the plates. The potential difference and the emf of the battery therefore depend on the force per unit charge provided by the external mechanisms and by the geometry of the battery, here represented by the distance s between the plates.

Chemical battery. Figure 14.3 illustrates a Daniell cell, which is a chemical battery named after the English physicist and chemist John Frederick Daniell (1790–1845). The battery consists of two electrodes, one with zinc (Zn) and one with copper (Cu) in a solution of SO_4^{2-} in water. In the solution, there are dissolved ions of Zn^{2+}, Cu^{2+}, and SO_4^{2-}. There is a membrane separating the two sides of the solution, as illustrated in the figure, and only SO_4^{2-} can pass through the membrane.

What happens in this system? When zinc dissolves from the zinc electrode, two electrons are released

$$\mathrm{Zn} \rightarrow \mathrm{Zn}^{2+} + 2\mathrm{e}^- . \tag{14.6}$$

The electrode reduction potential for this reaction is −0.76V. The charge of the two electrons travels to the Cu-electrode, where copper ions are absorbed from the solution onto the electrode:

$$\mathrm{Cu}^{2+} + 2\mathrm{e}^- \rightarrow \mathrm{Cu}. \tag{14.7}$$

The electrode reduction potential of this reaction is +0.34V. At the same time SO_4^{2-} diffuses through the membrane from the copper to the zinc side, closing the loop. The total voltage difference between the two electrodes due to this reaction is 1.1V per such cell. You will learn more about the physics underlying such processes in thermal and statistical physics.

Test your understanding
The figure shows a circuit with a chemical battery and a resistor. The potential along the x-axis is sketched. (a) What is the electric field $\mathbf{E}$ in the battery? (b) What is the net force on a charge Q in the battery? (c) What forces act on a charge Q inside the battery? (d) What is $\mathbf{f}_{NC}$, the non-Coulombic force per charge in the battery? (e) What is the emf of the battery? (f) What is Kirchhoff's law for this circuit?[1]

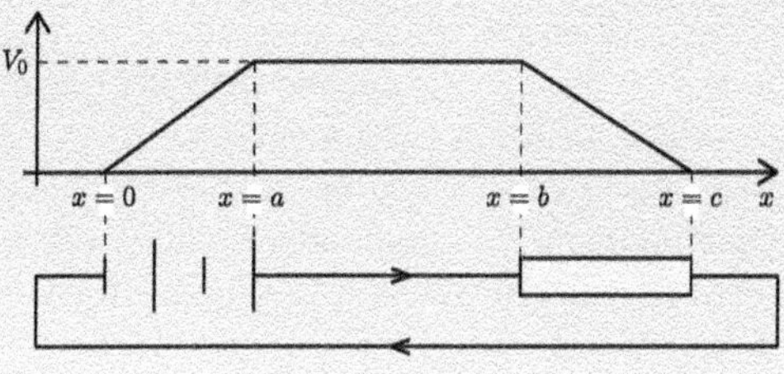

14.2 Motional Emf

What are possible sources of the non-Coulombic force? One possible source is the force from a magnetic field on a moving charge. If a conductor moves through a magnetic field, the conductor consists of charges q that are free to move around as well as charges $-q$ that are fixed. For example, for most conductors, electrons are free to move, whereas the atoms are fixed to the conductor. Figure 14.4 illustrates a circuit C in a magnetic field. The left side of the circuit is moving with a velocity $\mathbf{v}$ while the rest of the circuit is stationary. This can be realized for example by having a conducting rod slide along two conducting wires.

What is the force acting on a charge q in the rod? If the charge is at rest relative to the rod, the velocity of the charge is the same as the velocity of the rod, $\mathbf{v}$. The force on the charge is $\mathbf{F} = q\mathbf{v} \times \mathbf{B}$ for $\mathbf{B} = B\hat{\mathbf{z}}$ and $\mathbf{v} = v\hat{\mathbf{x}}$ as illustrated in Fig. 14.4, the resulting force is $\mathbf{F} = -qvB\hat{\mathbf{y}}$. The non-Coulombic force per charge is then $\mathbf{f}_{NC} = \mathbf{F}/q = -vB\hat{\mathbf{y}}$.

We find the corresponding emf by integrating the non-Coulombic force around the circuit. The only contribution is along the moving rod, which has a length L. The direction of the loop C is in the positive direction as illustrated in Fig. 14.4. The emf is

$$e = \oint_C (\mathbf{f}_{NC} + \mathbf{E}) \cdot \mathrm{d}\mathbf{l} = \oint_C \mathbf{f}_{NC} \cdot \mathrm{d}\mathbf{l} = -vB\hat{\mathbf{y}} \cdot (-L\hat{\mathbf{y}}) = vBL, \tag{14.8}$$

where we have used that $\oint_C \mathbf{E} \cdot \mathrm{d}\mathbf{l} = 0$. This demonstrates that we can calculate the emf from a moving or deforming circuit in a magnetic field. Notice that if we assume that the charge has a velocity $\mathbf{u}$ along the rod, the result does not change because

$\mathbf{w} = \mathbf{v} + \mathbf{u} = v\hat{\mathbf{x}} - u\hat{\mathbf{y}}$ is the net velocity and $\mathbf{f}_{NC} = \mathbf{w} \times \mathbf{B} = -vB\hat{\mathbf{y}} - uB\hat{\mathbf{x}}$ and it is only the component of $\mathbf{f}_{NC}$ along the y-axis that contributes to the integral.

14.3 Faraday's Law

The idea from motional emf above can be generalized to any circuit C which is moving with a velocity $\mathbf{v}(\mathbf{r})$ through a magnetic field $\mathbf{B}(\mathbf{r})$, where the velocity may vary with position, $\mathbf{r}$, which results in a deformed circuit as in the case of motional emf above. Figure 14.5 illustrates the situation. A charge q in the circuit moves with a velocity $\mathbf{u}$ along the circuit and the circuit moves with a velocity $\mathbf{v}$, so that the net velocity of the charge is $\mathbf{w} = \mathbf{u} + \mathbf{v}$. The non-Coulombic force on the charge q from the magnetic field $\mathbf{B}$ is

$$\mathbf{f}_{NC} = \frac{1}{q}\left(q\mathbf{w} \times \mathbf{B}\right) = \mathbf{w} \times \mathbf{B}. \tag{14.9}$$

The emf from this non-Coulombic force is given as the line integral along the circuit C:

$$e = \oint_C \mathbf{f}_{NC} \cdot \mathrm{d}\mathbf{l} = \oint_C (\mathbf{w} \times \mathbf{B}) \cdot \mathrm{d}\mathbf{l}. \tag{14.10}$$

We can rewrite this vector integral using the vector equality $\mathbf{a} \cdot (\mathbf{b} \times \mathbf{c}) = \mathbf{b} \cdot (\mathbf{c} \times \mathbf{a}) = \mathbf{c} \cdot (\mathbf{a} \times \mathbf{b})$:

$$\begin{aligned} e &= \oint_C (\mathbf{w} \times \mathbf{B}) \cdot \mathrm{d}\mathbf{l} = \oint_C \mathrm{d}\mathbf{l} \cdot (\mathbf{w} \times \mathbf{B}) \\ &= \oint_C \mathbf{B} \cdot (\mathrm{d}\mathbf{l} \times \mathbf{w}) = -\oint_C \mathbf{B} \cdot (\mathbf{w} \times \mathrm{d}\mathbf{l}). \end{aligned} \tag{14.11}$$

Here, we notice that $\mathbf{w} = \mathbf{u} + \mathbf{v}$ where $\mathbf{u}$ points along the circuit, that is, it is parallel to $\mathrm{d}\mathbf{l}$ and therefore $\mathbf{u} \times \mathrm{d}\mathbf{l} = 0$. Consequently, $\mathbf{w} \times \mathrm{d}\mathbf{l} = \mathbf{u} \times \mathrm{d}\mathbf{l} + \mathbf{v} \times \mathrm{d}\mathbf{l} = 0 + \mathbf{v} \times \mathrm{d}\mathbf{l}$. We have therefore found that

$$e = -\oint_C \mathbf{B} \cdot (\mathbf{v} \times \mathrm{d}\mathbf{l}). \tag{14.12}$$

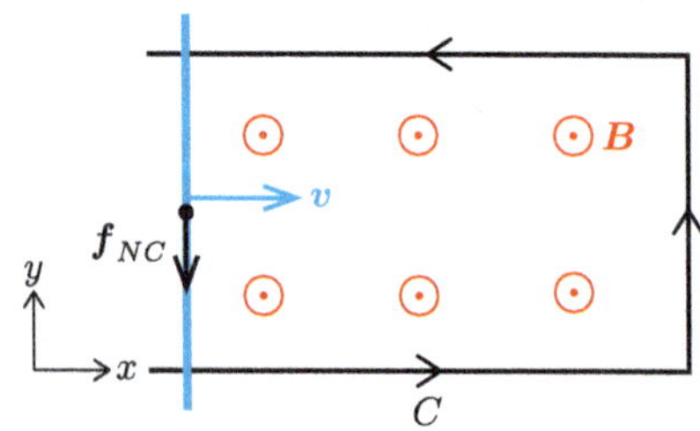

Fig. 14.4 Illustration of a circuit in a magnetic field $\mathbf{B} = B\hat{\mathbf{z}}$. The left side of the circuit consists of a rod that moves along conducting wires with a velocity $\mathbf{v} = v\hat{\mathbf{x}}$

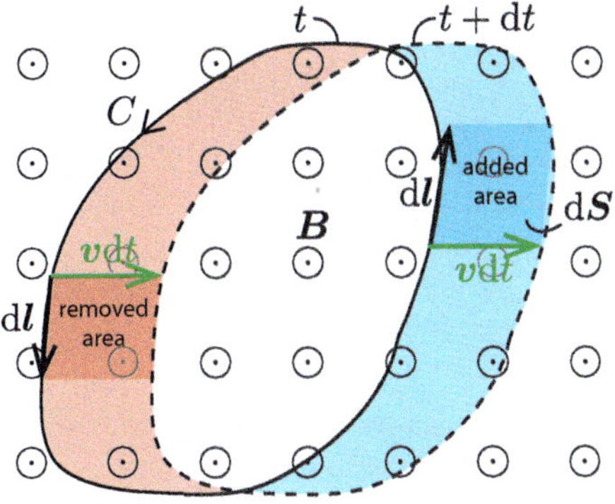

Fig. 14.5 Illustration of a circuit moving through a magnetic field at times t (solid line) and $t + \mathrm{d}t$ (dashed line). The element $\mathbf{v}\,\mathrm{d}t \times \mathrm{d}\mathbf{l} = \mathrm{d}\mathbf{S}$ corresponds to a change in the surface enclosed by the curve C. The red regions correspond to areas that are removed when the circuit moved from t to $t + \mathrm{d}t$. The blue regions correspond to areas that are added

How can we interpret the integral on the right-hand side of (14.12)? Figure 14.5 shows the circuit C at a time t and at a time $t + \mathrm{d}t$. During the time interval $\mathrm{d}t$ the circuit at $\mathbf{r}$ has moved to the position $\mathbf{r} + \mathbf{v}(\mathbf{r})\,\mathrm{d}t$. We see from Fig. 14.5 that we can interpret $\mathbf{v}\,\mathrm{d}t \times \mathrm{d}\mathbf{l}$ as an area $\mathrm{d}\mathbf{S}$ and $\mathbf{B} \cdot (\mathbf{v}\,\mathrm{d}t \times \mathrm{d}\mathbf{l})$ as the flux through this area. The red area on the left of Fig. 14.5 corresponds to the area that has been removed from the surface S spanned by the circuit C during the time interval $\mathrm{d}t$ and the blue area corresponds to the area that has been added. We can therefore interpret the integrand $\mathbf{B} \cdot (\mathbf{v}\,\mathrm{d}t \times \mathrm{d}\mathbf{l})$ as the contribution from the line segment $\mathrm{d}\mathbf{l}$ to the change in the flux of the magnetic field $\mathbf{B}$ through the circuit C during the time interval $\mathrm{d}t$. The total change, $\mathrm{d}\Phi$ in flux of $\mathbf{B}$ through the circuit is therefore the integral along the circuit:

$$\mathrm{d}\Phi = \oint_C \mathbf{B} \cdot (\mathbf{v}\,\mathrm{d}t \times \mathrm{d}\mathbf{l}). \tag{14.13}$$

We divide by $\mathrm{d}t$ and combine with (14.12) to find that

$$e = -\oint_C \mathbf{B} \cdot (\mathbf{v} \times \mathrm{d}\mathbf{l}) = -\frac{d\Phi}{\mathrm{d}t}. \tag{14.14}$$

This is called *Faraday's law* for motional emf, that is, when the circuit is moving through a stationary field.

Faraday's law

Faraday's law states that an emf is induced in a circuit C if the magnetic flux changes with time:

$$e = \oint_C (\mathbf{f} + \mathbf{E}) \cdot \mathrm{d}\mathbf{l} = -\frac{\mathrm{d}}{\mathrm{d}t} \int_S \mathbf{B} \cdot \mathrm{d}\mathbf{S} = -\frac{\mathrm{d}}{\mathrm{d}t}\Phi. \tag{14.15}$$

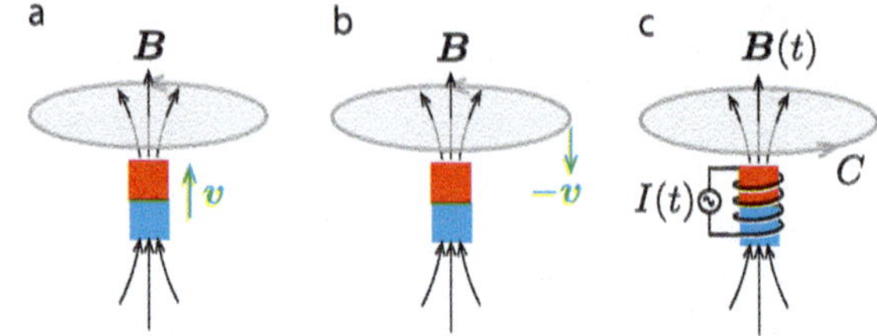

Fig. 14.6 **a** Illustration of a circuit moving with a velocity **v**. **b, c** Illustrations of a circuit moving with a velocity **v** in a stationary magnetic field or a magnetic field moving with a velocity −**v** through a stationary circuit

Faraday's law is valid independently of the source of change in the flux. The law applies both when a circuit is moving through a stationary magnetic field and when the magnetic field is varying in time through a stationary circuit. We call the emf generated by this effect an *induced emf.*

Faraday's Law is Independent of How the Flux Changes

Here, we have only proved Faraday's law for motional emf, that is for the case when a circuit is moving or is deformed in a magnetic field. However, Faraday executed a set of careful experiments to address three situations as shown in Fig. 14.6. In the first experiment in Fig. 14.6a, he moved the circuit with a velocity **v** into a stationary magnetic field. In the second experiment in Fig. 14.6b, he kept the circuit stationary and moved the magnet with a velocity −**v**. These situations clearly only depend on the reference system used to describe them, and we therefore expect the induced emf to be the same in both situations. This was also what Faraday found. However, he also performed a third experiment where the magnetic field was varying with time through a stationary circuit. The beauty of Faraday's law is that the induced emf can be described in the same way in all these situations:

$$e = -\frac{\mathrm{d}\Phi}{\mathrm{d}t}. \tag{14.16}$$

This means that the induced emf only depends on the change in flux through the circuit, independently of the reason for the change in flux.

The underlying symmetry in Faraday's law ensures that the emf induced in a circuit is the same if you take the circuit with you in a car and drive through a magnetic field which is localized in space or if you keep the circuit stationary and move the magnetic field, as Faraday did in the first two experiments. However, in the system inside the car, the circuit will be stationary, and the magnetic field will be changing in time. The behavior in all these situations must be the same if the physics is to be the same in all inertial systems. We will address these issues in more detail in the exercises.

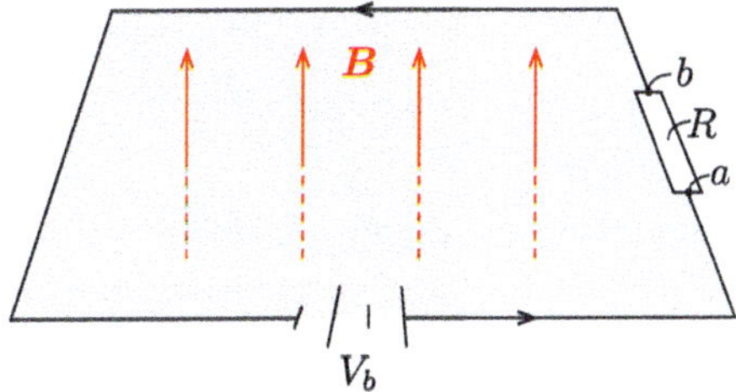

Fig. 14.7 Illustration of a circuit with a voltage source V_b, a resistor R and a magnetic field $\mathbf{B}$ that gives a flux Φ through the circuit

Test your understanding
(a) The figure shows a circuit of height and width $2a$. For $x > 0$ the magnetic field $\mathbf{B} = B_0\hat{\mathbf{z}}$ and for $x < 0$ the magnetic field is zero. What is the flux Φ_B through the circuit C? (b) The figure shows a circuit C in a magnetic field down into the plane. What is the sign of the flux Φ_B? (c) The figure shows a circuit C in a uniform magnetic field. The circuit is moving from the solid to the dashed position. What is the direction of the induced emf? (d) The figure shows an elliptical circuit rotating in a uniform magnetic field. What is the direction of the induced emf?[2]

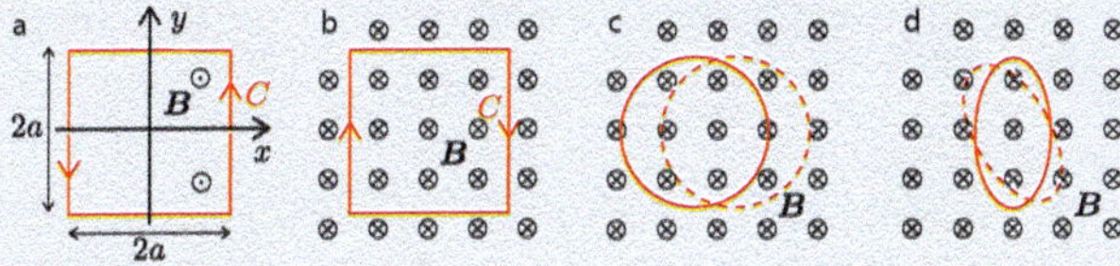

Emf in a Circuit

How do we include the magnetically induced emf in a circuit and how does it affect how we use Kirchhoff's voltage law? Figure 14.7 illustrates a circuit with a battery V_b, which represents one or more voltage sources, and a resistor R. There is a time-varying magnetic field $\mathbf{B}$ going through the circuit with a resulting flux $\Phi(t)$. How do we reformulate Kirchhoff's law to include the induced emf, the emf from the battery and the voltage drop across the resistor?

In this case, there are two non-Coulombic forces acting: a non-Coulombic force per unit charge, $\mathbf{f}_b$, inside the battery and the non-Coulombic magnetic force per unit charge, $\mathbf{f}_m$. The emf of the circuit is defined as

$$e = \oint_C (\mathbf{f}_b + \mathbf{f}_m + \mathbf{E}) \cdot d\mathbf{l} = \oint_C \mathbf{f}_b \cdot d\mathbf{l} + \oint_C (\mathbf{f}_m + \mathbf{E}) \cdot d\mathbf{l}. \tag{14.17}$$

We can add $\oint_C \mathbf{E} \cdot d\mathbf{l} = 0$ to $\oint_C \mathbf{f}_b \cdot d\mathbf{l}$ getting

$$e = \underbrace{\oint_C (\mathbf{f}_b + \mathbf{E}) \cdot \mathrm{d}\mathbf{l}}_{V_b} + \underbrace{\oint_C (\mathbf{f}_m + \mathbf{E}) \cdot \mathrm{d}\mathbf{l}}_{-\mathrm{d}\Phi/\,\mathrm{d}t} = V_b - \frac{\mathrm{d}\Phi}{\mathrm{d}t}. \tag{14.18}$$

The emf is therefore the sum of the emf of the battery and the induced emf. In general, we can add the emf contributions from different external forces such as for a circuit with several batteries.

However, if we look at the emf path integral

$$e = \oint_C (\mathbf{f}_b + \mathbf{f}_m + \mathbf{E}) \cdot \mathrm{d}\mathbf{l}, \tag{14.19}$$

we realize that $\mathbf{f}_b + \mathbf{f}_m + \mathbf{E} = 0$ everywhere except inside the resistor. Inside the wires, we have assumed that the resistance is effectively zero. This means that the voltage drop along the wire is zero and that the electric field is approximately zero. (This is of course an approximation.) Similarly, inside the battery, we assume that the force from the battery cancels the electric field. The only contribution to the integral in (14.19) is therefore from inside the resistor. Inside the resistor, the force from the battery is zero and we also assume that the contribution from $\mathbf{f}_m$ is small if the resistor is of limited extent. The emf is therefore

$$e = \int_a^b (\mathbf{f}_b + \mathbf{f}_m + \mathbf{E}) \cdot \mathrm{d}\mathbf{l} = \int_a^b \mathbf{E} \cdot \mathrm{d}\mathbf{l} = \Delta V_R = IR. \tag{14.20}$$

If we combine the results from (14.18) and (14.20), we get

$$e = V_b - \frac{\mathrm{d}\Phi}{\mathrm{d}t} = IR = \Delta V_R. \tag{14.21}$$

We found that the emf e is equal to the voltage drop ΔV_R across the resistor R. This argument will be the same if we replace the resistor with any other component i with a voltage drop ΔV_i. If we have a circuit with several sources e_j of emf and several components with voltage drops ΔV_i, we therefore get a version of Kirchhoff's law:

$$e = \sum_j e_j = \sum_i \Delta V_i. \tag{14.22}$$

This is one way to generalize Kirchhoff's law to include emfs. In this case, we put all the emfs on the left-hand side. These correspond to voltage increments. And we put all the voltage drops on the right-hand side. Notice that a battery with a voltage V_b contributes with a positive emf on the left-hand side. And a resistor with a voltage drop RI contributes with a positive voltage drop on the right-hand side. The wording may here be a bit confusing, but a voltage drop $\Delta V_R = RI$ means that the voltage drops by RI across the resistor in the direction of positive current I.

An alternative way to generalize Kirchhoff's law would be to include both the emfs, the voltage increments, and the voltage drops on the left-hand side:

$$\sum_j e_j - \sum_i \Delta V_i = 0. \tag{14.23}$$

This version may look more like a generalization of Kirchhoff's law, where we have included all the voltage changes on the left-hand side. Emfs contribute with positive voltage changes, whereas the other components contribute with negative voltage charges. For the situation at hand this would give us:

$$V_b - \frac{d\Phi}{dt} - RI = 0. \tag{14.24}$$

You may find that this is simply a small arithmetic adjustment. However, it is important that you ensure that you set the *signs* correctly!

Kirchhoff's voltage law

The generalized Kirchhoff's voltage law that includes both emfs e_j and voltage drops ΔV_i is

$$\sum_j e_j = \sum_i \Delta V_i, \tag{14.25}$$

or, equivalently,

$$\sum_j e_j - \sum_i \Delta V_i = 0. \tag{14.26}$$

Direction of the Induced Emf

What is the direction of the induced emf and the resulting induced current? Figure 14.8 illustrates a circuit and a uniform magnetic field. When we want to calculate the flux of **B** through a surface enclosed by circuit C, we need to determine an orientation of the surface. The flux integral is:

$$\Phi = \int_S \mathbf{B} \cdot d\mathbf{S}, \tag{14.27}$$

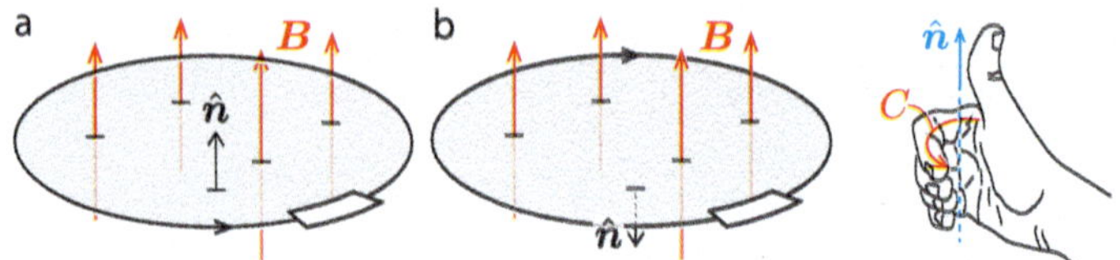

Fig. 14.8 Illustration of the direction of the circuit and the direction of the magnetic field

where the direction of the surface is given by the direction of d**S**, which is given by a normal vector to the surface. For simplicity, let us assume that the circuit is in a plane and we choose a surface in the same plane. We can then choose the normal vector to point either up from the plane or down into the plane. The choice of normal vector determines the positive direction of the circuit using the right-hand rule. For the circuit in Fig. 14.8a, you curve the four fingers from the index finger to your little finger on your right-hand in the direction of the circuit. Then your thumb will point in the direction of the surface normal. If the circuit is oriented in the positive direction, as in Fig. 14.8a, the normal vector points up. Whereas, if you choose the circuit to have the opposite orientation, as in Fig. 14.8b, the surface normal points down.

For positive orientation as in Fig. 14.8a, the flux is

$$\Phi = \int_S \mathbf{B} \cdot \mathrm{d}\mathbf{S} = BS, \tag{14.28}$$

If the magnetic field increases with time, $\mathrm{d}B/\mathrm{d}t > 0$, the flux also increases with time, $\mathrm{d}\Phi/\mathrm{d}t > 0$, which means that the emf will be $e = -\mathrm{d}\Phi/\mathrm{d}t < 0$. We find the associated current I in the circuit by applying Kirchhoff's voltage law: $e - RI = 0$, which gives $e = RI$ and $I = e/R$. Because e is negative, the current will be negative, that is, opposite the direction of the arrow.

What would happen if we instead chose the orientation in Fig. 14.8b? The flux would then be $\Phi = -BS$. If the magnetic field increases with time, we would get $\mathrm{d}B/\mathrm{d}t > 0$, and $\mathrm{d}\Phi/\mathrm{d}t < 0$, which would give $e = -\mathrm{d}\Phi/\mathrm{d}t > 0$. Again, Kirchhoff's voltage law gives $e - RI = 0$ and $I = e/R$, but now e is positive and the current is positive and in the direction of the arrow in Fig. 14.8b.

As expected, we get the same result for the direction of the emf and the direction of the current, independently of our choice of coordinate system and our choice of the positive direction for the circuit. We just have to be consistent and keep to our choices.

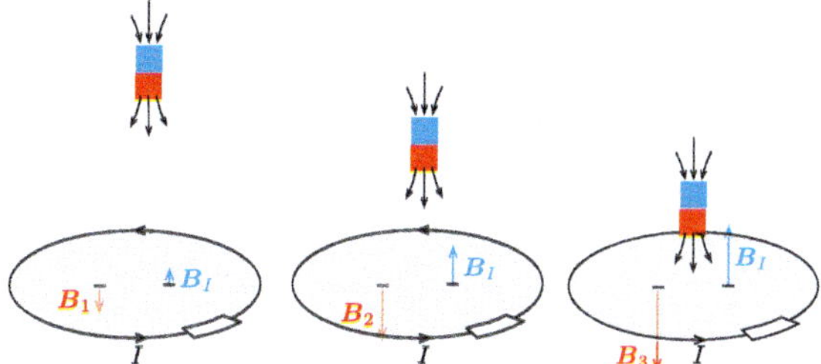

Fig. 14.9 Illustration of Lenz law. What is the direction of the magnetic field set up by the current induced by a permanent magnet moving towards the circuit?

Test your understanding
(a) A permanent magnet is placed under a circuit and pulled down. You friend Q states that since it is a permanent magnet, there cannot be any current induced in the circuit. If Q right? If there is an induced current, what is its direction? (b) A rectangular circuit is moving into an area with a uniform and constant magnetic field **B**. What is the direction of the induced current? What is the flux through the circuit? (c) Now, assume that the circuit is not moving, but the magnetic field is $\mathbf{B} = (B_0 + kt)\hat{\mathbf{z}}$. What is the induced emf? What is the direction of the magnetic field produced by the induced current?[3]

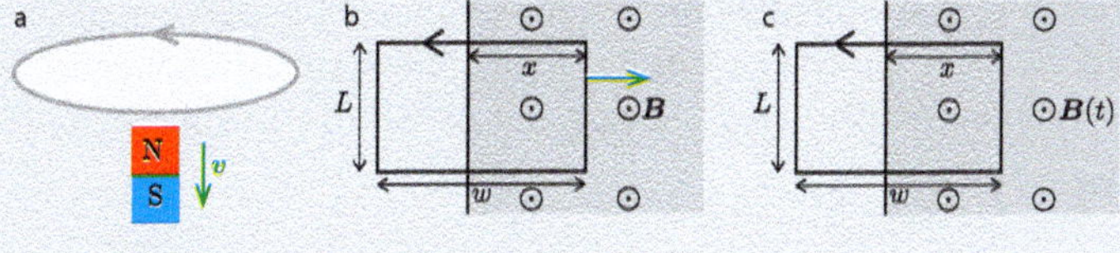

Lenz's Law

What happens if you move a permanent magnet towards an electric circuit? We have illustrated the field from a permanent magnet in Fig. 14.9. When the magnet is moved towards the circuit, the magnitude of the magnetic field through the surface of the circuit increases, and therefore the magnitude of the flux increases. This means that we induce a current in the circuit. However, this current will also set up a magnetic field. What is the direction of the magnetic field set up by the current, which is induced by the moving magnet?

That was a complicated question, but it has a surprisingly simple answer. Let us assume the permanent magnet and the circuit are oriented as in Fig. 14.9. When the permanent magnet is moved towards the circuit, the magnitude of the magnetic field on the surface S encompassed by the circuit increases with time. Because the flux is negative, this means that the flux decreases with time. The induced emf is $e = -\mathrm{d}\Phi/\mathrm{d}t$ and it increases with time. Therefore, the current I increases with time. The magnetic field $\mathbf{B}_I$ from this current is oriented in the direction opposite of the magnetic field from the permanent magnet and the magnitude of B_I increases

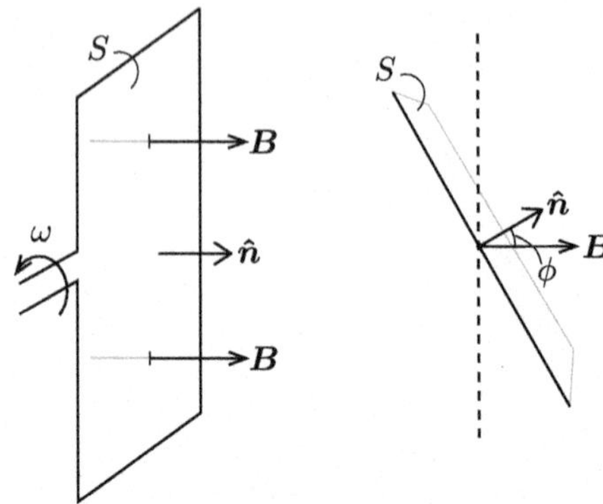

Fig. 14.10 Illustration a circuit rotating in an external magnetic field

with time. Therefore, the magnetic field set up by the current induced by the change in flux acts in a direction opposite of the change in the field that caused the current.

This result is general and is called *Lenz' law*: The direction of the current in the circuit is such that the magnetic field generated by the current will counteract the change in magnetic field that caused the current. This was a rather complex formulation. It can be simplified to "Nature counteracts changes in magnetic flux". If the flux is changed, the system will set up a current to generate a magnetic field, that will counteract the change. This is a very useful tool to check the sign of Faraday's law.

Example: Rotating Circuit in Static Magnetic Field

A circuit rotates with a constant angular velocity ω in a uniform, stationary magnetic field **B** *as illustrated in Fig. 14.10. The circuit is rectangular with sides a and b and has a resistance R. Find the induced emf and current in the circuit and the time-averaged power dissipated in the circuit. You can ignore the effect of the magnetic field generated by the current in the circuit.*

Solution : The magnetic flux through the circuit depends on the angle ϕ between the normal to the plane of the loop and the magnetic field as illustrated in the figure:

$$\Phi = \int_S \mathbf{B} \cdot \mathrm{d}\mathbf{S} = \mathbf{B} \cdot \mathbf{S} = BS\cos\phi, \tag{14.29}$$

where $\phi = \omega t$. The induced emf is therefore

$$e = -\frac{\mathrm{d}\Phi}{\mathrm{d}t} = -\frac{\mathrm{d}}{\mathrm{d}t}\left(BS\cos\omega t\right) = BS\omega\sin\omega t. \tag{14.30}$$

This means that to induce a large emf we need to have a large magnetic field B or a large surface area S of the circuit. A simple way to increase the surface area is to increase the number of windings, N, in the circuit. This is a simple model for a generator that generates an alternating voltage/current with a sinus-shape. To make a generator, you also have to find a smart way to connect two wires to the circuit. This is often done using brushes, which will also turn the polarity of the outgoing wire after a rotation of 180°.

From Kirchhoff's voltage law, we find that $e - IR = 0$ and $I = e/R$. The dissipated power is:

$$P = RI^2 = R\left(\frac{BS\omega \sin \omega t}{R}\right)^2 = \frac{\omega^2 B^2 S^2}{R} \sin^2 \omega t. \tag{14.31}$$

This is the instantaneous power. To find the time-averaged dissipated power we need to average over one rotational cycle. The average of the $\sin^2 \omega t$ term is

$$\langle \sin^2 \omega t \rangle = \frac{1}{T} \int_0^T \sin^2 \omega t \mathrm{d}t, \tag{14.32}$$

where $T = 2\pi/\omega$. We change variable to $u = \omega t$ so that the integral is from $u = 0$ to $u = T\omega = 2\pi$, where $\mathrm{d}u = \omega \mathrm{d}t$

$$\langle \sin^2 \omega t \rangle = \frac{1}{T\omega} \int_0^{2\pi} \sin^2 u \mathrm{d}u \tag{14.33}$$

This integral is simple to solve, and you may indeed remember it by heart, but we use `Sympy`:

```
import sympy as sp
u = sp.Symbol("u")
f = sp.sin(u)**2/(2*sp.pi)
sp.integrate(f, (u,0,2*sp.pi))
```

```
1/2
```

We therefore conclude that

$$\langle \sin^2 \omega t \rangle = \frac{1}{2}, \tag{14.34}$$

and that

$$\langle P \rangle = \frac{\omega^2 B^2 S^2}{R} \langle \sin^2 \omega t \rangle = \frac{\omega^2 B^2 S^2}{2R}. \tag{14.35}$$

14.4 Faraday's Law on Differential Form

Faraday's law relates the emf to the rate of change of flux through a circuit

$$e = \oint_C (\mathbf{f}_m + \mathbf{E}) \cdot \mathrm{d}\mathbf{l} = -\frac{\mathrm{d}}{\mathrm{d}t} \int_S \mathbf{B} \cdot \mathrm{d}\mathbf{S}. \tag{14.36}$$

If the circuit is stationary, then there are no magnetic forces on the charges, $\mathbf{f}_m = 0$, and we can move the derivative in (14.36) inside the integral, to get:

$$\oint_C \mathbf{E} \cdot \mathrm{d}\mathbf{l} = -\int_S \frac{\partial \mathbf{B}}{\partial t} \cdot \mathrm{d}\mathbf{S}. \tag{14.37}$$

where we can apply Stokes' theorem to get:

$$\nabla \times \mathbf{E} = -\frac{\partial \mathbf{B}}{\partial t}. \tag{14.38}$$

This is Faraday's law on differential form.

Faraday's law on differential form

$$\nabla \times \mathbf{E} = -\frac{\partial \mathbf{B}}{\partial t}. \tag{14.39}$$

Example: Circular Currents

A uniform, time-varying magnetic field, $\mathbf{B} = B_z(t)\hat{\mathbf{z}}$, passes through a circular conducting disc in the xy -plane. Find the associated electric field and current in the plane. You can assume that the field set up by the currents in the plane is negligible compared to $B_z(t)$.

Solution. We will attempt two types of solutions to this problem. First, we will solve the differential equation and second, we will use the integral formulation inspired by Ampere's law.

Solving the differential equation. First, we start from the partial differential equation for the electric field, Faraday's law:

$$\nabla \times \mathbf{E} = -\frac{\partial B_z}{\partial t}\hat{\mathbf{z}}. \tag{14.40}$$

We rewrite the curl in cylindrical coordinates, keeping only the z-component of the equations since the right-hand side only has a z-component:

$$\frac{1}{r}\left(\frac{\partial\left(rE_{\phi}\right)}{\partial r}-\frac{\partial E_{r}}{\partial\phi}\right)=-\frac{\partial B_{z}}{\partial t}. \tag{14.41}$$

From the symmetry of the system, we realize that E_r cannot have any ϕ-dependence because the system is rotationally symmetric around the z-axis, therefore $\partial E_r/\partial\phi = 0$. We are therefore left with:

$$\frac{1}{r}\frac{\partial\left(rE_{\phi}\right)}{\partial r}=-\frac{\partial B_{z}}{\partial t}\quad\Rightarrow\quad\frac{\partial\left(rE_{\phi}\right)}{\partial r}=-r\frac{\partial B_{z}}{\partial t}. \tag{14.42}$$

With solutions:

$$rE_{\phi}=-\frac{1}{2}r^{2}\frac{\partial B_{z}}{\partial t}+C\quad\Rightarrow\quad E_{\phi}=-\frac{1}{2}r\frac{\partial B_{z}}{\partial t}+\frac{C}{r}. \tag{14.43}$$

The electric field must be finite at $r = 0$, therefore $C = 0$. Consequently, the solution is

$$\mathbf{E}=-\frac{1}{2}r\frac{\partial B_{z}}{\partial t}\hat{\boldsymbol{\phi}}. \tag{14.44}$$

This is an electric field curling around the direction of the time-varying magnetic field. For a conducting material, we know from Ohm's law that $\mathbf{J} = \sigma\mathbf{E}$. The circular electric field induced by the time-varying magnetic field, will therefore induce circular currents. Inside a conducting volume exposed to a varying magnetic field, there will often be a combination of many such circular currents called *eddy currents*.

We visualize the electric field in the xy-plane. We insert that $\hat{\boldsymbol{\phi}} = (-y, x)/r$ where $\mathbf{r} = (x, y)$:

$$\mathbf{E}=-\frac{1}{2}r\frac{\partial B_{z}}{\partial t}\hat{\boldsymbol{\phi}}=-\frac{1}{2}\frac{\partial B_{z}}{\partial t}\frac{r(-y,x)}{r}=-\frac{1}{2}\frac{\partial B_{z}}{\partial t}(-y,x). \tag{14.45}$$

We visualize this field by calculating the field values on a $N \times N$ grid from $-5a$ to $5a$, where a is a length, and visualize using `quiver`. This is implemented in the following Python program and visualized in Fig. 14.11.

```
import numpy as np
import matplotlib.pyplot as plt
a = 1.0
L = 5*a
NL = 25
dBdt = 1.0
x = np.linspace(-L,L,NL)
y = np.linspace(-L,L,NL)
rx,ry = np.meshgrid(x,y,indexing="ij")
Ex = -(dBdt/2)*(-ry)
```

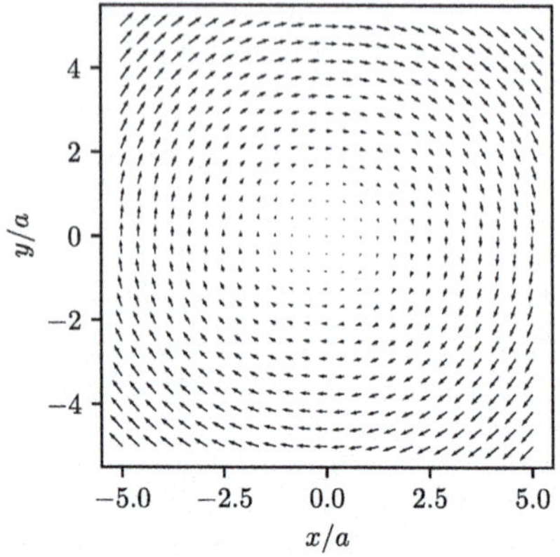

Fig. 14.11 Plot of the electric field (or the current density) associated with a time-varying magnetic field along the z-axis

```
Ey = -(dBdt/2)*(rx)
plt.quiver(rx,ry,Ex,Ey)
```

Solving by integration. We can also solve this by applying an approach similar to Ampere's law: The emf around a circular path with radius r around the z-axis is

$$e = \oint_C (\mathbf{E} + \mathbf{f}_{NC}) \cdot \mathrm{d}\mathbf{l} = \oint_C \mathbf{E} \cdot \mathrm{d}\mathbf{l}, \tag{14.46}$$

where there are no non-Coulombic forces, $\mathbf{f}_{NC} = 0$. Due to rotational symmetry, we expect $\mathbf{E}$ to only depend on r and z and not on ϕ. Only the component E_ϕ contributes to the curve integral around the circular path in (14.46), and since $E_\phi = E_\phi(r, z)$ does not depend on ϕ, we can place it outside the integral

$$e = \oint_C \mathbf{E} \cdot \mathrm{d}\mathbf{l} = E_\phi \oint_C \hat{\boldsymbol{\phi}} \cdot \mathrm{d}\mathbf{l} = E_\phi\, 2\pi r, \tag{14.47}$$

where E_ϕ is the azimuthal component of the eletric field. Faraday's law states that the emf e is related to the time derivative of the flux Φ through a surface S with C as its boundary:

$$e = -\frac{\mathrm{d}\Phi}{\mathrm{d}t} = -\frac{\mathrm{d}}{\mathrm{d}t} \int_S \mathbf{B} \cdot \mathrm{d}\mathbf{S}. \tag{14.48}$$

For a uniform magnetic field $\mathbf{B} = B_z(t)\hat{\mathbf{z}}$, the flux is

$$\Phi = \int_S \mathbf{B} \cdot \mathrm{d}\mathbf{S} = \mathbf{B} \cdot \mathbf{S} = B_z(t)\pi r^2. \tag{14.49}$$

We combine (14.47), (14.48) and (14.49), getting:

$$e = 2\pi r E_\phi = -\frac{\mathrm{d}}{\mathrm{d}t} B_z(t) \pi r^2. \tag{14.50}$$

Here, only $B_z(t)$ depends on time, and we get

$$E_\phi = -\frac{r}{2}\frac{\partial B_z}{\partial t}. \tag{14.51}$$

This is identical to the result in (14.44). This method is considered a curiosity, demonstrating how the methods we have developed for Ampere's law can be transferred to new areas if the equations are similar.

Summary

The **electromotoric force, emf**, is defined as

$$e = \oint_C (\mathbf{f}_{NC} + \mathbf{E}) \cdot \mathrm{d}\mathbf{l}.$$

The universal flux rule or **Faraday's law** states that

$$e = \oint_C (\mathbf{f}_{NC} + \mathbf{E}) \cdot \mathrm{d}\mathbf{l} = -\frac{\mathrm{d}}{\mathrm{d}t} \int_S \mathbf{B} \cdot \mathrm{d}\mathbf{S} = -\frac{\mathrm{d}}{\mathrm{d}t} \Phi.$$

This law is valid both when the change in flux is due to a change in the circuit and when the change in flux is due to a time-varying magnetic field.

We can include the emf from Faraday's law in the voltage drops in **Kirchhoff's voltage law** so that the sum of voltage drops along a circuit is

$$\sum_i \Delta V_i = 0$$

where we include $\Delta V_i = -\mathrm{d}\Phi/\mathrm{d}t$ for Faraday's law, $\Delta V_i = V_b$ for other sources such as a battery, $\Delta V_i = -RI$ for a resistor, and $\Delta V_i = -Q/C$ for a capacitor.

Faraday's law on differential form states:

$$\nabla \times \mathbf{E} = -\frac{\partial \mathbf{B}}{\partial t}.$$

Exercises

Discussion Exercises

14.1 Rotation of quadratic circuit. A quadratic circuit is in a region of space with a uniform (in space) and constant (in time) magnetic field. Is possible to rotate the circuit around one of the sides of the circuit without inducing an emf in the circuit? Discuss possible orientations of the rotational axis compared to the direction of the magnetic field.

14.2 Force on copper plate. A copper plate is placed between the poles of an electromagnet with a magnetic field perpendicular to the plate. When the plate is pulled out a noticeable force is necessary and the force required increases with the velocity of the plate. Explain this phenomenon.

14.3 Ring on the line. A long, straight conductor is placed in the center of a metallic ring. The conductor is directed along the symmetry axis of the ring. What happens if the current in the ring is changed? Explain.

Tutorials

14.4 Smart-emf. Figure 14.12 shows a circuit consisting of a smart capacitor, a resistor and a switch. The smart capacistor is constructed as a plate capacitor, as shown, with a distance L between the plates. But in addition, there is an external force $F = f\,q$ (where f is a constant) on charges q in the area between the two plates. You can assume that on the left plate in the capacitor, positive charges are released and can be transfered over the gap between the two plates to the right plate.

The switch is off. Assume first that the switch is off so that no current can pass through it.

(a) Explain what happens when you turn on the smart capacitor so that the force F starts working.

(b) After a long time, the system reaches a stationary state. What is the electric field in the capasitor then?

(c) What is the electric potential in the points A-F?

(d) What is $\oint_C \mathbf{E} \cdot d\mathbf{l}$ along the circuit? (Hint: The curve C may pass through areas where there is no wire.)

(e) What is the voltage drop across the capasitor?

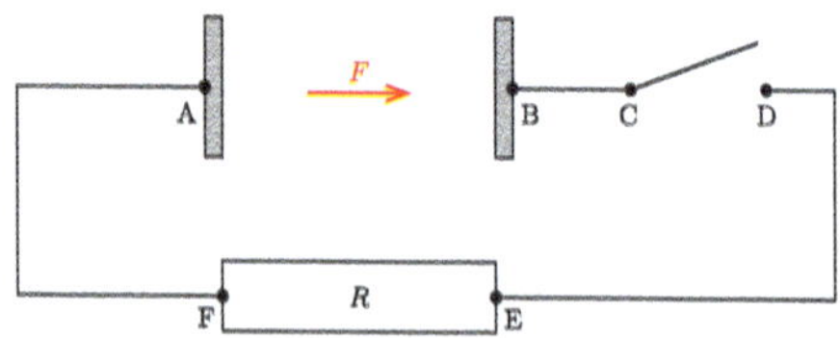

Fig. 14.12 Illustration of the smart capacitor

(f) What is the voltage drop across the switch, that is, between the left and the right part of the switch?

The switch is on. You turn the switch on, so that a current can pass through it.
(g) What is the voltage in the points A-F now?
(h) What is $\oint_C \mathbf{E} \cdot d\mathbf{l}$ along the circuit?
(i) What is the emf, e, in this circuit?
(j) What is the current, I, in the circuit?

14.5 Find the flux. Find the flux Φ_B through the surface S for the situations in Fig. 14.13. The orientation of the surface S is given by the direction of the curve or by the normal vector to the surface.
(a) (Figure A): A plane, square surface in the xy-plane from $-a$ to a in both the x- and the y-directions and a homogeneous magnetic field $\mathbf{B} = B_0\hat{\mathbf{z}}$.
(b) (Figure B): A plane, square surface in the xy-plane from $-a$ to a in both the x- and the y-directions and a magnetic field $\mathbf{B} = \frac{B_0}{b}x\hat{\mathbf{z}}$.
(c) (Figure C) A plane, rectangular surface in the xy-plane from $x = -a$ to $x = a$ and from $y = -a$ to $y = a$ and a magnetic field $\mathbf{B} = \frac{B_0}{b}x\hat{\mathbf{z}}$.
(d) (Figure D) A plane, square surface with sides $2a$ and rotated an angle θ around the y-axis. A homogeneous magnetic field $\mathbf{B} = B_0\hat{\mathbf{z}}$.
(e) (Figure E) A surface shaped as a half sphere with radius a and a homogeneous magnetic field $\mathbf{B} = B_0\hat{\mathbf{z}}$.
(f) (Figure F) A surface shaped as a half sphere with radius a and a magnetic field $\mathbf{B} = \frac{B_0}{b}r\hat{\mathbf{z}}$ where $r = \sqrt{x^2 + y^2}$.

14.6 Find the emf. Find the emf, both magnitude and direction, induced in the circuits shown in Fig. 14.14.

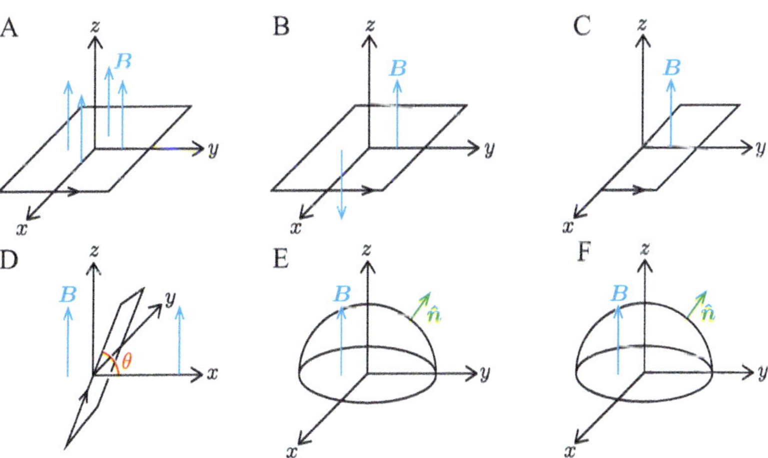

Fig. 14.13 Illustration of curves

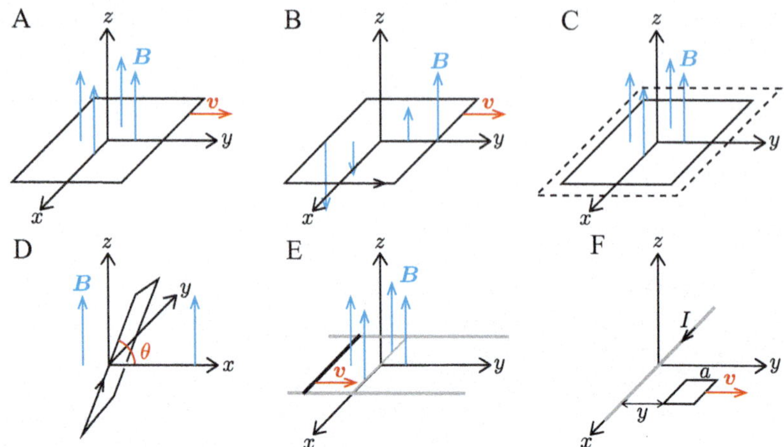

Fig. 14.14 Find the emf in the circuits

(a) (Figure A) A square circuit in the xy-plane with side $2a$ move at a constant velocity $\mathbf{v}$ along the x-axis through a homogeneous magnetic field $\mathbf{B} = B_0\hat{\mathbf{z}}$.
(b) (Figure B) A square circuit in the xy-plane with side $2a$ move with a constant velocity $\mathbf{v}$ along the x-axis through a magnetic field $\mathbf{B} = \frac{B_0}{b}x\hat{\mathbf{z}}$. The left side of the circuit is at a position y.
(c) (Figure C) A square circuit in the xy-plane expands in all direction with velocity v in a homogeneous magnetic field $\mathbf{B} = B_0\hat{\mathbf{z}}$. At the time $t = 0$ the side is $2a$.
(d) (Figure D) A place, square circuit with side $2a$ is rotated an angle θ around the y-axis. The angle θ increases with the angular velocity ω: $\theta = \omega t$. The circuit is in a homogeneous magnetic field $\mathbf{B} = B_0\hat{\mathbf{z}}$.
(e) (Figure E) A circuit consisting of two long conductors along the y-axis (in gray) and a conductor of length $2a$ that connects them. The last side in the circuit moves along the y-axis with a velocity $\mathbf{v}$ and pass the x-axis at $t = 0$. There is a homogeneous magnetic field $\mathbf{B} = B_0\hat{\mathbf{z}}$.
(f) (Figure F) There is a current I along a wire along the x-axis. A square circuit with side a is in the xy-plane at a distance y from the x-axis. The circuit moves in positive y-direction with the velocity $\mathbf{v}$.

14.7 Squeeze and charge. Your friend Q has come up with a brand new invention and proudly presents it to you. It is an elastic ball made of elastic foam. On two opposite sides of the ball, Q has glued to strong, permanent magnets. Q has wound a wire 100 times around the center of the ball. Figure 14.15 shows a sketch of the invention. *Look*, Q says, *when I squeeze this, I create a current I can use to charge my cellphone. Now I will no longer run out of battery*. You can assume that the ball has a diameter of 6 cm and that the magnetic field at the midpoint between the two magnets is approximately uniform with a strength of 0.01 mT. (It will not be uniform in reality, but for our purposes it is a reasonable approximation) (Fig. 14.15).

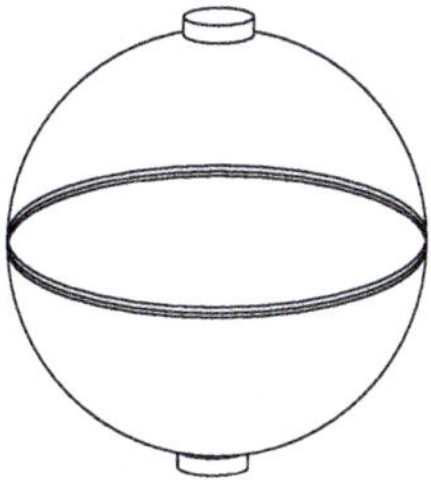

Fig. 14.15 A foam ball charger

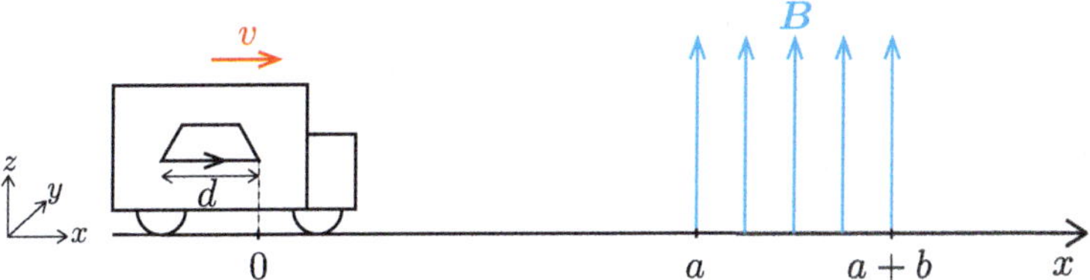

Fig. 14.16 Illustration of a truck passing a uniform magnetic field between $x = a$ and $x = a + b$

(a) Could this invention possibly work? Is any emf generated in this system?
(b) Estimate the magnitude of the emf, if you squeeze the ball so that the diameter around the equator is halved in one second. (You can assume that the magnetic field does not change in magnitude).
(c) How could you have made a more realistic model for the induced voltage?

14.8 Emf on the bus. Faraday's law is valid independently of how the flux varies–if it is the ciruit moving, if it is the field changing with time, or if it is the shape of the circuit changing with time. But how are these variant connected?

We construct a model system that we will study in detail. Figure 14.16 shows a truck driving along the x-axis with a constant velocity v. Inside the truck is a square circuit in the xy-plane with sides d. There is a constant magnetic field $\mathbf{B} = B_0\hat{\mathbf{z}}$ between $x = a$ and $x = a + b$. You can assume that $a > d$. At the time $t = 0$ the right hand side of the circuit pass $x = 0$.

Seen from the road. First, let us address how this process appears to a spectator standing on the road.
(a) Sketch the magnetic field as a function of x.
(b) Find the position of the right side of the circuit as a function of time, $x(t)$.
(c) Find the flux through the circuit as a function of time t.
(d) Find the induced emf in the circuit as a function of time t.

Seen from the truck. Then, let us see how this process appears to a spectator inside the truck.
(e) Sketch the magnetic field on the right hand side and on the left hand side of the circuit as a function of time.

(f) Sketch the magnetic flux in the circuit as a function of time.

14.9 Circular currents. What happens with a metallic plate in a magnetic field that varies in time? Here, we will first address a simplified model consisting of a circular circuit of radis a in the xy-plane. There is a uniform, time-varying magnetic field everywhere in space, $\mathbf{B} = B_0 \sin \omega t \hat{\mathbf{z}}$.
(a) What is the flux Φ_B through the circuit?
(b) What is the induced emf?
(c) If the conductor is made of a material with cross-section d^2 and conductivity σ, what is the resistance R of the circuit?
(d) What is the current I through the circuit?
(e) We assume that this is a model for a circular metallic plate consisting of many such circuits placed close to each other. What is the current as a function of the distance to the center of the plate?
(f) Sketch the current density in the plate.

Homework

14.10 Faraday's law. Faraday's law on differential form is

$$\nabla \times E = -\frac{\partial \mathbf{B}}{\partial t}. \tag{14.52}$$

(a) Use Stokes' theorem to rewrite Faraday's law onto its integral form.
(b) Consider a very long solenoid of radius R with n turns per length and current I. Compute the **B**-field everywhere. (You can assume that the **B**-field is zero outside the solenoid.)
(c) Suppose the current I in the solenoid is increasing at a steady rate $I(t) = Ct$, where C is a constant. Where do you think there is an **E**-field? (Inside the solenoid? Outside? Everywhere? Nowhere?) What do you think the **E**-field looks like? For now, just use your intuition, we'll check with calculations later.
(d) Use Faraday's Law on integral form to compute the electric field inside the solenoid. Specify the loop you chose for the integral.
(e) Use Faraday's law to compute the **E**-field outside the solenoid.

14.11 Rod on rails. A rod of mass m is running on rails in a magnetic field. The rails are connected with a wire at $x = 0$ so the rod and the rails together with the wire forms a closed circuit as illustrated in Fig. 14.17. The distance between the rails is l and the resistance in the rod is R. We will first assume that the rails and the wire are ideal conductors, so that the total resistance in the circuit is R. At the time $t = 0$ the rod moves in positive x-direction with velocity v_0.
(a) Find the emf in the circuit using Lorentz' law and indicate the contribution from the different parts of the circuit.
(b) Find the emf using Faraday's law. Did you get the same result?
(c) Find the magnitude and direction of the current in the circuit.

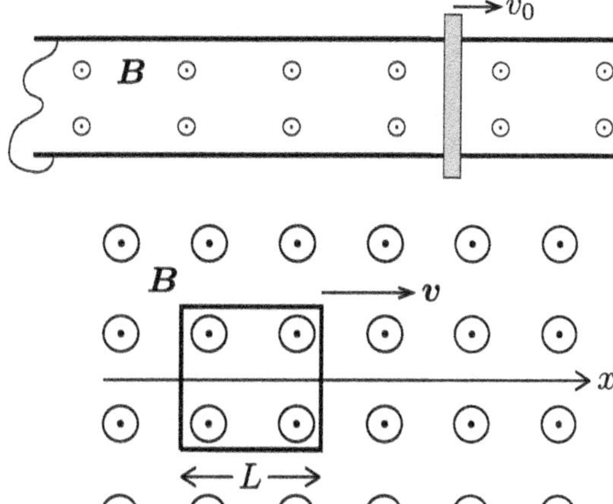

Fig. 14.17 A rod on rails

Fig. 14.18 A circuit moving through an inhomogeneous magnetic field

(d) Find the motion of the rod for $t > 0$ when you assume that is starts with a velocity v_0 at $t = 0$ and no other external forces is acting on the rod, except from the magnetic force. The rod has a mass m. Can you check your results using conservation of energy?
(e) In the rest of the exercise, we will assume that the rails are resistors with a resistivity ρ and a cross-sectional area A. The rod and the wire is now assumed to the ideal conductors. This is a good approximation if the rails are not very good conductors. What is the resistance of the circuit as a function of the position x of the rod?
(f) Write a program to find the motion of the rod when it starts at $x = 1\text{m}$ with the velocity v_0.
(g) You are charged with making a braking mechanism for a rollercoaster. You use electromagnets that allows you to construct an arbitrary magnetic field $B_z(x)$ along the last 100 m of the rollercoaster. You can choose to select where to place the wire that connects the two rails. Assume that the back wheels are connected by an ideal conductor. Perform simulations to determine what would be a reasonable magnetic field and where it is smart to place the wire. You can assume that the mass of the rollercoaster cart is $m = 500\text{kg}$ and that the starting velocity is $v_0 = 15\text{m/s}$.

14.12 Inhomogeneous field. You pull a quadratic current loop with side L with a constant velocity v_0 along the x-axis as illustrated in Fig. 14.18. The current loop is oriented in the xy-plane. There is a magnetic field $\mathbf{B}(x, y, z) = B(x, y, z)\hat{\mathbf{z}}$. The current loop has a resistance R. You may ignore the magnetic field generated by the current in the loop. At the time $t = 0$, the left side of the loop is in the position $x = 0$ (Fig. 14.18).
(a) Assume that the magnetic field is $\mathbf{B}(x, y, z) = B_0\hat{\mathbf{z}}$ where B_0 is a constant. What is the induced current I in the loop at a time t?
(b) Assume instead that the magnetic field is $\mathbf{B}(x, y, z) = B_0(x/L)\hat{\mathbf{z}}$. What is the induced current I in the loop at a time t?

(c) Assume that the magnetic field has the general form $\mathbf{B}(x, y, z) = B(x)\hat{\mathbf{z}}$. What is the induced current I in the loop at a time t?

14.13 An expanding circle. In this exercise we will address what happens with a flexible, conducting wire in a magnetic field. We assume the wire is shaped as a circle and made of a material with resistance R. The wire is in a homogeneous magnetic field and lies in a plane normal to the field.
(a) Sketch the system and introduce necessary symboles to describe the system.
(b) Assume that the diameter of the wire circle changes with a constant speed v_0. Find the current in the wire, both direction and magnitude.
(c) Check that your result is according to Lenz' law.
(d) How can you use an ampere-meter to find if the wire is expanding or shrinking?
(e) Assume that the magnetic field increases linearly with the distance r to the center of the circle, $B = B(r) = B_0(r/r_0)$, but that it has the same direction as before. Is this a physically possible magnetic field?
(f) What is now the current in the wire?

Chapter 15
Inductance

We will demonstrate how we can introduce a new element in a circuit, an inductor, to represent the inductance of the circuit, and we will show how you can calculate the inductance of a circuit based on the geometry of the current-carrying components and the magnetic materials. We will extend the concept of inductance to also include how one circuit can induce an emf in another circuit through the mutual inductance of the two circuits, which again only depends on the geometry of the system. Finally, we will introduce the energy of the magnetic field in an inductor in a way similar to how we introduced the energy of the electric field for a capacitor.

15.1 Inductance

Figure 15.1 illustrates a circuit made of a conductor of finite thickness. If a current runs in the circuit, a magnetic field will be generated, and the magnetic field will have a flux through the circuit itself. This self-induced flux is called a *self-inductance* and we introduce the term L to describe this effect a circuit has on itself:

$$\Phi = L\,I \quad , \quad L = \frac{\Phi}{I} \; . \tag{15.1}$$

When we calculate the flux Φ in this expression, it is only the flux from the magnetic field generated by the circuit itself that is to be included:

$$\Phi = \int_S \mathbf{B} \cdot \mathrm{d}\mathbf{S} \; , \tag{15.2}$$

where $\mathbf{B}$ is generated by I. There may be other sources of magnetic fields, such as a permanent magnet or another coil nearby, but these are not included in the flux Φ which is used to define the self-inductance.

© The Author(s), under exclusive license to Springer Nature Switzerland AG 2026

A. Malthe-Sørenssen, *Elementary Electromagnetism Using Python*, Undergraduate Texts in Physics, https://doi.org/10.1007/978-3-032-19876-1_15

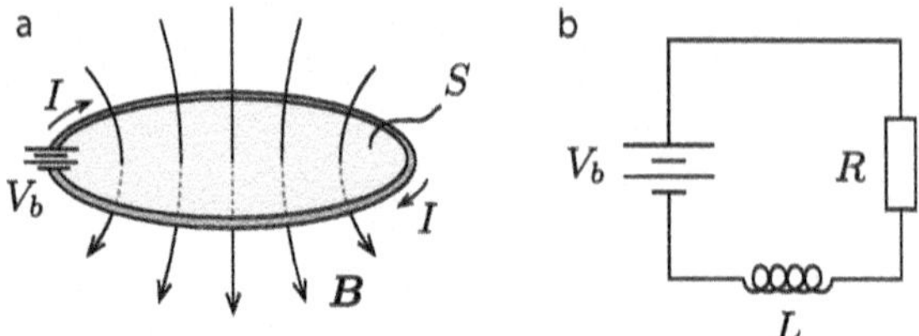

Fig. 15.1 **a** A circuit with a current I, a magnetic field **B** generated by I, and the resulting flux Φ through a surface S enclosed by the circuit C. **b** A simplified circuit diagram of the same system. The element L represents the self-inductance

Definition of self-inductance

Self-inductance

The **self-inductance** L of a system/circuit is defined as

$$L = \frac{\Phi}{I} , \tag{15.3}$$

where the flux Φ is the flux through the circuit from the magnetic field generated by the current I.

What determines the self-inductance L?

For a circuit as illustrated in Fig. 15.1, the magnetic field **B** is generated by the current as described by Biot-Savart's law:

$$\mathbf{B} = \int_C \frac{\mu_0 I \, \mathrm{d}\mathbf{l} \times \hat{\mathbf{R}}}{4\pi R^2} , \tag{15.4}$$

and the flux is

$$\Phi = \int_S \mathbf{B} \cdot \mathrm{d}\mathbf{S} , \tag{15.5}$$

integrated over a surface S enclosed by the circuit/path C. This shows that

- The magnetic field is proportional to the current and the flux is proportional to the magnetic field, $\Phi \propto B \propto I$. Consequently, there is a *linear* relationship between Φ and I. This motivates the definition of $\Phi = LI$ so that L describes this constant of proportionality.

- The inductance L is determined by the geometry of the system. It is the geometry of the circuit and the presence of magnetic materials that determine $\mathbf{B}$ and therefore the geometry of the circuit that determines the flux.
- The definition of L is similar to our previous definitions of the capacitance $C = Q/V$ or the resistance $R = V/I$ of a system.
- This also suggests a *method to find the inductance*: We assume a given I, calculate the resulting $\mathbf{B}$ and the resulting flux Φ and find L from $L = \Phi/I$. This method is effectively identical to the method we introduced for the capacitance and the resistance.

Method: Calculating the inductance

- Assume that a current I runs through the circuit. (Often you will only have to look at the part of the circuit that acts as a coil with many windings).
- Calculate the magnetic field $\mathbf{B}$ from the current I using Biot-Savart's law or Ampere's law. Include effects of magnetic materials.
- Calculate the flux of the magnetic field $\mathbf{B}$ set up by the current I through a convenient surface S enclosed by the circuit path C
- Find the self-inductance L from $L = \Phi/I$.

Test your understanding

The figure shows two circuits. Circuit 2 has the same shape as circuit 1 but consists of two loops on top of each other. (a) How is the magnetic field B_2 at the center of circuit 2 related to the magnetic field at the center of circuit 1? (b) How is the self-inductance of circuit 2 related to the self-inductance of circuit 1?[1]

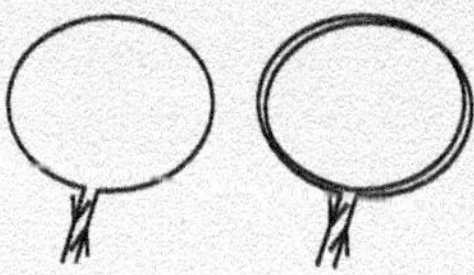

Test your understanding

The magnetic field inside a solenoid of length b, cross-sectional area S with N windings, the magnetic field is $B = \mu_0(N/b)I$ (a) What is the flux through a single winding? (b) What is the flux through the whole solenoid? (c) What is the inductance of the solenoid?[2]

---

[1] (a) $B_2 = 2B_1$; (b) $L_2 = 4L_1$.

[2] (a) $\Phi = \mu_0(N/b)IS$; (b) $\Phi = \mu_0(N^2/b)IS$; (c) $L = \mu_0 N^2 S/b$.

How does the self-inductance enter the circuit equations?

The self-inductance of a circuit describes the relation between the flux Φ and the current I in a circuit, $\Phi = LI$. But how does this affect the behavior of the circuit? If the current is *time-varying*, then the current will induce a time-varying flux:

$$I = I(t) \quad \Rightarrow \quad \Phi(t) = LI = LI(t) \; . \tag{15.6}$$

This will again introduce an emf to the circuit according to Faraday's law:

$$e = -\frac{\mathrm{d}}{\mathrm{d}t}\Phi(t) = -\frac{\mathrm{d}}{\mathrm{d}t}LI(t) \; . \tag{15.7}$$

If the geometry of the circuit does not change, then $\mathrm{d}L/\,\mathrm{d}t = 0$ and we get:

$$e = -L\frac{\mathrm{d}I}{\mathrm{d}t} \; . \tag{15.8}$$

To include this effect in a circuit, we introduce an element L in the circuit as illustrated in Fig. 15.1b. This element is drawn like a coil in a circuit diagram.

This element is in principle present in any circuit. For the circuit illustrated in Fig. 15.1, we introduce an inductance L to model the effect of the geometry of the circuit. We must also include the effect of the emf in Kirchhoff's voltage law for the circuit. We include the emf e in addition to the other voltage sources e_j in the circuit:

$$\sum e_j - \sum_i V_i = 0 \; , \tag{15.9}$$

where the V_i represents the voltage drops, such as $V_1 = RI$ for a resistor, whereas the emf's represents batteries, V_b and the emf due to inductance, $e = -\mathrm{d}\Phi/\,\mathrm{d}t$. For the illustrated circuit, we have the emf $V_b + e$ and

$$V_b + e - RI = 0 \; . \tag{15.10}$$

All circuits have some self-inductance, often as an undesirable side effect. We represent this inductance by a coil (inductor) circuit element. However, we also design and include inductors that play important roles of circuits, such as a coil. We call such circuit elements *inductors*. Inductors can store magnetic energy and play an important role in many circuits. In chapter 16, we will address how we construct circuits using the basic circuit elements resistor, capacitor and inductor, and how we can use circuits as models of model complex, real systems.

Notice that the model we have introduced for an inductor is a simplified model of the processes that goes on in the electromagnetic system, where we usually assume that the whole circuit can be described as a set of independent components. This is an approximation that sometimes breaks down. Then we have to apply the full machinery of an electromagnetic analysis.

Effect of self-inductance in a circuit

What is the effect of the self-inductance L in the circuit illustrated in Fig. 15.1? In this circuit, we found that

$$V_b + e - RI = V_b - L\frac{\mathrm{d}}{\mathrm{d}t}I - RI = 0\,. \tag{15.11}$$

What happens if a switch in the circuit is suddenly turned on? In this case, we expect the current to increase rapidly. The effect of the self-inductance is to induce an emf $e = -L\mathrm{d}I/\,\mathrm{d}t$. If the current increases rapidly, this will generate a large emf, which will counteract the change: If the current is increasing, the emf will be negative, whereas if the current is decreasing, the emf will be positive. The self-inductance will therefore act to reduce changes in current.

Similarly, if we very rapidly turn off a circuit, for example by unplugging a wire, there will be a very large emf due to the rapid change in current. This is what induces sparks when you unplug an electric device, because the electric field is so high that is causes dielectric breakdown in the air. We will address the effect of inductance in circuits in the next chapter. In this chapter, we will focus on how to calculate the inductance for a given circuit geometry, just like we did for capacitors and resistors.

Test your understanding

A part of a circuit has a self-inductance L as illustrated in the figure. (a) If the potential $V_A > V_B$, in what direction is the current flowing and is it increasing or decreasing? (b) If the current I is decreasing, what can we say about the voltage drop $V_A - V_B$ across the inductor L?[3]

Example: Toroid

Figure 15.2 illustrates a circuit wrapped N times around a toroid. The toroid has an inner radius a and an outer radius b, and a rectangular cross section of height h. Find the self-inductance L of this circuit.

[3] (a) Flow from A to B and is increasing; (b) $V_B > V_A$.

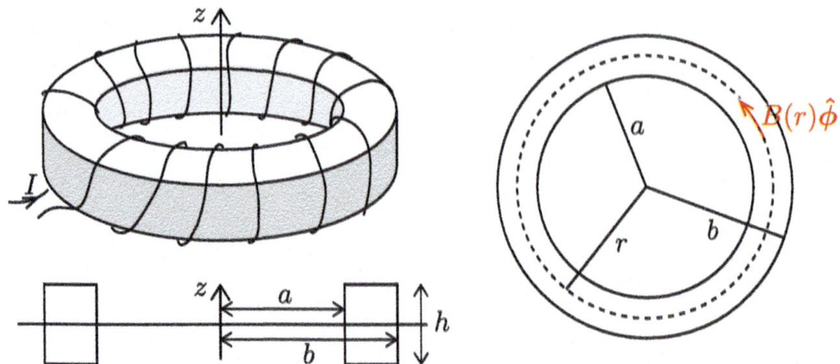

Fig. 15.2 Illustration of a toroid with rectangular cross section, inner radius a, outer radius b, height h, and N windings

Solution: Our plan is the use the method described above: We assume a current I in the wire, find the resulting magnetic field $\mathbf{B}$ in the region enclosed by the circuit, calculate the resulting flux Φ and find the (self) inductance from $L = \Phi/I$.

Finding the magnetic field. We find the magnetic field using Ampere's law. First, we realize from symmetry and that the divergence of the magnetic field is zero, that the magnetic field only has an azimuthal component, which only can depend on the distance r to the center of the toroid:

$$\mathbf{B} = B_\phi(r)\hat{\phi} = B(r)\hat{\phi} \,. \tag{15.12}$$

We apply Ampere's law on a circular loop C of radius r inside the toroid as illustrated in Fig. 15.2. The magnetic field is constant along this circle and Ampere's law gives:

$$\oint_C \mathbf{B} \cdot \mathrm{d}\mathbf{l} = B\, 2\pi r = \mu_0 I_{\mathrm{tot}} \,. \tag{15.13}$$

The total current going through the circle is I times the number of times the wire intersects the circle, which is N times, thus $I_{\mathrm{tot}} = NI$:

$$\mathbf{B} = \frac{\mu_0 N I}{2\pi r}\hat{\phi} \,. \tag{15.14}$$

Finding the flux. The flux through a single turn of the wire is

$$\Phi_1 = \int_S \mathbf{B} \cdot \mathrm{d}\mathbf{S} \,, \tag{15.15}$$

where the integration is over a cross-sectional area from $r = a$ to $r = b$. Notice that $\mathbf{B}$ is not uniform across this area. It points in the same direction across the whole area, in a direction normal to the wire loop, but the magnitude varies with r. A small surface element is $\mathrm{d}S = h\,\mathrm{d}r$. The flux of a single turn is therefore:

$$\begin{aligned} \Phi_1 &= \int_a^b Bh\,\mathrm{d}r = \int_a^b \frac{\mu_0 N I}{2\pi r} h\,\mathrm{d}r \\ &= h\frac{\mu_0 N I}{2\pi} \int_a^b \frac{1}{r}\,\mathrm{d}r = \frac{h\mu_0 N I}{2\pi} \ln\frac{b}{a} \,. \end{aligned} \tag{15.16}$$

This is the flux through a single winding. There are N windings, so that the total flux is:

$$\Phi_N = N\Phi_1 = \frac{h\mu_0 N^2 I}{2\pi} \ln\frac{b}{a} \,. \tag{15.17}$$

Notice that the number of windings enters this expression twice: Once for the magnetic field and once for the flux calculation.

Finding the inductance. We use this result to find the inductance:

$$L = \frac{\Phi}{I} = \frac{h\mu_0 N^2}{2\pi} \ln\frac{b}{a} \,. \tag{15.18}$$

We notice that this is indeed a geometric relation that only depends on geometric properties and of the circuit and not on the current I.

Approximation. What happens in the limit of a very thin toroid, where $b - a \ll a$? In this limit we can rewrite b as $b - a + a$, getting:

$$\ln\frac{b}{a} = \ln\frac{b-a+a}{a} = \ln\left(1 + \frac{b-a}{a}\right) \simeq \frac{b-a}{a} \,, \tag{15.19}$$

where we have used the Taylor expansion $\ln(1 + x) \simeq x$, which is valid when $x \ll 1$. This gives:

$$L \simeq \frac{h\mu_0 N^2}{2\pi} \frac{d}{a} \,, \tag{15.20}$$

where $d = b - a$ is the thickness of the solenoid. We can rewrite this as $dh = S$ is the cross-sectional area of the toroid:

$$L \simeq \frac{S\mu_0 N^2}{2\pi a} \,. \tag{15.21}$$

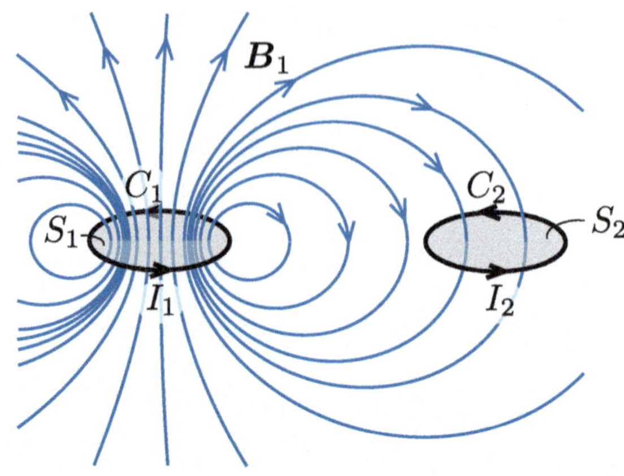

Fig. 15.3 Two circuits C_1 and C_2 interact through the magnetic fields generated by the currents in each circuit

15.2 Mutual Inductance

We have so far defined inductance by how a circuit induces a flux on itself. However, a circuit can also induce a flux in another circuit. We therefore generalize the term inductance to *mutual inductance*. Figure 15.3 illustrates two circuits C_1 and C_2. The current I_1 in C_1 generates a magnetic field $\mathbf{B}_1$, and the flux of the magnetic field $\mathbf{B}_1$ through C_2 is called

$$\Phi_{1,2} = \int_{S_2} \mathbf{B}_1 \cdot \mathrm{d}\mathbf{S} \,. \tag{15.22}$$

The first index refers to the source of the flux, which here is from circuit C_1. The second index refers to the circuit through which we measure the flux. Here it is a surface S_2 enclosed by C_2. This is the flux from circuit 1 through circuit 2. We define the mutual inductance $L_{1,2}$ as

$$L_{1,2} = \frac{\Phi_{1,2}}{I_1} \,, \tag{15.23}$$

using the same interpretation of the indexes as for the flux. We recognize the inductances $L_{1,1}$ and $L_{2,2}$ as the self-inductances of circuits C_1 and C_2 respectively. There are in general four mutual inductances $L_{1,1}$, $L_{1,2}$, $L_{2,1}$, and $L_{2,2}$.

Mutual inductance

We define the flux in a circuit C_j from a current I_i in circuit C_i to be

$$\Phi_{i,j} = \int_{S_j} \mathbf{B}_i \cdot \mathrm{d}\mathbf{S} \tag{15.24}$$

where S_j is a surface enclosed by C_j, and $\mathbf{B}_i$ is the magnetic field from the current I_i in circuit C_i. We define the **mutual inductance** $L_{i,j}$ as

$$L_{i,j} = \frac{\Phi_{i,j}}{I_i} . \tag{15.25}$$

Symmetry of mutual inductances

A very practical result, which we will often use, relates the two mutual inductances of two circuits C_1 and C_2:

$$L_{1,2} = L_{2,1} \tag{15.26}$$

This nice symmetry provides us with a very powerful tool: If we want to calculate the mutual inductance, we can choose to calculate the magnetic field from the circuit where this is the simplest and use this to find the flux through the other circuit. We will demonstrate how we use this principle below.

Proof. There is a very nice proof of this principle, which depends on the vector potential **A**. You may skip this proof without loss of continuity.

The flux through loop C_2 from the magnetic field set up by circuit C_1 is:

$$\Phi_{1,2} = \int_{S_2} \mathbf{B}_1 \cdot \mathrm{d}\mathbf{S} , \tag{15.27}$$

where we insert the vector potential $\mathbf{A}_1$ for $\mathbf{B}_1$:

$$\Phi_{1,2} = \int_{S_2} (\nabla \times \mathbf{A}_1) \cdot \mathrm{d}\mathbf{S} . \tag{15.28}$$

We apply Stokes' theorem to get:

$$\Phi_{1,2} = \int_{S_2} (\nabla \times \mathbf{A}_1) \cdot \mathrm{d}\mathbf{S} = \int_{C_2} \mathbf{A}_1 \cdot \mathrm{d}\mathbf{l}_2 . \tag{15.29}$$

The vector potential $\mathbf{A}_1$ is found from:

$$\mathbf{A}_1 = \frac{\mu_0 I_1}{4\pi} \oint_{C_1} \frac{\mathrm{d}\mathbf{l}_1}{R} \tag{15.30}$$

We insert this into the expression for the flux, and get:

$$\Phi_{1,2} = \frac{\mu_0 I_1}{4\pi} \oint_{C_2} \left(\oint_{C_1} \frac{d\mathbf{l}_1}{R} \right) \cdot d\mathbf{l}_2 \,. \tag{15.31}$$

We therefore find that the mutual inductance is

$$L_{1,2} = \frac{\mu_0}{4\pi} \oint_{C_1} \oint_{C_2} \frac{d\mathbf{l}_1 \cdot d\mathbf{l}_2}{R} \tag{15.32}$$

This is called the *Neumann formula*. The order in which we perform this double line integral is arbitrary. This proves that $L_{1,2} = L_{2,1}$. And it also demonstrates that the mutual inductance is a purely geometrical property: The line integrals depend on the shapes of the curves, but not on the currents in the circuits.

Test your understanding

The figure shows two identical circuits placed relative to each other in three possible ways (a-c). Which relative positioning gives the highest mutual inductance?[4]

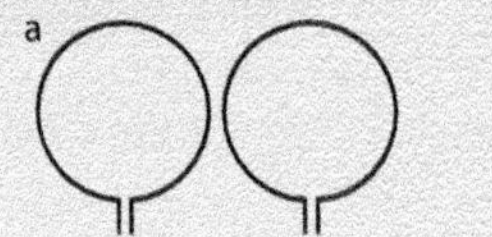

Example: Toroid with rectangular cross section

Figure 15.4 *shows a toroid with a rectangular cross section. A wire C_1 is wrapped N_1 times around the toroid. Another wire forms a loop C_2 around the toroid as illustrated in the figure. What is the mutual inductance $L_{2,1}$ of this system?*

Solution: To find $L_{2,1}$ we would need to find the magnetic field $\mathbf{B}_2$ generated by the current I_2 in circuit C_2. This is not simple. However, we can use that $L_{2,1} = L_{1,2}$. Finding $L_{1,2}$ is easier, because it is easier to find the magnetic field from the toroid circuit C_1.

Finding $\mathbf{B}_1$. We assume that a current I_1 runs in the toroid. What is the magnetic field inside the toroid? We have previously found that

$$\mathbf{B}_1 = \frac{\mu N_1 I_1}{2\pi r} \hat{\boldsymbol{\phi}} \,, \tag{15.33}$$

where N_1 is the number of times the wire is wrapped around the toroid and μ is the permeability of the magnetic material inside the toroid.

[4] The magnetic field decays with distance, so the closest placement gives the highest flux and the highest mutual inductance.

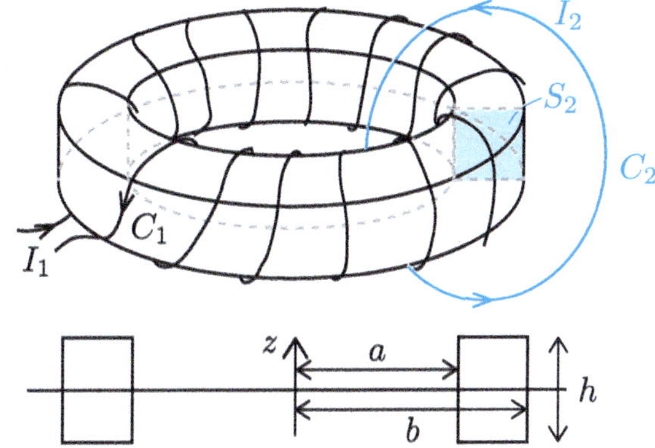

Fig. 15.4 A toroid circuit C_1 with a wire wrapped around a torus with rectangular cross section, interacts with a circuit C_2

Finding the flux. What is the flux of $\mathbf{B}_1$ through S_2? We notice that the magnetic field from the toroid is only non-zero inside the toroid and zero outside. It is therefore only the intersection of the surface S_2 and the toroid that contributes to the integral, corresponding to the shaded area in Fig. 15.4. We choose a surface which is normal to the toroid, so that the intersection is a rectangular intersection from $r = a$ to $r = b$ and with height h. (We can choose this particular surface, because we are free to choose *any* surface S_2 which is enclosed by C_2.) The flux of $\mathbf{B}_1$ through C_2 is therefore:

$$\Phi_{1,2} = \int_{C_2} \mathbf{B}_1 \cdot \mathrm{d}\mathbf{S} = \int_a^b Bh\,\mathrm{d}r = \int_a^b \frac{h\mu N_1 I_1}{2\pi r}\,\mathrm{d}r = \frac{\mu h N_1 I_1}{2\pi} \ln\frac{b}{a} . \qquad (15.34)$$

If circuit C_2 consisted of N_2 windings, we would get N_2 times the flux from one winding:

$$\Phi_{1,2} = \frac{\mu h N_1 I_1 N_2}{2\pi} \ln\frac{b}{a} . \qquad (15.35)$$

Mutual inductance. The mutual inductance is now

$$L_{1,2} = \frac{\Phi_{1,2}}{I_1} = \frac{\mu h N_1 N_2}{2\pi} \ln\frac{b}{a} \qquad (15.36)$$

We notice that this expression only depends on geometric properties of the system and not on any of the currents, as it indeed should.

15.3 Applications of Mutual Inductance

If a time-varying current I_1 is sent through circuit C_1 in Fig. 15.3, an emf $e_{1,2}$ will be induced in circuit C_2. This emf will be:

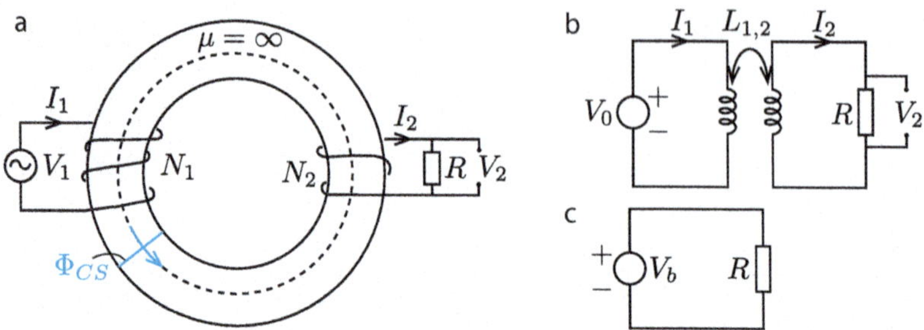

Fig. 15.5 **a** Illustration of a transformer consisting of two wires wrapped around a ferromagnetic toroid. **b** Circuit diagram model of the same system. **c** Simplified circuit diagram for one circuit

$$e_{1,2} = -\frac{\mathrm{d}\Phi_{1,2}}{\mathrm{d}t} = -\frac{\mathrm{d}\left(L_{1,2}I_1\right)}{\mathrm{d}t} . \tag{15.37}$$

If the loops do not change their size or position with time, then $\mathrm{d}L_{1,2}/\,\mathrm{d}t = 0$, and

$$e_{1,2} = -L_{1,2}\frac{\mathrm{d}I_1}{\mathrm{d}t} . \tag{15.38}$$

This opens for a *wireless transmission of voltage, current and energy between circuits*. We will here address a few examples of this interaction such as transformers, but this type of interaction is the basis of many technologies.

Example: Voltage and current transformation

Mutual inductance allows two circuits to be coupled and can be used to transform voltages or currents. Here, we will study a typical transformer geometry as illustrated in Fig. 15.5 and how it can be used to transform voltage and current. The system consists of a magnetic circuit formed by a thin toroid with a ferromagnetic core and two wires that are wound around the toroid, that is, two current loops, C_1 and C_2 with N_1 and N_2 windings respectively.

The emf induced in circuit 1 from I_1 is called $e_{1,1}$ and the emf induced in circuit C_2 from current I_1 is $e_{1,2}$. We follow the same indexing convention as before: The first index refers to the circuit that generates the magnetic field and the second index refers to the circuit where the flux is generated or the emf induced. From Faraday's law and the definition of the mutual inductances, we have that

$$\frac{e_{1,2}}{e_{1,1}} = \frac{-\frac{\mathrm{d}\Phi_{1,2}}{\mathrm{d}t}}{-\frac{\mathrm{d}\Phi_{1,1}}{\mathrm{d}t}} = \frac{L_{1,2}\frac{\mathrm{d}I_1}{\mathrm{d}t}}{L_{1,1}\frac{\mathrm{d}I_1}{\mathrm{d}t}} = \frac{L_{1,2}}{L_{1,1}} . \tag{15.39}$$

This is a general result valid for arbitrary circuits. We can use this result to address the ratio of the induced emfs based on calculated or experimentally measured values for the mutual and self-inductances.

Ideal transformer. In an ideal transformer, we assume that $\mu = \infty$ so that all the flux remains in the ferromagnet. We want to find the ratio between the voltage V_1 on the signal generator and the voltage V_2 across the resistor in circuit C_2.

In these systems it is often tricky to find the correct signs for the voltages. Let us look at the simplified circuit in Fig. 15.5c. For this circuit we know that $\sum e = RI$, which is

$$V_b + \left(-\frac{d\Phi}{dt} \right) = RI \ . \tag{15.40}$$

In the case when $R = 0$, which is the case on the left in the figure (circuit C_1), we get that $V_b = V_1 = d\Phi/dt$. In the case when $V_b = 0$, which is the case on the right in the figure (circuit C_2), we get that $RI = -d\Phi/dt$.

Voltage transformation. The ratio between V_2 and V_1 is therefore:

$$\frac{V_2}{V_1} = \frac{RI_2}{V_1} = \frac{-\frac{d\Phi_2}{dt}}{+\frac{d\Phi_1}{dt}} \tag{15.41}$$

What are the fluxes Φ_1 and Φ_2? These depend on the cross-sectional flux, Φ_{CS}, through a cross-section of the ferromagnetic material, which is the same for any cross section around the ferromagnet. The fluxes through circuit C_1 and C_2 are therefore $|\Phi_2| = N_2\Phi_{CS}$ and $|\Phi_1| = N_1\Phi_{CS}$, because each of the circuits wraps N_1 and N_2 times around the ferromagnet respectively. Let us also check the signs of these fluxes. The cross-sectional flux is positive as indicated in Fig. 15.5a. The flux Φ_1 should be positive, so that $\Phi_1 = N_1\Phi_{CS}$. However, we see that $\Phi_2 < 0$, so that $\Phi_2 = -N_2\Phi_{CS}$.

We insert this in (15.41) and find

$$\frac{V_2}{V_1} = \frac{-\frac{d\Phi_2}{dt}}{+\frac{d\Phi_1}{dt}} = \frac{\frac{dN_2\Phi_{CS}}{dt}}{\frac{dN_1\Phi_{CS}}{dt}} = \frac{N_2}{N_1} \ . \tag{15.42}$$

This system therefore acts as a transformer for a time-varying voltage and can change one voltage into another voltage, while transmitting the signal.

Current transformation. We can find the ratio between the two currents by applying Ampere's law to a circuit C as illustrated in Fig. 15.6. Ampere's law gives that:

$$\int_C \mathbf{H} \cdot d\mathbf{l} = N_1 I_1 - N_2 I_2 \ . \tag{15.43}$$

What is the value of the path integral? It is zero! However, we need a result we have not yet derived to demonstrate this. In Sect. 15.4 we show that the energy density in

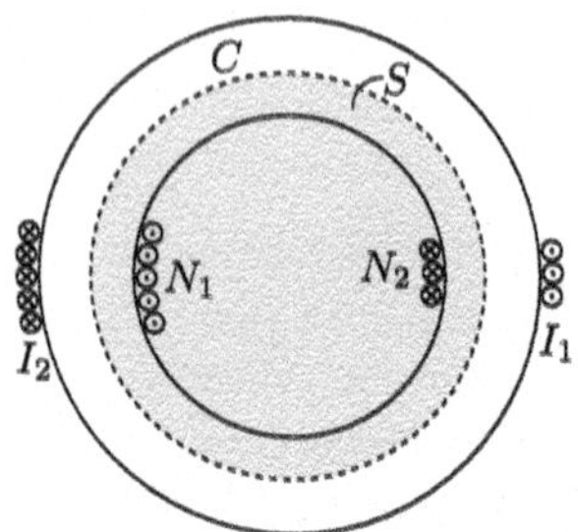

Fig. 15.6 Illustration of the integration path C and the surface S used for flux calculations. Only the currents on the *inside* of C contributes to the net current through C

space in a magnetic material is $\mu H^2/2$. In the case when $\mu \to \infty$ we must therefore have that $H \to 0$, because the energy density cannot be infinite. This is of course only an approximation, because the real μ is not infinite, but large. Notice that this does not imply that $\mathbf{B}$ is zero. We then get that:

$$N_1 I_1 = N_2 I_2 \quad \Rightarrow \quad \frac{I_2}{I_1} = \frac{N_1}{N_2} . \tag{15.44}$$

Notice that this is the opposite ratio of what we found for the voltages. This means that if a circuit transforms the voltage up, it will transform the current down. This is related to energy conservation: The power consumption must be the same on both sides of the magnetic circuit. The power consumption in the resistor is

$$P_2 = V_2 I_2 = \left(V_1 \frac{N_2}{N_1} \right) \left(I_1 \frac{N_1}{N_2} \right) = V_1 I_1 = P_1 . \tag{15.45}$$

What is the resistance of the circuit as seen from the source V_1? We find that:

$$\frac{V_1}{I_1} = \frac{V_2 \frac{N_1}{N_2}}{I_2 \frac{N_2}{N_1}} = \frac{V_2}{I_2} \left(\frac{N_1}{N_2} \right)^2 = R \left(\frac{N_1}{N_2} \right)^2 . \tag{15.46}$$

Notice that we have not made any assumptions about the time-dependence of the current. In principle, it is valid for any time-varying current, but in practice transformers are used for alternating currents.

Example: Numerical calculation of inductances

Find the self-inductance of a circular circuit of radius a, and the mutual inductance between two circular circuits of radius a at a distance b from each other.

Approach. Our plan is first to write a program to find the magnetic field from the circular circuit and then use the program to find the fluxes needed to find the inductances.

Find the magnetic field. We write a program to find the magnetic field by summing the contributions from individual current elements for a circle in the xy-plane with radius a and center in the origin. A current element runs from $\mathbf{r}_i$ to $\mathbf{r}_{i+1}$ with its center at $(\mathbf{r}_i + \mathbf{r}_{i+1})/2$, where $i = 0, \ldots, N$ enumerates points along the circuit. We use Biot-Savart's law to find the contribution to the magnetic field from this current element:

$$\mathrm{d}\mathbf{B}_i = \frac{\mu_0}{4\pi} \frac{I \mathrm{d}\mathbf{l}_i \times \mathbf{R}_i}{R_i^3} . \tag{15.47}$$

and the total magnetic field is the sum of the contributions from each current element. We define the circuit as the set of these current elements and write a function to find the magnetic field from the circuit described by the list. First, we create the list of positions for the circle:

```
import numpy as np
a, N = 1,25
I = 1.0 # Current, in ampere
list = np.zeros((N,3),float)
dtheta = 2*np.pi/N
for i in range(N):
    theta = i*dtheta
    rdv = np.array([a*np.cos(theta),a*np.sin(theta),0])
    list[i,:] = rdv
```

Then we write a function to find $\mathbf{B}(\mathbf{r})$ from this circuit

```
import scipy.constants as sc
def bfieldlist(r,list,I):
    # Find the magnetic field in the point r
    # list = array(N,3) of points, I = current
    B = np.array([0,0,0], float)
    N = np.shape(list)[0]
    for i in range(N):
        i0 = i
        i1 = i + 1
        if (i1>N-1):
            i1 = 0
        rdv = 0.5*(list[i1,:]+list[i0,:])
        R = r - rdv
        dlv = list[i1,:]-list[i0,:]
        dB = I*np.cross(dlv,R)/np.linalg.norm(R)**3
        B = B + dB
    return sc.mu_0/(4*np.pi)*B
```

To ensure that your solution looks reasonable, we recommend that you plot the magnetic field in the xy- and the xz-plane. However, to find the flux through the circle in the xy-plane, we divide the xy-plane into small square surface elements $\Delta \mathbf{S}_j = \Delta x^2 \hat{\mathbf{z}}$ and sum the contributions from all elements inside the circle:

```
def findflux(z0):
    Lx,Ly,Nx,Ny,I = a,a,25,25,1.0
    dA = (2*Lx/Nx)*(2*Ly/Ny)
    Phi = 0.0
    for ix in range(Nx):
        rxi = -Lx + (2*Lx/Nx)*(ix+0.5)
        for iy in range(Ny):
            ryi = -Ly + (2*Ly/Ny)*(iy+0.5)
            if (rxi*rxi+ryi*ryi<a*a):
                r = np.array([rxi,ryi,z0])
                Bx, By, Bz = bfieldlist(r,list,I)
                Phi = Phi + Bz*dA
    return Phi
Phi = findflux(0)
print("Phi = ",Phi)
```

We find the self-inductance by dividing the flux by the current, I, used in the calculation. Notice that this calculation requires a high resolution (a high N) to be precise, because the magnitude of the magnetic field is diverging close to the circuit.[5] However, we can use this approach to find the mutual inductance between two circular circuits by using the program to find the flux from the circle at height z through a circle of radius a at a height z as a function of height and plot the results:

```
z = np.linspace(0,5*a,10)
Phi = np.zeros_like(z)
for i in range(len(z)):
    Phi[i] = findflux(z[i])
import matplotlib.pyplot as plt
plt.plot(z,Phi)
plt.xlabel("$z$"), plt.ylabel("$\Phi$")
```

The mutual inductance of a circle of radius a in the origin and a circle with the same radius at z is found by dividing the flux by the current used in the calculation.

15.4 Energy in Magnetic Fields

For capacitors, we developed a theory for the energy needed to bring a capacitor to a given charge and used this to define the energy density of the electric field. Here, we will follow a similar approach to define the energy density of the magnetic field.

---

[5] To find the flux you should find the flux as a function of the resolution N, systematically increase N, and interpolate to $N \to \infty$.

Energy in a single circuit

What is the energy needed to get a current I flowing in the circuit in Fig. 15.7? You may think it is zero, because you can just turn a switch and then the current is running. However, if you start with zero current, you need to gradually change the current until you reach a current I. As you do this, you will experience an emf that will counteract the change in current. For the circuit in the figure, Kirchhoff's voltage law gives:

$$V_b = RI - e \;, \tag{15.48}$$

where V_b is the emf of the battery and e is the emf from the self-inductance of the circuit. The power P consumed in this circuit is

$$P = V_b I = RI^2 - eI \;, \tag{15.49}$$

and the work done during a short time interval $\mathrm{d}t$ is

$$\mathrm{d}W = P\,\mathrm{d}t = V_b I\,\mathrm{d}t = RI^2\,\mathrm{d}t - eI\,\mathrm{d}t \;. \tag{15.50}$$

The term $V_b I\,\mathrm{d}t$ represents the energy supplied by the battery. Some of this energy is dissipated in the resistor. This is the term $RI^2\,\mathrm{d}t$. The remaining energy is work against the emf due to the self-inductance of the circuit. This is the energy that is stored in the magnetic field in the circuit. The work done by external sources, the battery, on the magnetic field of the circuit during a short time interval $\mathrm{d}t$ is

$$\mathrm{d}W_m = -eI\,\mathrm{d}t = -\left(-\frac{\mathrm{d}\Phi}{\mathrm{d}t}\right) I\,\mathrm{d}t \tag{15.51}$$

We insert $\Phi = LI$, so that $\mathrm{d}\Phi/\mathrm{d}t = L\,\mathrm{d}I/\mathrm{d}t$ (when L is not changing with time), getting

$$\mathrm{d}W_m = LI\frac{\mathrm{d}I}{\mathrm{d}t}\,\mathrm{d}t = LI\,\mathrm{d}I \;. \tag{15.52}$$

This is the work done by the external source to change the current by an amount $\mathrm{d}I$. If we bring the system from zero current, where the magnetic energy is zero, up to a current I, the work done on the magnetic field is:

$$W_m = \int_0^I LI\,\mathrm{d}I = \frac{1}{2}LI^2 \;. \tag{15.53}$$

We can rewrite this using the flux $\Phi = LI$, getting:

$$W_m = \frac{1}{2}LI^2 = \frac{1}{2}\Phi I = \frac{\Phi^2}{2L} \;. \tag{15.54}$$

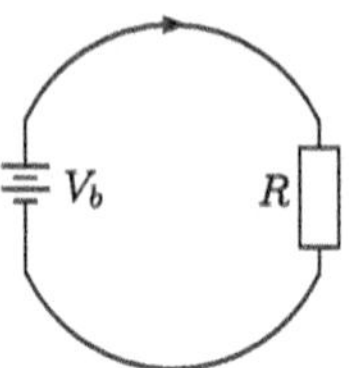

Fig. 15.7 Illustration of a simple circuit. What is the energy needed to get a current I flowing in this circuit?

We notice that this is similar to the expressions we found for a capacitor, where $W_e = QV/2 = CV^2/2$.

Energy stored in an inductor

The energy W_m stored in the magnetic field of an inductor with inductance L and a current I when it does not interact magnetically with any other circuits, is:

$$W_m = \frac{1}{2}LI^2 = \frac{1}{2}I\Phi \,. \tag{15.55}$$

Energy in coupled circuits

This result can be generalized to a system of N magnetically coupled circuit loops. The total energy is

$$W_m = \frac{1}{2}\sum_{i=1}^{N} I_i \Phi_i \,, \tag{15.56}$$

where the flux Φ_i through circuit i

$$\Phi_i = \sum_{j=1}^{N} \Phi_{j,i} \,, \tag{15.57}$$

is the sum of the fluxes from circuits $j = 1, \ldots, N$ on a circuit i, where $\Phi_{j,i} = L_{j,i} I_j$. The magnetic energy is therefore:

$$W_m = \frac{1}{2}\sum_{i=1}^{N}\sum_{j=1}^{N} L_{j,i} I_i I_j \,. \tag{15.58}$$

The magnetic energy of two coupled loops can therefore be written as:

$$W_m = \frac{1}{2}\left(\underbrace{L_{1,1}I_1I_1 + L_{2,1}I_2I_1}_{I_1\Phi_1} + \underbrace{L_{1,2}I_1I_2 + L_{2,2}I_2I_2}_{I_2\Phi_2}\right) \tag{15.59}$$

Notice that $L_{1,1}$ and $L_{2,2}$ are positive, but that $L_{1,2}$ can be both positive or negative. The total energy of a coupled system can therefore both be larger or smaller than the sum of the energies of the individual systems.

Energy stored in an set of coupled inductors

The energy W_m stored in the magnetic field in a set of N magnetically coupled inductors with mutual inductances $L_{i,j}$ and currents I_i is:

$$W_m = \frac{1}{2}\sum_i \sum_j L_{i,j} I_i I_j \;. \tag{15.60}$$

Proof for a set of coupled circuits. For completeness, we provide a proof of (15.56). For N coupled circuits Kirchhoff's voltage law for circuit j is

$$V_{b,j} + \left(-\frac{d\Phi_j}{dt}\right) = R_j I_j \;, \tag{15.61}$$

where $V_{b,j}$ is the emf of the voltage source in circuit j. The power dissipated in circuit j is:

$$P_j = V_{b,j}I_j = \left(R_j I_j + \frac{d\Phi_j}{dt}\right) I_j = R_j I_j^2 + \frac{d\Phi_j}{dt} I_j \;, \tag{15.62}$$

The work done by voltage source j in a time interval dt is therefore

$$dW_j = R_j I_j^2\, dt + I_j\, d\Phi_j \;. \tag{15.63}$$

where we recognize the first term as the power dissipation in the resistor in circuit j and the second term is related to the energy stored in the magnetic field. We call the second term the magnetic work. The total magnetic work done by all the voltage sources in a time interval dt is

$$dW_m = \sum_j I_j\, d\Phi_j \;, \tag{15.64}$$

where

$$\Phi_j = \sum_i \Phi_{i,j} = \sum_i L_{i,j} I_i \;. \tag{15.65}$$

When $L_{i,j}$ does not change with time we have that $\mathrm{d}(L_{i,j}I_i) = L_{i,j}\,\mathrm{d}I_i$:

$$\mathrm{d}W_m = \sum_i \sum_j L_{i,j} I_j \,\mathrm{d}I_i \ , \tag{15.66}$$

We now use the trick that $\mathrm{d}(I_i I_j) = I_i\,\mathrm{d}I_j + I_j\,\mathrm{d}I_i$ and that $L_{i,j} = L_{j,i}$ so that

$$\begin{aligned}
\mathrm{d}\left(\sum_i \sum_j L_{i,j} I_i I_j\right) &= \sum_i \sum_j L_{i,j}\left(I_j\,\mathrm{d}I_i + I_i\,\mathrm{d}I_j\right) \\
&= \sum_i \sum_j L_{i,j} I_j\,\mathrm{d}I_i + \sum_i \sum_j L_{j,i} I_i\,\mathrm{d}I_j \\
&= 2\sum_i \sum_j L_{i,j} I_j\,\mathrm{d}I_i \ ,
\end{aligned} \tag{15.67}$$

We therefore see that with

$$W_m = \frac{1}{2}\sum_i \sum_j L_{i,j} I_i I_j \ . \tag{15.68}$$

we get the relation in (15.66), which proves that this is the magnetic energy of the system.

Example: Energy density in a toroid

What is the energy in a thin toroid with N windings in Fig. 15.8 expressed in terms of the magnetic field B, H? The energy stored in this system is

$$W_m = \frac{1}{2}LI^2 = \frac{1}{2}I\Phi \ . \tag{15.69}$$

We can relate the current I to the magnetic field in the toroid by applying Ampere's law on a circuit C around the center of the toroid, where the magnetic **H**-vector is in the azimuthal direction due to symmetry, $\mathbf{H} = H(r)\hat{\boldsymbol{\phi}}$. Ampere's law gives

$$\oint_C \mathbf{H}\cdot \mathrm{d}\mathbf{l} = 2\pi r H = NI \quad \Rightarrow \quad I = \frac{2\pi r H}{N} \ . \tag{15.70}$$

The flux, Φ, of this magnetic field through the circuit C is N times the flux through a cross section with area S. If the toroid is thin compared to its radius, we can assume that the magnetic field is approximately uniform across the cross section and the flux through a cross-section S is:

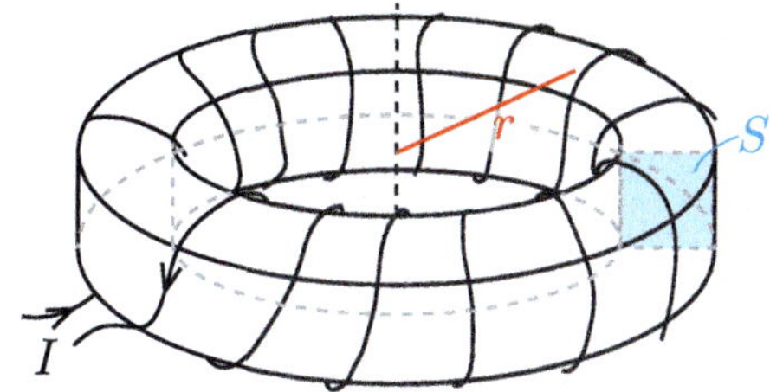

Fig. 15.8 A thin toroid with radius r, N windings and cross-sectional area S

$$\Phi_S = \int_S \mathbf{B} \cdot d\mathbf{S} = BS \tag{15.71}$$

and the total flux through N windings is $\Phi = N\Phi_S = NBS$. The magnetic energy stored in this system is therefore:

$$W = \frac{1}{2}I\Phi = \frac{1}{2}\left(\frac{2\pi r H}{N}\right) BSN = \frac{1}{2}2\pi r SBH = \frac{1}{2}BH\, 2\pi r S\,. \tag{15.72}$$

Here, we recognize the term $BH/2$ as the energy per unit volume and $2\pi r S$ as the volume of the toroid. We can therefore introduce the energy density u

$$u = \frac{W}{V} = \frac{1}{2}BH\,, \tag{15.73}$$

This result is general, even though we do not provide a general proof of this result here.

Energy density of a magnetic field

The **energy density** $u(\mathbf{r})$ of a magnetic field described by $\mathbf{B}$ and $\mathbf{H}$ is

$$u = \frac{1}{2}\mathbf{B} \cdot \mathbf{H}\,. \tag{15.74}$$

Test your understanding

Two solenoids are part of the same circuit. Solenoid 1 is twice as long, has twice as many windings and had twice the diameter of circuit 2. (The two solenoids are far from each other so you can disregard any interactions between them.) (a) What is the relation between the magnetic field in solenoid 1 and solenoid 2? (b) What is the relation between the energy densities in solenoid 1 and in solenoid 2? (c) What it the relation between the energy stored in solenoid 1 and solenoid 2?[6]

[6] (a) $B_1 = B_2$; (b) $u_1 = u_2$; (c) $W_1 = 8W_2$.

Magnetic energy in a non-linear medium

What is the energy stored in a non-linear magnetic material? This is an important practical question, because we often use non-linear ferromagnetic cores in magnetic circuits and in various coils (toroids) in electric circuits. The work performed by an external field (e.g. a voltage source or a battery) to change the magnetic flux in a magnetic circuit is:

$$\mathrm{d}W_m = I\,\mathrm{d}\Phi \ . \tag{15.75}$$

For a thin toroid of radius r we know from Ampere's law that the magnetic field H inside the toroid is

$$H = \frac{NI}{2\pi r} \ , \tag{15.76}$$

and that the total flux through the circuit wrapping around the toroid is $\Phi = NBS$. The contributions to the work is then:

$$\mathrm{d}W_m = \underbrace{\frac{2\pi r H}{N}}_{I} \underbrace{NS\,\mathrm{d}B}_{\mathrm{d}\Phi} = \underbrace{2\pi r S}_{\text{volume}} H\,\mathrm{d}B \ . \tag{15.77}$$

The change in energy density is the work done per unit volume, which is

$$\mathrm{d}u = H\,\mathrm{d}B \ . \tag{15.78}$$

So far, this is a general result. However, a non-linear medium has hysteresis. Figure 15.9 illustrates how the magnetic fields B and H are related during a complete cycle for an alternating current. The work per unit volume done on the magnetic system if we change the field from B to $B + \mathrm{d}B$ is the area of a rectangle with height $\mathrm{d}B$ and length H. In part 1 of the path, the changes in magnetic field are positive, but in part 2 the changes are negative, so that the work per unit volume also is negative. We therefore subtract the blue area in Fig. 15.9b from the red area in Fig. 15.9a, getting the dark gray area in Fig. 15.9c. The work done on the magnetic field over a whole period is the gray area enclosed by the hysteresis loop. This work corresponds to the energy dissipated as the orientations of the magnetic domains in the material changes and is transformed to thermal energy (heat) in the material.

The total area is the energy dissipated per cycle. If the cycle has a period T and a frequency $f = 1/T$, the dissipated energy, W, per unit time (and unit volume), that is the dissipated power per unit volume, is

$$p = \frac{P}{v} = \frac{W_m/T}{v} = \frac{1}{Tv}W_m = \frac{1}{T}\int \mathrm{d}u = \frac{1}{T}\int H\,\mathrm{d}B = A\,f \ , \tag{15.79}$$

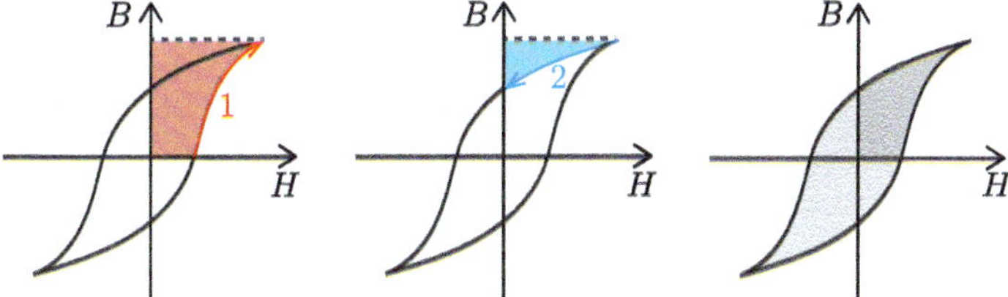

Fig. 15.9 Illustration of a hysteresis loop for the magnetic field in a circuit. **a** The integral of $H\,\mathrm{d}B$ along curve 1 is positive. **b** The integral of $H\,\mathrm{d}B$ along curve 2 is negative. **c** The net integral is the gray area enclosed by the hysteresis curve

where A is the area of the hysteresis curve and f is the frequency. To reduce power dissipation, we should therefore try to design a system with the smallest possible area of the hysteresis loop (and have low frequencies).

15.5 Magnetic Forces

We know that the force from a magnetic field on a current-carrying element is $\mathrm{d}\mathbf{F} = I\,\mathrm{d}\mathbf{l} \times \mathbf{B}$. However, this result is not practical in all situations. For example, if we want to find the force between two permanent magnets, we will need to know the magnetic field from one of them and the (bound) current density in the other to calculate the force. Here, we will introduce a more practical method based on energy considerations.

Magnetic force in a loss-free system

Figure 15.10 illustrates a loss-free system with a permanent magnet moving relative to an ideal conductor. The system is closed, so that the energy is conserved. Assume we displace the magnet in the z-direction, $z \rightarrow z + \mathrm{d}z$. The work performed by the magnetic force, $\mathbf{F}_m$, is then:

$$\mathrm{d}W = \mathbf{F}_m \cdot \boldsymbol{\Delta r} = \mathbf{F}_m \cdot \mathrm{d}z\,\hat{\mathbf{z}} = F_{m,z}\,\mathrm{d}z = -\,\mathrm{d}W_m \,, \tag{15.80}$$

where W_m is the magnetic energy. We interpret this to mean that the work done by the magnetic force must lead to a change in the magnetic energy. We can write this as:

$$F_{m,z} = -\frac{\partial W_m}{\partial z} \,. \tag{15.81}$$

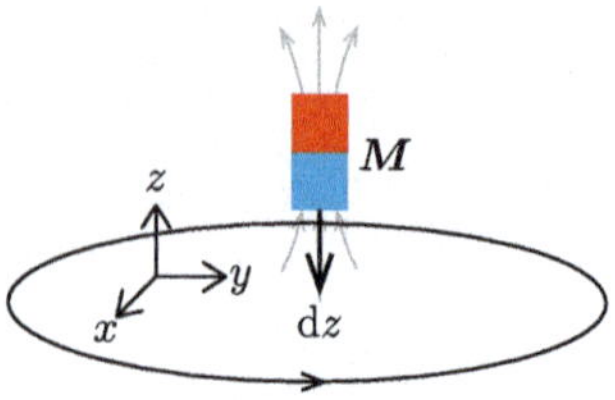

Fig. 15.10 An ideal conductor interacting with a permanent magnet

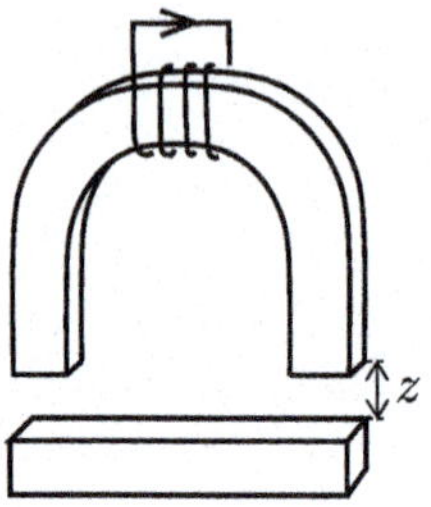

Fig. 15.11 A horseshoe shaped iron core electromagnet lifting an iron bar

We can introduce a similar argument for motion in the x and y directions. In general, we therefore get:

$$\mathbf{F}_m = -\nabla W_m \ , \tag{15.82}$$

where we must remember that W_m is an energy (and not work). This is a practical formula, because we may be able to find $W_m(x, y, z)$ for the system, which would allow us to find the magnetic force.

Example: Carrying force of an electromagnet

Figure 15.11 *illustrates a horseshoe-shaped iron electromagnet with a wire wrapped around its top. When a current I is running in the wire, the magnet can lift a weight G made of iron. What is the lifting force of the electromagnet?*

Solution: Our plan is to find an expression for the magnetic energy W_m of the system as function of the distance dz between the horseshoe magnet and the iron bar below it. We will assume that the permeability of the iron magnet and the iron bar is large ($\mu \to \infty$) so that all the flux remains inside the magnetic circuit consisting of the horseshoe magnet, the iron bar and the air gaps of thickness dz. The total energy will consist of three parts, the energy of the horseshoe, $W_{m,h}$, the air gaps, $W_{m,a}$ and the iron bar $W_{m,b}$. However, the energies of the horseshoe and the iron bar will remain constant during the process, it is only the magnetic energy in the air gaps that will depend on dz.

$$W_m = W_{m,h} + W_{m,a}(z) + W_{m,b} \ . \tag{15.83}$$

We find that the magnetic energy in the air gaps is the energy density in the air multiplied with the volume of the air gaps. Each air gap has a cross-sectional area S and a length z. The energy is therefore:

$$W_{m,a} = \underbrace{\frac{B^2}{2\mu_0}}_{u} \underbrace{2\,Sz}_{dv} = \frac{B^2\,Sz}{\mu_0} \,. \tag{15.84}$$

The resulting force is then

$$\mathbf{F} = F_z\hat{\mathbf{z}} = -\frac{\partial W_z}{\partial z}\hat{\mathbf{z}} = -\frac{B^2 S}{\mu_0}\hat{\mathbf{z}} \,. \tag{15.85}$$

The force is in the negative z-direction, that is, it is an attractive force pulling the two parts together. For a magnet with magnetic field $B = 1$ T and a surface area of 1 cm^2, we find that the force is $F \simeq 80$ N, which means that it can support about $80\,\mathrm{N}/9.8\,\mathrm{ms}^{-2} \simeq 8\,\mathrm{kg}$.

Summary

The **self-inductance** of a system with a flux Φ when a current I is running through it is $L = \Phi/I$.

We find the **self-inductance** by assuming a current I is running in the circuit, finding the resulting magnetic field **B** and the flux Φ of the magnetic field through the circuit. The self-inductance L is then found from $L = \Phi/I$.

The flux *in* a circuit C_j *from* a current I_i in circuit C_i is $\Phi_{i,j} = \int_{S_j} \mathbf{B}_i \cdot d\mathbf{S}$, where $\mathbf{B}_i$ is the magnetic field from current I_i. The **mutual inductance** is defined as:

$$L_{i,j} = \frac{\Phi_{i,j}}{I_i} \,.$$

The inductance $L_{i,i}$ is the self-inductance of circuit i.

The **mutual inductances are symmetric**: $L_{i,j} = L_{j,i}$.

The **magnetic energy** W_m stored in an inductor with inductance L and current I is

$$W_m = \frac{1}{2}LI^2 = \frac{1}{2}I\Phi \,.$$

The **energy density** u_m of a magnetic field is

$$u_m = \frac{1}{2}\mathbf{B}\cdot\mathbf{H} \,.$$

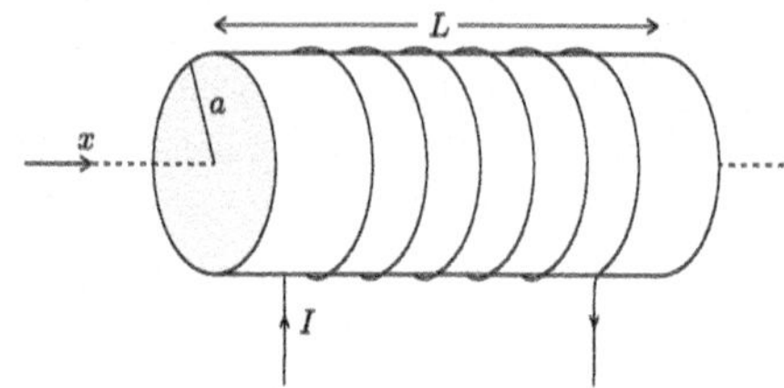

Fig. 15.12 Illustration of a solenoid

Exercises

Discussion exercises

15.1 Inductance. A tightly wound toroid is one of few configurations where it is simple to find the self-inductance. What properties of a toroid makes this simple?

15.2 Energy density in magnetic materials. For a given magnetic field strength, B, is the energy density higher in vacuum or in a magnetic material with permeability μ? Does the equation $u = (1/2)BH$ imply that for a long solenoid with current I, then the stored energy is proportional to $1/\mu$? And does this mean that for the same current, there is less energy stored in the solenoid when it is filled with a ferromagnetic material instead of with air? Explain.

15.3 Energy in a coil. We often call a solenoid a coil. Assume that you first connect a coil to a battery. How much energy is there in a coil with inductance L, if there is a current I running through it? You quickly disconnect the wire to the battery, so that the current goes to zero. Where did the energy in the coil go?

Tutorials

15.4 Introduction to self-inductance. Fig. 15.12 shows a coil, a solenoid, with diameter a, length L and N windings.
(a) What is the magnetic field inside the solenoid when a current I is running through it?
(b) What is the flux of the magnetic field through the circuit? Discuss both the part of the circuit that makes out the solenoid and the part of the circuit that ties the ends of the solenoid together.

15.5 Self-inductance per unit length. Two thin wires are handing in the air and carry a time varying current. The wires have a radius a and are hanging horizontally with a distance d from each other (d is the distance between the centers of the wires). We assume that this is a closed circuit, where there is current I one way in one of the wires and the other way in the other wire. We assume that the circuit is long, so that we can disregard edge effects at the end of the wires.

Fig. 15.13 Two coupled circuits

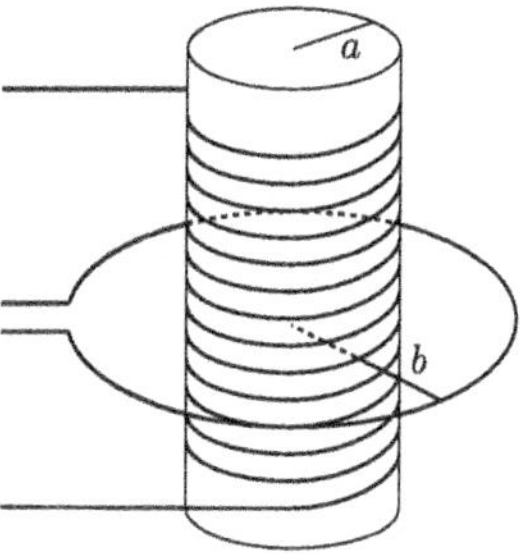

(a) Place a wire with its center in the origin and the other with its center at $(d, 0, 0)$. Assume that the wires are running in the y-direction. Make a drawing of the system.
(b) Find the magnetic field from the wire with its center in the origin as a function of x in the region between the wires.
(c) Find the magnetic field from the wire with its center at $(d, 0, 0)$ as a function of x in the region between the wires.
(d) Why do you think it was specified that the wire had a finite thickness, a?
(e) Find the flux of the total magnetic field for a length ℓ along the wires.
(f) Find the inductance per unit length for this system.
(g) If we want to reduce the inductance, how should we construct such as system?

15.6 Long and short circuit. A system consists of two circuits. One circuit consists of a very long, straight wire, which is then connected back to itself through a half-circle-shaped wire at the far distance. The other circuit consists of a small, quadratic circuit with sides a. The side of the quadratic circuit that is closes to the long wire is parallel to the long wire at a distance d from the long wire. We want to find the mutual inductance of this system.
(a) Make a drawing of the system.
(b) Through what circuit do you think that it is smart to let the current I_1 run through and through which circuit would like like to calculate the flux?
(c) Find the magnetic field at a distance x from the long wire.
(d) Find the flux through the quadratic circuit.
(e) Find the mutual inductance for the system.
(f) Why did we not need to include the magnetic field from the quadratic circuit, when we calculated the mutual inductance to this system?

15.7 Mutual inductance in a solenoid and a circular circuit. A solenoid consists of a wire wrapped N times around a cylinder of length L and radius a. A circular circuit of radius b is centered on the cylinder as illustrated in Fig. 15.13.
(a) What strategy would you choose to find the mutual inductance for this system? What symmetries would you use and how would you reason? In which system will you assume there is a current and in which system would you calculate the flux? Explain.

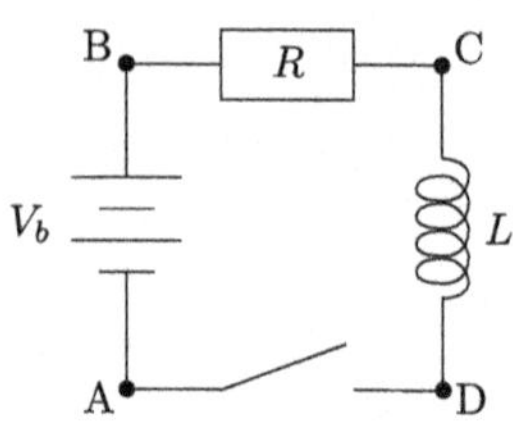

Fig. 15.14 Illustration of a circuit with a battery, a resistor, and a coil

(b) Based on your assumption, assume a current I_1 runs through your selected circuit 1. What is the magnetic field $\mathbf{B}_1$? Explain the steps in the argument used to find $\mathbf{B}_1$. How does the argument depend on what is inside the solenoid?
(c) Given the magnetic field $\mathbf{B}_1$, what is the flux $\Phi_{1,2}$ of $\mathbf{B}_1$ through circuit 2? Explain your reasoning.
(d) Finally, find the mutual inductance L_{12} for this system.
(e) Can you extend your argument to the case where $N = 1$ and $b \gg a$?

15.8 Energy densities. A solenoid consists of a wire wrapped N times around a cylinder of length h and radius a.
(a) What is the magnetic fields $\mathbf{H}$ and $\mathbf{B}$ inside the solenoid when a current I is running through the wire?
(b) What is the energy density inside the solenoid? And outside the solenoid?
(c) What is the total energy stored in the solenoid when a current I is running through the wire?
(d) Express the total energy in terms of the inductance L of the solenoid.

15.9 We turn the circuit on. Figure 15.14 shows a circuit consisting of a battery, a resistor and a coil. What is the voltage (potential) at the various components in the circuit when the switch is open as in the figure. You can assume that the potential in A is zero.
(a) In B, C, D?
(b) We close the switch at the time $t0$ so that it forms a closed circuit. What is now the potential at the various components at the time $t = 0$ immediately after the switch is closed? In B, C, D?
(c) Explain why you got this result at C.
(d) What is the potential drop across the coil at the time $t = 0$ immediately after the switch is closed?
(e) What is the current through the coil at the time $t = 0$ immediately after the switch is closed?
(f) Write down Kirchhoff's law of voltages for the circuit.
(g) What is the current in the circuit after a very long time?
(h) Find the current as a function of time.

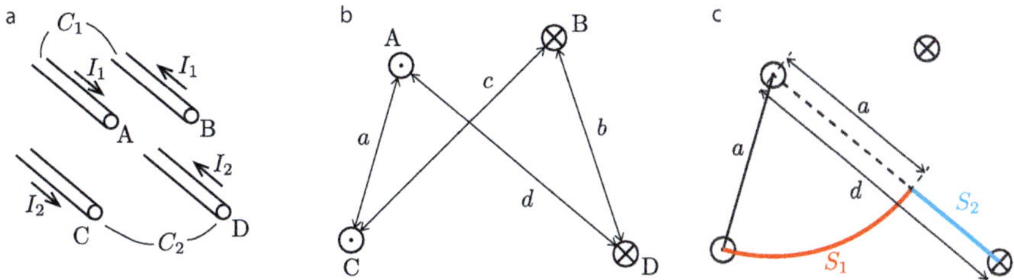

Fig. 15.15 **a** Two pairs of wires with two current I_1 and I_2 interact through a magnetic field. **b,c** Geometric aspects of the system

Homework

15.10 Self-inductance of a coaxial cable. A coaxial cable consists of an inner conductor with radius a and an outer, thin conductor at radius b.

(a) Make a sketch of the system. What do you define the circuit to be in this system?

(b) Find the magnetic field in the region between the inner and the outer conductor.

(c) Find the flux of the magnetic field through the surface, S, which is the flat surface spanned by the circuit in which the current I is running.

(d) Find the inductance per unit length of the system.

15.11 Cross-talk. Figure 15.15 illustrates two wire pairs. Each pair form loops with wires that are far away and outside the sketch. Loop C_1 carries a current I_1 and loop C_2 carries a current I_2.

(a) Find the magnetic field from a single wire in C_1 coming out of the plane in Fig. 15.3.

(b) Find the flux through C_2 from the magnetic field from the wires in C_1.

(c) Find the mutual inductance per unit length.

(d) Use this results to discuss the inductive coupling between two wires. How should you construct the wire pairs to reduce cross-talk?

15.12 Energy in coaxial cable. Find the energy per length stored in a coaxial cable carrying a current I. The inner radius of the cable is a and the outer radius is b.

15.13 Hairdryer. Assume that you bring your hairdryer from Europe to the US, where the voltage is 120 V instead of 230 V in Europe. In Europe you get an average power of 1600 W from the hair dryer.

(a) What can you do to get the same power from the hairdryer in the US as in Europe?

(b) How large current will you need from the outlet in the US?

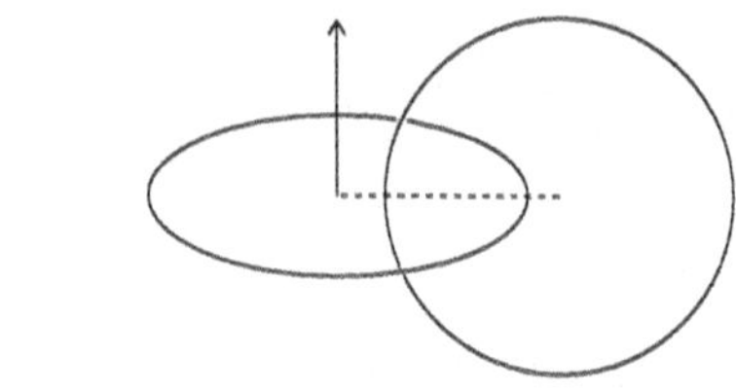

Fig. 15.16 Two intertwined conducting circles

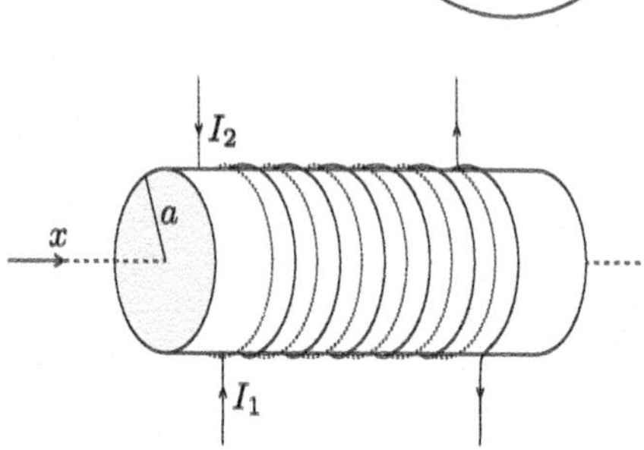

Fig. 15.17 Illustration of a solenoid

(c) What resistance will you experience that the US power grid will notice that your hairdryer has, when you have modified it to Europoean power consumption.

15.14 Force on a magnetic rod in a solenoid. A magnetic rod with circular cross section or radius a and length ℓ_0 is made of a magnetic material with permeability μ. The rod is partially inserted into a long solenoid with length ℓ_1 and radius b. The solenoid has n windings per unit length and initially a current I_0 is running through the wire.
(a) How large is the force on the rod? (Hint: Find an expression for the magnetic energy as a function of how far the rod has been inserted into the solenoid.)
(b) Is the rod pulled into or pushed out of the solenoid? What does you answer depend on?

15.15 Two circles. Two circular circuits are intertwined as shown in Fig. 15.16. A circle of radius a is in the xy-plane with its center in the origin.
(a) Another circle is placed in the yz-plane with its center in $(b, 0, 0)$ and radius R as shown in Fig. 15.16. What is the mutual inductance, L_{12}, of this system?
(b) Assume tha the other circle instead is parallel to the xy-plane and that its center is in $(0, 0, b)$ and that the radius of the circle is very small compared with a. Find an estimate of the mutual inductance L_{12} of this system. Argue for the approximations you have to do. (Hint: The circle in $(0, 0, b)$ is so small that you can assume that the magnetic field is uniform inside the circle.)

15.16 A double coil. In this exercise we will study a solenoide as shown in Fig. 15.17. The solenoid consists of a wire that is wrapped N times about a cylinder of radius a and length L, where $L \gg a$. The cylinder is filled with a magnetic material with permeability μ.

(a) Find the **H**-field, the **B**-field and the magnetization **M** inside the solenoid when a current I passes through the wire. You can assume that the **H**-field is zero outside the solenoid and that the **H**-field only has a component in the x-direction inside the coil.
(b) Find the self inductance L_{11} for the system.

Then we wrap another wire M times around the same cylinder as illustrated with a dashed line in Fig. 15.17. Notice that the indicated positive direction for current I_2 is opposite of the positive direction for current I_1. You may assume that a and L are the same for both wires, that the wires are insulated so that there is no contact, and that the wires cover the same area of the cylinder.
(c) What is the mutual inductance L_{12} between the two coils?
(d) We connect coil 1 to a time-varying voltage source with emf $V_1(t)$ and coil 2 to a resistor R_2. Find a set of equations for the currents I_1 and I_2 in coil 1 and coil 2. (You do not need to solve the equations).

15.17 Numerical inductance calculation. A square circuit with sides a is placed in the xy-plane.
(a) Write a program to find the self-inductance of this circuit.
(b) Write a program to find the mutual inductance between a square circuit in the xy-plane and an identical square circuit at a height z above the xy-plane. Plot the mutual induction as a function of z.

Chapter 16
Electric Circuits—Part 2

16.1 Circuit Components

We have already introduced the main components used in circuit diagrams. We now complete the description by introducing the *inductor* with inductance L. The inductor can either be thought of as modeling the effective inductance in a system, such as the inherent inductance in any current loop, or it can be thought of as a circuit element such as a coil with an iron core, which for example is used to damp out rapid changes in current. The circuit elements are illustrated in Fig. 16.1. The inductor is illustrated as a small spiral, imitating the look of a small solenoid.

An inductor is characterized by its inductance L. We know how to calculate or measure the inductance for a given circuit geometry, and we know that it only depends on the geometry of the system and the magnetic materials used. The flux through the circuit due to the inductor is

$$\Phi = LI, \tag{16.1}$$

and the emf associated with the inductor is

$$e = -\frac{\mathrm{d}}{\mathrm{d}t}\Phi = -L\frac{\mathrm{d}I}{\mathrm{d}t}, \tag{16.2}$$

where we have assumed that L does not vary with time. The emf from the inductor is included in Kirchhoff's voltage law for a circuit.

Description of Circuit Elements

Let us summarize the contributions to Kirchhoff's voltage law of the various components in the circuit illustrated in Fig. 16.2.

© The Author(s), under exclusive license to Springer Nature Switzerland AG 2026

A. Malthe-Sørenssen, *Elementary Electromagnetism Using Python*, Undergraduate Texts in Physics, https://doi.org/10.1007/978-3-032-19876-1_16

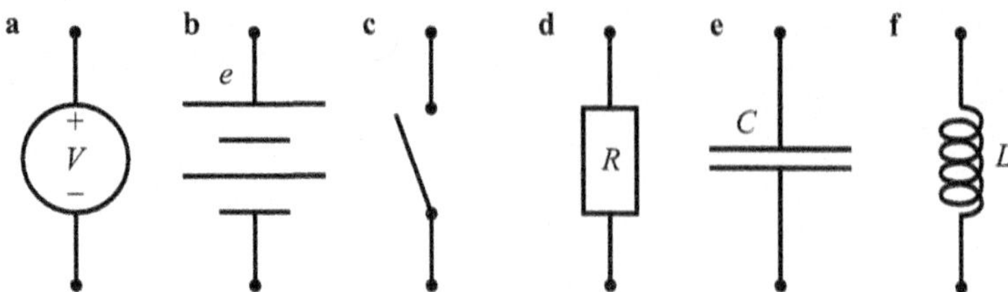

Fig. 16.1 Illustrations of the main circuit elements: **a** Voltage source, **b** battery, **c** switch, **d** resistor, **e** capacitor, **f** inductor

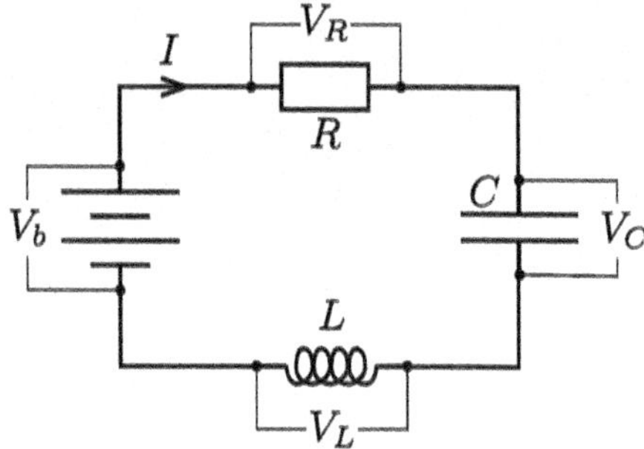

Fig. 16.2 Illustration of a circuit with a battery V_b, a resistor R, a capacitor C and an inductor L

Kirchhoff's voltage law for a complete loop around the circuit is

$$\sum_i e_i - \sum_j \Delta V_j = 0, \tag{16.3}$$

which in this case is

$$\underbrace{V_b}_{\text{emf}} - \underbrace{L\frac{\mathrm{d}I}{\mathrm{d}t}}_{V_L} - \underbrace{RI}_{V_R} - \underbrace{\frac{1}{C}Q}_{V_C} = 0 \tag{16.4}$$

Let us address the individual components.

Battery. A battery or a voltage source contributes with an emf $V_b(t)$ to the circuit. This corresponds to a voltage increase where the other components correspond to voltage drops. We recall that a battery may be thought of as a mechanical battery that lifts charges by an external, non-Coulombic force against the electric field.

Resistor. The resistor R is drawn either as a small rectangle, illustrating the cylindrical shape often used, or as a zig-zag line. The voltage drop across the resistor is $\Delta V = RI$, where R is the resistance of the resistor. We recall that we combine resistors in series by adding them: $R = \sum_i R_i$ and that we combine resistors in parallel by adding their inverse: $1/R = \sum_i 1/R_i$. The power dissipation in a resistor is $P = RI^2$.

Capacitor. The capacitor C acts as a charge storage device and is characterized by its capacitance $C = Q/\Delta V$, where Q is the charge on each side of the capacitor when the voltage difference is ΔV. The current flowing into a capacitor is related to the change in charge Q on the capacitor: $I = \mathrm{d}Q/\mathrm{d}t$. If we insert $Q = C\Delta V$, we get

$$\frac{\mathrm{d}Q}{\mathrm{d}t} = I = C\frac{\mathrm{d}V}{\mathrm{d}t}, \tag{16.5}$$

when the capacitance C does not vary with time. The capacitor can be thought of as a component that resists changes in voltage in the system and stores charges. We recall that we combine capacitors in series by adding their inverse capacitances: $1/C = \sum_i 1/C_i$ and that we combine capacitors in parallel by adding their capacitances: $C = \sum_i C_i$.

Inductor. The inductor L introduces an additional emf $e = -\mathrm{d}\Phi/\mathrm{d}t = -L\mathrm{d}I/\mathrm{d}t$. The emf is included as a voltage jump, similar to the battery, or we include $-e$ as a voltage drop in the circuit. The voltage drop across the inductor is then $\Delta V = -e = L\mathrm{d}I/\mathrm{d}t$.

Test your understanding
The figure shows a circuit with four elements. (a) If the current is constant, over which components does the voltage drop or increase? (b) If the current is decreasing with time, over which components do the voltage drop or increase?[1]

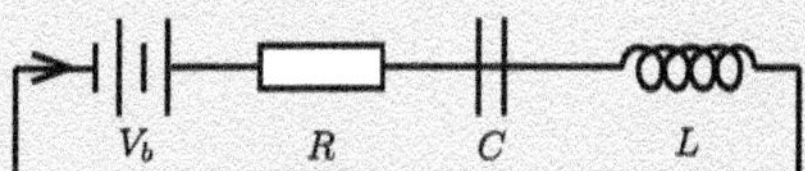

Combining Inductors

We have found rules for combining both resistors and capacitors in series and parallel, and this is actively used to analyze circuits. Can we find similar laws for the combination of inductors?

Inductors in series. Figure 16.3a illustrates a series coupling of two inductors L_1 and L_2. Can we replace these two elements by a single inductor L? The same current I runs through both inductors. The voltage drops across each of the two inductors are $V_1 = L_1\mathrm{d}I/\mathrm{d}t$ and $V_2 = L_2\mathrm{d}I/\mathrm{d}t$. The total voltage drop across the two inductors are

$$V = V_1 + V_2 = L_1\frac{\mathrm{d}I}{\mathrm{d}t} + L_2\frac{\mathrm{d}I}{\mathrm{d}t} = \underbrace{(L_1 + L_2)}_{L}\frac{\mathrm{d}I}{\mathrm{d}t} = L\frac{\mathrm{d}I}{\mathrm{d}t}. \tag{16.6}$$

[1] (a) Increase over V_b, decrease over R, C, constant over L; (b) Increase over V_b, and L, decrease over R and C.

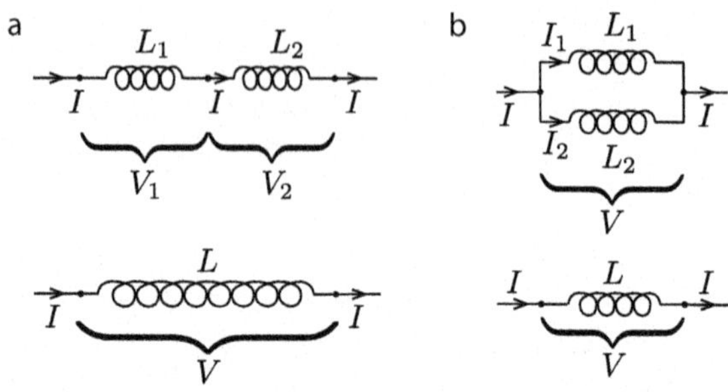

Fig. 16.3 **a** Illustration of a series coupling of two inductors. **b** Illustration of a parallel coupling of two inductors

We can therefore replace the two inductors L_1 and L_2 with a single inductor with inductance $L = L_1 + L_2$. We therefore conclude that we *add inductances in series*.

Inductors in parallel. Figure 16.3b illustrates a parallel coupling of two inductors L_1 and L_2. The voltage drop over both inductors are V so that

$$V = L_1 \frac{dI_1}{dt} = L_2 \frac{dI_2}{dt}, \tag{16.7}$$

giving

$$\frac{V}{L_1} = \frac{dI_1}{dt} \text{ and } \frac{V}{L_2} = \frac{dI_2}{dt}, \tag{16.8}$$

where $I = I_1 + I_2$ so that

$$\frac{V}{L} = \frac{dI}{dt} = \frac{dI_1}{dt} + \frac{dI_2}{dt} = \frac{V}{L_1} + \frac{V}{L_2} = V\left(\frac{1}{L_1} + \frac{1}{L_2}\right). \tag{16.9}$$

We can therefore replace the two inductors L_1 and L_2 with one inductor with inductance $1/L = 1/L_1 + 1/L_2$ and we conclude that we add the inverses of inductors in parallel.

Test your understanding
Three inductors, all with the same inductance L, are connected as in the figure. If we want to replace this circuit with a single component, what should be its inductance?[2]

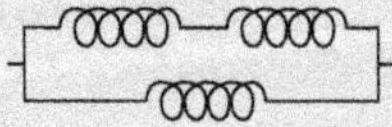

[2] $2L/3$.

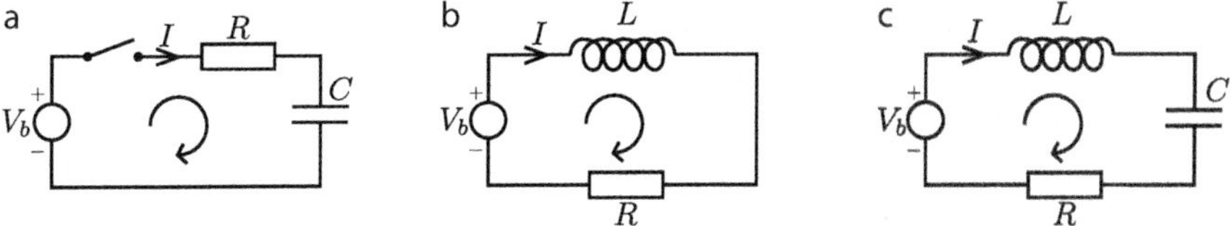

Fig. 16.4 Illustrations of **a** an RC circuit, **b** a RL circuit, and **c** an RLC circuit

16.2 Circuits in the Time Domain

To analyze the time dynamics of circuits, we apply Kirchhoff's voltage and current laws to form a set of (differential) equations describing the currents and voltages in the circuits. Let us address some classical examples.

Example: An RC Circuit

We have analyzed an example of an RC-circuit previously. Let us quickly repeat here. Figure 16.4a illustrates a circuit consisting of a voltage source $V_b(t)$ connected in series with a resistor R and a capacitor C. We apply Kirchhoff's voltage law to the complete loop:

$$V_b - V_R - V_C = 0, \tag{16.10}$$

where V_b is the time-varying voltage source, $V_R = RI(t)$ is the voltage drop across the resistor, and $V_C = Q/C$ is the voltage drop across the capacitor. We assume that the capacitor initially is without any charge.

Approach 1. To solve this we take the time derivative of all the voltages

$$\frac{\mathrm{d}V_b}{\mathrm{d}t} - \frac{\mathrm{d}V_R}{\mathrm{d}t} - \frac{\mathrm{d}V_C}{\mathrm{d}t} = \frac{\mathrm{d}V_b}{\mathrm{d}t} - R\frac{\mathrm{d}I}{\mathrm{d}t} - \frac{1}{C}\underbrace{\frac{\mathrm{d}Q}{\mathrm{d}t}}_{I} = 0, \tag{16.11}$$

which gives a differential equation for the current $I(t)$:

$$\frac{\mathrm{d}V_b}{\mathrm{d}t} - R\frac{\mathrm{d}I}{\mathrm{d}t} - \frac{1}{C}I = 0, \tag{16.12}$$

Approach 2. Alternatively, we can use that $Q = V_C C$ so that

$$I(t) = \frac{\mathrm{d}Q}{\mathrm{d}t} = C\frac{\mathrm{d}V_C}{\mathrm{d}t}, \tag{16.13}$$

we insert this into $V_R = RI(t)$, getting a differential equation in V_C:

$$V_b(t) - RC\frac{\mathrm{d}V_C}{\mathrm{d}t} - V_C = 0. \tag{16.14}$$

Characteristic time τ. In both cases we introduce a characteristic time $\tau = RC$. The second approach can then be rewritten as:

$$V_b(t) - \tau\frac{\mathrm{d}V_C}{\mathrm{d}t} - V_C = 0. \tag{16.15}$$

Analytical solution. If the V_b is a constant and the circuit is closed at $t = 0$, we expect $V_C(0) = 0$. After an infinite time, we expect that the system has reached a stationary state, so that the time derivative is zero and $V_b - V_C = 0$, that is, $V_C(\infty) = V_b$. The solution to this equation is then

$$V_C(t) = V_b\left(1 - e^{-t/\tau}\right) , \tag{16.16}$$

which you can check by insertion.

Numerical solution. The differential equation can also be solved by e.g. Euler's method for numerical integration of the differential equation

$$\frac{\mathrm{d}V_C}{\mathrm{d}t} = \frac{1}{\tau}\left(V_b - V_C\right). \tag{16.17}$$

which is done by the stepwise algorithm:

$$V_C(t + \Delta t) = V_C(t) + \frac{\Delta t}{\tau}\left(V_b(t) - V_C(t)\right), \tag{16.18}$$

with initial condition $V_C(0) = 0$. This is implemented in the following Python program

```
import numpy as np
import matplotlib.pyplot as plt
tau = 1.0
dt = 0.01
Vb = 1.0
time = 10.0
N = int(time/dt)
VC = np.zeros(N)
t = np.zeros(N)
VC[0] = 0.0
for i in range(N-1):
    VC[i+1] = VC[i] + dt/tau*(Vb-VC[i])
    t[i+1] = t[i] + dt
plt.plot(t,VC)
plt.xlabel("$t$")
plt.ylabel("$V_C$")
```

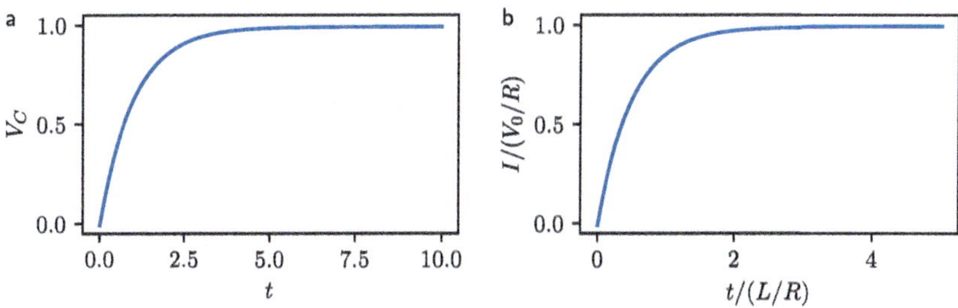

Fig. 16.5 **a** Plot of $V_C(t)$ for the RC-circuit solved by Euler's method. **b** Plot of $I(t)$ for the RL-circuit

The resulting plot is shown in Fig. 16.5a. It may seem unnecessary to solve this equation numerically all the time we have an exact solution. However, the numerical solution opens for studying perturbations or more realistic variants of the system. For example, what if the voltage source $V_b(t)$ had a different shape, such as a square pulse?

$$V_b(t) = \begin{cases} 0 & \text{for } t < 0 \\ V_0 & \text{for } 0 \le t < t_0 \\ 0 & \text{for } t_0 \le t \end{cases}, \tag{16.19}$$

or if the capacitor or resistor has a time dependent value or a non-linear behavior? These cases are easily solved by the same numerical approach but generally cannot be solved by analytical methods.

Example: The RL Circuit

Figure 16.4b illustrates a circuit consisting of a voltage source $V_b(t)$, an inductor L and a resistor R in series. We call such a circuit an RL-circuit. Kirchhoff's voltage law around the circuit gives:

$$V_b - V_L - V_R = V_b - L\frac{\mathrm{d}I}{\mathrm{d}t} - IR = 0, \tag{16.20}$$

which gives a differential equation for the current I in the circuit:

$$L\frac{\mathrm{d}I}{\mathrm{d}t} = V_b - RI. \tag{16.21}$$

We introduce the characteristic time $\tau = L/R$:

$$\frac{L}{R}\frac{\mathrm{d}I}{\mathrm{d}t} = \tau\frac{\mathrm{d}I}{\mathrm{d}t} = \frac{V_b}{R} - I. \tag{16.22}$$

If we turn the circuit on at $t = 0$, we expect the current to start from zero, that is, the initial condition is $I(0) = 0$. The exact solution to this equation is then

$$I(t) = \frac{V_b}{R}\left(1 - e^{-t/\tau}\right). \tag{16.23}$$

The resulting plot is shown in Fig. 16.5b and shows that the effect of the inductor L is to dampen the rapid onset of current when the circuit is turned on with a characteristic time scale $\tau = L/R$, which depends on L: The larger L, the longer time it takes the current to reach its stationary value. This equation can also be solved numerically using Euler's scheme as we did above.

Example: The RLC Circuit

Figure 16.4c illustrates a circuit consisting of a capacitor C, an inductor L and a resistor R in series. We call such a circuit an RLC-circuit. Kirchhoff's voltage law around the circuit, when the voltage source, $V_b(t) = 0$, is zero, gives:

$$-L\frac{\mathrm{d}I}{\mathrm{d}t} - \frac{Q}{C} - RI = 0. \tag{16.24}$$

We can relate Q and the current I by $I = \mathrm{d}Q/\mathrm{d}t$. Taking the time derivative of (16.24) therefore gives:

$$-L\frac{\mathrm{d}^2 I}{\mathrm{d}t^2} - \frac{1}{C}I - R\frac{\mathrm{d}I}{\mathrm{d}t} = 0. \tag{16.25}$$

We can rewrite this equation as

$$\frac{\mathrm{d}^2 I}{\mathrm{d}t^2} + \frac{1}{\tau}\frac{\mathrm{d}I}{\mathrm{d}t} + \omega_0^2 I = 0, \tag{16.26}$$

where $\tau = \frac{L}{R}$ and $\omega_0^2 = \frac{1}{LC}$. We recognize this equation from mechanics as the equation for damped, harmonic oscillations. We also need two initial conditions. The general solution to this equation is

$$I(t) = I_0 e^{-t/(2\tau)} \cos\left(\omega_0' t\right), \tag{16.27}$$

where

$$\omega_0' = \sqrt{\omega_0^2 - \frac{1}{4\tau^2}}. \tag{16.28}$$

In the limit when $\omega_0^2 \gg 1/\tau^2$, that is, when R is small, we get that $\omega_0' \simeq \omega_0$. We therefore get damped oscillations with an angular frequency ω_0 and a damping with

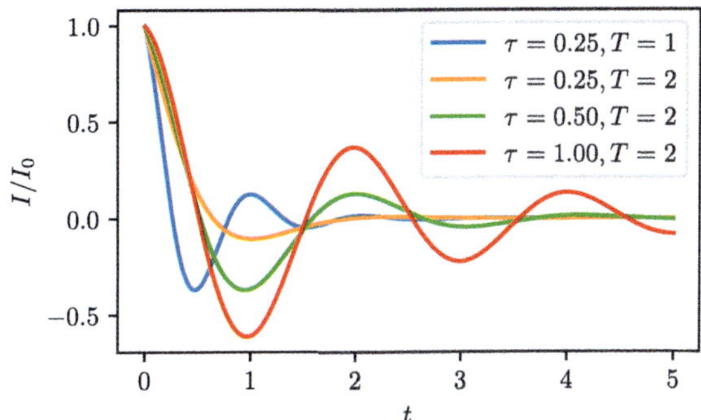

Fig. 16.6 Plot of $I(t)$ for the RLC-circuit

a characteristic time 2τ. Figure 16.6 illustrates a few cases for the dynamics of the circuit. Notice both the damping and the oscillatory behavior. We will address the various types of damping and applications of such circuits in more detail when we study waves and oscillations.

Test your understanding
The figure shows an RL circuit. The switch is closed so that current may run through the system at $t = 0$. (a) What is the current at $t = 0$ immediately after the switch is closed? (b) Which graph best represents the time development of the current?[3]

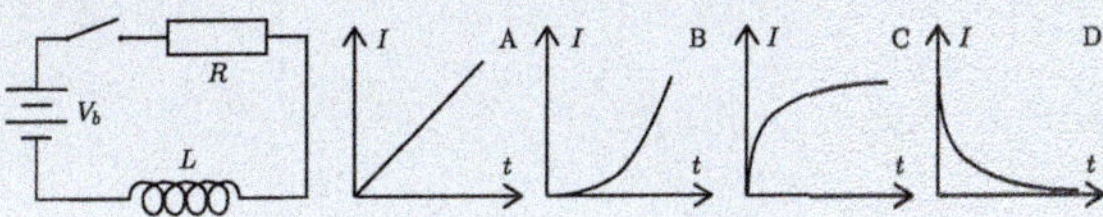

Example: Periodic Signal

Figure 16.4 a illustrates a simple circuit consisting of a voltage source $V_b(t) = V_0 \sin \omega t$, a capacitor C and a resistor R. Assume that the system starts with the voltage $V(0) = 0$ across the capacitor. Find the behavior of the system. Find $V(t)$ numerically in the case when $V_b(t) = V_0 \text{sign}(\sin \omega t)$, that is, when $V_b(t)$ is a square pulse. Test your numerical solution by comparison with the exact solution for the sinus-pulse.

We apply Kirchhoff's voltage law to a loop around the circuit:

$$V_b - V_C - IR = 0. \tag{16.29}$$

To relate V_C and I we use that $V_C = Q/C$ so that

---

[3] (a) $I(0) = 0$; (b) C.

$$\frac{dV_C}{dt} = \frac{1}{C}\frac{dQ}{dt} = \frac{1}{C}I \quad \Rightarrow \quad I = C\frac{dV_C}{dt}. \tag{16.30}$$

We can therefore rewrite (16.29) by inserting this expression for I:

$$V_b(t) - V_C - RC\frac{dV_C}{dt} = 0. \tag{16.31}$$

Analytical solution. We look for the long-term solution after potential initial transients. We expect this solution to have the same periodicity as the driving voltage source: $V_C(t) = A\sin(\omega t + \phi_0)$, where A and ϕ_0 must be determined. We insert this and $V_b = V_0 \sin\omega t$ into (16.31):

$$V_0 \sin\omega t - A\sin(\omega t + \phi_0) - RCA\omega\cos(\omega t + \phi_0). \tag{16.32}$$

We write $RC = \tau$. (16.32) must be true when $t = 0$:

$$-A\sin\phi_0 - \tau\omega A\cos\phi_0 = 0 \quad \Rightarrow \quad \tan\phi_0 = -\omega\tau\,. \tag{16.33}$$

And, similarly, (16.32) must be true when $\omega t + \phi_0 = \pi/2$ so that

$$V_0 \sin\left(\frac{\pi}{2} - \phi_0\right) - A = 0 \quad \Rightarrow \quad V_0 = \frac{A}{\cos\phi_0}. \tag{16.34}$$

Numerical solution. We rewrite (16.31)

$$\tau\frac{dV_C}{dt} = V_b(t) - V_C(t)\,, \tag{16.35}$$

and introduce a dimensionless time $t' = t/\tau$ and a dimensionless potential $V_C' = V_C/V_0$ so that

$$\tau\frac{dV_C'}{dt'}\frac{dt'}{dt} = \frac{dV_C'}{dt'} = \frac{V_b(t'\tau)}{V_0} - V_C'\,. \tag{16.36}$$

We solve this equation numerically using a forward Euler integrator:

$$V_C'(t' + \Delta t') = V_C'(t') + \Delta t'\left(V_b'(t'\tau) - V_C'(t')\right). \tag{16.37}$$

The voltage source is

$$V_b'(t'\tau) = \frac{V_0 \sin\omega(t'\tau)}{V_0} = \sin\omega' t', \tag{16.38}$$

where $\omega' = \omega\tau$. We rewrite the exact solution in terms of t' and V', getting

$$V_C(t) = V_0\cos\phi_0 \sin(\omega t + \phi_0) \quad \Rightarrow \quad V_C'(t) = \cos\phi_0 \sin(\omega' t + \phi_0) \tag{16.39}$$

where $\phi_0 = \arctan(-\omega')$. This is implemented in the following program:

```
import numpy as np
import matplotlib.pyplot as plt
dt = 0.001
time = 20.0
omegam = 1.0 # omega*tau
N = int(time/dt)
VC = np.zeros(N)
t = np.zeros(N)
for i in range(N-1):
    Vb = np.sin(omegam*t[i])
    VC[i+1] = VC[i] + dt*(Vb-VC[i])
    t[i+1] = t[i] + dt
# Plot result
plt.plot(t,VC)
# Compare with analytical solution
phi0 = np.arctan(-omegam)
A =  1.0*np.cos(phi0)
Va = A*np.sin(omegam*t+phi0)
plt.plot(t,Va,"-r")
```

The resulting plot in Fig. 16.7 shows that there is a short transient time when the analytical solution and the numerical solution differ, but that the long-term behavior coincides. This provides a test of the numerical solution method. We apply the method also to the square-pulse signal by replacing `Vb = np.sin(omegam*t[i])` with `Vb = np.sign(np.sin(omegam*t[i]))`. The resulting plot is also shown in Fig. 16.7. You can change the parameters to see what happens to the resulting signal.

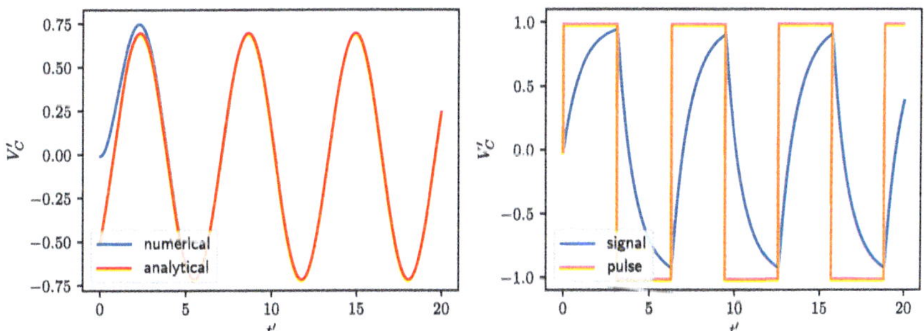

Fig. 16.7 Plots of the comparison of the analytical and numerical solution and comparison of the driving voltage pulse and the resulting voltage signal

16.3 Circuits in the Frequency Domain

We have demonstrated how we can find the behavior of a circuit by applying Kirchhoff's voltage law in combination with the description of individual components and that this leads to a set of (differential) equations for the current and voltages in the circuit. In the last example, we saw that if a circuit is driven by a harmonic voltage source, the voltage and current in the circuit also follows the same periodic behavior after a transient time. Here, we will introduce methods to effectively describe the stationary (non-transient) behavior of circuits driven by a harmonic voltage source. We will study circuits that are driven by periodic, harmonic signals with an angular frequency ω and then look for solutions with the same frequency ω. This is a good approximation for the stationary behavior and will provide us with very efficient methods for describing their behavior and with new concepts that are useful to *understand* the behavior of circuits with alternating currents.

Complex Representation of a Signal

We represent a harmonic signal by two values, the frequency ω and the phase ϕ as illustrated in Fig. 16.8:

$$V(t) = V_0 \cos(\omega t + \phi), \tag{16.40}$$

where ω is related to the period T through $\omega = 2\pi/T$. The phase ϕ determines how the signal is shifted. The sign of the phase can sometimes be confusing: A negative phase, $\phi < 0$ means that the curve $V(t)$ is shifted to the right, while a positive phase, $\phi > 0$, means that the curve $V(t)$ is shifted to the left. However, a negative phase also means that the signal is lagging: It takes a small time ϕ/ω before the signal from $\cos \omega t + \phi$ has the same value as $\cos \omega t$.

Mathematically, it is more convenient to work with complex numbers to represent the signal because this allows us to work with exponential functions instead of sines and cosines. We will therefore introduce a complex representation of a signal

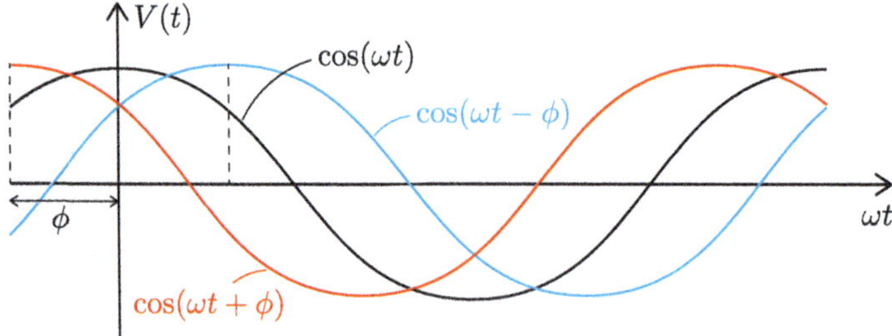

Fig. 16.8 Illustration of an harmonic signal $V(t) = V_0 \cos \omega t$ and two signals with two different phases, $V_0 \cos(\omega t - \phi)$ and $V_0 \cos(\omega t + \phi)$

$$V(t) = \hat{V} e^{i\omega t}, \tag{16.41}$$

where $\hat{V}$ is a complex number and

$$e^{i\omega t} = \cos \omega t + i \sin \omega t. \tag{16.42}$$

We will use this to simplify our calculations, realizing that it is only the real part of the expression that represents the physical part of the signal, so that

$$V(t) = \text{Re}\left\{ \hat{V} e^{i\omega t} \right\}, \tag{16.43}$$

is the physical voltage. The complex amplitude $\hat{V}$ represents both the real amplitude and the phase of the signal. In general, we can write a complex number $\hat{V}$

$$\hat{V} = |\hat{V}| e^{i\phi}, \tag{16.44}$$

where ϕ is the phase. The physical voltage $V(t)$ is the real part:

$$V(t) = \text{Re}\left\{ |\hat{V}| e^{i\phi} e^{i\omega t} \right\} = |\hat{V}| \cos(\omega t + \phi). \tag{16.45}$$

We call $\hat{V}$ the complex amplitude or the *phasor* of the signal. Figure 16.9 illustrates a signal with a phase $\phi = -\pi/4$ as a function of time and how the phasor is illustrated in the complex plane.

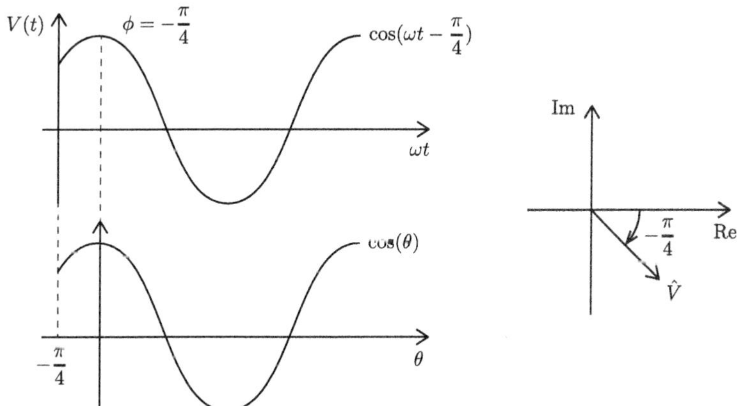

Fig. 16.9 Illustration of a signal with a phase $\phi = -\pi/4$

Test your understanding
Repeating complex numbers: (a) What is $|2+i|$? (b) Draw $2+i$ in the complex plane (c) Write $2+i$ as $A^{i\phi}$[4]

Test your understanding
The figure shows the voltage and current in a circuit. (a) What is the phase for the two signals? (b) Write down the signals on complex form.[5]

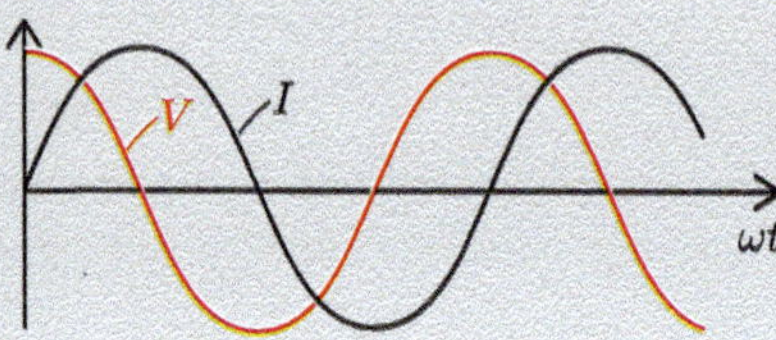

Circuits: Complex Representation of a Resistor

How can we use this representation to describe a circuit with a signal generator, the voltage source V_b, and a resistor R? We represent the voltage and the current using the complex representation:

$$V(t) = \hat{V} e^{i\omega t} \quad , \quad I(t) = \hat{I} e^{i\omega t}. \tag{16.46}$$

Kirchhoff's voltage law describes relations between the real components of the voltages and currents:

$$\text{Re}\{V(t)\} = R\,\text{Re}\{I(t)\} = \text{Re}\{R\,I(t)\}, \tag{16.47}$$

and

$$\text{Re}\left\{\hat{V} e^{i\omega t}\right\} = \text{Re}\left\{R\,\hat{I} e^{i\omega t}\right\}. \tag{16.48}$$

This means that we have the relation

$$\hat{V} = R\hat{I}, \tag{16.49}$$

for a resistor. We can therefore use Ohm's law to relate the complex amplitudes of the voltages and currents. To understand the impact of this relation, let us address how we can represent a capacitor.

[4] (a) $\sqrt{5}$; (c) $\sqrt{5}e^{i\phi}$, $\phi = \arctan(1/2)$.
[5] (a) $\phi_V = 0$, $\phi_I = -\pi/2$; (b) $V = V_0 e^{i\omega t}$, $I = I_0 e^{i(\omega t - \pi/2)}$.

Circuits: Complex Representation of a Capacitor

In a circuit consisting of a voltage source V_b and a single capacitor C, we know that the voltage drop over the capacitor is related to the charge Q on the capacitor, $C = Q/V$, so that $Q = CV$ and the current $I = \mathrm{d}Q/\mathrm{d}t$ is

$$I = \frac{\mathrm{d}Q}{\mathrm{d}t} = C\frac{\mathrm{d}V}{\mathrm{d}t}. \tag{16.50}$$

The current I is the real part of the complex current $\hat{I}e^{i\omega t}$ and the voltage is the real part of the complex voltage, $\hat{V}e^{i\omega t}$:

$$\mathrm{Re}\left\{\hat{I}e^{i\omega t}\right\} = C\frac{\mathrm{d}}{\mathrm{d}t}\mathrm{Re}\left\{\hat{V}e^{i\omega t}\right\} = \mathrm{Re}\left\{C\hat{V}i\omega e^{i\omega t}\right\}. \tag{16.51}$$

This equation implies that the prefactors also must be equal:

$$\hat{I} = Ci\omega\hat{V}, \tag{16.52}$$

and

$$\hat{V} = \frac{1}{i\omega C}\hat{I} = \hat{Z}\hat{I}. \tag{16.53}$$

We call the term $\hat{Z}$ the **impedance**. The impedance acts like a complex resistance.

Interpretation of the complex impedance. The resistor R had a real impedance $\hat{Z} = R$, while the capacitor has a complex impedance $\hat{Z} = 1/(i\omega C)$. How does a complex impedance affect the current or voltage in a circuit? Let us assume that the current I is real, that is, that the phase of the current in zero and $\hat{I} = |\hat{I}|$:

$$I(t) = \mathrm{Re}\left\{\hat{I}e^{i\omega t}\right\} = \hat{I}\cos\omega t. \tag{16.54}$$

For a circuit with an impedance $\hat{Z}$, the voltage is $\hat{V} = \hat{Z}\hat{I}$:

$$V(t) = \mathrm{Re}\left\{\hat{Z}\hat{I}e^{i\omega t}\right\} = |\hat{Z}|\hat{I}\,\mathrm{Re}\left\{e^{i(\omega t+\phi)}\right\} = |\hat{Z}|\hat{I}\cos(\omega t+\phi). \tag{16.55}$$

This allows us to interpret the role of a complex impedance $\hat{Z}$:

- The norm $|\hat{Z}|$ describes the ratio between the voltage and current amplitudes.
- The phase ϕ describes the phase shift of the voltage relative to the current: It describes by how much the voltage is before (in time) the current.

For the capacitor the complex impedance is:

$$\hat{Z} = \frac{1}{i\omega C} = \frac{-i}{\omega C} = \frac{1}{\omega C}e^{-i\frac{\pi}{2}}. \tag{16.56}$$

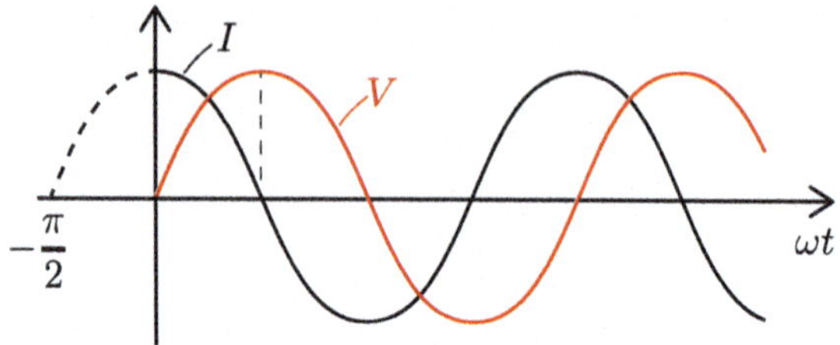

Fig. 16.10 Illustration of the current, $I(t)$, and voltage, $V(t)$, for a circuit with a capacitor C and an impedance $\hat{Z} = 1/(i\omega C)$

which corresponds to a phase shift of $-\pi/2$. The corresponding $I(t)$ and $V(t)$ curves are illustrated in Fig. 16.10.

Circuits: Complex Representation of an Inductor

For a circuit with a voltage source V_b and an inductor L, Kirchhoff's law for the circuit gives

$$V - L\frac{\mathrm{d}I}{\mathrm{d}t} = 0 \ \Rightarrow \ V = L\frac{\mathrm{d}I}{\mathrm{d}t}. \tag{16.57}$$

We rewrite this on complex form:

$$\mathrm{Re}\left\{\hat{V}e^{i\omega t}\right\} = L\frac{\mathrm{d}}{\mathrm{d}t}\mathrm{Re}\left\{\hat{I}e^{i\omega t}\right\} = \mathrm{Re}\left\{i\omega L\hat{I}e^{i\omega t}\right\}. \tag{16.58}$$

This means that we again have the relation $\hat{V} = \hat{Z}\hat{I}$ where $\hat{Z} = i\omega L$ for the inductor.

Circuits with Impedances

We have found that we can describe the relation between the complex amplitudes of the current and voltage with an impedance for resistors, capacitors, and inductors.

Impedances for the main components

We describe a harmonic signal in complex notation as

$$I(t) = \mathrm{Re}\left\{\hat{I}e^{i\omega t}\right\} \ , \ \ V(t) = \mathrm{Re}\left\{\hat{V}e^{i\omega t}\right\}, \tag{16.59}$$

where $\hat{I}$ and $\hat{V}$ are called the complex amplitudes or the **phasors** of the signals. The phasor includes both an amplitude and phase information:

$$\hat{I} = |\hat{I}|e^{i\phi_I} \quad , \quad \hat{V} = |\hat{V}|e^{i\phi_V} \tag{16.60}$$

For the main components resistor, capacitor and inductor, we find a generalized Ohm's law relating the complex amplitudes to the **impedance** $\hat{Z}$ of the component:

$$\hat{V} = \hat{Z}\,\hat{I}, \tag{16.61}$$

where the impedances are

- $\hat{Z} = R$ for a **resistor**
- $\hat{Z} = \frac{1}{i\omega C}$ for a **capacitor**
- $\hat{Z} = i\omega L$ for an **inductor**

We can prove[6] that we can use the same rules for combining impedances as we already know for resistors:

- **Two impedances** $\hat{Z}_1$ and $\hat{Z}_2$ **in series** can be replaced by a single component with impedance $\hat{Z} = \hat{Z}_1 + \hat{Z}_2$.
- **Two impedances** $\hat{Z}_1$ and $\hat{Z}_2$ **in parallel** can be replaced by a single component with impedance $1/\hat{Z} = 1/\hat{Z}_1 + 1/\hat{Z}_2$.

However, we must remember that impedances are *complex numbers* and we must use correct algebra for complex numbers when we perform these calculations.

Example: RC-Circuit in the Frequency Domain

Figure 16.11 *illustrates two possible circuits consisting of an input signal $V_S(t)$ (a signal generator), a resistor R and a capacitor C. Determine how $\hat{V}$ measured as shown in the figure is related to $\hat{V}_S$ in the two cases.*

Solution. In both cases the two components act in series so that

$$\hat{V}_S = \hat{I}\left(\hat{Z}_C + \hat{Z}_R\right), \tag{16.62}$$

and

$$\hat{I} = \frac{\hat{V}_S}{\hat{Z}_R + \hat{Z}_C}. \tag{16.63}$$

Low-pass filter. For the circuit in Fig. 16.11a, we measure the voltage across the capacitor, giving

[6] We leave this proof for the exercises.

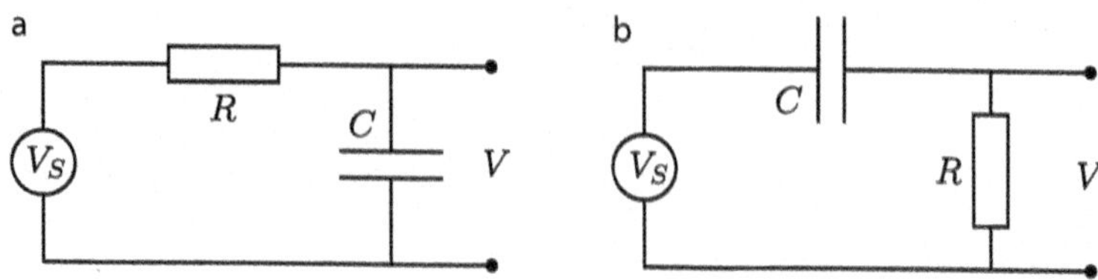

Fig. 16.11 Illustration of two circuits with a resistor R and a capacitor C corresponding to **a** a low-pass filter and **b** a high-pass filter

$$\hat{V} = \hat{Z}_C\,\hat{I} = \frac{\hat{Z}}{R+\hat{Z}}\hat{V}_S = \frac{\frac{1}{i\omega C}}{R+\frac{1}{i\omega C}}\hat{V}_S = \frac{1}{1+i\omega RC}\hat{V}_S = \frac{1}{1+i\omega\tau}\hat{V}_S. \quad (16.64)$$

where we have inserted the characteristic time $\tau = RC$. To find the amplitude and phase, we insert $\hat{V} = |\hat{V}|e^{i\phi}$. We get that

$$|\hat{V}| = \frac{1}{\sqrt{1+(\omega\tau)^2}}|\hat{V}_S| \quad , \quad \phi = -\arctan(\omega\tau), \quad (16.65)$$

We see that in the limit when $(\omega\tau) \gg 1$ we get that $\sqrt{1+(\omega\tau)^2} \simeq \omega\tau$, and (16.65) becomes:

$$|\hat{V}| \simeq \frac{1}{\omega\tau}|\hat{V}_S|, \quad (16.66)$$

which implies that $|\hat{V}| \ll |\hat{V}_S|$. Whereas in the limit when $(\omega\tau) \ll 1$ we get that $\sqrt{1+(\omega\tau)^2} \simeq 1$ and $|\hat{V}| \simeq |\hat{V}_S|$. This means that frequencies much larger than $1/\tau$ are not let through, but frequencies smaller than this are let through without any significant damping. We call such a circuit a *low-pass filter*.

High-pass filter. For the circuit in Fig. 16.11b, we measure voltage across the resistor, giving

$$\hat{V} = \hat{Z}_R\,\hat{I} = \frac{R}{R+\hat{Z}}\hat{V}_S = \frac{R}{R+\frac{1}{i\omega C}}\hat{V}_S = \frac{i\omega RC}{1+i\omega RC}\hat{V}_S = \frac{i\omega\tau}{1+i\omega\tau}\hat{V}_S. \quad (16.67)$$

Again, we insert $\hat{V} = |\hat{V}|e^{i\phi}$ and find that

$$|\hat{V}| = \frac{\omega\tau}{\sqrt{1+(\omega\tau)^2}}|\hat{V}_S| \quad , \quad \phi = \frac{\pi}{2} - \arctan(\omega\tau), \quad (16.68)$$

In this case, when $(\omega\tau) \gg 1$ we get that $|\hat{V}| \simeq |\hat{V}_S|$ whereas when $(\omega\tau) \ll 1$ we get that $|\hat{V}| \simeq \omega\tau|\hat{V}_S|$. This means that frequencies much larger than $1/\tau$ are let through, but frequencies smaller than this are damped out. We call such a circuit a *high-pass filter*.

Power Dissipation in Complex Circuits

The power dissipated in a resistor is $P = RI^2 = V(t)I(t)$. For a system described using complex voltages and currents this is

$$P(t) = V(t)\,I(t) = \mathrm{Re}\left\{\hat{V}e^{i\omega t}\right\}\mathrm{Re}\left\{\hat{I}e^{i\omega t}\right\}. \tag{16.69}$$

Notice that this is *not* the same as $\mathrm{Re}\{\hat{V}e^{i\omega t}\,\hat{I}e^{i\omega t}\}$. Here, we are interested in finding the average power dissipated over a period of the oscillating signal. If the current has a phase ϕ relative to the voltage, then $V(t) = |\hat{V}|\cos\omega t$ and $I(t) = |\hat{I}|\cos(\omega t - \phi)$ so that

$$P(t) = |\hat{V}|\cos\omega t\,|\hat{I}|\cos(\omega t - \phi). \tag{16.70}$$

We rewrite this using $\cos(a+b) = \cos a\,\cos b - \sin a\,\sin b$:

$$P(t) = |\hat{V}|\,|\hat{I}|\left(\cos^2\omega t\,\cos\phi + \cos\omega t\,\sin\omega t\,\sin\phi\right). \tag{16.71}$$

If we average this over one period, $2\pi/\omega$, then the average of $\cos^2\omega t$ is $1/2$ and the average of $\cos\omega t\,\sin\omega t$ is 0. The average power dissipated is therefore

$$\langle P\rangle = \frac{1}{2}|\hat{V}|\,|\hat{I}|\cos\phi. \tag{16.72}$$

This means that for a resistor $\phi = 0$ and the dissipated power is $|\hat{V}|\,|\hat{I}|/2$. For a capacitor or a inductor $\phi = \pi/2$ or $-\pi/2$ and $\cos\phi = 0$. This means that there is on average no power dissipated in capacitors or inductors, they only store energy during parts of the cycle.

Root-mean-square values. It is engineering practice to use the root-mean-square averaged values and not the directly averaged values in circuits with alternating currents. The root-mean-square value is calculated over a complete cycle. The root-mean-square value of the voltage is therefore

$$V_{\mathrm{rms}} = \left[\frac{\omega}{2\pi}\int_0^{2\pi/\omega}|\hat{V}|^2\cos^2\omega t\,\mathrm{d}t\right]^{1/2} = \frac{|\hat{V}|}{\sqrt{2}}. \tag{16.73}$$

and similarly for the current, $I_{\mathrm{rms}} = |\hat{I}|/\sqrt{2}$. The power-dissipation is therefore

$$\langle P\rangle = V_{\mathrm{rms}}\,I_{\mathrm{rms}}\cos\phi. \tag{16.74}$$

Notice that the peak voltage for US household circuits is approximately 170 V so that $V_{\mathrm{rms}} = (170\,\mathrm{V})/\sqrt{2} = 120\,\mathrm{V}$. Similarly for European household circuits the voltages provided are the root-mean-square voltages and not the amplitudes.

16.4 Assumptions Underlying Circuit Models

Previously, we discussed how our representation of circuits was an approximation to the physical system. We assume that we can describe the system as a set of separate components that do not interact. This is one of several underlying assumptions that we have implicitly made when we model a real system as a circuit, but also in our treatment of circuits we have built from circuit components. Modeling a system as a circuit is an *approximation* to the full electromagnetic theory described by Maxwell's equation—an approximation we use because it allows us to understand, predict and explain behavior, and to develop a simplified theory that may guide our intuition. What are the underlying assumptions, and how can we compensate or modify the model to accommodate these deviations?

- We have assumed that conductors (wires) are ideal conductors. In practice, they are not. We can compensate for this by modeling a wire as an ideal conductor and a resistor in series. However, in many cases, other resistors in the circuit are much larger than the resistance from the wires, and we can therefore assume that the resistance from the wire is negligible.
- We have assumed that wires and junctions do not accumulate any charges. This conservation of charges flowing into and out of any junction was the basis for Kirchhoff's law of currents. We can introduce capacitors in the circuit to model the effect of local charge accumulation.
- We have assumed that various components do not interact through their electric or magnetic fields. That is, we have assumed that components are far away from each other compared to how quickly the fields decay or that the mutual inductances are small. This is usually a good approximation. We can compensate for this by introducing additional components in the circuits.
- In addition, we have implicitely assumed that changes in the system are sufficiently slow. This means that we have assumed that frequencies are sufficiently small so that the wavelength $\lambda = c/f$, where c is the speed of light, is long compared with the circuit. You will learn more about this when you learn about waves and oscillations.

Summary

We described circuits using a set of components: a battery with emf V_b, a resistor with voltage drop RI, a capacitor with voltage drop Q/C, and an inductor with voltage drop $L\,\mathrm{d}I/\,\mathrm{d}t$. Kirchhoff's voltage law for a circuit with all components in series is $V_b - RI - Q/C - L\,\mathrm{d}I/\,\mathrm{d}t = 0$.

The inductance of **inductors in series** is the sum of the inductances: $L = \sum_i L_i$. Whereas the inductance of **inductors in parallel** is the inverse sum of the inductances: $1/L = \sum_i 1/L_i$.

The harmonic signal in a circuit with an alternating voltage source is $V(t) = V_0 \cos(\omega t + \phi)$, where V_0 is the amplitude of the voltage, ω is the frequency, and ϕ is the phase. For simplicity, we describe the signal using complex numbers, $V = \hat{V} e^{i\omega t}$ where $\hat{V} = |\hat{V}| e^{i\phi}$ is called a phasor and includes both the magnitude and the phase. The physical voltage is the real value of the complex voltage: $V(t) = \text{Re}\{V\}$.

Components can be described with their complex **impedance**, $\hat{Z}$. The impedance of a resistor is $\hat{Z}_R = R$. The impedance of a capacitor is $\hat{Z}_C = 1/(i\omega C)$. The impedance for an inductor is $\hat{Z}_L = i\omega L$. We can use Ohm's law for complex impedances to relate the voltage and current through a component: $\hat{V} = \hat{Z}\,\hat{I}$.

The power dissipated in a component averaged over one period of oscillation is $P = (1/2)|\hat{V}|\,|\hat{I}|\,\cos\phi$. For a resistor $\phi = 0$, $\cos\phi = 1$, and the dissipated power is $(1/2)|\hat{V}|\,|\hat{I}| = (1/2)R\,|\hat{I}|^2$. A capacitor and an inductor do not dissipate any power.

Exercises

Discussion Exercises

16.1 Winding resistor. You are making a resistor by winding a wire around a cylinder. To make the inductance as small as possible, you have been suggested to wind half of the wire one way and the other half the other way. Will this give the effect you are after? Explain why or why not.

16.2 A coil plus a coil equald four coils. Your friend Q comes to you with a question: *If you place two coils with inductances L after each other in a circuit, they will act as one inductor with inductance* $2L$. *But if you place one coil with inductance L on top of a coil with inductance L, the total inductance is* $4L$. *How can this be possible?* What would you tell Q? Has he made a mistake or is something else wrong with the argument or the assumptions?

16.3 Energy transfer. The current in a wire with alternating current changes direction a given number of times per second and the average current is zero. Explain how it is possible to transport energy (power) along such a system.

16.4 Cancelling components. A circuit consists of an alternating current source, a light bulb, a capacitor and a coil, all connected in series. Is it possible that light in the bulb is the same if you remove both the capacitor and the coil? Describe the behavior of the system using the impedence of the capacitor ($\hat{Z} = 1/(i\omega C)$) and the coil ($\hat{Z} = i\omega L$).

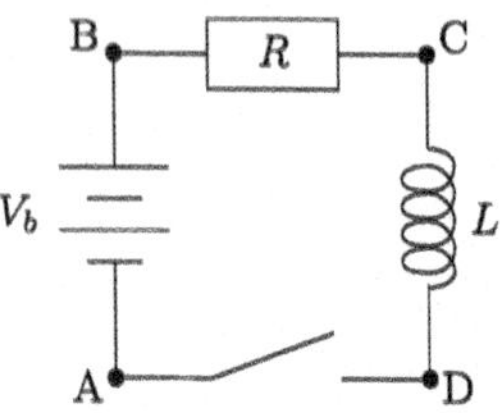

Fig. 16.12 Circuit with coil and resistor

Tutorials

16.5 Circuit with coil and resistor. Figure 16.12 shows a circuit consisting of a battery, a resistor and an inductor (coil). At the time $t = 0$, we close the switch and current may start running in the circuit.

(a) What is the current at $t = 0$?

(b) What is the Kirchhoff's voltage law for this circuit?

(c) What is the current after a very long time?

(d) After a very long time, the battery suddenly stops working so that it no longer provides a voltage V_b, but a current may still pass through it. What happens in the circuit?

(e) Let us now reset the time and assume that the battery stops working at $t = 0$. What is the current as a function of time for $t > 0$?

16.6 A simple oscillator circuit. A circuit consists of an inductor L and a capacitor C.

(a) What is Kirchhoff's voltage law for this system?

(b) What is the equation for the current in this system?

(c) What initial conditions do we need to find the behavior of this circuit?

(d) Assume that we know that at $t = 0$, $Q = Q(0) = Q_0$ and that we turn on a switch the connects it to the inductor at $t = 0$. What are the initial conditions?

(e) Does the sign on $I'(0)$ seem reasonable? (Hint: Check the change in charge for a small time interval Δt.)

(f) Find the current as a function of time for this system.

16.7 Model for a switch. In this exercise we will develop a model for what happens when we turn on a switch in a circuit that consists of a battery, a resistor and a switch. The switch is shaped like a plate capacitor with two circular plates of radius a and distance d. When the plates are in contact, that is when $d = 0$, the switch is closed and a current can run in the circuit.

(a) Make a sketch of the circuit.

(b) Assume that the switch is open, so that the distance between the plates is $d > 0$ and that the system has been like this for a long time. What is then the voltage across the different components?

(c) Assume that we suddenly (in zero time) halve the distance d. What is now the voltage across the different components immediately after this change?

(d) We keep the plates at this distance. Describe what happens in the system.

(e) If we move the plates very slowly toward each other, what happens to the electric field between the two plates?
(f) What will you eventually observe in the region between the two plates?

16.8 Periodically driven circuits. A circuit consists of a resistor R and a capacitor C and is driven by a period voltage source $U(t)$.
(a) What is the Kirchhoff's voltage law?
(b) We want to describe the system with the voltage V_C across the capacitor. How can you express the current I through the resistor using V_C?
(c) What is now Kirchhoff's voltage law expressed in terms of U and V_C?
(d) We want to solve this equation numerically. You can assume that $RC = 0.1$ and that $U(t) = U_0 \sin \omega t$ where $\omega = 1$ and $U_0 = 1$. Choose different values for $V_C(0)$ and visualize the result. Does it matter what $V_C(0)$ is?

16.9 LR circuit. Consider a circuit consisting of a battery with emf V_0, a resistance R and an inductance L connected in series in a loop. A current I is flowing due to the emf.
(a) Draw the circuit using standard symbols for the components. Illustrate the positive direction for I.
(b) What is the voltage drop V_R across the resistor? (The voltage drop is a positive number in the direction of the current).
(c) For the inductance L the flux is $\Phi = LI$. What is the emf across the inductance L?
(d) Write down the sum of emf and potential drops around the circuit.
(e) What is the potential drop V_L over the inductance L? Ensure that you get the same sum of emf and potential drops around the circuit when you represent the inductance as a component with a voltage drop as when you represented it by an emf.
(f) Find a differential equation for the current, I.
(g) First, find the solution to this equation when $V_0 = 0$ and $I(0) = I_0$.
(h) Second, find the solution to this equation when $V_0 > 0$ and $I(0) = 0$.

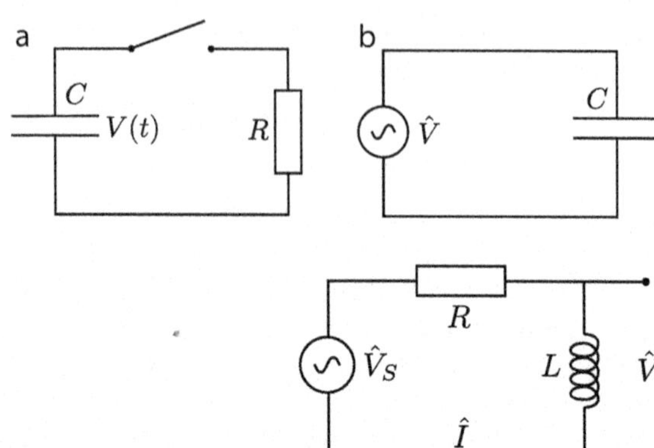

Fig. 16.13 Illustration of a circuit with a capacitor

Fig. 16.14 Illustration of a circuit with an inductance

Homework

16.10 Complex capacitor.
(a) A capacitor C is first charged to the voltage V_0. After $t = 0$ it is discharged through the resistor R as illustrated in Fig. 16.13a. Sketch the voltage $V(t)$ across the capacitor as a function of time. How long time, approximately, does it take to charge or discharge a capacitor?
(b) What is the impedance of the capacitor?
(c) We now connect the capacitor directly to an AC source with complex amplitude $\hat{V}$ as illustrated in Fig. 16.13b. What is the complex current $\hat{I}$ through the capacitor?
(d) Give an interpretation of the complex circuit equation (the connection between current and voltage) you found in the previous exercise. You should interpret both the absolute value and the imaginary unit i in the equation. (Hint: i is $e^{i\pi/2}$ in polar coordinates.)

16.11 Complex RL-circuit. An AC voltage source with angular frequency ω is connected in series with a resistance R and an inductance L as shown in Fig. 16.14.
(a) What is the physical, time-varying voltage $V_s(t)$ that corresponds to the complex amplitude $\hat{V}_s$?
(b) Find the complex voltage $\hat{V}$ expressed in terms of the source $\hat{V}_s$, the time constant $\tau = L/R$ and the angular frequency ω. Also find an expression for $|\hat{V}|/|\hat{V}_s|$, plot it and provide an interpretation.

Chapter 17
Maxwell's Equations

17.1 Displacement Current

There is Something Wrong with the Equations

We have introduced a set of equations to describe the coupling between the electric and magnetic fields. On differential form, the equations are:

$$\nabla \times \mathbf{E} = -\frac{\partial \mathbf{B}}{\partial t}, \tag{17.1}$$

$$\nabla \times \mathbf{H} = \mathbf{J}, \tag{17.2}$$

$$\nabla \cdot \mathbf{D} = \rho, \tag{17.3}$$

$$\nabla \cdot \mathbf{B} = 0. \tag{17.4}$$

In addition, we have conservation of charge, which is formulated as:

$$\nabla \cdot \mathbf{J} + \frac{\partial \rho}{\partial t} = 0. \tag{17.5}$$

Unfortunately, these equations are not consistent with each other. We can see this by realizing that the divergence of a curl is always zero. This means that

$$\nabla \cdot (\nabla \times \mathbf{H}) = 0. \tag{17.6}$$

Which would imply that $\nabla \cdot \mathbf{J} = 0$. But this is not always true, it is only true for static systems. In general we know from the conservation of charge that

© The Author(s), under exclusive license to Springer Nature Switzerland AG 2026
A. Malthe-Sørenssen, *Elementary Electromagnetism Using Python*, Undergraduate Texts in Physics, https://doi.org/10.1007/978-3-032-19876-1_17

$$\nabla \cdot \mathbf{J} = -\frac{\partial \rho}{\partial t}. \tag{17.7}$$

This implies that there must be something wrong with or lacking from the equation $\nabla \times \mathbf{H} = \mathbf{J}$. In what situations would this problem appear, and how can we fix this?

Example: Flow Into a Capacitor

Let us examine a problem where charge conservation is essential by addressing how charge flows into a capacitor. Figure 17.1 illustrates a simple circuit with a battery, a resistor and a capacitor. When the battery is connected, there will be a current flowing into the capacitor. Over time, the current will decrease as charges are building up across the capacitor, increasing the potential drop $V = Q/C$ across the capacitor until the potential drop equals the emf of the battery.

If we apply Ampere's law to a loop C_1 around the wire leading into the capacitor, as illustrated in the figure, we know that

$$\oint_{C_1} \mathbf{H} \cdot \mathrm{d}\mathbf{l} = I_S \tag{17.8}$$

where the current I is the current through a surface S_1, which has the loop C_1 as its boundary. This equation should hold for any such surface. We have illustrated two surfaces in the figure. For surface S_1, the current I_{S_1} is the current $I(t)$ in the wire. However, if we choose a surface S_2 which passes through the gap between the capacitor plates, there is no current through this surface, and $I_{S_2} = 0$. Hmmm. This means that also Ampere's law is inconsistent. How can we fix this?

We know that the current $I(t)$ in the wire is related to the charge $Q(t)$ on the capacitor, which again is related to the electric field $\mathbf{D}$ in the space between the capacitors:

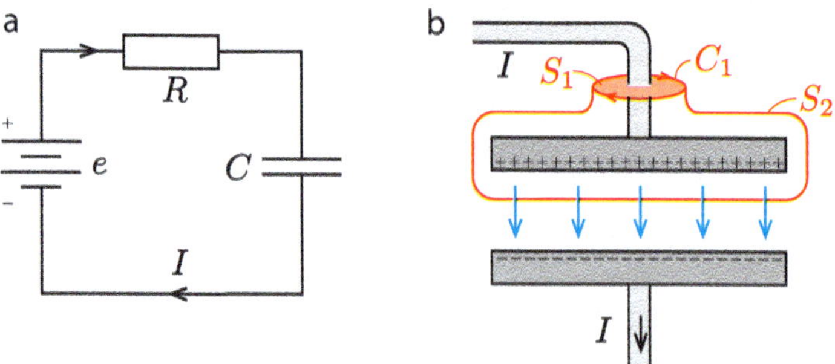

Fig. 17.1 **a** Illustration of a circuit with a capacitor, **b** Illustration of the capacitor

$$I(t) = \frac{\mathrm{d}Q}{\mathrm{d}t}, \tag{17.9}$$

We can relate $Q(t)$ to the electric field through Gauss' law:

$$\oint_S \mathbf{D}(t) \cdot \mathrm{d}\mathbf{S} = Q(t). \tag{17.10}$$

The current I is therefore:

$$I(t) = \frac{\mathrm{d}Q}{\mathrm{d}t} = \frac{\mathrm{d}}{\mathrm{d}t} \oint_S \mathbf{D}(t) \cdot \mathrm{d}\mathbf{S}. \tag{17.11}$$

If the surface S does not change with time, the time variation only comes from the time dependence of the electric field $\mathbf{D}(t)$, and we can move the time derivative into the integral:

$$I(t) = \oint_S \frac{\partial \mathbf{D}}{\partial t} \cdot \mathrm{d}\mathbf{S}. \tag{17.12}$$

It is useful to consider the right-hand side of this expression to be a current that flows across the gap between the capacitor plates, even though there is no actual transport of charges across the gap. This allows us to consider $I(t)$ a current that flows continuously around the whole circuit, including through the capacitor gap.

Displacement Current

We introduce the term displacement current based on this definition:

Displacement current

The displacement current I_d through a surface is defined as

$$I_d = \int_S \frac{\partial \mathbf{D}}{\partial t} \cdot \mathrm{d}\mathbf{S}\,. \tag{17.13}$$

Notice that the displacement current comes in addition to the conductor current. For example, if a capacitor is filled with a dielectric with a very low conductivity, there would be both an ordinary (conduction) current I and a displacement current I_d.

This resolves the apparent inconsistency: On the surface S_2 the conduction current is zero, but the displacement current is

$$I_d = \int_{S_2} \frac{\partial \mathbf{D}}{\partial t} \cdot \mathrm{d}\mathbf{S} = \frac{\mathrm{d}}{\mathrm{d}t} \int_{S_2} \mathbf{D} \cdot \mathrm{d}\mathbf{S} = \frac{\mathrm{d}Q}{\mathrm{d}t} = I\,, \tag{17.14}$$

so now both surfaces gives the same answer: S_1 gives $I + 0$, and S_2 gives $0 + I_d$, with $I_d = I$.

Fixing Maxwell's Equations

We can now fix Ampere's law by including both types of currents.

$$\oint_C \mathbf{H} \cdot \mathrm{d}\mathbf{l} = I + I_d \tag{17.15}$$

We write the conduction current $I(t)$ as the surface integral of the current density $\mathbf{J}$:

$$I(t) = \int_S \mathbf{J} \cdot \mathrm{d}\mathbf{S}, \tag{17.16}$$

where the surface S is a (any) surface with the loop C as its boundary. Similarly, we write the displacement current I_d as:

$$I_d = \int_S \frac{\partial \mathbf{D}}{\partial t} \cdot \mathrm{d}\mathbf{S}. \tag{17.17}$$

Ampere's law therefore becomes:

$$\oint_C \mathbf{H} \cdot \mathrm{d}\mathbf{l} = \int_S \left(\mathbf{J} + \frac{\partial \mathbf{D}}{\partial t} \right) \cdot \mathrm{d}\mathbf{S}. \tag{17.18}$$

This fixes the problem in the example above. But does it also fix the problem we discovered in Maxwell's equations on differential form? We use Stokes' theorem to rewrite Ampere's law to:

$$\int_S \nabla \times \mathbf{H} \cdot \mathrm{d}\mathbf{S} = \int_S \left(\mathbf{J} + \frac{\partial \mathbf{D}}{\partial t} \right) \cdot \mathrm{d}\mathbf{S}. \tag{17.19}$$

This is valid for any surface S, and therefore, the equation also holds for the arguments of the integral:

$$\nabla \times \mathbf{H} = \mathbf{J} + \frac{\partial \mathbf{D}}{\partial t}. \tag{17.20}$$

Let us check if this now fixes the problem. The divergence of a curl is always zero, so the divergence of the left side is zero:

$$\nabla \cdot (\nabla \times \mathbf{H}) = 0. \tag{17.21}$$

What about the right-hand side? We take the divergence and find:

$$\nabla \cdot \left(\mathbf{J} + \frac{\partial \mathbf{D}}{\partial t} \right) = \nabla \cdot \mathbf{J} + \frac{\partial}{\partial t} \nabla \cdot \mathbf{D}. \tag{17.22}$$

We can now use Gauss' law on differential form, $\nabla \cdot \mathbf{D} = \rho$, giving:

$$\nabla \cdot \mathbf{J} + \frac{\partial}{\partial t} \nabla \cdot \mathbf{D} = \nabla \cdot \mathbf{J} + \frac{\partial \rho}{\partial t}. \tag{17.23}$$

We recognize this as the conservation of charge from (17.5) and that this is always equal to zero. Hence, we have found that the new formulation at least provides us with a consistent set of the equations: Maxwell's equations, and a new formulation of Ampere's law, which is called Ampere-Maxwell's law:

Ampere-Maxwell's law

Ampere-Maxwell's law on *integral form* states that for any closed curve C which encloses a surface S:

$$\oint_C \mathbf{H} \cdot d\mathbf{l} = \int_S \left(\mathbf{J} + \frac{\partial \mathbf{D}}{\partial t} \right) \cdot d\mathbf{S}. \tag{17.24}$$

Ampere-Maxwell's law on *differential form* states that:

$$\nabla \times \mathbf{H} = \mathbf{J} + \frac{\partial \mathbf{D}}{\partial t}. \tag{17.25}$$

We call the term

$$\mathbf{J}_D = \frac{\partial \mathbf{D}}{\partial t}, \tag{17.26}$$

the *displacement current density*.

17.2 Maxwell's Equations

We now have a complete set of equations, which are called Maxwell's equations, that provide a complete description of all electromagnetic phenomena:

Maxwell's equations
Maxwell's equations are:

$$\nabla \times \mathbf{E} = -\frac{\partial \mathbf{B}}{\partial t}, \tag{17.27}$$

$$\nabla \times \mathbf{H} = \mathbf{J} + \frac{\partial \mathbf{D}}{\partial t}, \tag{17.28}$$

$$\nabla \cdot \mathbf{D} = \rho, \tag{17.29}$$

$$\nabla \cdot \mathbf{B} = 0. \tag{17.30}$$

These equations, together with the continuity equation and the definitions of **D** and **H** provide a complete description of the physics of electromagnetism:

Charge continuity

Charge continuity is formulated as

$$\nabla \cdot \mathbf{J} + \frac{\partial \rho}{\partial t} = 0. \tag{17.31}$$

Dielectric and magnetic materials

For **dielectric materials** we have defined **D** as

$$\mathbf{D} = \epsilon_0 \mathbf{E} + \mathbf{P}. \tag{17.32}$$

For *linear dielectric materials* we have that $\mathbf{P} = \chi_e \epsilon_0 \mathbf{E}$ and

$$\mathbf{D} = \epsilon \mathbf{E}, \tag{17.33}$$

where $\epsilon = \epsilon_r \epsilon_0$.

For **magnetic materials** we have defined **H** as

$$\mathbf{H} = \frac{1}{\mu_0} \mathbf{B} - \mathbf{M}. \tag{17.34}$$

For a *linear magnetic material* we have that $\mathbf{M} = \chi_m \mathbf{H}$ and

$$\mathbf{B} = \mu \mathbf{H} = \mu_r \mu_0 \mathbf{H}. \tag{17.35}$$

Boundary conditions

The boundary conditions at an interface between two materials (1,2) for the electric and magnetic fields are:

$$E_{1,t} = E_{2,t} \quad , \quad \mathbf{D}_1 \cdot \hat{\mathbf{n}} - \mathbf{D}_2 \cdot \hat{\mathbf{n}} = \rho_s . \tag{17.36}$$

$$\mathbf{B}_1 \cdot \hat{\mathbf{n}} = \mathbf{B}_2 \cdot \hat{\mathbf{n}} \quad , \quad \hat{\mathbf{n}} \times \mathbf{H}_1 - \hat{\mathbf{n}} \times \mathbf{H}_2 = \mathbf{J}_s . \tag{17.37}$$

Appendix
Answers

1.2 **Vector sums**
(b) 0 **(c)** $\sqrt{2}\,\mathrm{m}$ **(e)** $\sqrt{3}/2\,\mathrm{m}^2$ **(f)** $\sqrt{2+\sqrt{3}/2}\,\mathrm{m}$ **(g)** $\mathbf{c}\cdot\hat{\mathbf{x}} = 1\,\mathrm{m}$, $\mathbf{c}\cdot\hat{\mathbf{y}} = 0$, $\mathbf{d}\cdot\hat{\mathbf{x}} = \sqrt{3}/2\,\mathrm{m}$, $\mathbf{d}\cdot\hat{\mathbf{y}} = (1/2)\,\mathrm{m}$. **(h)** $\mathbf{c} = (1, 0)\,\mathrm{m}$, $\mathbf{d} = (\sqrt{3}/2, 1/2)\,\mathrm{m}$.

1.3 **Decomposition**
(a) 8 **(c)** $(8/5)(2, 1)$ **(d)** $(\mathbf{b} - \mathbf{b}\cdot(\mathbf{a}/a))\cdot\mathbf{a} = \mathbf{b}\cdot\mathbf{a} - \mathbf{b}\cdot\mathbf{a} = 0$

1.4 **Cross products of a unit vector**
(a) $\hat{\mathbf{x}}\cdot\hat{\mathbf{y}} = (1, 0, 0)\cdot(0, 1, 0) = 0 + 0 + 0 = 0$ and similar for the other. **(b)** $(1, 0, 0) \times (0, 1, 0) = (0\,0 - 0\,1, 0\,0 - 1\,0, 1\,1 - 0\,0) = (0, 0, 1)$ **(c)** $(\hat{\mathbf{y}}, \hat{\mathbf{z}}, \hat{\mathbf{x}})$, $(\hat{\mathbf{z}}, \hat{\mathbf{x}}, \hat{\mathbf{y}})$.

1.5 **Sketching a vector field**
(b)

```
import numpy as np
import matplotlib.pyplot as plt
def efield(r):
    return r/np.linalg.norm(r)**3
x = np.linspace(-2,2,20)
y = np.linspace(-2,2,20)
rx,ry = np.meshgrid(x,y,indexing="ij")
Ex = rx.copy()
Ey = ry.copy()
for i in range(len(rx.flat)):
    r = np.array([rx.flat[i],ry.flat[i]])
    Ex.flat[i],Ey.flat[i] = efield(r)
plt.quiver(rx,ry,Ex,Ey)
```

(c)

```
p = np.array([1,0])
def efield2(r,p):
    return 3*np.dot(r,p)*r/np.linalg.norm(r)**5-p/np.linalg.norm(r)**3
x = np.linspace(-2,2,20)
y = np.linspace(-2,2,20)
```

© The Editor(s) (if applicable) and The Author(s), under exclusive license to Springer Nature Switzerland AG 2026

A. Malthe-Sørenssen, *Elementary Electromagnetism Using Python*, Undergraduate Texts in Physics, https://doi.org/10.1007/978-3-032-19876-1

```
rx,ry = np.meshgrid(x,y,indexing="ij")
Ex2 = rx.copy()
Ey2 = ry.copy()
for i in range(len(rx.flat)):
    r = np.array([rx.flat[i],ry.flat[i]])
    Ex2.flat[i],Ey2.flat[i] = efield2(r,p)
plt.quiver(rx,ry,Ex2,Ey2)
```

1.6 **Fluxes**
(a) $\Phi = 4E_0$. **(b)** $\Phi = 4E_0$ **(c)** $\Phi = 4E_0$ **(d)** $\Phi = 4E_0$ **(e)** $\Phi = 0$ **(f)** $\Phi = 4E_0$

1.7 **Gradients**
(a) $\nabla V = V_0\hat{\mathbf{x}}$ **(b)** $\nabla V = V_0 \cos \hat{\mathbf{x}}\,\hat{\mathbf{x}}$ **(c)** $\nabla V = V_0(1, 1, -1)$ **(d)** $\nabla V = (2x, 2y, 2z)$ **(e)** $\nabla V = \mathbf{r}/|\mathbf{r}|$ **(f)** $\nabla V = -V_0\frac{\mathbf{r}}{r^3}$

1.8 **Divergence and curl**
(a) $\nabla \cdot \mathbf{E} = 0$, $\nabla \times \mathbf{E} = \mathbf{0}$ **(b)** $\nabla \cdot \mathbf{E} = 3$, $\nabla \times \mathbf{E} = \mathbf{0}$ **(c)** $\nabla \cdot \mathbf{E} = 0$, $\nabla \times \mathbf{E} = \mathbf{0}$ **(d)** $\nabla \cdot \mathbf{E} = 0$, $\nabla \times \mathbf{E} = (0, 0, 2)$ **(e)** $\nabla \cdot \mathbf{E} = 0$, $\nabla \times \mathbf{E} = \mathbf{0}$ **(f)** $\nabla \cdot \mathbf{E} = 0$, $\nabla \times \mathbf{E} = \mathbf{0}$

1.9 **The field in a tornado**
(b) The velocity will approach zero in the middle of the tornado and far away. **(c)** $\mathbf{v}(\mathbf{r}) = e^{-r/a}(-y, x)$ **(d)**

```
import numpy as np
import matplotlib.pyplot as plt
x0,x1,y0,y1 = -4,4,20,20
x = np.linspace(x0,x1,Nx)
y = np.linspace(y0,y1,Ny)
rx,ry = np.meshgrid(x,y,indexing="ij")
a = 1.75
vx = np.zeros((Nx,Ny),float)
vy = np.zeros((Nx,Ny),float)
for i in range(Nx):
    for j in range(Ny):
        rnorm = np.sqrt(rx[i,j]**2+ry[i,j]**2)
        vx[i,j] = np.exp(-rnorm/a)*(-ry[i,j])
        vy[i,j] = np.exp(-rnorm/a)*( rx[i,j])
plt.quiver(rx,ry,vx,vy)
plt.axis("equal")
```

1.11 **Along a line or along a curve?**
(a) $5(y_0 + 2)$ **(b)** 7

1.12 **Against the current**
(b) $\Phi = vS$ **(c)** $\Phi = 0$

1.14 **Do not unleash the divergence**
(b) $2C$ **(c)** No **(d)** $\Phi = \frac{C}{2\pi r}$

1.15 **Exploring a scalar field**
(b) $\nabla h = \hat{\mathbf{x}}e^{-y}\cos x - \hat{\mathbf{y}}e^{-y}\sin x$

1.16 **Volume integrals**
(a) $a^2(1-e^{-a})$ **(b)** $(4\pi/3)a^3\rho_0$

1.17 **Line integrals**
(b) $2E_0$ **(c)** 0 **(d)** $2E_0$

1.18 **Fluxes of curved surfaces**
(a) $\Phi = 2\pi a L E_0$ **(b)** $E(r) = C/r$ **(c)** $\Phi = 4\pi a^2 E_0$ **(d)** $E(r) = C/r^2$

1.19 **Cylindrical coordinates**
(a) $2\pi L V_0(b^3 - a^3)/3$ **(b)** $\Phi = 0$ **(c)** $2\pi a E_0$ **(d)** $\nabla V = (V_0/r)\hat{\mathbf{r}}$ **(e)** $\mathbf{E} = r\hat{\phi}$

2.7 **Field and R-vector for a single charge**
(c) $\mathbf{E} = \frac{-Q}{4\pi\epsilon_0 a^2 2^{3/2}}(1,1)$

2.9 **Three unknown charges**
$q_1 > 0, q_2 < 0, q_3 > 0$

2.10 **Net charge of a charge distribution**
(a)

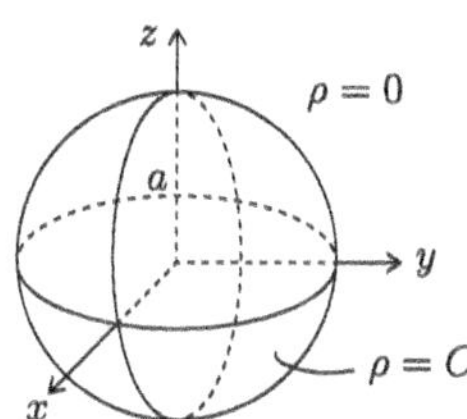

(b)

$$Q(r) = \begin{cases} \frac{4\pi}{3}r^3 C & \text{for } r < a \\ \frac{4\pi}{3}a^3 C & \text{for } r \geq a \end{cases}$$

2.11 **Balloon**
(a) A surface charge density $\rho_S = Q/(4\pi a^2)$. **(b) 0**

2.12 **A long line of charge**
(a)

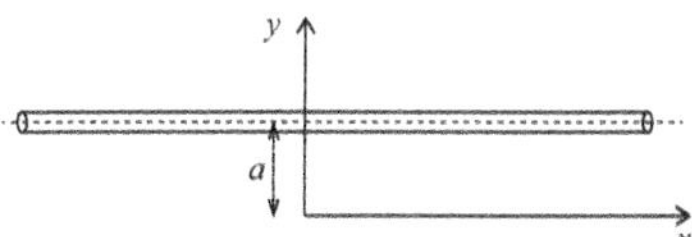

(b) Normal to the rod and away from the rod, that is, in negative y-direction. **(c)** Q is right. **(d)** $\mathrm{d}q = (Q/L)\,\mathrm{d}l$ **(e)** $\mathbf{R} = (-x', -a, 0)$ **(f)**

$$dE = \frac{(Q/L)\,dx'}{4\pi\epsilon_0}\frac{(-x', -a, 0)}{((x')^2 + a^2)^{3/2}}$$

(g)

$$E_x = \int_{-\infty}^{\infty} \frac{(Q/L)\,dx'}{4\pi\epsilon_0}\frac{-x'\hat{\mathbf{x}}}{((x')^2 + a^2)^{3/2}}$$

$$E_y = \int_{-\infty}^{\infty} \frac{(Q/L)\,dx'}{4\pi\epsilon_0}\frac{-a\hat{\mathbf{y}}}{((x')^2 + a^2)^{3/2}}$$

2.13 **Electric dipole**
(a) $-\hat{\mathbf{x}}$ **(b)** $\mathbf{R}_1 = (-a, y, 0)$, $\mathbf{R}_2 = (a, y, 0)$ **(c)** $\mathbf{R}_1 = (0, y, z)$, $\mathbf{R}_2 = (2a, y, z)$ **(d)** $\mathbf{E} = Q/(4\pi\epsilon_0)\left((0, y, z)/(y^2 + z^2)^{3/2} - (2a, y, z)/(4a^2 + y^2 + z^2)^{3/2}\right)$ **(e)** Cylindrical around x.
(f) $\mathbf{E} = Q/(4\pi\epsilon_0)\left[\left((x - a)\hat{\mathbf{x}} + \rho\hat{\boldsymbol{\rho}}\right)/((x - a)^2 + \rho^2)^{3/2} - \left((x + a)\hat{\mathbf{x}} + \rho\hat{\boldsymbol{\rho}}\right)/((x + a)^2 + \rho^2)^{3/2}\right]$

2.14 **Line charge**
(a) $\rho_l = Q/L$ **(b)** $dq = (Q/L)\,dx$ **(c)** $d\mathbf{E} = \rho_l\,dx'/(4\pi\epsilon_0)(-x', y, 0)/\left((x')^2 + y^2\right)^{3/2}$, $\mathbf{R} = (-x', y, 0)$. **(d)** $\mathbf{E} = \int_{-L/2}^{L/2} \rho_l/(4\pi\epsilon_0)(-x', y, 0)/((x')^2 + y^2)^{3/2}\,dx'$

2.15 **Plane charge**
(a) $\rho_s(\mathbf{r}) = \rho_{s,0}$ **(b)** Infinite **(d)** $\mathbf{R} = (x - x', y - y', z)$, $d\mathbf{E} = \rho_s dx'\,dy'/(4\pi\epsilon_0)\,\mathbf{R}/R^3$ **(e)** $\mathbf{E} = \int_0^a \int_0^b \rho_A/(4\pi\epsilon_0)(x - x', y - y', z)/\left((x - x')^2 + (y - y')^2 + z^2\right)^{3/2} dy'\,dx'$

2.16 **Angle between suspended charges**
(b) $\theta_1 = \theta_2 \simeq q^2/(\pi\epsilon_0 L^2 mg)$

2.17 **Adding and subtracting using superposition**
(a) 0 **(b)** $\mathbf{F} = Qq/(4\pi\epsilon_0 L^2)\hat{u}$, where $\hat{u}$ points from the center to the removed charge. **(c)** 0 and $Q/(4\pi\epsilon_0 r^2)\hat{\mathbf{x}}$ **(d)** $\mathbf{E}_{\text{shell}} = 0$ for $r < b$; $\mathbf{E}_{\text{shell}} = \rho/(3\epsilon_0)\left(r - b^3/r^2\right)\hat{\mathbf{r}}$ for $b < r < a$.

2.18 **Point charges**
(b) $\mathbf{F}_2 = (3.85 \cdot 10^{-8}, -1.92 \cdot 10^{-8})$N **(c)** $\mathbf{F}_2 = (-3.85 \cdot 10^{-8}, 1.92 \cdot 10^{-8})$N

2.19 **The electric field from groups of point charges**
(d) $E(r) \propto r^{-3}$ **(e)** $E(r) \propto r^{-2}$ **(g)** $E(r) \propto r^{-4}$

2.20 **The electric field above a finite disk**
(a) $d\mathbf{E} = \rho_A r\,dr\,d\theta/(4\pi\epsilon_0)(-r\cos\theta, -r\sin\theta, z)/(r^2+z^2)^{3/2}$ **(b)** $\mathbf{E} = \rho_A/(2\epsilon_0)\left(1 - z/(a^2+z^2)^{1/2}\right)\hat{\mathbf{z}}$ **(c)** $\mathbf{E} = \rho_A/(2\epsilon_0)\hat{\mathbf{z}}$

2.21 **Work on point charges**
$W = Q^2/(2\pi\epsilon_0 a)$

2.22 **Plate with a hole**
$E_z = \rho_s/(2\epsilon_0)\,z/(z^2+a^2)^{1/2}$

2.23 **Half circle**
(a) $Q/(2\pi^2\epsilon_0 a^2)\hat{\mathbf{y}}$ **(c)** $\mathbf{E} = -\frac{Q}{16\epsilon_0 a^2}$

2.24 **The electric field above an infinite plane**
(b) $dq = \rho_S r\,dr\,d\theta$ **(c)** $d\mathbf{E} = \rho_A r\,dr\,d\theta/(4\pi\epsilon_0)(-r\cos\theta, -r\sin\theta, z)/(r^2+z^2)^{3/2}$
(d) $dE_z = \frac{2\pi\rho_A r\,dr}{4\pi\epsilon_0}\frac{z}{(r^2+z^2)^{3/2}}$ **(e)** $E_z = \frac{\rho_A}{2\epsilon_0}\frac{z}{|z|}$

2.25 **Dipole in a uniform electric field**
(a) 0 **(b)** $\boldsymbol{\phi} = \mathbf{p}\times\mathbf{E}$

2.26 **Field from a hemisphere shell**
$\mathbf{E}_z = \frac{\rho}{4\epsilon_0}\hat{\mathbf{z}}$

2.27 **Visualize electric field**
(a) $\mathbf{E} = Q/(4\pi\epsilon_0)(x-x_1, y-y_1)/((x-x_1)^2+(y-y_1)^2)^{3/2}$

2.29 **Crossing lines**
(a) $\mathbf{E} = -(Q/L)/(2\pi\epsilon_0 y)\,\hat{\mathbf{y}}$ **(b)** $\mathbf{E} = \frac{Q}{2\pi\epsilon_0 L}\frac{(y-a)}{|y-a|^2}\hat{\mathbf{y}}$

```
import numpy as np
import matplotlib.pyplot as plt
def E1(y):
    return 1/y
def E2(y,a):
    return (y-a)/(y-a)**2
a = 1.0
y = np.linspace(-a,2*a,1000)
E = E1(y)+E2(y,a)
plt.plot(y,E,"-k")
plt.ylim((-20,20))
plt.xlabel("$x/a$")
plt.ylabel("$E/E_0$")
```

(c) $\frac{Q}{2\pi\epsilon_0 L}(-1/x, 1/y)$

3.7 **Electric potential and potential energy**
(a) Downward, in negative y-direction. **(b)** Negative **(c)** Positive **(d)** $U_p(y=d) = eE_0 d$ **(e)** $V = E_0 y$ **(f)** Downward (negative y-direction) which is toward lower elec-

tric potential. **(g)** $V(d/2) = Ed/2$. Toward higher electric potential.

3.8 **Optimal path**
The work is the same along all paths.

3.9 **From potential to field**

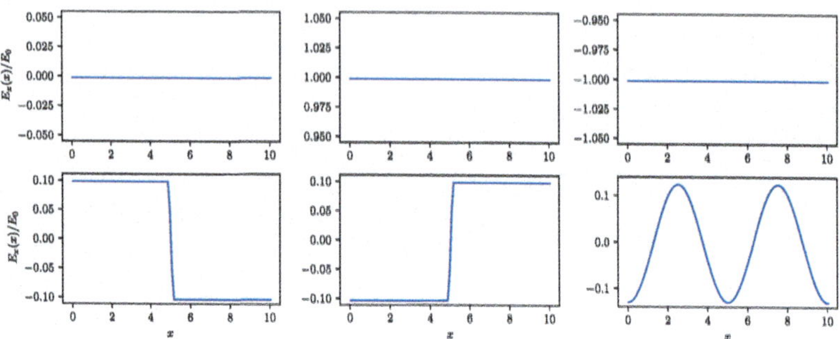

3.10 **Potential, field and forces**
(a) a: to the right; b: zero; c: to the left. **(c)** a: to the left; b: will not move; c: to the right.

3.11 **Units for electric potential and energy**
(a) 15km **(b)** 1 eV = 1.60218 10^{-19} J

3.12 **Quadrupole**
(b) The electric potential is zero along the x and y axes. The electric field is zero in the origin. **(c)** The electric potential is zero along the x-axis. **(e)**

```
Ex,Ey = np.gradient(-V)
plt.contourf(rx,ry,V)
plt.quiver(rx,ry,Ex,Ey)
plt.axis("equal")
```

(f)

```
xp = np.linspace(-3,3,1000)
Vx = xp.copy()
for i in range(len(xp)):
    r = np.array([xp[i],xp[i]])
    Vx[i] = epotlist(r,Q,R)
plt.plot(xp*np.sqrt(2),Vx)
plt.xlabel("r")
plt.ylabel("V(x)")
```

(h) $\mathbf{p} = \mathbf{0}$

3.13 **Two point charges**
(a) $U = -1.35\,10^{-6}$ J **(c)** $V = -359.5$ V

3.14 **Four charges**
(a) $U = 3.60\,10^{-5}$ J **(b)** $V = 50841$ V

3.15 **Two charges again**
(a) At a distance of 2 cm from Q_1 toward Q_2 and at a distance 3.33 cm from Q_1 away from Q_2.

3.16 **A distance away**
(a) 10 m **(b)** $Q = 1.1$ nC

3.17 **Potential in the plane**
(a) $\mathbf{E} = (-Cy, -Cx)$ **(c)** $\mathbf{r}(t) = (x_0, y_0) + (-\frac{\partial V}{\partial y}, \frac{\partial V}{\partial x})t$

3.18 **Accelerate an electron**
(b) $m_0c^2 = 0.5110$ MeV **(c)** $v/c = 0.98$

3.19 **Electric fields and electric potential**
(a) $V_A = -30$ V, $V_B = 60$ V, $V_C = 60$ V. **(c)** $\mathbf{F}(x, -2\,\text{mm}) = (1, 0)\,10^{-1}$ N for -3 mm $< x < 0$ and $(2, 0)\,10^{-1}$ N for $0 < x < 3$ mm **(d)** Along the y-axis.

3.20 **Gradients**
(a) $V(x) = \frac{q}{4\pi\epsilon_0}\left(\frac{1}{|x-a|} + \frac{1}{|x+a|}\right)$ **(d)** $V(y)q/(2\pi\epsilon_0(y^2 + a^2)^{1/2})$

3.21 **Line charge**
(a) $dV = (Q/L)dx'/(4\pi\epsilon_0((x')^2 + y^2)^{1/2})$ **(b)** $V = \int_{-L/2}^{L/2}(Q/L)/(4\pi\epsilon_0 \left((x')^2 + y^2\right)^{1/2})dx'$ **(c)** No

3.22 **Three-dimensional visualization**
(a)

```
import numpy as np
import scipy.constants as sc
def Vfield(r,Q,ri):
    R = r - ri
    Rnorm = np.linalg.norm(R)
    return 1.0/(4.0*np.pi*sc.epsilon_0)*(1/Rnorm)
```

(b)

```
# Generate a ring charge
qilist = []
rilist = []
```

```
a = 1.0
Q = 1.0
Nring = 100
qi = Q/Nring
for ic in range(Nring):
    thetai = ic/Nring*2*np.pi
    ri = np.array([a*np.cos(thetai),a*np.sin(thetai),0])
    qilist.append(qi)
    rilist.append(ri)
```

3.24 **Half a line**
(a) $\mathrm{d}V = \frac{\rho_l \mathrm{d}x'}{4\pi\epsilon_0 |x-x'|}$ **(b)** $V(x) = \frac{\rho_l}{4\pi\epsilon_0}(\ln(x+L) - \ln(x))$ **(c)** $E_x = -\frac{\rho_l}{4\pi\epsilon_0}\left(\frac{1}{x+L} - \frac{1}{x}\right)$
(d) Because we only know $V(x, 0)$ and not $V(x, y)$ needed to find the derivative along the y-direction.

3.25 **Potential and field above a disk**
(a) $\mathrm{d}V = \frac{\rho_s r \, \mathrm{d}r \, \mathrm{d}\phi}{4\pi\epsilon_0 (r^2+z^2)^{1/2}}$ **(b)** $V(z) = \frac{rho_s}{2\epsilon_0}\left((z^2+a^2)^{1/2} - z\right)$ **(c)** $-\frac{rho_s}{2\epsilon_0}\left(\frac{z}{(z^2+a^2)^{1/2}} - 1\right)$
(d) $V_{z\to 0} = \frac{\rho_s}{2\epsilon_0}$

3.26 **Far away from charges**
$V \simeq \frac{Q}{4\pi\epsilon_0}\frac{2a^2}{x^3}$

3.27 **Plate with hole**
$\mathbf{E} = \frac{\rho_s}{2\epsilon_0}\frac{z}{\sqrt{z^2+a^2}}\hat{\mathbf{z}}$

3.29 **Half spherical shell**
(a) $V(x) = \frac{\rho_s a^2}{2\epsilon_0}\frac{(|x+a| - (a^2+x^2)^{1/2}}{2ax}$ $(x \neq 0)$; $V(x) = \frac{\rho_s a^2}{2\epsilon_0 a}$ $(x = 0)$ **(c)** $E_x = \frac{\rho_s a^2}{2\epsilon_0}\left(-\frac{1}{x^2} - \frac{1}{a(a^2+x^2)^{1/2}} + \frac{(a^2+x^2)^{1/2}}{ax^2}\right)$

3.30 **Electric potential from an oxygen molecule**
(b)

```
theta = 104.5/180.0*np.pi # degrees -> rad
a = 0.95 # AAngstrom
QO = -0.7 # e
QH = -QO/2
# Setup charges
rilist = []
qilist = []
ri = np.array([0,0])
rilist.append(ri)
qilist.append(QO)
ri = a*np.array([np.sin(theta/2),np.cos(theta/2)])
rilist.append(ri)
qilist.append(QH)
ri = a*np.array([-np.sin(theta/2),np.cos(theta/2)])
rilist.append(ri)
qilist.append(QH)
```

(c)

```
Lx, Ly, Nx, Ny = 3, 3, 30, 30
x = np.linspace(-Lx,Lx,Nx)
y = np.linspace(-Ly,Ly,Ny)
rx,ry = np.meshgrid(x,y,indexing="ij")
# Set up electric potential
V = np.zeros((Nx,Ny),float)
# Calculate the potential
for ix in range(Nx):
    for iy in range(Ny):
        r = np.array([rx[ix,iy],ry[ix,iy]])
        for ic in range(len(qilist)):
            ri = rilist[ic]
            qi = qilist[ic]
            V[ix,iy] = V[ix,iy] + Vfield(r,qi,ri)
# Visualize potential and field in xy-plane
import matplotlib.pyplot as plt
Ex,Ey = np.gradient(-V)
plt.figure(figsize=(6,6))
plt.contourf(rx,ry,V,20,cmap="seismic")
plt.quiver(rx,ry,Ex,Ey)
plt.xlabel("$x (Ang) $"), plt.ylabel("$y (Ang) $"), plt.axis("equal")
```

3.31 **Dipole of two rings**
(b) $E_z = \frac{Q}{4\pi\epsilon_0}\frac{z}{(z^2+a^2)^{3/2}}$

4.6 **Maximum flux**
(a) $E_0 S$ **(b)** $\theta = 0$ **(c)** $\theta = \pi$. **(d)** $\Phi = E_0 S \cos\theta$

4.7 **Smartsurface**
(b) Yes **(c)** Yes **(d)** No

4.8 **Waterflux**
(a) $\Phi_A = \Phi_B$, $\Phi_C = 0$ **(b)** Equal amounts are flowing out of surfaces A and B and there is no net flow out of surface C. **(c)** Negative charges corresponds to a sink or a tube that sucks water. **(d)** Zero.

4.9 **Unknown charge**
(a) $\Phi = 4\pi a^2 E_0$ **(b)** $Q = 4\pi\epsilon_0 a^2 E_0$.

4.10 **A plane charge**
(a) That the surface charge density is the same everywhere in the plane. **(k)** $E_z(z) = \rho_S/(2\epsilon_0)$ for $z > 0$ and $E_z(z) = -\rho_S/(2\epsilon_0)$ for $z < 0$ **(m)** Yes **(n)** No

4.11 **Multiple planes**
(b)

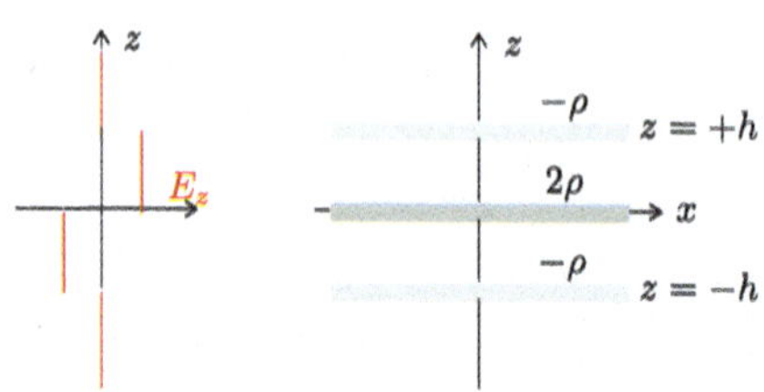

(c) $E_z = \frac{2\rho}{2\epsilon_0} = \frac{\rho}{\epsilon_0}$ for $-h < z < h$, zero otherwise.

4.12 **Flux-trix**
(a) $\hat{\mathbf{n}}_A = (0, -1, 0)$, $\hat{\mathbf{n}}_B = (0, 0, -1)$, $\hat{\mathbf{n}}_C = (1, 0, 0)$, $\hat{\mathbf{n}}_D = (-1, 0, 0)$, and $\hat{\mathbf{n}}_E = (0, y, z)/\sqrt{y^2 + z^2}$. **(b)** $\Phi_A = -4a^2 E_0$, $\Phi_B = \Phi_C = \Phi_D = 0$. **(c)** $\Phi_E = 4a^2 E_0$. **(d)** $\Phi = -(\pi a^2/2)E(a)$

4.13 **Overlapping clouds of charge**
(a)

$$\mathbf{E} = \begin{cases} \frac{1}{3\epsilon_0}\rho_0 r\hat{\mathbf{r}} & \text{when } r < a \\ \frac{1}{3\epsilon_0}\rho_0 \frac{a^3}{r^2}\hat{\mathbf{r}} & \text{when } r > a \end{cases}$$

(b) $\mathbf{E} = \mathbf{E}_+ + \mathbf{E}_- = 2C_+ (\mathbf{r}_- - \mathbf{r}_+)$

4.14 **Fluxes from point charges**
(a) $\Phi_E = 0$ **(b)** $\Phi_{1.5\,\text{m}} = -677.6\,\text{Vm}$, $\Phi_{2.5\,\text{m}} = -225.8\,\text{Vm}$ **(c)** No

4.15 **Field from charge distributions**
(a) $\mathbf{E}(r) = \frac{Q}{4\pi\epsilon_0 r^2}\hat{\mathbf{r}}$ **(b)** $\mathbf{E}(r) = \frac{Q}{4\pi\epsilon_0 a^3} r\hat{\mathbf{r}}$ for $r < a$ and $\mathbf{E}(r) = \frac{Q}{4\pi\epsilon_0 r^2}\hat{\mathbf{r}}$ for $r \geq a$. **(c)** $E(r) = Q/(4\pi\epsilon_0 r^2)\hat{\mathbf{r}}$ for $r > a$ and $E(r) = 0$ for $r < a$. **(d)** $k = Q/(\pi a^4)$, $\mathbf{E}(r) = Qr^2/(4\pi\epsilon_0 a^4)\hat{\mathbf{r}}$ for $r \leq a$.

4.16 **Two wires (A)**
(b) $E = \rho/\pi\epsilon_0 x$ **(d)** $|\mathbf{E}| = \frac{9\rho}{4\pi\epsilon_0 x}$ towards the negative wire.

4.17 **Two wires (B)**
(d) $|E| = \frac{3\rho}{4\pi\epsilon_0 x}$

4.18 **Electrons in a box**
2000

4.19 **Two line charges**
(a) $\mathbf{E} = \rho_l/(2\pi\epsilon_0)(x - a, y)/((x - a)^2 + y^2)$ **(b)** $\mathbf{E} = \rho_l/(2\pi\epsilon_0)(x - a, y)/((x - a)^2 + y^2) - \rho_l/(2\pi\epsilon_0)(x + a, y)/((x + a)^2 + y^2)$ **(c)** $\mathbf{E}(x, y = 0) \simeq -\mathbf{p}/(2\pi\epsilon_0 x^2)$

4.20 **Cylinder**
$E(r) = \rho r/(2\epsilon_0)$ for $r < a$, $E(r) = \rho a^2/(2\epsilon_0 r)$ for $r > a$

5.4 **Electric field through water**
(a) $E_z = \frac{\rho}{\epsilon_0}$ from the positive to the negative plane. **(b)** $E_z = \frac{1}{\epsilon_r}\frac{\rho}{\epsilon_0}$ **(d)** No.

5.5 **Dielectric plate in uniform field**
(b) A possible solution is: top at $y = d/2$ and bottom at $y = -d/2$. **(c)** $\rho_{s,b,\text{top}} = -\rho_{s,b,\text{bottom}} > 0$ **(d)** No **(e)** $E_{z,T} = E_0$ $(y < -d/2, y > d/2)$, $E_0 - \rho_{s,b}/\epsilon_0$ $(-d/2 < y < d/2)$ **(f)** $P_z = (\epsilon - \epsilon_0)(E_0 - \rho_{s,b}/\epsilon_0)$ **(g)** $\rho_{s,b} = (1 - \epsilon_0/\epsilon)\,\epsilon_0 E_0$, $E_T = (\epsilon_0/\epsilon)E_0$, $P = (\epsilon - \epsilon_0)(\epsilon_0/\epsilon)\,E_0$ **(h)** $\mathbf{D} = \epsilon_0 \mathbf{E}_0$ **(i)** $E_{t,1} = E_{t,2}$, $D_{n,1} - D_{n,2} = \rho_{s,f} = 0$

5.6 **Torque on dipole**
(a) $\mathbf{p} \parallel \mathbf{E}$ **(b)** $\mathbf{p} \perp \mathbf{E}$ **(c)** No

5.7 **Bound charges**
(a) 0 **(b)** $\rho_{s,b} = P\sin\theta$ **(c)** 0

5.8 **Charge in dielectric**
(a) $D_1 = D_2$ **(b)** $E_1 > E_2$ **(c)** $P_2 > P_1$

5.9 **Charge in hole**
(a) $E_r(r) = q/(4\pi\epsilon_0 r^2)$ **(b)** $D_r(r) = q/(4\pi r^2)$ **(c)** $D_r(r) = q/(4\pi r^2)$ **(d)** $E_r(r) = q/(4\pi\epsilon r^2)$ for $r > a$.

5.11 **A simple model for polarization**
(a) $E(z) = \frac{\sigma}{2\epsilon_0}\text{sign}(z)$. **(b)** $\rho_+ = \rho_0\Delta z$, $\rho_- = -\rho_0\Delta z$ **(c)** $E_i = -\rho_0\Delta z/\epsilon_0$ **(d)** $E_i = -\rho_0\chi E_0/\epsilon_0$ **(e)** $E_{tot} = E_0/(1 + \rho_0\chi/\epsilon_0)$

5.12 **Polarized cube**
(a) $\rho_{v,b} = 0$, $\rho_{s,b} = P_0$ for $x = a$, $\rho_{s,b} = -P_0$ for $x = 0$, and $\rho_{s,b} = 0$ otherwise. **(b)** $\rho_{v,b} = P_0 x/a^2$, $\rho_{s,b} = P_0 y/a$ for $x = a$, and $\rho_{s,b} = 0$ otherwise.

5.13 **A cell membrane**
(a) Spherical symmetry with $\mathbf{D} = D_r(r)\hat{\mathbf{r}}$ **(b)** $E_r = \frac{Q}{4\pi\epsilon r^2}$ **(c)** $V(b) - V(a) = \frac{Q}{4\pi\epsilon}\left(\frac{1}{b} - \frac{1}{a}\right)$ **(d)** About 800k ions.

5.14 **Long cylinder**
(a) $E_r = (r\rho)/(2r\epsilon)$ for $r < a$ and $E_r = (a\rho)/(2r^2\epsilon_0)$ for $r > a$ **(b)** $\rho_{v,b} = -(\epsilon - \epsilon_)\rho/\epsilon$, $\rho_{s,b} = (\epsilon - \epsilon_0)a\rho/(2\epsilon)$.

5.15 **Hollow sphere**
(a) $D_r = \rho r/3$ for $r < a$, $D_r = \rho a(a/r)^2/3$ for $r \geq a$ **(b)** $E_r = \rho r/(3\epsilon)$ for $r < a$, $E_r = \rho a(a/r)^2/(3\epsilon)$ for $a < r < b$, $E_r = \rho a(a/r)^2/(3\epsilon_0)$ for $b < r$ **(c)** $\rho_{v,b} = -\nabla \cdot \mathbf{P} = -(\epsilon - \epsilon_0)\rho r^2/\epsilon_0$.

6.3 **Laplace operator in one dimension**
(a)

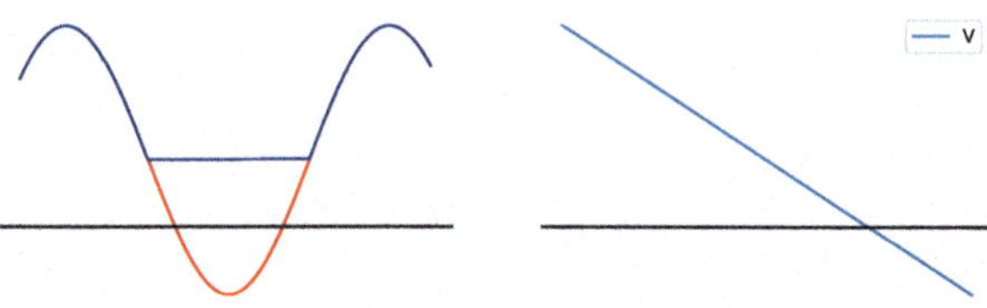

Red: positive, Blue: negative **(b)** No **(c)** Yes. The charge distribution is minus the shown distribution. **(d)** Yes

6.4 **Boundary conditions in one dimension**
(a) $V(0) = 1.5\,\text{V}$, $V(1\,\text{cm}) = 0\,\text{V}$. Infinitely far away we do not yet know what the potential is, but we expect it to be a constant. **(b)** Because $\rho = 0$. **(c)** $V(x) = Ax + B$ **(d)** $V(x) = -1.5\,\text{V/cm}x + 1.5\,\text{V}$. **(e)** $E_x = 1.5\,\text{V/cm}$ for $0 < x < 1$. Outside this region $E_x = 0$. **(f)** Two infinite, parallell plates.

6.5 **The Laplace operator in two and three dimensions**
(a) 0; Negative; Negative; Positive; Negative; Negative. **(b)** Yes; No; Yes.

6.8 **The discrete Laplace operator**
(a) -2 **(b)** $+2$ **(c)** 0

6.9 **Manual Jacobi iteration**
(a) $4, 0, 0, 0, 0$ **(b)** $4, 2, 0, 0, 0$ **(c)** $4, 11/4, 3/2, 3/4, 0$ **(d)** $4, 3, 2, 1, 0$

6.11 **Laplace operator**
(a) Laplace's equation. **(b)** Poisson's equation for $\rho(x) = \epsilon_0\omega^2 \sin \omega x$ **(c)** Laplace's equation. **(d)** Laplace's equation. **(f)** Poisson's equation for $\rho = -2ay\epsilon_0$.

6.12 **Spatial variation of potentials (Laplace equation)**
(b) $V(x, y) = 0$ **(d)** $V_1(x) = 2V_1x/L$ and $E_1(x) = -2V_1/L$ for $0 < x < L/2$; $V_2(x) = 2(V_1/L)(L/2 - x) + V_1$ and $E_2(x) = 2V_1/L$ for $L/2 < x < L$. **(f)** $\rho_a = 2V_1\epsilon/L$ on each side of $x = L/2$ and $\rho_a = -2V_1\epsilon/L$ at $x = 0$ and $x = L$.

6.13 **Surface charges**
(a) $E = 0$ for $r < a$ and $\mathbf{E} = Q/(4\pi\epsilon r^2)\hat{r}$ for $r > a$. **(b)** $V(r) = Q/(4\pi\epsilon r)$ for $r > a$ and $V(r) = Q/(4\pi\epsilon a)$ for $r \leq a$. **(d)** $\rho_a = Q/(4\pi a^2)$

6.15 **Laplace's equation for a long cylinger**

6.16 **Charge distribution between metal plates**
(a) $V(x) = -\rho x^2/(2\epsilon) + (\rho d/(2\epsilon) - V_0/d)\,x + V_0$ **(b)** $\mathbf{E} \approx (-1.13x + 6.65)\hat{\mathbf{x}}$ V/mm **(c)** $V_{\min} = -9.57$ V

6.17 **Poisson's equation in one dimension**
(a)

$$A = \frac{1}{\Delta x^2}\begin{bmatrix} 2 & -1 & 0 & 0 & 0 & 0 & \dots \\ -1 & 2 & -1 & 0 & 0 & 0 & \dots \\ 0 & -1 & 2 & -1 & 0 & 0 & \dots \\ \vdots & 0 & -1 & 2 & -1 & 0 & \dots \\ \vdots & \vdots & \ddots & \ddots & \ddots & \ddots & \ddots \\ \ddots & \ddots & \ddots & \ddots & \ddots & \ddots & \ddots \\ 0 & & \dots & \dots & 0 & -1 & 2 \end{bmatrix}$$

6.18 **Poisson's equation**
(a) $V(h) = (-10\,\text{V/m})h + 10\,\text{V}$

6.19 **Potential**
(a) $V(x) = V_0(x/L)$ **(b)** $V(x) = \frac{V_0}{L}x + \frac{\rho_0}{2\epsilon}xL - \frac{\rho_0}{2\epsilon}x^2$

6.20 **Earnshaw's theorem**
No, there is no *stable* equilibrium point inside the box, but there is an *unstable* equilibrium.

6.23 **Modelling a 3D ion trap**
(b) $V(x, y, z) = A(x^2 + y^2 - 2z^2)$

7.4 **Conductors and electric field**
(a) 0 **(b)** Positive charges are on the top surface and negative charges are on the bottom surface. There are no charges inside the cube. **(c)** No **(d)** 0 **(e)** $+Q$ on the uppermost surface and $-Q$ on the lowermost surface. **(f)** A charge $+Q$ on the lower surface of the upper half cube and a charge $-Q$ on the upper surface of the lower half cube.

7.6 **Secret briefcase**
No

7.7 **Millions of electrons**
(b) $E = 18.75$ V/m **(c)** $1.04\ 10^7$

7.8 **Coaxial cable**
(e) 0

7.9 **Shielding**

7.10 **Faraday cage**
(a) 0 **(b)** $\mathbf{E} = 0$

7.11 **Point charge near conducting sphere**

7.12 **Point charge above conducting plane**
(a) $E_z = -\frac{Q}{4\pi\epsilon_0}\left(\frac{1}{(h-z)^2} + \frac{1}{(h+z)^2}\right)$ **(b)** $\rho_s = -\frac{Qh}{2\pi(r^2+h^2)^{\frac{3}{2}}}$ **(e)** $\mathbf{F} = -Q^2/(16\pi\epsilon_0 h^2)\,\hat{\mathbf{z}}$

7.13 **A lightning rod**
(c) $E_r = V_0 a/r^2$

7.14 **A lightning rod in two dimensions**
(b) $E_r = -V_0/(r\ln(a/b))$

8.4 **Cylinder**
(a) Uniformly on the surface of the cylinder. **(b)** Uniformly on the inner surface of the outer shell. **(c)** $\frac{Q/L}{2\pi\epsilon_0}\ln b/a$ **(d)** $C = 2\pi\epsilon_0/(\ln(b/a))$

8.5 **Double trouble**
No

8.6 **Sandwich**
(a) $E_z = \rho_S/(2\epsilon_0)$ for $z > 0$ and $E_z = -\rho_S/(2\epsilon_0)$ for $z < 0$. **(c)**

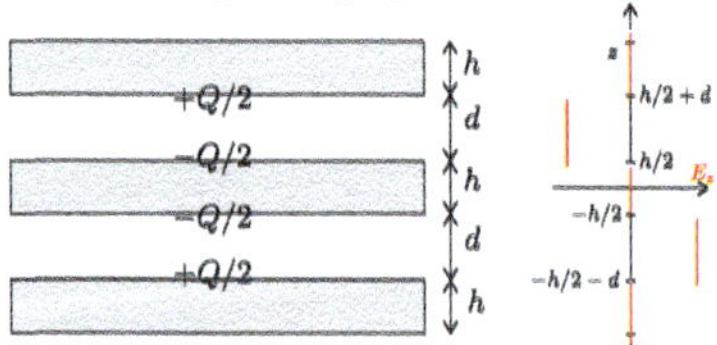

(d) $\frac{Qd}{2A\epsilon_0}$ **(e)** $C = \frac{2A\epsilon_0}{d}$ **(f)** We interpret this as the capacitor having an effective area of $2A$.

8.7 **Capacitance of the axon**
(a) No **(b)** $\mathbf{E} = (Q/L)/(2\pi\epsilon r)\hat{\mathbf{r}}$ for $a < r < b$ and zero otherwise. **(c)** $V(a) = (Q/L)/(2\pi\epsilon)\ln(b/a)$ **(d)** $C = (2\pi\epsilon L)/(\ln(b/a))$

8.9 **Parallel plates**
(a) No change **(b)** No change **(c)** Decreases **(d)** Increases

8.10 **Capacitance**
(a) Parallel **(b)** $C_{max} = 8.0\,\mu$F, $C_{min} = 0.5\,\mu$F **(c)** 0.2 mJ

8.11 **Coaxial cable**
(a) $\mathbf{E} = V_0/(r\ln(b/a))\hat{\mathbf{r}}$, $C' = 2\pi\epsilon/\ln(b/a)$ **(b)** $C' = 8.58 \times 10^{-11}$ F/m **(c)** $U' = \frac{\pi\epsilon}{\ln\frac{b}{a}}V_0^2$ **(d)** Zero.

8.12 **Capacitor of two spheres**
(a) Uniformly on the surface

8.13 **Capacitor from two circuits**
(a) $\mathbf{E}(0) = 0$, $V(0) > V(\infty)$. **(e)** $C \simeq 1.1 \cdot 4\pi\epsilon_0 a$

9.5 **Moving charges**
(a) 2 **(b)** 1,3 and 4

9.6 **Different velocities**
(a) $I_1 = 1.602\,10^{-19}$ A **(b)** $I = 1.60210^{-7}$ A. **(c)** $J = 1.602\,10^{-1}$ A/m^2 along the conductor. **(d)** Mb/L **(e)** $Mv\Delta t/L$ **(f)** $I_1 = MvQ/L = Mve/L$ **(g)** $I = N^2Mve/L$ **(h)** $J = Mve/(a^2L)$

9.7 **Current through inclined surface**
(a) $\mathbf{J} = I/a^2\hat{\mathbf{x}}$ **(c)** The same, but negative, $-I$.

9.8 **Ohm's law for charges in a liquid**
(a) The Coulomb-force, $q\mathbf{E}$, and a viscous force from the liquid: **D**. **(b)** $v = \frac{qE}{6\pi\eta a}$ **(c)** $N = nAv\Delta t$ **(d)** $J = \frac{nq^2}{6\pi\eta a}E$

9.9 **Quadratic cross section**
(a) $J_x = I/A = I/a^2$ **(b)** $E_x = J_x/\sigma = I/(a^2\sigma)$ **(c)** $V(x) = V_0 - xI/(a^2\sigma)$ **(d)** $\Delta V = I\frac{L}{a^2\sigma}$ **(e)** $R = \frac{\Delta V}{I} = \frac{L}{a^2\sigma}$ **(f)** $V(x) = V_0 + x\frac{V_1 - V_0}{L}$ **(g)** $E_x = -(V_1 - V_0)/L$ **(h)** $J_x = \sigma(V_0 - V_1)/L$ **(i)** $I = \frac{\sigma a^2(V_0 - V_1)}{L}$, $R = (V_0 - V_1)/I = L/(\sigma a^2)$ **(j)** Half.

9.10 **Resistance gymnastics**
(a) $R = L/(\sigma a^2)$. **(b)** $R = \int_0^L \mathrm{d}x/(\sigma a^2)$ **(c)** $\mathrm{d}R = \mathrm{d}r/(\sigma r\theta h)$. **(d)** $R = \frac{1}{\sigma h\theta}\ln\frac{b}{a}$ **(f)** $R = L/(\sigma a^2\cos^2\theta)$.

9.11 **Bent resistor**
(a) The electric field will only have a component along the conductor, that is, $\mathbf{E} = E_\phi\hat{\boldsymbol{\phi}}$. **(b)** $V = V(\phi)$ (No radial dependence because the electric field does not have a radial component). **(c)** $\frac{1}{r^2}\frac{\partial^2 V}{\partial\phi^2} = 0$ and $V(\phi = 0) = V_0$, $V(\phi = \pi) = V_1$. **(d)** $V(\phi) =$

$V_1(\phi/\pi)$ (Where $\phi = 0$ along the positive x-axis) **(e)** $E_\phi = -\frac{V_1}{\pi r}$ **(f)** $J_\phi = -\frac{\sigma V_1}{\pi r}$ **(g)** $I = -\frac{a\sigma V_1}{\pi}(\ln(b+a) - \ln b)$ (Negative means that the current goes from V_1 to $V_0 = 0$.) **(h)** $R = \frac{\pi}{a\sigma \ln((b+a)/a)}$ **(i)** There is no current in the air outside the conductor, because there are not (enough) charges there that can contribute to the current.

9.12 **Current density**
(a) $I = \pm J_0 4a^2$ for the surfaces through $(\pm 1, 0, 0)$, and zero for the other surfaces. **(b)** $\mathbf{J} = \rho v_0 \hat{z}$.

9.13 **Cylindrical resistor**
(c) $\mathbf{E} = I/(2\pi r L\sigma)\,\hat{r}$ when $a < r < b$ and 0 otherwise. **(d)** $V = I/(2\pi L\sigma)\ln(b/a)$ **(e)** $R = \ln(b/a)/(2\pi L\sigma)$

9.14 **Current of ions**
Yes, to the right.

9.15 **Lightning**
(a) 1.1 kV **(b)** 2.1 kJ

9.16 **Spherical current**
(a) $J(r) = \sigma Q/(4\pi\epsilon_0 r^2)$ **(b)** $I = \sigma Q/\epsilon_0$ **(c)** $R = ((1/a) - (1/b))/(4\pi\sigma)$

9.17 **Leaky coaxial cable**
(a) $\mathbf{J} = \frac{\sigma V_0}{r \ln \frac{b}{a}}\hat{\mathbf{r}}$ **(b)** $P_J = 98$ GW **(c)** $R = 2.76\,\mu\Omega$.

9.18 **A cylindrical component**
(a) $\mathbf{E} = \frac{(Q/L)}{2\pi\epsilon r}\hat{\mathbf{r}}$ **(b)** $V(r) = \frac{(Q/L)}{2\pi\epsilon}\ln\left(\frac{a+d}{r}\right)$ **(c)** $C = \frac{2\pi\epsilon L}{\ln(1+\frac{d}{a})}$ **(d)** $J(r) = \frac{I}{2\pi r L}$ **(e)** $\mathbf{E} = \frac{I}{2\pi\sigma L r}\hat{\mathbf{r}}$

9.19 **Two resistors**
(a) $E_1 = I/(\pi a^2\sigma_1)$, $E_2 = I/(\pi a^2\sigma_2)$ **(b)** $V(x) = \frac{I(L-x)}{\pi a^2\sigma_2}$ for $x > L/2$ and $I\frac{(L/2)-x}{\pi a^2\sigma_1} + I\frac{(L/2)}{\pi a^2\sigma_2}$ for $x < L/2$. **(c)** $R = \frac{L}{2\pi a^2}\left(\frac{1}{\sigma_1} + \frac{1}{\sigma_2}\right)$

9.21 **Resistance of two cylinders**
(a) $R = L/(\sigma\pi r_0^2)$ **(b)** $R = 5L/(2\sigma\pi a^2)$ **(c)** $R = L/(\sigma\pi a^2)\,(61/27)$

10.4 **Current and eletric potential in a poor conductor**
(a) It is homogeneous and points from high to low potential. **(b)** There is a positive charge approximately uniformly distributed on the edge at V_1 and similarly a uniformly distributed negative charge at V_0. **(c)** The electric field will point along the conductor inside the conductor, from high to low potential. **(d)** The charge will be distributed on the edges of the conductor to ensure that the electric field points along the conductor. **(e)** No, and it will generally not be zero. **(f)** $E_\phi = \frac{V_0 - V_1}{r(\pi/2)}$.

10.5 **Model system**

(a)

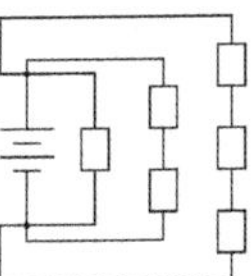

(b) No

10.6 **Twin circuits**

The same circuit.

10.7 **Potential drop of resistor**

(c)

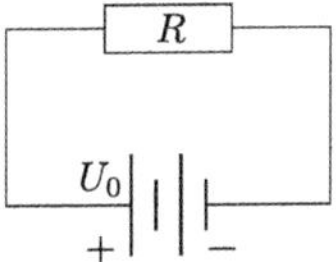

(d) We assume that the current is toward the right through the resistor. The voltage drop is IR. (Notice that this is a drop, that is, a reduction across the resistor.) **(e)** $\sum \Delta V = U_0 - IR = 0$ **(f)** $\sum \Delta V = -U_0 + IR = 0$ **(g)** $I = \frac{U_0}{R} = \frac{1.5\,\text{V}}{1000\,\text{ohm}} = 1.5\,\text{mA}$ **(h)** The potential drop is $-IR$. (Notice the word potential *drop*. It means that in Kirchhoff's law, the contribution is $\sum \Delta V = +RI$. Negative potential drop means an increase in potential. **(i)** No. Assuming that the current goes in the opposite direction does not imply any physical change in the systemm while turning the battery around is a physical change. **(j)** $\sum \Delta V = U_0 + IR = 0$ **(k)** $I = -\frac{U_0}{R} = -1.5\,\text{mA}$

10.8 **Potential drop of capacitor**

(a) The potential on the side of the capacitor that is connected to the battery is U_0 higher than the other side. **(b)** The electric field in the capacitor is uniform. **(c)** The charge is $+Q = C(V_1 - V_0)$ on the positive side (to the right) and correspondingly $-Q$ on the left side. Calculation gives $C = \frac{A\epsilon_0}{d} = \frac{(10^{-2}\,\text{m})^2\, 8.854\, 10^{-12}\,\text{Fm}^{-1}}{10^{-3}\,\text{m}} = 8.854\, 10^{-13}\,\text{F}$. **(d)**

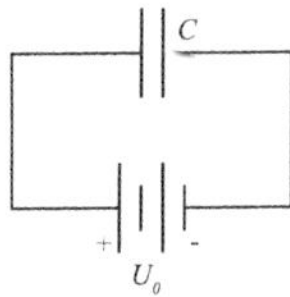

(e) The current is positive.

10.9 **Kirchhoff's law**

(**a**) The potential difference across the battery. (**b**) We chose positive current in counter-clockwise direction: $\sum \Delta V = U_0 - IR - IR = 0$ (**c**) $I = \frac{U_0}{2R}$ (**d**) Positive direction for the current is clockwise, but Kirchhoff's law along a counter-clockwise path: $\sum \Delta V = U_0 + IR + IR = 0$ (**e**) $I = -\frac{U_0}{2R}$ (The same as before!) (**f**) In the connecting points.

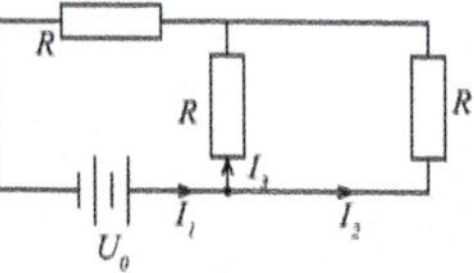

(**g**) The point marked with a circle was chosen. We find the net current into this point: $I_1 - I_2 - I_3 = 0$ (**h**) Net current into this point is: $I_3 + I_2 - I_1 = 0$ (**i**) 3

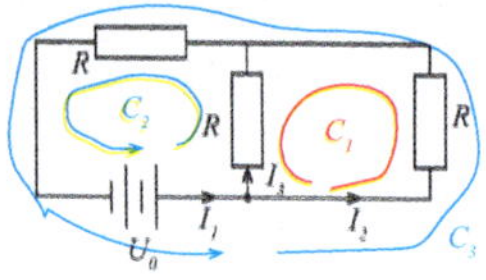

(**j**) For C_1: $I_2 R - I_3 R = 0$, for C_2: $U_0 - I_3 R - I_1 R = 0$, for C_3: $-I_2 R - I_1 R + U_0 = 0$. (**k**) $I_1 = \frac{2}{3}\frac{U_0}{R}$, $I_2 = \frac{1}{3}\frac{U_0}{R}$ (**l**)

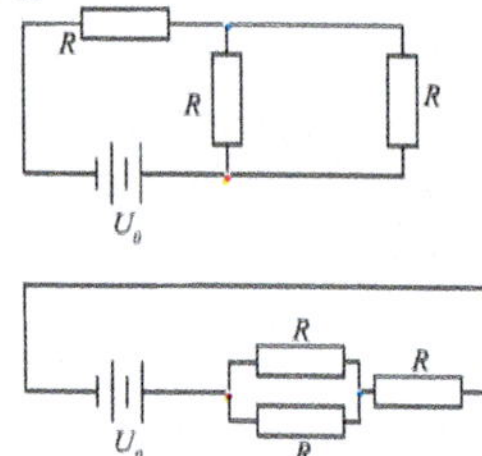

(**m**) No, we only used one of the equations from Kirchhoff's current law, because the two equations are identical. Similarly, we only used two of the equations from Kirchhoff's voltage law, because the third law is linearly dependent on the two others.

10.10 **Equal currents**

(**a**) $I_1 = I_2 = I_3 = I_4 = I_5 = I_6$ (**b**) $I_{12} = I_7 = I_8 = -I_9 = -I_{10} = I_{11}$ (**c**) In general, none.

10.11 **Battery or capacitor**

(**a**)

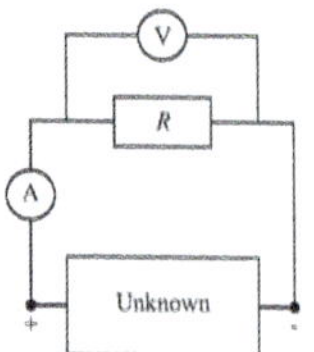

(**b**) We expect V and I to be constant, and that V is the voltage of the battery. (**c**)

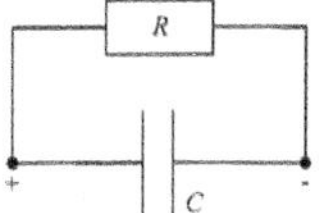

(**d**) When the current is out from the capacitor, in positive direction, the charge is decreasing so that $I = -\mathrm{d}Q/\mathrm{d}t$. (**e**) $\frac{\mathrm{d}V}{\mathrm{d}t} = \frac{1}{C}\frac{\mathrm{d}Q}{\mathrm{d}t} = -\frac{1}{C}I$ (**f**) $V - IR = 0 \Rightarrow I = \frac{V}{R}$ and $\frac{\mathrm{d}V}{\mathrm{d}t} = -\frac{1}{C}I = -\frac{1}{C}\frac{V}{R} = -\frac{V}{RC}$ (**h**) The time constant $\tau = RC$. (**i**) $C = \frac{A\epsilon}{d} = \frac{1\,\mathrm{m}^2\, 8.854\, 10^{-12}\,\mathrm{Fm}^{-1}}{10^{-4}\,\mathrm{m}} = 8.85\, 10^{-8}\,\mathrm{F}$ and therefore $\tau = 100\,\mathrm{ohm} \cdot 8.85\, 10^{-8}\,\mathrm{F} = 10^{-5}\,\mathrm{s}$. This is a very small number. The figure shows that the discharge is much slower than this. This may be because there is a large resistor inside the suitcase that makes the time constant much longer and therefore the discharge slower.

10.12 **Is the circuit closed**
(**a**) Yes, there is a current. This is effectively a capacitor. (**b**) Eventually the electric field in the space between the two cylinders becomes so large that the air experiences dielectric breakdown and a current will pass through the air in the form of a spark.

10.13 **A simple circuit**
(**a**) $I = \frac{R_1+R_2}{R_1 R_2} U_0$. (**b**) $P = \frac{R_1+R_2}{R_1 R_2} U_0^2$.

10.14 **From model to circuit**
(**b**)

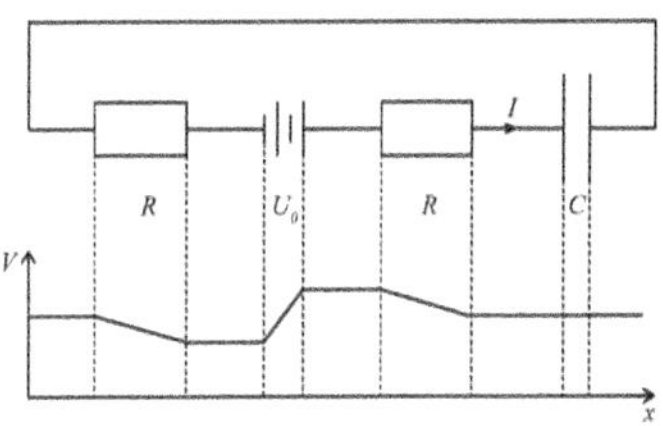

(**c**) $I = e/(2R)$ (**d**) $I = 0$. No drop across the resistor. A drop e across the capacitor equal to the increase e across the battery. (**e**) $\frac{\mathrm{d}I}{\mathrm{d}t} = -\frac{1}{2RC}I$, $I(t) = I_0 e^{-t/t^*}$

10.15 **A circuit-model for a cell**
(**a**) $I(t = 0) = e/R$ (**b**) $Q = Q_\infty = eC$

10.16 **Wheatstone bridge**
(**a**) $V_a - V_b = \left(\frac{R_1}{R_1+R_3} - \frac{R_2}{R_2+R_x}\right) V_s = \frac{R_1 R_x - R_2 R_3}{(R_1+R_3)(R_2+R_x)} V_s$. (**b**) $R_x = \frac{R_2 R_3}{R_1}$

10.17 **Lightning strike**
(**a**) $Q(0) = 0, V(0) = 0$. (**b**) $V = V_0, Q = CV_0$ (**c**) $V(t) = V_0(1 - \exp(-t/(R_0 C)))$

10.18 **Kirchhoff's circuit laws**
(**a**) $I_1 \approx 0.318\mathrm{A}$, $I_2 \approx 0.091\mathrm{A}$ and $I_3 \approx 0.227\mathrm{A}$. (**b**) $I_1 \approx 0.318\mathrm{A}$, $I_2 \approx 0.091\mathrm{A}$, $I_3 \approx 0.227\mathrm{A}$, $I_4 = -0.6\mathrm{A}$, $I_5 = -0.48\mathrm{A}$ and $I_6 \approx -0.517\mathrm{A}$.

10.19 **Circuit model of conducting sphere**

(b) $R = 1/(4\pi\sigma)(1/a - 1/b)$ **(c)** $C = 4\pi\epsilon/(1/a - 1/b)$. **(d)**

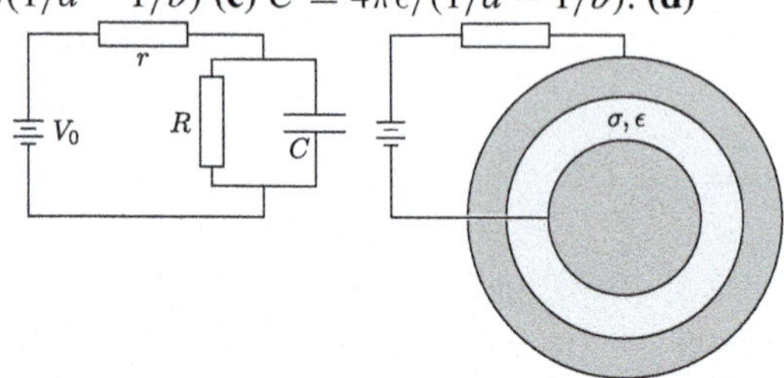

(e) $I = V_0/r$ **(f)** $I = V_0/(r + R)$ **(g)**

(h) $I = I_R + I_C$

10.20 **Voltage gated ion channels**

(d)

```
import numpy as np
R, r, C, v = 10.0, 1.0, 1.0, 2.0
tmax, dt = 10.0, 0.01
nt = int(tmax/dt)
VC = np.zeros(nt)
t =  np.zeros(nt)
def VS(t):
    return 5.0
for i in range(nt-1):
    if (VC[i]>v):
        dVC = (1/C)*(VS(t[i])-VC[i]/R-VC[i]/r)
    else:
        dVC = (1/C)*(VS(t)-VC[i]/R)
    VC[i+1] = VC[i] + dVC*dt
    t[i+1] = t[i]+dt
plt.subplot(2,1,1)
plt.plot(t,VC)
plt.ylabel("$V_C$")
plt.subplot(2,1,2)
plt.plot(t,np.gradient(VC)*C)
plt.ylabel("$I_C$")
plt.xlabel("$t$")
```

11.6 **Magnetic field from a current element**

(a) A: $-\hat{\mathbf{z}}$, B: $-\hat{\mathbf{z}}$, C: $\mathbf{0}$, D: $+\hat{\mathbf{z}}$, E: $\hat{\mathbf{z}}$, F: $\hat{\mathbf{z}}$, G: $\mathbf{0}$, H: $-\hat{\mathbf{z}}$ **(b)** A: $\hat{\mathbf{y}}$, B: $\frac{1}{\sqrt{2}}(-\hat{\mathbf{x}} + \hat{\mathbf{y}})$, C: $-\hat{\mathbf{x}}$, D: $\frac{1}{\sqrt{2}}(-\hat{\mathbf{x}} - \hat{\mathbf{y}})$, E: $-\hat{\mathbf{y}}$, F: $\frac{1}{\sqrt{2}}(+\hat{\mathbf{x}} - \hat{\mathbf{y}})$, G: $+\hat{\mathbf{x}}$, H: $\frac{1}{\sqrt{2}}(+\hat{\mathbf{x}} + \hat{\mathbf{y}})$ **(c)** A: $\frac{\mu_0}{4\pi}\frac{I\,\mathrm{d}l(+\hat{\mathbf{y}})}{a^2}$,

B: $\frac{\mu_0}{4\pi}\frac{I\,dl\frac{1}{\sqrt{2}}(-\hat{\mathbf{x}}+\hat{\mathbf{y}})}{2a^2}$, C: $\frac{\mu_0}{4\pi}\frac{I\,dl(-\hat{\mathbf{x}})}{a^2}$, D: $\frac{\mu_0}{4\pi}\frac{I\,dl\frac{1}{\sqrt{2}}(-\hat{\mathbf{x}}-\hat{\mathbf{y}})}{2a^2}$, E: $\frac{\mu_0}{4\pi}\frac{I\,dl(-\hat{\mathbf{y}})}{a^2}$, F: $\frac{\mu_0}{4\pi}\frac{I\,dl\frac{1}{\sqrt{2}}(\hat{\mathbf{x}}-\hat{\mathbf{y}})}{2a^2}$, G: $\frac{\mu_0}{4\pi}\frac{I\,dl(+\hat{\mathbf{x}})}{a^2}$, H: $\frac{\mu_0}{4\pi}\frac{I\,dl\frac{1}{\sqrt{2}}(+\hat{\mathbf{x}}+\hat{\mathbf{y}})}{2a^2}$

11.7 **Two current elements**
(a) $+\hat{\mathbf{z}}$ **(b)** For example as circle around the midpoint between the two elements, but there are many possible solutions. **(c)** 0 **(d)** One solution is to extend each element to an infinitely long line. Another solutions is to make two quadradic or rectangular circuits symmetrically placed around the midpoint. **(e)** 0 **(f)** We extend each of the two circuits to quadratic or rectangular circuits with a side equal to the current element. The circuits are placed symmetrically around the midpoint.

11.8 **Comparing a line charge and a line current**
(a) $\mathbf{R} = (r, 0, -z)$ **(b)** For each charge element ρdl there will also be a charge element symmetrically placed around the xy-plane, and the z-contributions from each of these elements will cancel. **(c)** $\mathrm{d}E_x = \frac{\rho \mathrm{d}l}{4\pi\epsilon_0}\frac{R_x}{R^3} = \frac{\rho \mathrm{d}l}{4\pi\epsilon_0}\frac{r}{(z^2+r^2)^{3/2}}$ **(d)** $E_x = \int_{-\infty}^{\infty}\frac{\rho}{4\pi\epsilon_0}\frac{r}{(z^2+r^2)^{3/2}}\,\mathrm{d}z = \frac{\rho}{2\pi\epsilon_0 r}$ **(e)** $\mathbf{R} = (r, 0, -z)$ **(f)** We see from the right-hand-rule that all the contributions will be directed along the y-axis. **(g)** $\mathrm{d}\mathbf{B} = \frac{\mu_0}{4\pi}\frac{I\mathrm{d}l r\hat{\mathbf{y}}}{(r^2+z^2)^{3/2}}$ **(h)** $\mathbf{B} = \int_{-\infty}^{\infty}\frac{I\mu_0}{4\pi}\frac{r\hat{\mathbf{y}}}{(z^2+r^2)^{3/2}}\,\mathrm{d}z$ **(i)** $\mathbf{B} = \frac{I\mu_0}{2\pi r}\hat{\mathbf{y}}$ **(j)** The most important difference is between using $\mathbf{R}$ directly to find the electric field and $\mathbf{dl} \times \mathbf{R}$ to find the magnetic field.

11.9 **Current loop in a magnetic field**
(a) 0 **(b)** $\boldsymbol{\tau} = -Ia^2 B_0\hat{\mathbf{x}}$ **(c)** To measure the direction, you let the element rotate freely around the x-axis. The current loop will then orient so that its normal vector points in the direction of the field. To measure the magnitude, you direct the loop so that it is normal to the field, using the method you developing in the first part. Then you let the loop rotate around x-axis while it is attached to a torsional spring. The extent of the rotation will then be proportional to the torque.

11.10 **Charges in a tokamak**
(a) Nothing, the force from the magnetic field is zero. **(b)** It will move in a circular orbit with radius $r = mv/(QB_0)$. **(c)** No **(d)**

```
for i in range(nstep-1):
    B = B_vec(B0,r[i])
    F = np.cross(Q*v[i],B)
    a = F/m
    v[i+1] = v[i] + a*dt
    r[i+1] = r[i] + v[i+1]*dt
```

11.11 **An explicitely parameterized curve integral**
(a) $\mathbf{r}(\phi) = a(\cos\phi, \sin\phi, 0)$ **(b)** $\mathbf{R} = (0, 0, z) - a(\cos\phi, a\sin\phi, 0) = (-a\cos\phi, -a\sin\phi, z)$ **(c)** $\mathbf{dl} = \frac{\mathrm{d}\mathbf{r}'(\phi)}{\mathrm{d}\phi}\mathrm{d}\phi = a(-\sin\phi, \cos\phi)\mathrm{d}\phi$ **(d)** $\mathrm{d}\mathbf{B} = \frac{\mu_0}{4\pi}\frac{Ia\,\mathrm{d}\phi(z\cos\phi, z\sin\phi, a)}{(z^2+a^2)^{3/2}}$ **(e)** $\mathbf{B} = \frac{\mu_0}{4\pi}\frac{Ia2\pi\hat{\mathbf{z}}}{(z^2+a^2)^{3/2}} = \frac{\mu_0 Ia}{2(z^2+a^2)^{3/2}}\hat{\mathbf{z}}$ **(f)** The calculations are nearly identical, but the integral is now over s from 0 to $2\pi a$.

11.12 **Finding B from a line segment**
(b) $dB_y = -\mu_0 I \, dl/(4\pi)(z/R^3)$

11.13 **Force on a current element**
$F = \frac{\mu_0 I_1 I_2 L}{\pi r}$

11.14 **Magnetic field from a thin sheet**
(a) Because the system has translational symmetry in the xy-direction.

11.16 **Infinite bent wire**
(a) $\mathbf{B}(\mathbf{0}) = -\frac{\mu_0 I}{4a}\left(\frac{2}{\pi} + 1\right)\hat{\mathbf{z}}$.

11.17 **Field from a rotating circle**
(a) $a\omega\rho$ **(b)** $B_z = (\mu_0 \rho \omega a^3)/(2(z^2 + a^2)^{3/2})$

11.19 **Two semi-infinite line currents**
(b) $\mathbf{dB} = (I\mu_0)/(4\pi)\,(-y\hat{\mathbf{z}})/((x - x')^2 + y^2)^3$ **(c)** $\mathbf{B} = (-I\mu_0\hat{\mathbf{z}})/(4\pi y)\left(1 + \frac{x}{(x^2+y^2)^{1/2}}\right)$ **(d)** $\mathbf{B} = (-I\mu_0\hat{\mathbf{z}})/(4\pi x)\left(1 + \frac{y}{(y^2+x^2)^{1/2}}\right)$ **(e)** $\mathbf{B} = \frac{-I\mu_0\hat{\mathbf{z}}}{4\pi}\left(\frac{1}{x}\left(1 + \frac{y}{(y^2+x^2)^{1/2}}\right) + \frac{1}{y}\left(1 + \frac{x}{(y^2+x^2)^{1/2}}\right)\right)$

11.20 **Force from magnetic field on curved wire**
$\mathbf{F} = -2IBa\hat{\mathbf{y}}$

11.23 **A finite solenoid**
(b)

```
import numpy as np
def Bcircle(r,r0,a,I,N):
    mu0 = 1.257e-6
    K = mu0/(4*np.pi)
    B = np.array([0,0,0])
    Idl = I*a*2*np.pi/N
    for j in range(N):
        theta = (j/N)*2*np.pi
        rj = r0 + np.array([a*np.cos(theta),a*np.sin(theta),0])
        dlvec = np.array([-np.sin(theta),np.cos(theta),0])
        R = r-rj
        dB = K*Idl*np.cross(dlvec,R)/np.linalg.norm(R)**3
        B = B + dB
    return B
```

(c)

```
# We test by setting r0 = (0,0,0)
r0 = np.array([0,0,0])
I = 1.0
a = 0.1
N = 100
```

```
r = np.array([0,0,0])
Bnum = Bcircle(r,r0,a,I,N)
print("Numerical = ",Bnum)
mu0 = 1.257e-6
Bexact = mu0*I/(2.0*a)*np.array([0,0,1])
print("Exact = ",Bexact)
```

```
Numerical = [0.000e+00 0.000e+00 6.285e-06]
Exact = [0.000e+00 0.000e+00 6.285e-06]
```

(d)

```
a = 0.1
I = 1.0
Lx, Lz, Nx, Nz = 2*a,2*a,20,20
x = np.linspace(-Lx,Lx,Nx)
z = np.linspace(-Lz,Lz,Nz)
rx,rz = np.meshgrid(x,z,indexing="ij")
Bx = np.zeros_like(rx)
Bz = np.zeros_like(rz)
r0 = np.array([0,0,0])
N = 100
for ix in range(Nx):
    for iz in range(Nz):
        r = np.array([rx[ix,iz],0,rz[ix,iz]])
        Bx[ix,iz],By,Bz[ix,iz] = Bcircle(r,r0,a,I,N)
import matplotlib.pyplot as plt
plt.figure(figsize=(8,4))
plt.subplot(1,2,1)
plt.quiver(rx,rz,Bx,Bz)
plt.xlabel("$x$ (m)"), plt.ylabel("$z$ (m)")
plt.subplot(1,2,2)
plt.streamplot(rx.T,rz.T,Bx.T,Bz.T)
plt.xlabel("$x$ (m)")
```

(e)

```
import numpy as np
def Bsolenoide(r,a,I,N,M):
    B = np.array([0,0,0])
# Make M circular currents and sum the contributions
    for i in range(M):
        zi = -a + (2*a)/(M-1)*i
        r0 = np.array([0,0,zi])
        B = B + Bcircle(r,r0,a,I,N)
    return B
```

(f)

```
a = 0.1
I = 1.0
Lx, Lz, Nx, Nz = 2*a,2*a,20,20
x = np.linspace(-Lx,Lx,Nx)
z = np.linspace(-Lz,Lz,Nz)
```

```
rx,rz = np.meshgrid(x,z,indexing="ij")
Bx = np.zeros_like(rx)
Bz = np.zeros_like(rz)
r0 = np.array([0,0,0])
N = 100 # Antall elementer i hver krets
M = 5 # Antall æsirkulre kretser
for ix in range(Nx):
    for iz in range(Nz):
        r = np.array([rx[ix,iz],0,rz[ix,iz]])
        Bx[ix,iz],By,Bz[ix,iz] = Bsolenoide(r,a,I,N,M)
plt.figure(figsize=(12,6))
plt.subplot(1,2,1)
plt.quiver(rx,rz,Bx,Bz)
plt.xlabel("$x$ (m)"), plt.ylabel("$z$ (m)")
plt.subplot(1,2,2)
plt.streamplot(rx.T,rz.T,Bx.T,Bz.T)
plt.xlabel("$x$ (m)")
```

(g)

```
N = 100
a = 0.1
M = 5
Ni = 100
zp = np.linspace(-a,a,Ni)
Bz1 = np.zeros_like(zp)
Bz2 = np.zeros_like(zp)
Bz3 = np.zeros_like(zp)
for i in range(Ni):
    zi = zp[i]
    r1 = np.array([0,0,zi])
    Bxi,Byz,Bzi = Bsolenoide(r1,a,I,N,M)
    Bz1[i] = Bzi
    r2 = np.array([a/2,0,zi])
    Bxi,Byz,Bzi = Bsolenoide(r2,a,I,N,M)
    Bz2[i] = Bzi
    r3 = np.array([2*a,0,zi])
    Bxi,Byz,Bzi = Bsolenoide(r3,a,I,N,M)
    Bz3[i] = Bzi
plt.plot(zp,Bz1,label="$B_z(0,0,z)$")
plt.plot(zp,Bz2,label="$B_z(a/2,0,z)$")
plt.plot(zp,Bz3,label="$B_z(2a,0,z)$")
plt.legend()
plt.xlabel("$z$ (m)"), plt.ylabel("$B_z$")
```

12.4 **Possible and impossible fields**
(a) Yes. **(b)** No. **(c)** Yes, if $B_\phi(r, z)$ is *not* at function of ϕ. **(d)** Yes

12.5 **It cannot possibly be a field**
For example $\mathbf{B} = x\hat{\mathbf{x}}$.

12.6 **Is something wrong with the law**
No

12.7 **Net current**
(a) $I = 2J_0a^2$. Positive sign. **(b)** $I = 2J_0a^2$. Positive sign. **(c)** For a plane surface it is the cross-sectional area, the area normal to the current density, which is included in the $I = \mathbf{J} \cdot \mathbf{S} = J S_{\text{cross-section}}$

12.8 **Strange Amperian loops**
A: $I_S = 0$, B: $I_S = 0$, C: $I_S = 0$, D: $I_S = +6I$, E: $I_S = -2I$

12.9 **Ampere's law and symmetries**
(a) Cartersian. **(b)** $\mathbf{B} = \mathbf{B}(z)$ **(c)** No **(d)** $\mathbf{B}(z) = -\mathbf{B}(-z)$. **(e)** No **(f)** A rectangular loop symmetrically around the origin such as: $(0, y, z)$, $(0, -y, z)$, $(0, -y, -z)$, $(0, y, -z)$. **(g)** $\oint_C \mathbf{B} \cdot \mathbf{dl} = -4yB_y(z) = \mu_0 I = \mu_0(2y)J_0$ **(h)** $B_y = -\frac{\mu_0 J_0}{2}$ $(z > 0)$ and $B_y = \frac{\mu_0 J_0}{2}$ $(z < 0)$

12.10 **Surface current on a cylinder**
(a) $\mathbf{B} = \mathbf{B}(r)$ **(b)** No **(c)** No **(d)** B_z is constant outside the cylinder. **(e)** $B_z = 0$ **(f)** That B_z is constant for $r < a$. **(g)** $B_z = \mu_0 J_s$

12.11 **Coaxial cable**
(a) $\mathbf{B}(r) = \frac{\mu_0 I r}{2\pi a^2}\hat{\phi}$ for $r < a$ and $\mathbf{B}(r) = \frac{\mu_0 I}{2\pi r}\hat{\phi}$ for $r > a$. **(b)** $\mathbf{B}(r) = \mu_0 I \frac{r^2 - b^2}{2\pi r(c^2 - b^2)}\hat{\phi}$ for $b < r < c$ and $\mathbf{B}(r) = \frac{\mu_0 I}{2\pi r}\hat{\phi}$ for $r > c$.

12.14 **The field from a rotating charged cylinder**
(d) $B_z = \mu_0 \rho_s \omega a$. **(e)** $\mathbf{J} = B_0/(\mu_0 r)\hat{\mathbf{z}}$.

13.6 **Cylindrical magnet**
(a) $\mathbf{H} = \mathbf{B}_0/\mu_0$ **(d)** $H = B_0/\mu_0$ **(f)** $\mathbf{M} = \chi_m \mathbf{B}_0/\mu_0$ **(g)** $\mathbf{B} = (1 + \chi_m)\mathbf{B}_0$

13.7 **Similarities and differences between Gauss and Ampere**
(b) $Q = v\rho = \pi a^2 L\rho$, $\mathbf{D} = \frac{\rho a^2}{2r}\hat{\mathbf{r}}$ **(e)** $\mathbf{H}(r) = (1/2)r|\mathbf{J}|\hat{\phi}$

13.8 **Magnetic circuits**
(b) $\oint_C \mathbf{E} \cdot \mathbf{dl}$, $\oint_C \mathbf{H} \cdot \mathbf{dl}$ **(d)** $\mathbf{J} = \sigma\mathbf{E}$ and $\mathbf{B} = \mu\mathbf{H}$ **(e)** $I_1 = J_1 S_1 = \sigma E_1 S_1$ $I_2 = J_2 S_2 = \sigma E_2 S_2$, $\Phi_1 = B_1 S_1 = \mu H_1 S_1$ $I_2 = B_2 S_2 = \mu H_2 S_2$ **(g)** $E_i = \frac{I}{\sigma S_i}$ $H_i = \frac{\Phi}{\mu S_i}$

13.9 **A non-uniformly magnetized cylinder**
(a) $\mathbf{J}_b = 2M_0 r/a^2\hat{\phi}$, $\mathbf{J}_{b,s} = 0$ **(b)** $B_z = \mu_0 M_0$

13.10 **A magnetized sphere**
(a) $\mathbf{J}_{b,s} = M_0 \sin\theta\hat{\phi}$ **(b)** $B_z = (2/3)\mu_0 M$

13.11 Surface current

(a) $\mathbf{B} = \mu J_S/2\hat{\mathbf{y}}$ for $z > 0$, and $-\mu J_S/2\hat{\mathbf{y}}$ for $z < 0$ **(b)** $\mathbf{H} = J_0\hat{\mathbf{z}}$, $\mathbf{B} = \mu_0 J_0\hat{\mathbf{z}}$ inside the cylinder and zero outside. **(c)** $B = H = 0$

13.12 A simple magnetic trap

(b) $\nabla \cdot \mathbf{B} = 0$ **(c)** 0 **(d)** $e = \frac{B_0}{b} 2Lv$

13.13 Rectangular magentic circuit

(b) $R_{m,1} = l_1/(\mu_1 S_1)$, $R_{m,2} = l_2/(\mu_2 S_2)$ **(c)** $R_m = \sum_i R_{m,i}$

13.14 A quadratic circuit

(a) $\mathbf{B}(x, y, 0) = \frac{\mu_0 I}{2\pi y}\hat{\mathbf{z}}$ **(b)** $\mathrm{d}\mathbf{B} = \frac{\mu_0}{4\pi}\frac{I\,\mathrm{d}x' y\hat{\mathbf{z}}}{((x')^2+y^2)^{3/2}}$ **(c)** $\int \mathrm{d}\mathbf{B} = \frac{I\mu_0}{2\pi y}\frac{1}{\sqrt{1+\frac{y^2}{(a/2)^2}}}$ **(d)**

```
# Felt fra linje
mu0 = 1.257e-6
K = mu0/(4*np.pi)
def Bline(r,a_vec,a_0,I,N):
    B = np.array([0,0,0])
    Idl_vec = I*a_vec/N
    for j in range(N):
        rj = a_0 + a_vec*j/N
        R = r-rj
        dB = K*np.cross(Idl_vec,R)/np.linalg.norm(R)**3
        B = B + dB
    return B
```

(e)

```
# Find the field from a line along the x-axis
a = 1.0
a_0 = np.array([-a/2,0,0])
a_vec = np.array([a,0,0])
I = 1.0 # Current (A)
N = 100 # Number of points in the plot
x = np.zeros((N,1))
y = 0.1*a
B = np.zeros((N,3))
for ix in range(N):
    xix = -a+2*a*ix/N
    r = np.array([xix,y,0])
    x[ix] = xix
    B[ix] = Bline(r,a_vec,a_0,I,30)
plt.plot(x,B[:,2]*1e6)
B_theory = K*I*2/y*np.ones((N,1))
plt.plot(x,B_theory*1e6)
B_line = K*I*2/y/np.sqrt(1+(y/(0.5*a))**2)
plt.plot(0,B_line*1e6,"o")
plt.xlabel("$z$ (m)")
plt.ylabel("$B_z$ ($\mu$T)")
```

(f)

```
# Find the field inside the circuit
mu0 = 1.257e-6
K = mu0/(4*np.pi)
def Bsquare(r,a,I,N):
    B = np.array([0,0,0])
    # Four lines, each with N/4 points
    N4 = int(N/4)
    # Line 1
    a_0 = np.array([a/2,a/2,0])
    a_vec = np.array([-a,0,0])
    B = B + Bline(r,a_vec,a_0,I,N4)
    # Line 2
    a_0 = np.array([-a/2,a/2,0])
    a_vec = np.array([0,-a,0])
    B = B + Bline(r,a_vec,a_0,I,N4)
    # Line 3
    a_0 = np.array([-a/2,-a/2,0])
    a_vec = np.array([a,0,0])
    B = B + Bline(r,a_vec,a_0,I,N4)
    # Line 4
    a_0 = np.array([a/2,-a/2,0])
    a_vec = np.array([0,a,0])
    B = B + Bline(r,a_vec,a_0,I,N4)
    return B
a = 1.0
L = 1.0
I = 1.0
N = 100
Nx = 20
Ny = 20
Bz = np.zeros((Nx+1,Ny+1))
for ix in range(Nx+1):
    for iy in range(Ny+1):
        xi = -L/2+L*ix/Nx
        yi = -L/2+L*iy/Ny
        r = np.array([xi,yi,0])
        Bi = Bsquare(r,a,I,N)
        Bz[ix,iy] = Bi[2]
plt.contourf(Bz,cmap="jet")
plt.colorbar()
```

14.4 **Smart-emf**
(a) Positive charges will move from left to right in the capacitor so that a positive charge builds up on the right hand side and a negative charge on the left hand side. **(b)** $E = F/q = f$. **(c)** $V_A = 0$, $V_B = fL$, $V_C = V_B$, $V_D = V_A = 0$, $V_E = V_A = 0$, $V_F = V_A = 0$ **(d)** $\oint_C \mathbf{E} \cdot d\mathbf{l} = 0$. **(e)** The voltage increase fL across the capacitor, so that the voltage drop is $-fL$. **(f)** The voltage drop is fL. **(h)** 0 **(i)** $e = fL$ **(j)** $I = fL/R$

14.5 **Find the flux**
(a) $\Phi_B = 4a^2 B_0$. **(b)** $\Phi_B = 0$ **(c)** $\Phi_B = \frac{B_0 a^3}{b}$ **(d)** $\Phi_B = B_0 4a^2 \cos\theta$ **(e)** $\Phi_B = \pi a^2 B_0$

(f) $\Phi_B = \frac{2\pi a^3 B_0}{3b}$

14.6 **Find the emf**
(a) $e = -\frac{d\Phi_B}{dt} = 0$ **(b)** $e = -\frac{4a^2 v B_0}{b}$ **(c)** $e = -B_0(4av + 2vt)$ **(d)** $e = B_0 4a^2 \omega \sin\theta$
(e) $e = -B_0 v 2a \quad (t > 0) \qquad B_0 v 2a \quad (t < 0)$ **(f)** $e = \frac{I\mu_0 a}{2\pi}\left(\frac{1}{y+a} - \frac{1}{y}\right) v$

14.7 **Squeeze and charge**
(a) Yes. The flux will change when you squeeze, so there will be induced an emf. **(b)** $e = 2.0\,\mu$V **(c)** You may use a better approximation for the magnetic field from a permanent magnetic, for example in the form of a magnetic dipole field.

14.8 **Emf on the bus**
(b) $x(t) = vt$ **(c)** $\Phi_B = 0$ when $t < a/v$; $B_0 d(vt - a)$ when $a/v < t < (d+a)/v$; $B_0 d^2$ when $(d+a)/v < t < (a+b)/v$; $B_0 d(a + b - (vt - d))$ when $(a+b)/v < t < (a+b+d)/v$; and 0 when $(a+b+d)/v < t$ **(d)** $e = 0$ when $t < a/v$; $-B_0 dv$ when $a/v < t < (d+a)/v$; 0 when $(d+a)/v < t < (a+b)/v$; $B_0 dv$ when $(a+b)/v < t < (a+b+d)/v$; and 0 when $(a+b+d)/v < t$. **(e)**

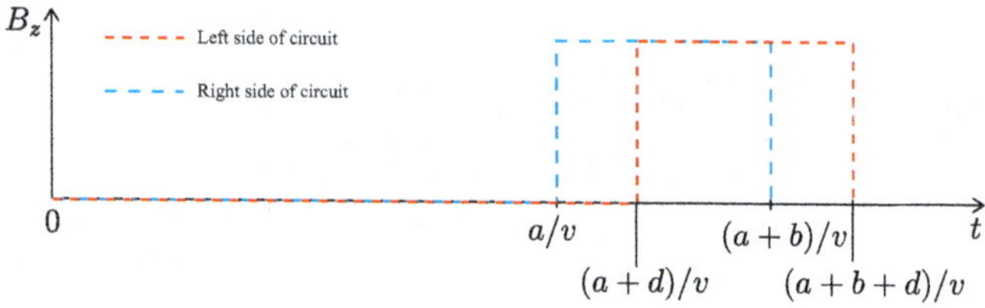

14.9 **Circular currents**
(a) $\Phi_B = B_0 \sin\omega t \pi a^2$ **(b)** $e = -\pi a^2 B_0 \omega \cos\omega t$ **(c)** $R = 2\pi a/(d^2\sigma)$ **(d)** $I = -\frac{ad^2\sigma B_0 \omega \cos\omega t}{2}$ **(e)** $I = I_0 a$ where I_0 depends on the time and the magnetic field, but not on a. **(f)** $\mathbf{J} = J_0 r \hat{\phi}$

14.10 **Faraday's law**
(b) $B = \mu_0 I n$

14.11 **Rod on rails**
(a) $e = vBl$ **(b)** $e = -vBl$ **(c)** $I = -vBl/R$ **(d)** $v = v_0 e^{-t/t_0}$ where $t_0 = mR/(Bl^2)$
(e) $R = 2\rho x/A$

14.12 **Inhomogeneous field**
(a) $I = 0$ **(b)** $I = -B_0 v_0 L/R$ **(c)** $I = v_0 L\,(B(v_0 t + L) - B(v_0 t))\,/R$

14.13 **An expanding circle**
(b) $I = -\frac{2\pi a b_0 B_0}{R}$ **(e)** Yes **(f)** $I = \frac{e}{R} = -\frac{2\pi B_0 a^2 v_0}{R r_0}$

15.4 **Introduction to self-inductance**
(a) $B = \frac{\mu_0 N I}{L}$ **(b)** $\Phi_B = \frac{\mu_0 N^2 \pi a^2}{L} I$

15.5 **Self-inductance per unit length**
(a)

(b) $B_z = \frac{\mu_0 I}{2\pi x}$ **(c)** $B_z = \frac{\mu_0 I}{2\pi(d-x)}$ **(d)** Otherwise the magnetic field would diverge in the region where we need to find the flux. **(e)** $\Phi_B = \frac{\mu_0 I \ell}{\pi} \ln \frac{d-a}{a}$ **(f)** $L/\ell = \frac{\mu_0}{\pi} \ln \frac{d-a}{a}$

15.6 **Long and short circuit**
(b) The line wire. **(c)** $B_y = \frac{\mu_0 I}{2\pi x}$ **(d)** $\Phi_B = \frac{\mu_0 I a}{2\pi} \ln \frac{d+a}{d}$ **(e)** $L_{1,2} = \frac{\mu_0 a}{2\pi} \ln \frac{d+a}{d}$ **(f)** This is only included in the self-inductance, which we also need to include if we want to model the system.

15.8 **Energy densities**
(a) $\mathbf{H} = NI/h\hat{z}$, $\mathbf{B} = \mu\mathbf{H}$. **(b)** $u = (1/2)\mathbf{B}\cdot\mathbf{H} = (1/2)\mu(NI/h)^2$ **(c)** $W_m = uV = u\pi a^2 h = (1/2)\mu\pi a^2 (NI)^2/h$ **(d)** $L = \Phi/I = N\pi a^2 \mu N/h$, $W_m = (1/2)LI^2$.

15.9 **We turn the circuit on**
(a) B: V_b; C: V_b; D: V_b **(b)** B: V_b; C: V_b; D: 0. **(c)** The potential does not change in C when you close the circuit. **(d)** V_b **(e)** $I = 0$ **(f)** $\sum \Delta V = 0 = V_b - IR - L\frac{dI}{dt} = 0$ **(g)** $I = V_b/R$ **(h)** $I(t) = V_b/R(1 - e^{-tRL})$

15.10 **Self-inductance of a coaxial cable**
(a)

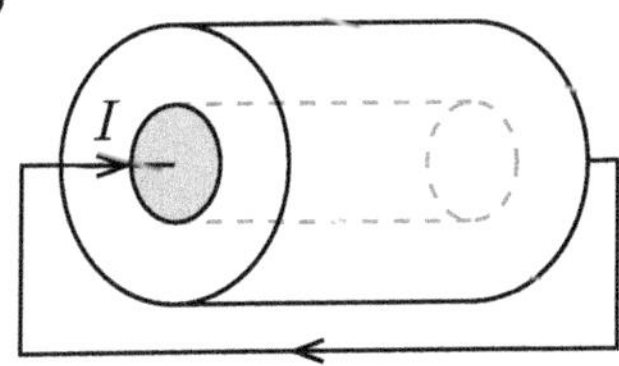

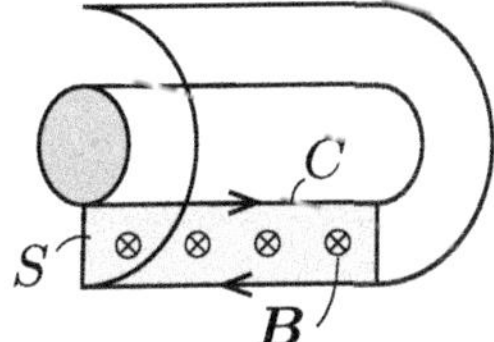

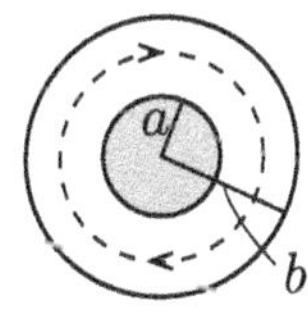

(b) $\mathbf{B} = \frac{\mu_0 I}{2\pi r}\hat{\phi}$ **(c)** $\Phi = \frac{\mu_0 I}{2\pi} l \ln \frac{b}{a}$ **(d)** $\frac{L}{l} = \frac{\mu_0}{2\pi} \ln \frac{b}{a}$

15.11 **Cross-talk**
(a) $B_1 = \frac{\mu_0 I_1}{2\pi r}$ **(b)** $\Phi = \frac{\mu_0 I_1 l}{2\pi} \ln \frac{dc}{ab}$ **(c)** $L_{1,2} = \frac{\mu_0}{2\pi} \ln \frac{dc}{ab}$

15.12 **Energy in coaxial cable**
$W_m = \frac{1}{2} l \frac{\mu}{2\pi} \ln \frac{b}{a} I^2$

15.13 **Hairdryer**
(b) 13 A **(c)** 9.0 Ω

15.15 **Two circles**
(a) $L_{12} = 0$

15.16 **A double coil**
(a) $\mathbf{H} = NI/L\hat{\mathbf{x}}$, $\mathbf{B} = \mu NI/L\hat{\mathbf{x}}$, $\mathbf{M} = (\mu/\mu_0 - 1)\mathbf{H}$ **(b)** $L_{11} = \frac{\mu N^2 \pi a^2}{L}$ **(c)** $L_{12} = \frac{\mu N M \pi a^2}{L}$

16.5 **Circuit with coil and resistor**
(a) $I(0) = 0$ **(b)** $+V_b - IR - L\frac{dI}{dt} = 0$ **(c)** $I = V_b/R$ **(e)** $I(t) = (U_0/R)e^{-t/(L/R)}$

16.6 **A simple oscillator circuit**
(a) $-\frac{Q}{C} - L\frac{dI}{dt} = 0$ **(b)** $\frac{d^2 I}{dt^2} = -\frac{1}{LC}I$ where $I = dQ/dt$. **(c)** We need to initial conditions, e.g. $I(t_0)$ and $dI/dt\ (t_0)$. **(d)** $I(0) = 0$ and $I'(0) = -Q_0/(LC)$ **(e)** After Δt some charge will have flowed out from the positive side of the capacitor. This corresponds to a negative chrage with the chosen positive direction of current ($dQ/dt = I$). Therefore, $I'(0)$ must be negative. **(f)** $I(t) = -Q_0\omega \sin \omega t, \quad \omega = \frac{1}{\sqrt{LC}}$

16.7 **Model for a switch**
(a) This is a circuit consisting of a battery, a resistor and a capacitor in series. **(b)** The voltage increases with V_b across the battery and decreases by V_b across the switch. **(c)** The voltage increases by V_b across the battery, falls by $V_b/2$ across the resistor and with $V_b/2$ across the capacitor. **(d)** The voltage across the capacitor will gradually build up to V_b again. **(e)** If the voltage drop is constant at V_b, the electric field will be $E = V_b/d$, which becomes larger as d becomes smaller. **(f)** A spark when the electric field increases beyond a limit, the air is ionized and will become conducting.

16.8 **Periodically driven circuits**
(a) $U - V_C - IR = 0$ **(b)** $I = C dV_C/dt$ **(c)** $U - V_C - RC\frac{dV_C}{dt} = 0$ **(d)**

```
import numpy as np
import matplotlib.pyplot as plt
RC = 0.1
U0 = 1.0
VC0 = 0.0
dt = 0.001
time = 20.0
omega = 1.0
N = int(time/dt)
VC = np.zeros(N)
VC[0] = VC0
t = np.zeros(N)
for i in range(N-1):
    Vb = U0*np.sin(omega*t[i])
    VC[i+1] = VC[i] + dt*(Vb-VC[i])/RC
```

```
    t[i+1] = t[i] + dt
plt.plot(t,VC)
```

16.9 **LR circuit**
(b) IR **(c)** $-LdI/dt$ **(d)** $V_0 + e - RI = V_0 + (-d\Phi/dt) - RI = 0$ **(e)** $V_0 - RI - V_L = V_0 - RI - d\Phi/dt$ **(f)** $\frac{dI}{dt} = \frac{V_0}{L} - \frac{1}{\tau}I$ **(g)** $I(t) = I_0 e^{-t/\tau}$ **(h)** $I(t) = (V_0/R)\left(1 - e^{-t/\tau}\right)$

16.10 **Complex capacitor**
(b) $\hat{Z} = 1/(i\omega C)$ **(c)** $\hat{I} = i\omega C\hat{V}$

References

A. G. Csaszar, G. Czako, T. Furtenbacher, J. Tennyson, V. Szalay, S. V. Shirin, N. F. Zobov, and O. L. Polyansky. On equilibrium structures of the water molecule. *Journal of Chemical Physics*, 122:214305, 2005.

Lawrence C. Evans. *Partial Differential Equations, Second Edition*. American Mathematical Society, 2010.

Nathan Ida. *Engineering Electromagnetics*. Springer, 2015.

Anders Malthe-Sorenssen. *Percolation Theory Using Python*. Springer, 2024.

C. Martin and H. Zipse. Charge distribution in the water molecule—a comparison of methods. *Journal of Computational Chemistry*, 26:97–105, 2005.

Branislav M. Notaros. *Electromagnetics*. Pearson, 2011.

Johannes Skaar. *Elektromagnetisme*. University of Oslo, 2019.

Johannes Skaar. *Kretsanalyse basert pa elektromagnetisme*. University of Oslo, 2019.

Index

© The Editor(s) (if applicable) and The Author(s), under exclusive license to Springer Nature Switzerland AG 2026

A. Malthe-Sørenssen, *Elementary Electromagnetism Using Python*, Undergraduate Texts in Physics, https://doi.org/10.1007/978-3-032-19876-1

N

O

P

R

S

T

The manufacturer's authorised representative in the EU is Springer
Nature Customer Service Centre GmbH, Europaplatz 3, 69115 Heidelberg,
Germany. If you have any concerns regarding our products, please
contact ProductSafety@springernature.com

Printed and bound by CPI Group (UK) Ltd, Croydon, CR0 4YY
21/07/2026
02173373-0001